MULTIPHASE CHEMICAL REACTORS
VOLUME II – DESIGN METHODS

NATO ADVANCED STUDY INSTITUTES SERIES

Proceedings of the Advanced Study Institute Programme, which aims at the dissemination of advanced knowledge and the formation of contacts among scientists from different countries.

The series is published by an international board of publishers in conjunction with NATO Scientific Affairs Division

A	Life Sciences	Plenum Publishing Corporation
B	Physics	London and New York
C	Mathematical and Physical Sciences	D. Reidel Publishing Company Dordrecht and Boston
D	Behavioural and Social Sciences	Sijthoff & Noordhoff International Publishers B.V.
E	Applied Sciences	Alphen aan den Rijn, The Netherlands and Rockville, Md., U.S.A.

Series E: Applied Sciences – No. 52

MULTIPHASE CHEMICAL REACTORS
VOLUME II - DESIGN METHODS

edited by

ALIRIO E. RODRIGUES
University of Porto,
Department of Chemical Engineering
Porto, Portugal

JOSEPH M. CALO
Princeton University
Princeton, N.J., U.S.A.

and

NORMAN H. SWEED
Oxirane International
Princeton, N.J., U.S.A.

Springer-Science+Business Media, B.V. 1981

Proceedings of the NATO Advanced Study Institute on
Multiphase Chemical Reactors
Hotel Golf Mar, Vimeiro, Portugal
August 18-30, 1980.

ISBN 978-94-011-8128-0 ISBN 978-94-011-8126-6 (eBook)
DOI 10.1007/978-94-011-8126-6

Originally published by Sijthoff & Noordhoff International Publishers B.V. in 1981
Softcover reprint of the hardcover 1st edition 1981

PREFACE

Multiphase reactors are becoming an important subject as a result of numerous applications in various fields ranging from chemical and petrochemical to food, biological and pharmaceutical (FBP) engineering, namely in the hydrodesulfurization of petroleum, production of antibiotics and wastewater treatment.
The importance of the subject is measured by the increasing number of publications and meeting sections dedicated to it. Therefore an Advanced Study Institute on Multiphase Chemical Reactors was held in Vimeiro, Portugal from 18 to 30 August 1980, following an International Symposium on Chemical Engineering of Gas-Liquid-Solid Catalyst Reactions, held in Liège, 1978.
The purpose of this NATO ASI was to present a didacted approach to multiphase reactor design, through the interaction among hydrodynamics, reaction and transport processes.
The lectures delivered there are compiled in two volumes.
In the first volume an overview of the field is presented, followed by a detailed analysis of the fundamental concepts (interphase mass transfer models, mass transfer with chemical reaction, theory of residence time distributions and micromixing phenomena). Numerical methods are often required to solve the mathematical models involved; so a chapter is dedicated to this subject followed by a discussion of steady-state multiplicity and stability.
Finally, a section with special applications of multiphase reactors is included, presenting topics on biological reactors, emulsion polymerization, supported liquid phase catalyst and enhanced oil re-

covery.

In the second volume modeling of multiphase chemical reactors is discussed; then the problem of intraparticle and interphase transport resistances in multiphase systems is tackled. The general characteristics of multiphase reactors (hydrodynamics, heat and mass transfer) follows, with particular emphasis on trickle-beds. A chapter is dedicated to three-phase fluidization. Finally laboratory reactors and scale-up methods are treated in some detail.

We want to acknowledge the assistance of our co-workers, especially the secretary of the course, J.M. Loureido and M. Idalina, who typed some of the manuscript.

This NATO Advanced Study Institute was made possible through the NATO ASI Programme. The participants and we ourselves are much indebted to this organization.

A.E. Rodrigues
J.M. Calo
N.H. Sweed

TABLE OF CONTENTS

DESIGN METHODS

THE MODELING OF MULTIPHASE CHEMICAL REACTORS

J. M. Calo

Department of Chemical Engineering, Princeton University, Princeton, New Jersey 08544

"Take up this mangled matter at the best:
Men do their broken weapons rather use
Than their bare hands."

Othello, Act I, Scene 3
William Shakespeare

"When we mean to build,
We first survey the plot, then draw the model;"

Henry the Fourth, Part Two
Act I, Scene 3
William Shakespeare

1. INTRODUCTION

The modeling of heterogeneous chemical reactors has progressed considerably during the past two decades or so to the point where the *a priori* design of the two-phase, gas-solid catalytic fixed bed reactor, as envisioned by Wilhelm [1], has become a reality (e.g., see [2-4] and even more significantly some recent standard textbooks [5-8]). However, few practitioners of the art of chemical reactor engineering would venture to make a similar statement concerning the design of what are commonly known as multiphase chemical reactors. Although the adjective "multiphase" literally means "having many phases", here it is used in the reactor engineering vernacular as explicitly meaning reactors with three phases: gas-liquid-solid. In reflection, the label "multiphase" is not as inappropriate as it may logically seem, since once gas, liquid, and solid phases are all present together in a chemical reactor, the introduction or presence of additional phases has at most only an incremental effect on the considerable degree of complexity which already exists.

The modeling of multiphase chemical reactors differs qualitatively from that of the two-phase, catalytic fixed bed reactor, as represented by the almost invariable necessity of at least one (and perhaps more) intermediate scale-up steps between the laboratory and commercial operation. In this respect, multiphase reactors are more akin to the two-phase fluidized bed. Of course, generically speaking, multiphase slurry reactors share similarities with the two-phase fluidized bed, while the multiphase fixed bed reactors are derivatives of the two-phase fixed bed. However, the effects of interphase mass interchange and hydrodynamics on reactor behavior shared by all multiphase reactors are quite reminiscent of similar phenomena in fluidized beds.

One way to more fully describe the difficulties involved in modeling multiphase chemical reactors derives from the idea of the chemical reactor model promulgated by Levenspiel [9] schematized in Figure 1. As depicted, the fundamental idea of developing a

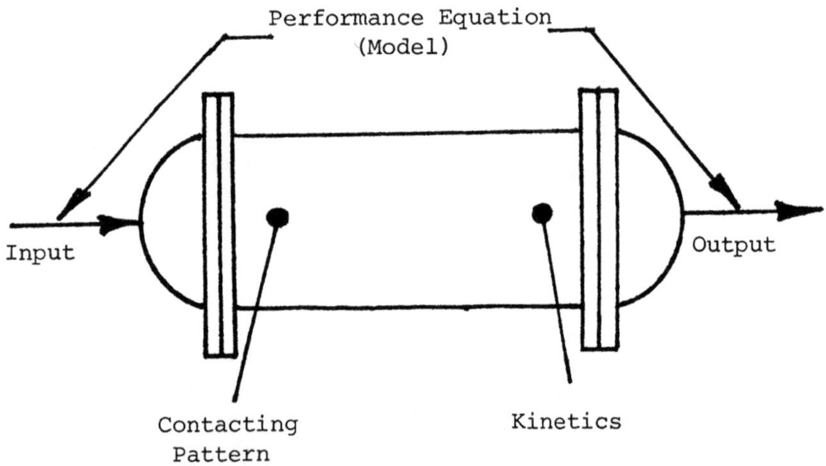

Figure 1. Chemical reactor model concept [9].

performance equation or model for a chemical reactor is actually quite simple and can be expressed mathematically as

$$\text{Output} = f[\text{Input, Kinetics, Contacting}] \qquad (1.1)$$

Once the three basic elements in this functionality have been determined, the chemical reactor model follows quite naturally. What then is the particular difficulty with multiphase reactors? In order to answer this question, let us consider the elements of kinetics and contacting, assuming the input is relatively well-defined.

Contacting. The contacting pattern describes how material flows through the reactor. In homogeneous chemical reactors when the flow deviates from the plug flow or well-mixed ideals, contacting can be characterized using well-developed stimulus-response tracer techniques, for example. In the two-phase, catalytic fixed bed reactor, the single fluid phase can exhibit flow regimes ranging from laminar through transition to turbulent, which are all reasonably well-characterized by the properties and velocity of the fluid and the size of the solid packing. In multiphase chemical reactors, however, the flow regime depends not only on the properties and flow rates of two fluid phases, but also on their relative flow orientation (cocurrent-upflow or downflow or countercurrent-flow), the nature of the fluid distributors, and the size of the reactor. In addition, the distribution of fluid within and even between different multiphase reactors can vary significantly depending on size and operating mode. Thus, obviously the number of parameters required to describe the contacting, flow pattern, or flow regime increases significantly in proceeding from even the already modestly complex two phase, catalytic fixed bed reactor to multiphase chemical reactors. But surely, even a relatively large number of parameters represents no truly significant problem for numerical solution -- evidence the success with large industrial models with many parameters for fluid catalytic cracking [10, 11] and reforming [12, 13]. The qualitative difference with multiphase reactors is that currently we do not even know what all the appropriate parameters are and their complex interrelationships -- irrespective of their number.

Kinetics. The kinetic processes describe the rate of chemical change within the reactor. In homogeneous systems this is obviously synonymous with the intrinsic chemical rate, but in the two-phase catalytic fixed bed reactor, interphase, interparticle, and intraphase transport of mass and heat can also participate in determining the observed or global rate. Even so, the interfacial area for transport remains relatively constant, the variation of the interphase transport coefficients with fluid properties and velocity is relatively well-characterized, at least empirically, and intraphase transport phenomena in the porous solid catalyst, although complex, involve only one fluid phase and have been successfully described by single effective diffusivities. In multiphase chemical reactors, however, interphase transport is intimately coupled with the flow regime which affects both the transport coefficients and the interfacial transport area. In addition, intraphase or intraparticle transport can involve more than one fluid phase.

Thus in summary, the primary qualitative differences which make the modeling of multiphase reactors difficult and challenging are: (1) proper identification of a large number of incompletely characterized parameters and their interrelationships; and (2) the strong coupling between flow regime (hydrodynamics) and interphase transport which directly affects the global kinetics.

Given this current state of affairs, it is perhaps fair to ask the following questions: What then, if anything, can be expected to be gained by attempting to model these reactors? And, assuming at least an encouraging answer to the previous question, what approaches and techniques seem appropriate? The first question is actually quite a fundamental one which directly addresses the philosophy of modeling. At least a partial answer has been provided by Shinnar [14]. In addition to the rather obvious objective of predicting operating performance, models can be used heuristically to indicate possible structural relationships among variables which may prove fruitful in unraveling the complexities of the phenomena involved in multiphase chemical reactors, for example. Also, models, no matter how simple, must always be constructed to explain experimental observations and their interrelationships. In doing so, theories concerning the observed phenomena are created and provide additional guidance and feedback for further experimentation, eventually leading to a more complete understanding of the particular system. For these and many other reasons, the modeling of physicochemical processes and systems, such as multiphase chemical reactors, is widely accepted today even when the resultant models are not precise or robust enough to accurately predict operating behavior under all conditions of interest.

Having thus reaffirmed what amounts to the *raison d'etre* for modeling, we turn to the approaches and techniques which have already been used with relative success and which show promise for the modeling of multiphase reactors. In view of the success in modeling homogeneous and simple two-phase heterogeneous chemical reactors, it is only natural to apply the same ideas and philosophy to more complex reactors. This approach is strongly imbued with the *a priori* concept advocated by Wilhelm [1]. Simply stated, this approach maintains that the individual physicochemical phenomena which comprise a chemical reactor can be identified, characterized, and measured independently in the laboratory and rationally combined in a mathematical model to accurately predict the behavior of a chemical reactor. Thus, as specifically applies to multiphase reactors, phenomena such as gas-liquid and liquid-solid mass transfer, heat transfer, hydrodynamics of the fluid phases, solids (catalyst) wetting, interparticle and intraparticle transport, and intrinsic chemical kinetics would be characterized as independently as possible to provide numerical parameters, and combined with appropriate mathematical formulations to describe overall reactor behavior. It is implicitly presupposed that if this is done correctly, the resultant model, although admittedly complex, will be an accurate representation of the reactor. While this objective has been achieved in a sense for the two-phase, catalytic fixed bed reactor, the *a posteriori* approach, as identified by Hlavacek, et al. [4], which has been an integral part of chemical reactor engineering since its inception, is still quite necessary in the development of multiphase reactors. In this approach, initially conceived models are tested experimentally in gradually larger-scale reactors and adjusted in an iterative fashion in as many steps as necessary until a commercial-scale design is achieved. The *a priori* approach is not yet directly applicable to multiphase chemical reactor modeling and will not be until all the major appropriate parameters and their interrelationships have been firmly identified.

Therefore, what approaches, techniques, and models are available and appear to be fruitful for characterizing the behavior of multiphase chemical reactors? Again, quite naturally almost all known approaches are based on methods that have proven successful in modeling homogeneous and simple heterogeneous reactors. Consequently, since the detailed hydrodynamics or fluid flow pattern seems to be of fundamental concern,it is not surprising that many models have been based on residence time distributions (RTDs) of the fluid phases. This approach attempts to characterize the complex hydrodynamics of each fluid phase by determining the contact time, or, more precisely, the macroscopic distribution of residence times of fluid elements in each phase within the reactor. The requisite experimental RTDs are usually determined with stimulus-response tracer techniques. In principle, once this information is known, it can be combined with the intrinsic kinetics, interphase, interparticle, and intraphase transport to formulate a phenomenological model in the spirit of the schematic in Figure 1. A manifold of other models and approaches have also been used which focus on other aspects of multiphase reactor phenomena such as contacting effectiveness between liquid and solid phases or the roles of interphase mass transport resistances. In the present paper, both the spectrum of approaches and specific models will be examined in order to provide a comprehensive view of the current status of this area.

The focus of this review is on the *modeling and models* of multiphase chemical reactors. Thus the detailed nature, measurements, and correlations pertaining to individual phenomena and properties which characterize multiphase reactors, although of paramount importance in determining model parameters, are not specifically covered, except when pertinent to the discussion. These topics are explicitly addressed in other papers in this Institute and in recent reviews by Satterfield [15], Charpentier [16], Goto, Levec, and Smith [17], Hofmann [18, 19], Gianetto, et al. [20], Chaudhari and Ramachandran [21], Shah, Stiegel, and Sharma [22], the classic review by Østergaard [23], and the recent excellent monograph by Shah [24].

2. REACTOR TYPES, CLASSIFICATION, AND FLOW REGIMES

A brief categorization of multiphase chemical reactors is useful at this point to define exactly the types of reactors and operating modes to which the models to be discussed here apply. Østergaard [23] grouped multiphase processes into two main classes:

2.1 Fixed Bed

The solid (particulate) phase forms a fixed bed and the fluid phases may be in cocurrent-downflow, cocurrent-upflow, or countercurrent-flow as shown in Figure 2. The flow regimes are usually depicted on flow regime maps patterned after those devised for two-phase flow in pipes (e.g., see Wallis [25] and Govier and Aziz [26]). However, as may be expected, the boundaries on these maps are complex functions of the fluid properties and flow rates and the properties of the solid, and therefore are either usually in dispute or specific to certain systems. Thus rather than become involved in the intricacies and nuances of flow regime maps, which are a complete subject of discussion unto themselves (e.g., see Gianetto, et al. [20] or Shah [24]), a qualitative description of the predominant flow regimes will suffice for our purposes here.

Essentially two types of flow regimes are predominant in cocurrent-downflow: trickle flow where the liquid trickles over the packing in discontinuous films, rivulets, and drops near a stagnant continuous gas phase; and bubble-flow where the liquid phase is continuous and the gas phase is dispersed as separate bubbles, at low enough gas flow rates. Spray flow is also a gas-continuous flow regime which occurs at high gas rates. The liquid is suspended primarily as a heavy mist but also covers the packing in a liquid layer, the thickness of which decreases with increasing gas flow rate. Pulsed flow is an intermediate flow regime where slugs are formed due to liquid bridging of gaps between adjacent particles. These liquid slugs are pushed down the bed by the gas such that discernible regions of high and low liquid loading propagate down the bed as pulses or waves [27, 28]. The pulse flow regime is of current interest since some commercial operations are approaching this operating mode [24, 29]. It should also be noted that on the

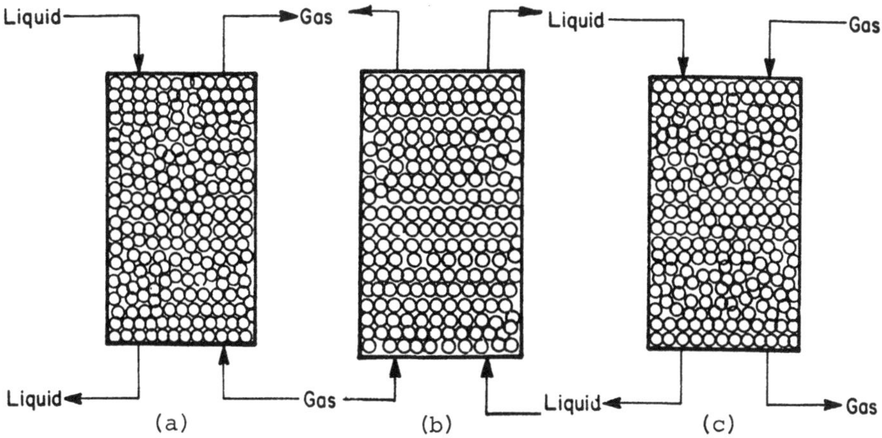

Figure 2. Three-phase fixed bed reactor types [22]: (a) countercurrent-flow; (b) cocurrent-upflow; (c) cocurrent-downflow.

basis of gas-liquid interaction, Middoux, et al. [30] and Specchia and Baldi [31] have suggested that there are actually only two flow regimes: a low interaction regime which includes only trickle flow; and a high interaction regime, comprising all other flow regimes.

The predominant flow regimes in cocurrent-upflow and countercurrent-flow are qualitatively the same as described for cocurrent-downflow, except of course that the flow regime boundaries are quantitatively different (i.e., liquid-continuous, gas-continuous, slug flow, etc.).

2.2 Slurry

The solid phase particles are suspended in the liquid phase and a large degree of relative motion exists between particles. Depending on the nature of transport of momentum, the following subtypes were distinguished by Østergaard [23]:

1. Stirred or agitated batch slurry reactor (particles suspended by bubble movement and mechanical agitation).

2. Bubble-column or nonagitated batch slurry reactor (particles suspended by bubble movement only).

3. Gas-liquid fluidized or cocurrent three-phase fluidized-bed reactor (particles suspended by bubble movement and cocurrent liquid flow).

These three are the most commonly used in practice and are schematized in Figure 3. In addition to these, Shah [24] separately

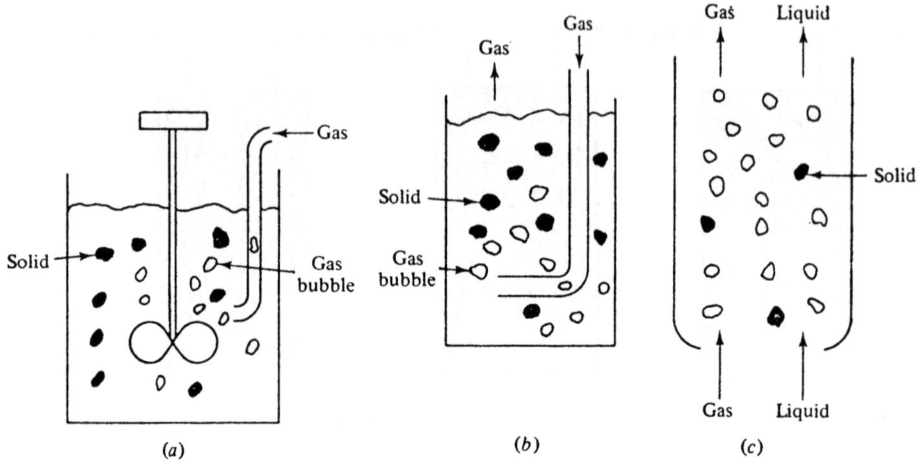

Figure 3. Three-phase slurry reactor types [24]: (a) agitated; (b) nonagitated; (c) fluidized bed.

lists two reactor types which have been operated only on a laboratory scale:

4. Countercurrent (liquid, solid downflow, and gas upflow) flow reactor or three-phase spouted bed reactor.

5. Batch pulsating reactor.

Unlike the fixed bed multiphase reactors which are classified primarily by the flow directions of the fluids, the three-phase slurry reactors are classified basically according to flow characteristics, and thus more detailed considerations concerning flow regimes will not be addressed here.

In addition to the preceeding, three-phase chemical reactors can be classified according to the nature of the reaction:

I. Reactions where all three phases are reactants or products.

II. Gas-liquid-solid reactions where the solid is a catalyst.

III. Two reacting phases with the third phase inert.

Here we will be concerned only with category II, where the solid or packing is a catalyst.

3. HYDRODYNAMIC OR RTD MODELS

The general reactor performance equation (1.1) indicates that the fundamental basis of reactor design is the characterization of the coupling between intrinsic chemical kinetics and the contacting pattern or hydrodynamics. The latter affects not only the local concentrations of reactants and products, but also, and of at least equal importance in heterogeneous reactors, the physical transport rates between phases. The hydrodynamic models which have been developed to describe fluid mixing in multiphase reactors are all naturally derived from single phase considerations. However, in general, the extent of mixing in each fluid phase will be different and should be considered separately.

The hydrodynamic description of fluid mixing in chemical reactors usually assumes the form of a residence time distribution (RTD) which defines the residence time of fluid elements within the reactor. This formulation is convenient because the RTD can usually be determined experimentally by well-established, stimulus-response techniques. The RTD of a fluid phase defines only the convective or bulk component of fluid mixing known as macromixing. The resultant RTD reflects bulk phenomena such as nonuniform velocity fluctuations due to molecular and eddy diffusivity, reactor shape and internals, backflow of fluid due to velocity differences between phases, and internal recycle due to mechanical agitation. The RTD does not reflect exchange of mass between fluid elements due to molecular diffusion, sometimes referred to as micromixing.

The fractional conversion of a reactant, X, is then related to the RTD, E(t), by [9]:

$$X = 1 - \int_0^\infty \frac{C(t)}{C_o} E(t)dt \qquad , \qquad (3.1)$$

where E(t)dt is the fraction of the exit stream that has resided in the reactor for a time between t and t + dt, C(t) is the reactant concentration as a function of time according to the intrinsic chemical kinetics, and C_o is the reactant concentration at the inlet.

Schwartz and Roberts [32] reviewed existing RTD models for trickle bed reactors and defined the proper RTD applicable to a heterogeneous reactor. Although directed specifically at trickle bed reactors, the results have some implications for RTD models in general. The reactor was visualized as a system of nonidentical, parallel plug flow tubular reactors with no communication between tubes. It was shown that this segregated-flow (RTD) model is appropriate provided that (1) liquid holdup is due only to the packing, i.e., holdup on the reactor walls is negligible; (2) the liquid film thickness around each particle is constant; (3) the bed is uniformly packed; and that the RTD used in the calculation is that of the *external* liquid; i.e., the RTD which would be observed with nonporous packing. For a first-order irreversible reaction in an isothermal reactor,

$$C_j = C_o \exp(-k_1 \tau_j) \quad , \tag{3.2}$$

where C_j is the reactant concentration in tube j at residence time τ_j and k_1 is the effective first order rate constant. Substitution of Eq. (3.2) into Eq. (3.1) following the development of Schwartz and Roberts [32] yields

$$X = 1 - \int_0^\infty e^{-(k_1 \rho_c / h_{dL} \rho_L) t'} E(t') dt' \quad , \tag{3.3}$$

where ρ_c is the catalyst density based on reactor volume, ρ_L is the liquid density, E(t') is the RTD of liquid external to catalyst particles, and

$$h_{dL} = \bar{t}_L Q_L / V \tag{3.4}$$

is the dynamic liquid holdup, for which

$$\bar{t}_L = \int_0^\infty t' E(t') dt' \tag{3.5}$$

is the mean liquid residence time, Q_L is the volumetric liquid flow rate, and V is the total reactor volume.

While the majority of the RTD models presented here have been specifically developed for the liquid phase in trickle beds, some are also applicable to liquid and solid phase mixing in slurry reactors. Gas phase mixing, especially in trickle beds, has not received as much attention as that in the liquid phase, primarily due to the minimal influence of axial gas dispersion on reactor performance. However, Hochman and Effron [33] and Dunn, et al. [34] have satisfactorily interpreted the gas phase RTD in a trickle bed in terms of the single-parameter dispersion model and found that dispersion increased with increasing liquid and gas flow rates. Of course, the effect of the gas phase on liquid phase hydrodynamics and mass transfer is quite important in all multiphase reactors.

RTD models are usually classified according to whether mixing is described by a Fickian diffusion-like formulation, which results in differential models, or by perfectly mixed staged regions, which results in stagewise or discrete models. The RTD models have also been classified according to the number of parameters required to characterize them. This number is subject to some interpretation, however, and can become confusing when the same model is referred to in different sources as having a different number of parameters. For example, Gianetto, et al. [20] contend that the fluid phase holdup must be defined from the RTD and hence must be included as a parameter. However, Schwartz and Roberts [32], Shah [24], and Shah, et al. [22] exclude holdup from the number of characteristic parameters. Here we follow the latter convention.

3.1 Single-parameter models.

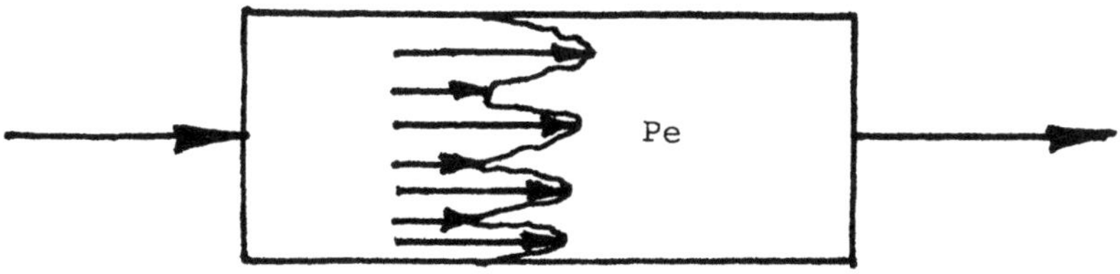

Figure 4a. Dispersion or piston-dispersion (PD) model.

3.1.1 The dispersion model (Figure 4a). The differential mass balance for a delta function input of inert tracer for the one-dimensional dispersion model [9] or the piston-dispersion (PD) model [20] is:

$$\frac{\partial C}{\partial t} = D_Z \frac{\partial^2 C}{\partial Z^2} - \bar{u} \frac{\partial C}{\partial Z} \qquad (3.6)$$

where C is the tracer concentration, D_Z is the axial dispersion coefficient, and $\bar{u}$ is the mean axial fluid velocity. This model assumes that the fluid has a flat velocity profile and fluid elements are dispersed axially according to a Fickian diffusion formulation. Since the fundamental steady-state version of the mass balance is a differential equation, this model forms the basis of all differential models.

In dimensionless form, Eq. (3.6) becomes

$$\frac{\partial v}{\partial \theta} = \frac{1}{Pe} \frac{\partial^2 v}{\partial z^2} - \frac{\partial v}{\partial z} \qquad (3.7)$$

where $v = C/C_o$ (C_o is a reference concentration), $z = Z/L$, where L is a characteristic length scale (usually the reactor diameter or length, or the characteristic packing size), $\theta = t/\bar{t}$ ($\bar{t}=\bar{u}/L$), and $Pe = \bar{u}L/D_Z$. Thus the dispersion model describes the degree of backmixing solely by the axial dispersion coefficient, D_Z, or equivalently, in dimensionless form, by the Peclet number, Pe. (When the characteristic particle size is used as the length scale, the Peclet number is sometimes known as the Bodenstein number.) The extreme values of Pe = 0 and Pe = ∞ define the ideal flow patterns of complete macromixing and plug flow (no mixing), respectively.

The single characteristic parameter of the dispersion model, Pe, can be obtained by various techniques such as fitting the experimentally determined RTD to solutions of the dispersion model [35, 36], or the method of moments [37] (for a more complete discussion see Himmelblau and Bischoff [38]). For example, the Peclet number can be obtained from the second central moment or the variance of the RTD curve. For a closed system reactor (no dispersive fluxes at the inlet and outlet):

$$\sigma_\theta^2 = \frac{1}{\bar{t}^2}\int_0^\infty (t - \bar{t})^2 E(\theta)\,d\theta = \frac{2}{Pe} - \frac{2}{Pe^2}(1 - e^{-Pe}) \qquad (3.8)$$

where $E(\theta)$ is the RTD curve or the response to a delta function input of tracer, and $\bar{t}$ is the average residence time determined by the first non-central moment of the RTD. Other analytical results for the variance of the dispersion model with other boundary conditions can be found in Himmelblau and Bischoff [38].

3.1.2 <u>N-stirred tanks-in-series model</u> (Figure 4b).

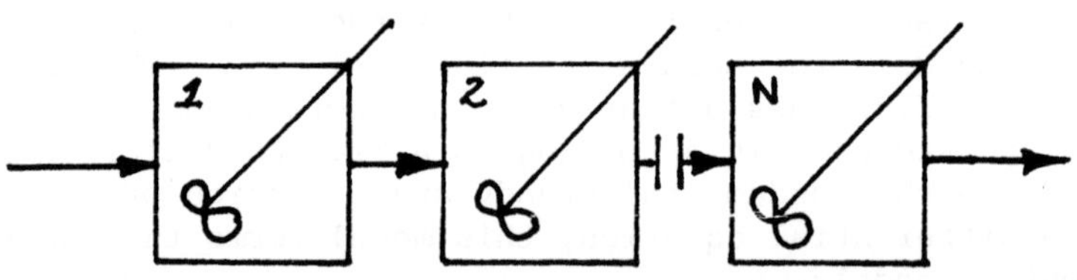

Figure 4b. N-stirred tanks-in-series model.

The N-stirred tanks-in-series model characterizes fluid back-mixing as occurring in a series of perfectly mixed compartments, cells, or tanks [9]. The single parameter is simply the number of stirred tanks, N, where the extreme values of N = 1 and N = ∞ represent the ideal flow patterns of complete macromixing and plug flow, respectively. The simple nature of this model (ordinary differential equations for transient response and algebraic equations for steady-state behavior) readily lends itself to more complex embellishments such as backflow, dead volume, mass interchange between different fluid regions, etc., with the introduction of additional parameters. Thus this model serves as the basis for all stagewise models.

The N-stirred tanks-in-series model can be used whenever the dispersion model is applicable, and for relatively small deviations from plug flow (i.e., large N) yields similar results. The parameter N can be obtained using the same techniques as for obtaining the Peclet number for the dispersion model. The RTD for this model is [38]:

$$E(\theta) = \frac{N^N}{(N - 1)!} \theta^{N-1} \exp(-N\theta) \quad , \qquad (3.9)$$

the variance of which is simply,

$$\sigma_\theta^2 = \frac{1}{N} \qquad (3.10)$$

A comparison of Eqs. (3.10) and (3.8) for small departures from plug flow (i.e., large N and Pe) yields

$$N \approx \frac{2}{Pe} \quad , \qquad (3.11)$$

which represents the fundamental relationship between the two models.

3.1.3 <u>Laminar film-mixing-bypass model</u> (Figure 4c). In order for the dispersion or N-stirred tanks models to fit experimental RTDs accurately, they must describe the physical situation adequately. Michell and Furzer [39] have claimed that the dispersion model does not explain the relatively long tailing in RTD data from the liquid phase in trickle beds for example. Therefore, they proposed a model whereby liquid flow in a trickle bed is viewed as laminar film flow over the packing between junctions, and at each junction a fraction, χ, of the flow enters a perfectly mixed region, while the remainder completely bypasses the mixed region. In effect, mixing is due to hydrodynamic phenomena at the static junctions rather than by diffusion-like processes. This model is most easily visualized from the schematic in Figure 4c.

The total mean residence time of the liquid is given by:

$$\bar{t} = n\bar{t}_f + \chi n\bar{t}_m \qquad (3.12)$$

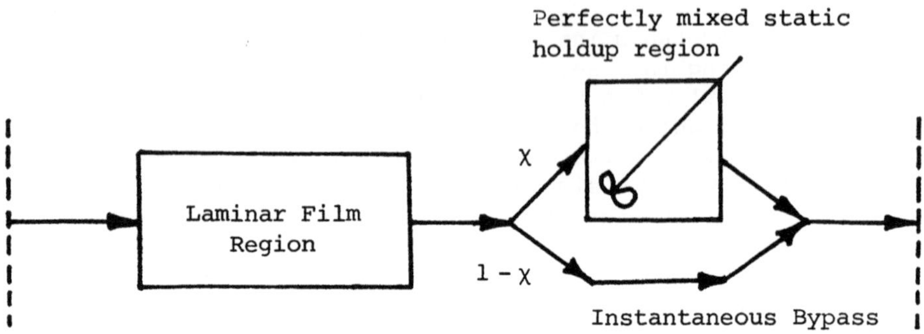

Figure 4c. Element of laminar film-mixing-bypass model [39].

where n is the number of laminar flow elements in the series, $\bar{t}_f$ is the mean residence time of the fluid in each flow element, and $\bar{t}_m$ is the mean residence time in a single mixer. It is assumed that the length of each laminar flow element is on the order of the diameter of a solid particle such that $n = L/d_p$, where L is the total packed height and d_p is the mean diameter of the packing. Therefore, $\bar{t}_f = d_p/\bar{u}_f$, where $\bar{u}_f$ is the mean flow velocity of the films. The perfectly mixed junction regions correspond to the static liquid holdup, h_{SL}, such that $\bar{t}_m = \rho_L h_{SL} d_p/\chi G_L$, where G_L is the superficial liquid mass velocity. Assuming h_{SL}, G_L and ρ_L are known or can be measured separately, this leaves two parameters, χ and $\bar{t}_f$. However, $\bar{u}_f$ can be related to the wetted area, a_w, by known relations such as [40]:

$$\bar{u}_f = (4G_L/a_w\mu_L)^{2/3}(\mu_L g/4\delta\rho_L)^{1/3} \quad , \tag{3.13}$$

where μ_L is the liquid viscosity and g is the gravitational constant. Assuming a_w can be measured or estimated, the model reduces to the single parameter, χ.

Michell and Furzer [39] correlated χ with the film Reynolds number as

$$\chi = 0.0187 \, Re_f^{0.32} \quad , \tag{3.14}$$

where $Re_f = 4\delta\rho_L\bar{u}_f/\mu_L$ and δ is the film thickness. Also, by comparing the variances obtained from this model with those from an axial dispersion model for the same system, the following approximate relationship between the two models resulted:

$$Pe = \chi(\bar{t}_m + \bar{t}_f)^2/\bar{t}_m^{\,2} \qquad (3.15)$$

which was in turn used to correlate Pe with the film Reynolds number as

$$Pe = 0.039\ Re_f^{\,0.5} \qquad (3.16)$$

with a considerable amount of data on nonporous packing from several investigators [39]. Although (3.16) correlates the data well, it has not been tested against data for porous packing where a significant portion of the static liquid holdup can be due to liquid within the porous catalyst particles.

3.2 Two-parameter models.

Schwartz and Roberts [32] have shown that the following three, two-parameter models are mathematically equivalent insofar as they all lead to the same RTD expression upon appropriate redefinition of parameters.

3.2.1 Crossflow model (Figure 5a).

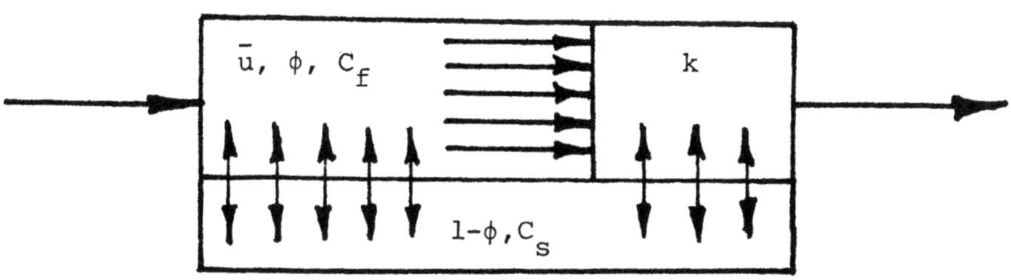

Figure 5a. Crossflow or piston-exchange (PE) model.

In the two-parameter crossflow, or piston-exchange model [20] the fluid is assumed to be divided into a plug flow region and a stagnant region in contact with the flowing fluid with which mass can be exchanged. This model has been successfully applied to liquid phase backmixing in trickle bed reactors by Hochman and Effron [33] and Hochman and McCord [41]. The physical interpretation is that stagnant fluid can accumulate in trickle beds on top of the packing, in certain interstices between tightly packed particles, and between the reactor wall and the packing.

This physical description can be expressed mathematically by a mass balance on the flowing fluid,

$$\phi \frac{\partial C_f}{\partial t} + \bar{u} \frac{\partial C_f}{\partial z} + k(C_f - C_s) = 0 \qquad (3.17)$$

and the stagnant fluid:

$$(1 - \phi) \frac{\partial C_s}{\partial t} - k(C_f - C_s) = 0 \quad , \qquad (3.18)$$

where ϕ is the fraction of liquid in plug flow, k is the mass transfer or interchange coefficient between stagnant and flowing fluid, $\bar{u}$ is the mean interstitial fluid velocity, and C_f and C_s are the tracer concentrations in the flowing and stagnant fluids, respectively.

Since the crossflow model has two parameters, ϕ and k, they can be determined in principle by the first three moments. However, due to uncertainties in measuring third moments, Hochman and Effron [33] recommended the technique used by Hoogendorn and Lips [42]. The latter authors showed that if the flowing liquid is in plug flow, ϕ can be determined from the initial breakthrough time of the tracer, t_i, by

$$\phi = t_i/\bar{t} \quad , \qquad (3.19)$$

where $\bar{t}$ is the mean residence time. k can then be calculated from:

$$k = 2(1 - \phi)^2/\sigma^2\bar{t} \quad , \qquad (3.20)$$

where σ is the variance of the RTD curve.

Values of ϕ were correlated [33] by:

$$\phi = 0.52 \ Re_L^{0.08} \ 10^{0.001 Re_G} \quad , \qquad (3.21)$$

where Re_L and Re_G are the liquid and gas phase Reynolds numbers, respectively. Also, values of ϕ estimated from operating and total liquid holdup agreed well with values calculated from the model, which lends a physical basis to the crossflow model.

Values of k were correlated [33] by:

$$k = 0.010 \ Re_L^{0.6} \quad , \qquad (3.22)$$

which is also reasonable. In the same study, a correlation of the Peclet number, calculated for an axial dispersion model, with Re_L exhibited significant scatter, but yielded

$$Pe_L = 0.042 \ Re_L^{0.5} \qquad (3.23)$$

which is remarkably similar to Eq. (3.16) obtained by Michell and Furzer [39]. These results indicate that at least for trickle beds, the crossflow model has some physical justification and seems to fit the data better than the one-parameter axial dispersion model.

3.2.2 Time-delay model (Figure 5b).

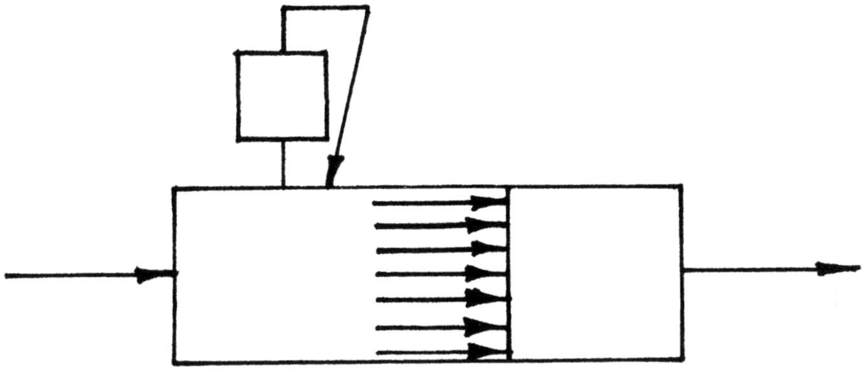

Figure 5b. Time-delay model [43].

The probabilistic time delay model of Buffham, et al. (43] assumes a main fluid in plug flow with fluid elements randomly delayed on their passage through a fixed bed of solids. The delayed fluid elements rejoin the flow after a time delay. Any particular element may experience a number of such delays in its passage through the bed. Assuming an experimental distribution of delay times, the RTD for this model is

$$E(t) = 0 \quad , \quad t < t_i \tag{3.24a}$$

$$E(t) = \exp\left[\frac{\bar{t} - t_i}{t_d}\right]\delta(t - t_i) + \frac{1}{t - t_i} \cdot \exp\left[\frac{-\bar{t} - t + 2t_i}{t_d}\right] \cdot \sum_{n=1}^{\infty} \frac{[(\bar{t} - t_i)(t - t_i)/t_d^{\,2}]^{2n-1}}{n!(n-1)!} \quad , \quad t \geq t_i \tag{3.24b}$$

where n is the number of time delays, $\bar{t}$ is the mean liquid residence time, and the two parameters required to define this model are t_i, the initial tracer breakthrough time (or the time taken to attain a normalized concentration of 0.05 [43]), and t_d, the mean delay time, which can be determined from the peak height or variance of the RTD curve [43].

3.2.3 Modified mixing-cell model (Figure 5c).

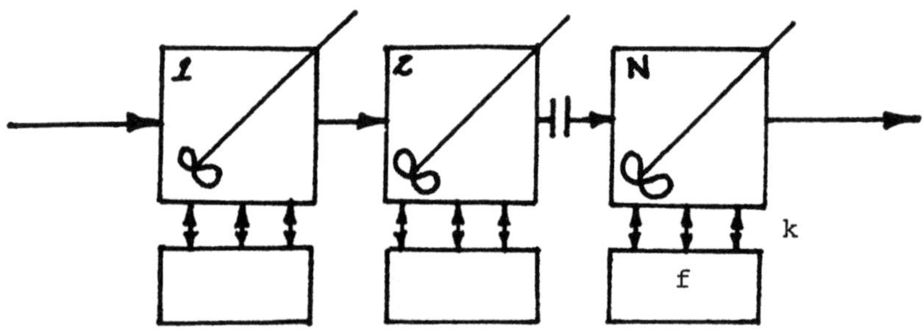

Figure 5c. Modified mixing-cell model [44].

Deans [44] proposed that the mixing cell model of Deans and Lapidus [45] could be modified to describe flow through porous media by allowing for the possibility of mass transfer between stagnant and flowing regions in each cell. The equivalence between this model and the later crossflow model [39] becomes quite evident in the limit of a large number of cells, for which the modified mixing-cell model may be represented by the partial differential equations (3.17) and (3.18). The resultant RTD for this limiting form of the model is:

$$E(t) = 0 \quad , \qquad t < t_i \tag{3.25a}$$

$$E(t) = \exp(-k\bar{t})\,\delta(t - t_i) + \left(\frac{k^2\bar{t}}{f}\right)\exp\left[-k\bar{t}\left(2 + \frac{t - \bar{t}}{f\bar{t}}\right)\right]$$

$$\left[\frac{I_1\left(2k\bar{t}\sqrt{1+(t-\bar{t})/(f\bar{t})}\right)}{k\bar{t}\sqrt{1+(t-\bar{t})/f\bar{t}}}\right] , \qquad t \geq t_i \tag{3.25b}$$

where $f = (1 - \phi)$ is the fraction of liquid which is stagnant. Equations (3.24) and (3.25) can be shown to be identical by writing $I_1(z)$ in its infinite series form, setting

$$k\bar{t} = \frac{\bar{t} - t_i}{t_d} \tag{3.26}$$

and

$$f = 1 - (t_i/\bar{t}) \tag{3.27}$$

The two parameters required to define this model are, of course, f and k, which can be respectively determined from Eq. (3.27) and the variance of the RTD,

$$\sigma^2 = 2f^2/k\bar{t} \ . \tag{3.28}$$

3.3 Three-Parameter Models.

3.3.1 Piston-Dispersion-Exchange (PDE) model (Figure 6a).

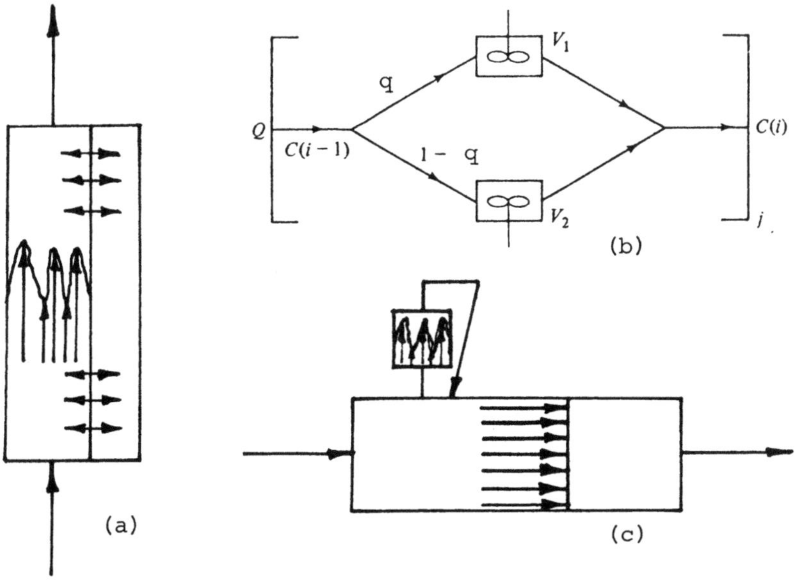

Figure 6. Three-parameter models: (a) PDE [24]; (b) parallel mixing cell [50]; (c) modified time-delay [52].

The PDE model is essentially the two-parameter crossflow model modified to take into account dispersion in the flowing liquid phase. The three parameters required to characterize this model are the Peclet number, $Pe = \bar{u}L/D_Z$, the fraction of total liquid flowing, ϕ, and the mass interchange coefficient between the dynamic and stagnant regions, which can be expressed in dimensionless form as [24],

$$\bar{N} = K_m a_s/\varepsilon h_L \phi \quad , \tag{3.29}$$

where K_m is the mass transfer coefficient, a_s is the packing surface area per unit volume, ε is the void fraction of the dry bed, and h_L is the total liquid holdup based on total void volume. Obviously, when $\bar{N} = 0$ and $\phi = 1$, the PDE model must reduce to the dispersion model, and when $D_Z = 0$ to the crossflow model. The liquid static holdup and mass interchange values for the PDE model are obviously different than the corresponding values for the crossflow model due to the effect of dispersion.

The analytical solution of the PDE model for an impulse of tracer has been obtained for closed-open [46, 47] and closed-closed [48] boundary conditions. For the latter,

$$\sigma^2 = \frac{2}{Pe} + \frac{3}{Pe^2} + \frac{2(1 - \phi)^2}{\bar{N}}\left(\frac{1}{Pe} + 1\right). \qquad (3.30)$$

The mechanisms of axial dispersion in the liquid phase and mass interchange between flowing and stagnant liquid, both at particle contact points and within porous particles, enable the PDE model to fit the long tailing observed in RTD curves from trickle bed reactors better than the single-parameter dispersion model, as shown in Figure 7, or the two-parameter PD (piston-dispersion, $\bar{N}$ = 0) model of Matuura, et al. [49], or the crossflow model.

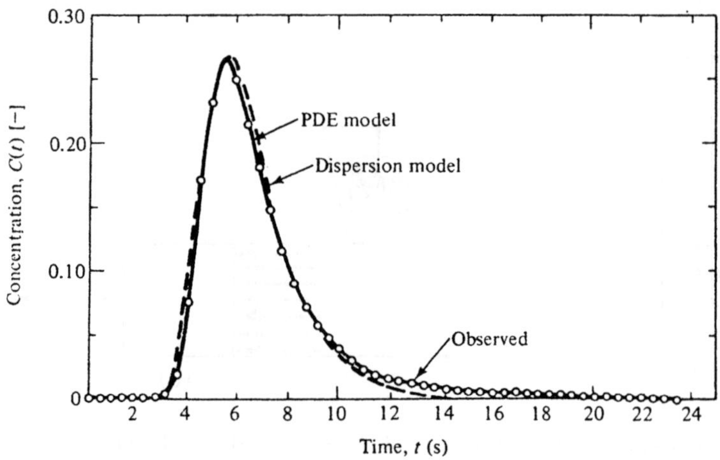

Figure 7. Comparison of measured tracer output with the PDE and single-parameter dispersion models [49].

3.3.2 Parallel mixing-cell model (Figure 6b). As shown in Figure 6b, the parallel mixing-cell model assumes the reactor can be represented by a series of perfectly mixed elements arranged in parallel. The liquid flow rate is distributed between the parallel elements in the ratio q/(1-q), where q is the fraction of the flow through the mixing cell representing the stagnant zone. The volumes of the two mixers are assumed to be in the ratio of the static and dynamic holdups. For q = 0, this model reduces to the modified mixing cell model. For q > 0, Van Swaaij, et al. [50] derived the first three moments from which the two parameters, q and N (number of stages), can be evaluated. The number of stages can be found from the variance of the RTD curve, i.e.,

$$\sigma_\theta^2 = \frac{2}{N}\frac{[1 + q(A^2 - 1)]}{[1 + q(A - 1)]^2} - \frac{1}{N} \quad (3.31)$$

where $A = V_1(1 - q)/V_2q$, under conditions where the stagnant zone is negligible $(q = 0)$, for which $\sigma_\theta^2 = 1/N$. q can then be determined from the variance under conditions where $q \neq 0$. The third parameter for this model is the ratio of the two volumes, V_1/V_2. The primary uncertainty in this model is the relationship between the dynamic and static liquid holdups and the mobile and stagnant zones.

3.3.3 Modified time-delay model. There are at least two modifications of the probabilistic time delay model of Buffham, et al. [43] which give rise to an additional parameter. The first is due to Buffham and Gibilaro [51] whereby a gamma distribution of time delays,

$$F_i(t) = \left(\frac{m}{t_d}\right)^{mi} \frac{t^{mi-1}\, e^{-mt/t_d}}{\Gamma(mi)} \quad (3.32)$$

is proposed in lieu of the exponential time delay distribution. The additional parameter is the constant m, and i is the cell index. For m = 1, Eq. (3.32) reduces to the exponential time distribution.

A second modified form, due to Oorts and Hellinckx [52], takes into account imperfect mixing in the delayed lateral streams with an additional parameter (Figure 6c). This model has been fit to experimental data with a lower mean square deviation than the PDE model. However, its validity at fast response times has not been verified.

3.4 Four-Parameter Models.

3.4.1 Rao and Varma [53] model (Figure 8a). As shown in Figure 8a, in this model the total liquid volume in a stage, V_t, is divided into a region with axial dispersion, V_p, to model the flowing liquid film on the particle surface, followed by a perfectly mixed region, V_m, which exchanges mass with a stagnant region, V_s. Thus generically this model is a combination of the bypass and PDE models. The four parameters in this model are the Peclet number of the axial dispersion region, Pe, the crossflow ratio, q, and $\alpha = V_p/V_t$ and $\beta = V_s/(V_m + V_s)$. The number of stages, N, is taken as the ratio of bed length to particle size. Rao and Varma have shown that this model fits experimental data quite well. The most significant parameters are Pe and β. The parameter α and q remained approximately constant at 0.5 and 10^{-5}, respectively. The stagnant zone typically represents 5 to 14% of the total holdup.

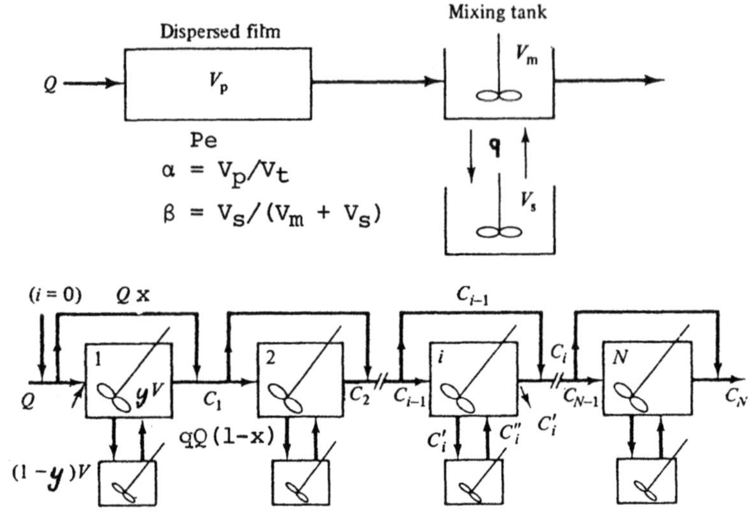

Figure 8. Four-parameter models: (a) Raoand Varma [53]; (b) Raghuraman and Varma [54].

3.4.2 Raghuraman and Varma [54] model (Figure 8b). This model is an extension of the stagewise PDE model with instantaneous bypass, as schematized in Figure 8b. The four parameters in this model are the bypass fraction, x, the interchange coefficient or crossflow ratio, q, the fractional stagnant volume, y, and the number of stages, N, which may or may not be given by the simple relationship $N = L/d_p$, or equivalently, the volume of each stage, V. This model would be useful for modeling large degrees of bypassing in fixed beds. For $x \to 0$ this model reduces to the modified mixing-cell model, while for $N \to \infty$, $x \to 0$, it reduces to the two parameter crossflow or time-delay model.

3.5 Recycle-Flow Model with Crossmixing (Figure 9).

This model has been analyzed by Hochman and McCord [41] in the distributed form and Mah [55] in the stagewise form. It is especially useful for describing RTDs for bubble columns or vigorously bubbling fluidized beds where the flow at the center rises rapidly and flows down at the walls, or for agitated reactors which exhibit departures from perfect mixing. The circulation loop can be divided into two flow regions joining at the inlet and separating at the outlet (i.e., recycle), with crossmixing or mass interchange between regions as schematized in Figure 9. In most cases the dispersion in each phase is negligible in comparison to the sufficiently large crossflow, which produces a similar effect. Thus the model is described by:

$$\frac{\partial C_1}{\partial t} + \bar{u}_1 \frac{\partial C_1}{\partial z} = - \frac{k_g a_s}{V_1/V} (C_1 - C_2) = -k_1' \, (C_1 - C_2) \qquad (3.33)$$

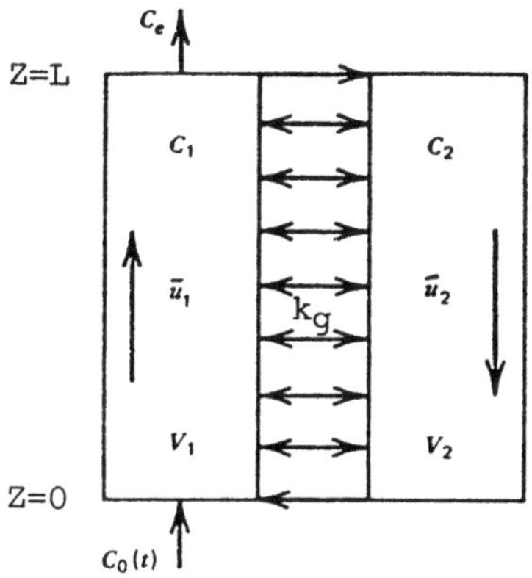

Figure 9. Recycle-flow model with crossmixing.

$$\frac{\partial C_2}{\partial t} - \bar{u}_2 \frac{\partial C_2}{\partial z} = \frac{k_g a_s}{V_2/V} (C_1 - C_2) = k_2' (C_1 - C_2) \tag{3.34}$$

with the two boundary conditions,

$$C_1(0,t) = \frac{1}{1+r}[C_o(t) + rC_2(0,t)] \quad , \tag{3.35}$$

where $r = Q_2/Q_o$ is the recycle ratio, and

$$C_1(L,t) = C_2(L,t)\,;\; C_1(L,0) = C_2(L,0) = 0, \text{ for an impulse} \tag{3.36}$$

As would be expected, for limiting cases this model reduces to other familiar forms; i.e.: as $r \to \infty$, the model approaches perfectly mixed behavior; as $r \to 0$ it becomes the distributed limit of the modified mixing-cell model, crossflow model, or time delay model; as $k_g \to \infty$ the model approaches plug flow, with net flow rate $Q_o = Q_1 - Q_2$; and as $k_g \to 0$ it becomes the ordinary recycle model with time delay in both fluid regions.

The solution to Eqns. (3.33) and (3.34) for an impulse tracer input in Laplace transform space is given by Hochman and McCord [41]:

$$E(s) = \frac{\alpha \exp[(t_m - t_1)s + \beta]}{[(2r + 1)t_m s]\sinh\alpha + \alpha\cosh\alpha} \quad , \tag{3.37}$$

where

$$t_i = L/\bar{u}_i = V_i/Q_i \quad ,$$

$$t_m = (t_1 + t_2)/2 \quad ,$$

$$\alpha = [t_m^2 s^2 + 2(2r + 1)\beta t_m s + \beta^2]^{\frac{1}{2}} \quad , \text{ and}$$

$$\beta = k_g a_s V/Q_o 2r(r + 1) \quad .$$

The first moment and variance of the RTD are:

$$\mu_1 = t_1 + 2rt_m = V/Q_o = \bar{t} \tag{3.38}$$

and

$$\sigma_\theta^2 = \frac{4t_m^2 r(r + 1)}{\bar{t}^2 \beta} \quad 1 - \frac{1}{2\beta}(1 - e^{-2\beta}). \tag{3.39}$$

From Eq. (3.38) it can be seen that of the three parameters, t_1, t_2, and r, only two are independent. Thus this model usually has three parameters: $k_g a_s$, r, and t_1 (or t_2), for example. Hochman and McCord [41] proposed a two tracer technique whereby t_1 is determined from an impulse at Z = 0, and t_2 from an impulse at Z=L. r can then be found from Eq. (3.38), and $k_g a_s$ from Eq. (3.39). The stagewise form of this model analyzed by Mah [55] has two additional parameters -- the number of cells in each region, N_1 and N_2.

3.6 Pulsed Flow Model.

Lerou, Glasser, and Luss [29] developed an RTD model which accounts for the major factors responsible for liquid phase dispersion in pulsed, cocurrent gas-liquid downflow in fixed beds. In this flow regime slugs are formed due to liquid bridging of gaps between adjacent particles. These liquid slugs are pushed down the bed in a piston-like fashion by the gas flow. The slug continually sheds a fraction of its liquid, which forms a slow-moving film while picking up additional liquid shed by previous slugs. Inert tracer experiments indicate that tracer is picked up and shed by several consecutive slugs, and that spreading of the tracer is small upon transit of the bed. These results suggest that the slugs do not remain integral, but rather exchange material rapidly with the liquid film. Thus the primary cause of liquid phase dispersion is the difference in velocity and rapid mixing between slugs and films.

This behavior has been described by Lerou, et al.[29] using a model which views the liquid phase as two continuous, cocurrently flowing liquid regions with different velocities and continuous mass exchange between them. The slug phase can be approximated as continuous since the mean residence time of the liquid in the column is significantly greater than that of a slug. The inert tracer mass balances for the two phases are given exactly by Eqns. (3.33) and (3.34) for the recycle-flow model with crossmixing [41, 55]. The appropriate boundary conditions, however, are not the same; e.g., for an impulse input:

$$C_1(0,t) = C_2(0,t) = \delta(t) \quad 1 = \text{liquid film phase} \tag{3.40a}$$

$$C_1(L,0) = C_2(L,0) = 0 \,. \quad 2 = \text{liquid slug phase} \tag{3.40b}$$

Lerou, et al. transformed Eqns. (3.33) and (3.34) to the chromatography (i.e., moving front) equation and obtained the variance of the RTD, which is:

$$\sigma_t^2(Z) = \frac{2k_1'k_2'(\bar{u}_2 - \bar{u}_1)^2 Z}{(k_1'\bar{u}_2 + k_2'\bar{u}_1)^3}\left\{1 + \frac{\exp\left[-\left(\frac{k_1'}{\bar{u}_1} + \frac{k_2'}{\bar{u}_2}\right)Z\right] - 1}{\left(\frac{k_1'}{\bar{u}_1} + \frac{k_2'}{\bar{u}_2}\right)Z}\right\} \tag{3.41}$$

where $k_n' = k_g a/h_{Li}$.

The single-parameter dispersion model erroneously predicts that tracer spreads upstream of the injection point and that it propagates at infinite speed in the bed. This model does not exhibit either of these properties, but it contains four parameters (k_1', k_2', $\bar{u}_1$, $\bar{u}_2$) of which only $\bar{u}_2$, the mean slug velocity, can be measured independently. However, Lerou, et al. have shown that for pulsed flow,

$$\frac{\sigma_t^2(Z)}{\bar{t}} \simeq \frac{2}{k_1'} \,, \tag{3.42}$$

where $\bar{t}$ is the total liquid volume divided by the total liquid volumetric flow rate. Thus, actually k_1' is the single parameter which characterizes dispersion, and it can be determined from the variance and the mean residence time. The relationship between this model and the single-parameter dispersion model was shown [29] to be:

$$D_{Z_L} = \bar{u}_L^2/k_1'. \tag{3.43}$$

The predictions of the model agreed with impulse response experiments. An important conclusion derived from the results of Lerou, et al. [29], as well as those of Matsuura, et al. [49], is that liquid phase dispersion in pulsed flow is much smaller than in the trickle flow regime. This has been attributed to a high rate of mass interchange between the films and slugs, and the mixing between different parallel flowing liquid streams in the bed caused by the pulses [24]. This effect increased with liquid flow rate, and liquid phase dispersion attained a constant asymptotic value for sufficiently high gas velocity [29]. Thus pulsed flow is preferred to trickle flow when plug flow is favored to attain high conversion. Additional advantages cited for pulsed flow operation are increased wetting of the catalyst pellets, high interphase transport coefficients, and improved flow distribution throughout the bed.

3.7 General Comments on RTD Models and Backmixing.

As indicated previously, most of the RTD models have been applied to describing liquid phase backmixing in multiphase reactors. However, in addition to gas phase dispersion in trickle beds [33, 34], the gas phase RTDs in three-phase fluidized beds have been measured by Schügerl [56], Michelsen and Østergaard [57], and Østergaard and Michelsen [58], and discussed by Massimilla [59]. Schügerl [56] reported that mixing intensity decreases in both gas and liquid phases from the top of the bed to the bottom. Also, the gas phase Peclet number increases with gas flow rate at low liquid flow rates, and exhibits a maximum with gas flow rate at high liquid flow rates in three-phase fluidized beds [56]. The gas phase Peclet number appears to decrease with increasing liquid flow rate, although this effect is unclear [56]. The effect of suspended particles on liquid phase backmixing has also been studied [56-58], and Vail, et al. [60] showed that the presence of solids serves to reduce liquid phase backmixing. The backmixing of suspended particles in a three-phase fluidized bed has been studied by Kato, et al. [61] and Imafuku, et al. [62]. For small particles in small diameter columns, the axial dispersion coefficient for the solids was found to be the same as that of the liquid.

Most of the available RTD models are strongly related to each other, even when they differ as to form (i.e., differential or stagewise), primarily because they represent various degrees and combinations of similar phenomena such as mass interchange with stagnant regions, bypass, recycle, crossflow, mixing, dispersion, etc. Thus the models discussed here are by no means intended to be an exhaustive compilation but rather a generically representative group of the numerous models which have appeared in the literature. The optimum model for a particular reactor type or mode of operation is probably the simplest one which most closely approximates the actual physical situation. Thus differential models may be better for trickle beds, for example, whereas combinations of differential and stagewise models may better approximate physically compartmentalized or segmented reactors. In many cases, the choice of a particular RTD model becomes more a matter of taste or convenience than one based on purely technical grounds.

In order for the RTD approach to be useful in predicting the performance of a multiphase chemical reactor, it must be combined with the intrinsic chemical kinetics and the mass and heat transport which occur within a reactor. Even though in principle the chemical kinetics may be available, Satterfield [15] points out that the conversion according to Eq. (3.3) is strictly true only for first order kinetics. For other than first order kinetics, the conversion becomes a function of the RTD and micromixing between fluid elements, even though bounds on the relative importance of this effect can be estimated. Also, most of the RTD models have been developed for bubble and trickle flow conditions, and their applicability to other flow regimes for the most part remains to be investigated.

The most fundamental problem in using RTD models to predict multiphase reactor performance, however, is the characterization of the effectiveness of contacting between phases and its relationship to the RTD. For example, although there is no doubt that local liquid velocity and holdup are important in liquid-solid contacting, the fundamental relationships are not well known. Even if all the appropriate relationships were well characterized, which they are not, it is obvious that the resultant mathematical model describing the performance of a multiphase chemical reactor could become quite complex, even when the RTD is described by a single-parameter model. For this reason, Shah, et al. [22] believe that RTD models with more than two parameters will find very limited use for reactor design and modeling. This is especially true in view of the fact that the single-parameter dispersion model has proven adequate for modeling trickle bed reactors, at least for simple reactions.

Thus having described RTD models and their role in multiphase chemical reactor modeling, attention is now focussed on more specific reactor models which consider contacting effectiveness and interphase mass transport explicitly.

4. FIXED BED REACTORS

4.1 General Distributed Isothermal Trickle Bed Model.

Here a general isothermal model of a trickle bed reactor, following Shah [24], is presented for the solid-catalyzed reaction:

$$A(g) + C(\ell) \longrightarrow B(\ell) \tag{4.1}$$

The principal features of this model are schematized in Figure 10.

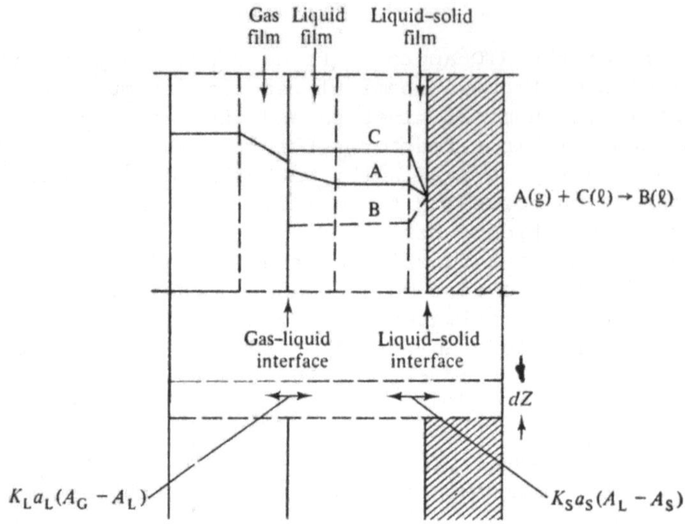

Figure 10. General distributed isothermal trickle bed model [24].

The mass balances, including interphase and intraparticle mass transport resistance and dispersion, are:

A balances

$$D_{ZG} \frac{d^2 A_G}{dZ^2} - U_G \frac{dA_G}{dZ} - K_L a_L (A_G - A_L) = 0 \quad (4.2)$$

$$D_{ZL} \frac{d^2 A_L}{dZ^2} - U_L \frac{dA_L}{dZ} + K_L a_L (A_G - A_L) - K_{SA} a_s (A_L - A_S) = 0 \quad (4.3)$$

$$K_{SA} a_s (A_L - A_S) = k_2 a_s A_s C_s \eta \quad (4.4)$$

C balances

$$D_{ZL} \frac{d^2 C_L}{dZ^2} - U_L \frac{dC_L}{dZ} - K_{SC} a_s (C_L - C_S) = 0 \quad (4.5)$$

$$K_{SC} a_s (C_L - C_S) = k_2 a_s A_s C_s \eta \quad (4.6)$$

B balances

$$D_{ZL} \frac{d^2 B_L}{dZ^2} - U_L \frac{dB_L}{dZ} + K_{SB} a_s (B_S - B_L) = 0 \quad (4.7)$$

$$K_{SB} a_s (B_S - B_L) = k_2 a_s A_s C_s \eta \quad (4.8)$$

The eight boundary conditions required to complete the mathematical definition of this model are taken as the Danckwerts boundary conditions for Eqns. (4.2, 4.3, 4.5, and 4.7). The holdups of the gas and liquid phases are assumed constant throughout the reactor; gas and liquid phase dispersion are considered separately, but the dispersion coefficients of all the species in each phase are assumed equal; $K_L a_L$ is the overall gas-liquid mass transfer coefficient, and the individual species liquid-solid mass transfer coefficients are denoted by $K_s a_s$; η is the appropriate isothermal intraparticle effectiveness factor; k_2 is the intrinsic second order surface reaction rate constant; and U_G and U_L are the superficial gas and liquid phase velocities, respectively.

Although the full model is best solved numerically, various simplifying assumptions lead to limiting forms convenient for analytical solution. For example, Goto, et al. [17, 63] obtained an analytical solution for the case where the gas phase concentration is constant and the liquid phase reactant C is in excess. In this case Eqns. (4.2, 4.5, and 4.6) are unnecessary, and the remaining four equations are solved with the corresponding Danckwerts boundary conditions for Eqns. (4.3) and (4.7). The analytical results were used to calculate the efficiency of the trickle bed reactor, η_t, defined as the ratio of the actual production rate to that possible under conditions of zero mass transfer resistance (i.e., at the intrinsic reaction rate); viz.,

$$\eta_t = \frac{U_L\left(B_{L_e} - B_{L_o}\right)/L}{k_2 C_s a_s A_L^*} = \frac{\left(y_{B_e} - y_{B_o}\right)\eta\eta_L}{Da} , \qquad (4.9)$$

where A_L^* is the saturated liquid concentration of the gaseous reactant, $Da = k_2 C_s \eta\eta_L L/U_L$, $\eta_L = (1 + k_2 C_s \eta/K_s)^{-1}$, and $y_B = B_L/A_L^*$. The overall effectiveness factor for this model, η_o, is given by:

$$\eta_o = \left|1 + k_2 C_s \eta\eta_L / K_L a_L \eta_G\right|^{-1} , \qquad (4.10)$$

assuming the gas film resistance is negligible. Goto, et al. [17, 63] found good agreement with experimental results from three studies involving the oxidation of ethanol [64], the hydrogenation of α-methyl styrene [65], and the hydrogenation of aniline [66]. In all three cases, the gas film resistance was negligible, and the liquid-solid effectiveness, η_L, was in the range 0.67 to 0.86, while η was in the range 0.005 to 0.025. Thus the overall efficiency of the trickle bed, η_t, was controlled by intraparticle diffusion. These results suggest that the performance of trickle bed reactors can be improved by using smaller, more accessible catalyst particles, consistent with pressure drop limitations [17, 63].

In the same work, Goto, et al. [63] distinguished three regimes of the dependence of $\eta_t/\eta\eta_L$ on Da as shown in Figure 11: (1) small Da -- $\eta_t/\eta\eta_L \simeq y_{AL,o}$ independent of Da, Pe, and η_o, and thus reactor efficiency can be improved by presaturating liquid feed with the gaseous reactant, A; (2) intermediate Da -- $\eta_t/\eta\eta_L$ dependent on all the parameters. In this region plug flow is preferable to complete mixing if $y_{AL,o} > \eta_o$ and vice versa for $y_{AL,o} < \eta_o$; (3) large Da -- $\eta_t/\eta\eta_L \simeq \eta_o$ independently of all the parameters. The role of gas-liquid mass transfer can become significant.

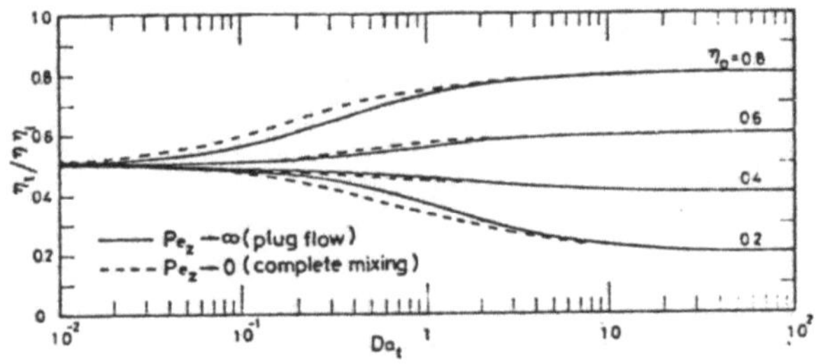

Figure 11. Trickle bed reactor efficiency for $y_{AL_o} = 0.5$ [63].

In another study, Goto and Smith [67] also analyzed an isothermal model quite similar to Eqns. (4.2) - (4.9) for the reversible, solid-catalyzed reaction,

$$A(g) + C(\ell) \rightleftharpoons B(g) \tag{4.11}$$

(i.e., the product can be transported back to the gas phase) for the effects of cocurrent and countercurrent gas-liquid operation. The model was reduced to an eighth-order, linear, differential equation for y_{Ag} as a function of dimensionless reactor length. The eight roots of the differential operator polynomial and the pre-exponential constants were solved for numerically [67]. Some typical results calculated with this model are presented in Figure 12, which shows how increasing the liquid rate (decreasing Da_ℓ) and/or the gas rate (decreasing Da_g) increase the reactor efficiency for conversion to products. Also, even though the effect is not very marked for the intermediate values of the Peclet numbers used, cocurrent operation is shown to be superior to countercurrent. For gas purification, however, countercurrent operation is preferable as long as the liquid phase feed contains no gaseous impurity. This was demonstrated for the removal of sulfur dioxide by Goto and Smith [68]. In any case however, the differences due to flow direction are not very large, except for plug flow conditions [67].

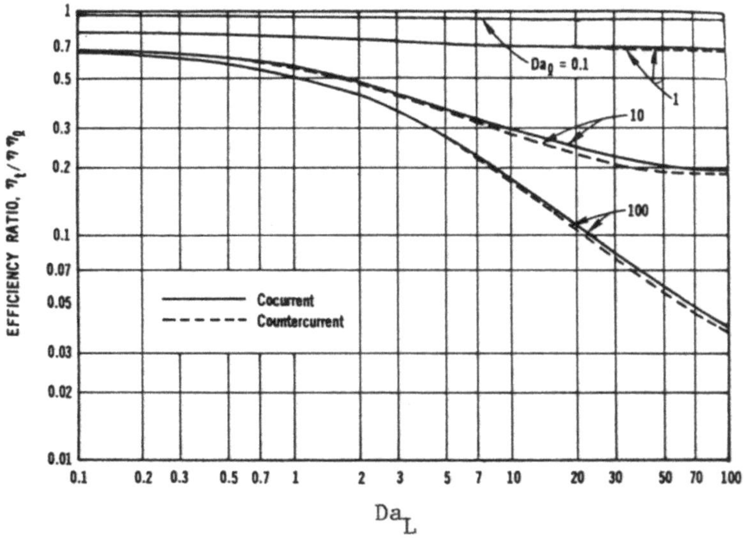

Figure 12. Trickle bed reactor efficiency as a function of liquid flow rate for cocurrent and countercurrent operation [67].

Shah and Paraskos [69] extended the analysis of Mhaskar [70] for an isothermal fixed bed model for the case of reaction (4.1) with no interphase and intraparticle transport resistance and no dispersion in the gas phase. Thus this model reduces to Eqns. (4.2 - 4.6) with $D_{ZG} = 0$ and the appropriate interphase transport terms eliminated; i.e.,

$$\frac{dy_A}{dz} = - P^* y_A y_B \tag{4.12}$$

$$\frac{d^2 y_B}{dz^2} = \pm \frac{dy_B}{dz} + P^* Q^* y_A y_B \quad Pe_L \tag{4.13}$$

where the + sign is for cocurrent operation and the - sign for countercurrent operation, and

$$P^* = \frac{a_s L}{\bar{u}_G \varepsilon_G A_o} \sqrt{\nu k_2 D_A B_o A^*_L{}^2} \quad ; \quad Q^* = \frac{\nu \bar{u}_G \varepsilon_G A_o}{\bar{u}_L (1 - \varepsilon_G) B_o} \quad ; \; Pe_L = \frac{\bar{u}_L L}{D_{ZL}},$$

ε_G is the fractional cross sectional area occupied by the gas, A_o and B_o are the inlet concentrations of A and B, D_A is the diffusivity of A in the liquid phase, A_L^* is the concentration of A in the liquid at saturation, ν is the stoichiometric ratio between A and B, and $\bar{u}_G$ and $\bar{u}_L$ are the mean velocities of the gas and liquid phases, respectively. Equations (4.12) and (4.13) along with the appropriate boundary conditions (the inlet condition for Eq. (4.12) and the Danckwerts boundary conditions for Eq. (4.13)) were solved by a collocation technique which assumes that the approximate solutions for monotonic nth order kinetics are of the same form as the analytical solution for a first order reaction, under otherwise similar reaction conditions. For the second order reaction considered here, the approximate solution was found to be within 10% of the exact numerical solution for most of the reaction conditions investigated. It is interesting to note, however, that unlike Goto and Smith [67], Shah and Paraskos [69] found that countercurrent operation is preferable to cocurrent operation for conversion of reactants to products for the Peclet numbers and reaction conditions examined. Since the primary difference between these two models is resistance to interphase transport, the importance of this effect is underscored by these results.

4.2 Trickle Bed Models and Plug Flow.

The general isothermal distributed model of the previous section can be considerably reduced in complexity under conditions where additional simplifying assumptions are warranted. For example, Satterfield [15] derived an ideal isothermal trickle bed reactor model by assuming:

1. liquid phase in plug flow;
2. no interphase or intraparticle mass and heat transport resistances; the liquid is saturated with the gas phase reactant;
3. first order, irreversible reaction at the liquid phase-catalyst interface;
4. completely wetted particles; and
5. no vaporization or condensation.

The first assumption eliminates the dispersion term, and actually in many situations trickle bed operation can be approximated by plug flow in the liquid phase without serious error [18,19,32] (see subsequent discussion on axial dispersion). The second assumption eliminates all the interphase transport terms and sets $\eta = 1$. The remaining assumptions are already present in the general isothermal distributed model.

A differential mass balance yields:

$$Q \frac{dC}{dV} = - k_1 C(1 - \varepsilon) \qquad (4.14)$$

where Q is the volumetric liquid flow rate, C is the limiting reactant concentration in the liquid phase, V is the reactor volume,

k_1 is defined as volume of liquid/volume of catalyst·time, and ε is the bed void fraction. Eq. (4.14) can be integrated to:

$$\ln \frac{C_o}{C_e} = \frac{V}{Q} k_1(1 - \varepsilon) = \frac{k_1(1 - \varepsilon)}{\text{LHSV}} , \tag{4.15}$$

where C_o and C_e are the feed and exit concentrations, respectively, and LHSV is the liquid hourly space velocity. For second order kinetics (i.e., $-r = k_2C^2$), this model yields:

$$\frac{1}{C_o} - \frac{1}{C_e} = \frac{k_2(1 - \varepsilon)}{\text{LHSV}} . \tag{4.16}$$

The assumption of no intraparticle diffusion resistance can readily be relaxed by incorporation of the appropriate isothermal effectiveness factor, η. However, overall catalyst effectiveness may also depend on liquid flow rate. Henry and Gilbert [71] proposed to account for this by multiplying Eq. (4.15), for example, by the total liquid holdup, h_L, i.e.,

$$\ln \frac{C_o}{C_e} = \frac{k_1(1 - \varepsilon)\eta h_L}{\text{LHSV}} \tag{4.17}$$

(It should be noted that liquid holdup is implicitly included in the liquid phase dispersion in terms of the model in Section 4.1 [72].) Using the correlation proposed by Satterfield, et al. [65]; viz.,

$$h_L \propto [Fr_L/Re_L]^{1/3} , \tag{4.18}$$

where $Fr_L = G_L^2/\rho_L^2 d_p g$ is the liquid Froude number, $Re_L = G_L d_p/\mu_L$ is the liquid Reynolds number, and the other parameters are defined as previously. Substituting Eq. (4.18) into Eq. (4.17) yields:

$$\ln \frac{C_o}{C_e} \propto L^{1/3}(\text{LHSV})^{-2/3}(d_p)^{-2/3}(\mu_L/\rho_L) , \tag{4.19}$$

where L is the reactor length. Henry and Gilbert [71] demonstrated the conversion dependence predicted by Eq. (4.19) for hydrocracking, hydrodenitrogenation, and hydrodesulfurization in trickle bed reactors.

Mears [72], however, has criticized the process model of Henry and Gilbert [71] on the grounds that: (1) the correlation of Satterfield, et al. [65] (Eq. 4.18) applied to liquid holdup on a vertical string of glass beads and a special definition of the Reynolds number for that system; and (2) the model suggests unrealistic improvement in conversion for large reactor scaleups. As an alternate explanation to the observed dependence of conversion on L and LHSV Mears [72] proposed that the reaction rate is directly proportional to the fraction of external catalyst surface effectively wetted by the flowing liquid; i.e., Eq. (4.15) should be written as

$$\ell n \frac{C_o}{C_e} = \frac{k_1(1-\varepsilon)\eta(a_w/a_t)}{\text{LHSV}} , \tag{4.20}$$

where a_w is the effective wetted surface area, and a_t is total surface area of the catalyst. Mears invoked two correlations for a_w/a_t: that of Puranik and Vogelpohl [73] for pilot scale reactors:

$$\frac{a_w}{a_t} = 1.05\ \text{Re}^{0.047}\ \text{We}^{0.135}(\sigma_c/\sigma)^{0.206} , \tag{4.21}$$

where σ is the liquid surface tension, σ_c is critical liquid surface tension (at zero contact angle), and $\text{We} = G_L{}^2 d_p/\sigma\rho_L$ is the Weber number; and the correlation of Onda, et al. [74] for larger liquid mass velocities:

$$\frac{a_w}{a_t} = 1 - \exp\ - 1.36\ \text{Ga}^{0.05}\ \text{We}^{0.2} \left(\frac{\sigma_c}{\sigma}\right)^{0.75} , \tag{4.22}$$

where $\text{Ga} = d_p{}^3 g\rho_L{}^2/\ {}_L{}^2$ is the Galileo number. Eqns. (4.21) and (4.22) coincide over the range $0.136 < G_L < 1.36$ kg/s m^2. Eq. (4.21) when substituted into Eq. (4.20) yields

$$\ell n \frac{C_o}{C_e}\ \alpha\ L^{0.32}\ (\text{LHSV})^{-0.68}\ d_p{}^{-.18} \left[\mu_L/\rho_L\right]^{-0.05} \left(\frac{\sigma_c}{\sigma}\right)^{0.21} \eta d_p \tag{4.23}$$

It is quite obvious that Eqns. (4.19) and (4.23) predict the same dependence of conversion on reactor length and liquid space velocity, while differing slightly on the effects of d_p, μ_L/ρ_L, and σ. The correlation of Onda, et al. [74], when combined with Eq. (4.20) yields:

$$\ln \frac{C_o}{C_e} = \frac{k_1(1-\varepsilon)\eta}{\text{LHSV}} \left\{1 - \exp\left[\gamma L^{0.4}(\text{LHSV})^{0.4}\right]\right\}, \tag{4.24}$$

where the constant γ includes the dependence on μ_L, ρ_L, σ, and d_p.

Paraskos, et al. [75] found that logarithmic plots of $\ln(C_o/C_e)$ vs. $(\text{LHSV})^{-1}$ and bed length, L, for hydrotreating 53% reduced Kuwait crude (Kuwait atmospheric tower bottoms, KATB) and various gas oils were indeed linear. However, the slopes were found to be dependent on reactor temperature and the nature of the reaction, and varied between 0.532 to 0.922 for the $(\text{LHSV})^{-1}$ dependence, and 0.498 to 0.55 for the L dependence. Also, Montagna and Shah [76] obtained a slope of 0.2 for removal of sulfur, vanadium, nickel, carbon residue, nitrogen, and pentane insolubles from 36% reduced Kuwait crude. Thus these results substantiate the liquid space velocity and bed length dependence of Henry and Gilbert [71] and Mears [72]. Of the two models, however, the effective catalyst wetting model has been shown to better correlate experimental data on hydrodesulfurization of 22% reduced Kuwait vacuum tower bottoms and 36% reduced KATB [77]. The catalyst wetting model is also more reasonable in that it predicts conversion should decrease with increasing viscosity, than the increase with viscosity implied by Eq. (4.19) [17, 18], and that the reaction rate is proportional to the wetted catalyst area, rather than the liquid volume (holdup).

The effect of catalyst bed length on conversion can also be due to axial dispersion in the liquid phase, as shown by Mears [72, 78]. In order to assess the relative importance of this effect, and to establish conditions under which the plug flow assumption is valid, Mears [78] developed a simple perturbation criterion for the minimum reactor length to be within 5% of that required for plug flow, for monotonic nth order kinetics; i.e.,

$$L/d_p > \frac{20n}{Bo_L} \ln \frac{C_o}{C_e}, \tag{4.25}$$

where $Bo_L = U_L d_p / D_{ZL}$ is the Bodenstein (Peclet) number. Condition (4.25) correctly gives a minimum length of zero for zero-order reactions, which are not affected by dispersion. For a first order reaction condition (4.25) yields

$$Pe > 20\ Da, \tag{4.26}$$

where $Pe = U_L L / D_{ZL}$ and $Da = k_1 L / U_L$. However, Goto, et al. [63] have noted that Mears' criterion does not take interphase mass transport resistances into account, and is valid only if the reaction rate is independent of gas solute but dependent on the liquid reactant. Applying the same 5% criterion, they derived:

$$Pe > (Pe)_{\delta=0.05}, \qquad \text{for } y_{A_o} > \eta_o \tag{4.27a}$$

$$Pe > (Pe)_{\delta=-0.05}, \qquad \text{for } y_{A_o} < \eta_o \tag{4.27b}$$

where

$$(Pe)_\delta = \frac{(y_{A_o} - \eta_o)h^2}{\delta h \eta_o e^h + (y_{A_o} - \eta_o)(e^{\delta h} - 1)} \tag{4.28a}$$

$$h = Da\eta_t/(1 - \eta_o) \tag{4.28b}$$

$$\delta = \left(Da\eta_t - Da\eta_{t\ \text{plug flow}}\right)/Da\eta_t \,. \tag{4.28c}$$

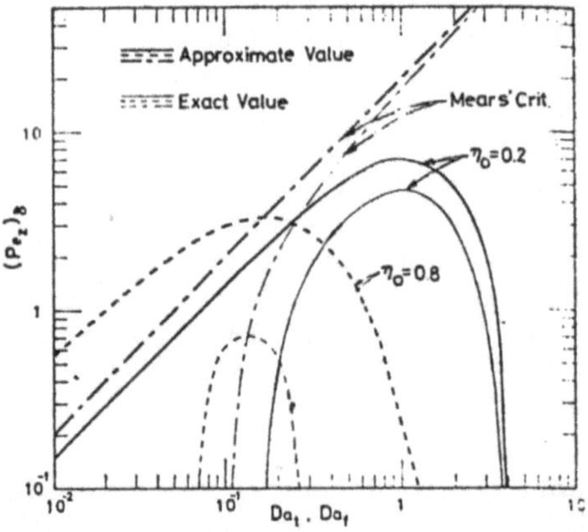

Figure 13. Axial dispersion criteria of Mears [78] and Goto et al. [63].

Criteria (4.25) and (4.26) are shown in Figure 13 from Goto, et al. [63]. Two curves each were calculated for Mears' criterion, $\eta_o = 0.2$ and $\eta_o = 0.8$ since approximation (4.28a) is not valid at low Pe_δ and thus was calculated numerically for the lower curve in each case. However, it is obvious that the criteria differ qualitatively. Mears' criterion is monotonic with Da, whereas that of Goto, et al. exhibits a different maximum for each η_o which guarantees no axial dispersion effect for Peclet numbers greater than that at the maximum.

Although only isothermal operation has been explicitly considered thus far, many hydroprocessing trickle bed reactors normally operate under near-adiabatic conditions. Shah and Paraskos [79] have extended Mears' analysis to adiabatic trickle bed reactors. Approximate solutions to a model for the special cases of plug flow and of equal Peclet numbers for the dispersion of mass and heat in the near-plug flow regime were obtained and compared to numerical solutions. Resultant criteria for significant axial dispersion effects are presented in Figure 14, the parameters in which are defined:

$$\delta' = (E_a/RT_o)\times\left((-\Delta H)C_o/\rho_L C_{p_L} T_o\alpha\right) \quad (4.29a)$$

$$\alpha = 1 + U_G\rho_G C_{p_G}/U_L\rho_L C_{p_L} \quad (4.29b)$$

$$R_n' = A_o(1-\varepsilon)\tau\, C_o^{\,n-1}\exp(-E_a/RT_o) \quad , \quad (4.29c)$$

where E_a is the activation energy, $(-\Delta H)$ is the heat of reaction, A_o is the frequency factor, τ is the space time, R is the universal gas constant, and the other parameters are as previously defined. The most significant aspect of the criteria is that for low values of R_n', lower Peclet numbers are required to minimize axial dispersion in adiabatic reactors than in isothermal reactors, while the converse holds true for high values of R_n'. It was found that dispersion effects could be significant in hydrodesulfurization and hydrocracking of gas-oils [79].

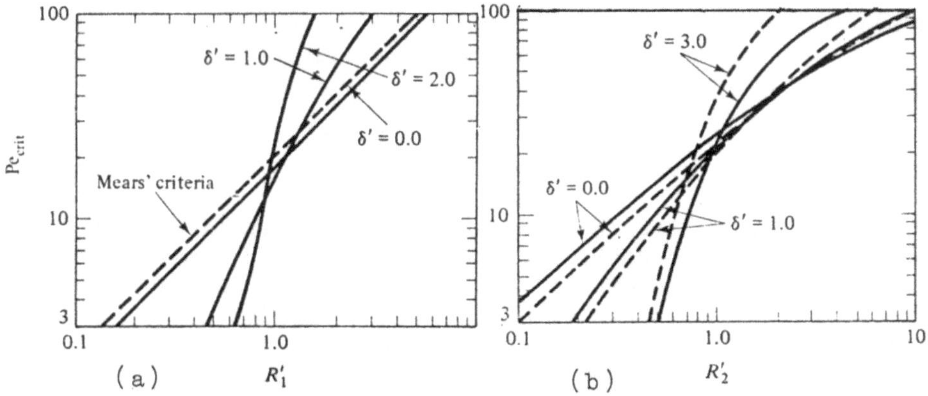

Figure 14. Nonisothermal axial dispersion criteria of Shah and Paraskos [79] and Mears [78] (δ'=0): (a) n=1; (b) n=1.5(---), 2(——).

4.3 Nonisothermal Trickle Bed Reactor Models.

The various trickle bed reactor models discussed here have all assumed isothermal operation. However, many industrial hydroprocessing reactions have a relatively large exothermic heat of reaction ($-\Delta H_r \simeq 30 - 50$ kcal/mol). Therefore, rigorous trickle bed models should include both mass and energy balances. Moreover, due to the relatively large volume-to-external surface area ratio, commercial hydroprocessing trickle bed reactors usually approximate adiabatic operation.

In studying axial dispersion effects in adiabatic trickle bed hydroprocessing reactors, Shah and Paraskos [79] developed a non-isothermal model for nth order monotonic kinetics for the case where the limiting reactant is either in the liquid phase or is being continuously transferred from the gas phase such that the gas and liquid phase concentrations are always in equilibrium. Under these conditions the mass balance is given in dimensionless form by:

$$\frac{1}{Pe_m}\frac{d^2v}{dz^2} - \frac{dv}{dz} - R_n \exp[-\gamma(1 - \theta)]v^n = 0 , \qquad (4.30)$$

where

$$R_n = k_n \tau C_o^{n-1}(1 - \varepsilon_c) \qquad (4.31a)$$

$$\tau = \frac{L(1 - \varepsilon)}{U_L(1 + b)(1 - \varepsilon_c)} = \frac{L_u}{U_L} \qquad (4.31b)$$

$$\theta = (T - T_o)/T_c \qquad , \qquad (4.31c)$$

ε_c is the void fraction of undiluted catalyst, L_u is the length of undiluted bed, b is the dilution ratio of inerts to catalyst, T and T_o are the temperature and feed temperature, respectively, and all the other symbols are as given previously. In Eq. (4.30) it was assumed that:

$$\exp(-E_a/RT) \simeq \exp[-\gamma(1 - \theta)] \quad ; \quad \theta << 1. \qquad (4.32)$$

In developing the energy balances it was assumed that: (1) radial gradients are negligible and the dispersion model adequately represents heat transport in the axial direction; (2) gas, liquid and solid temperature are equilibrated at all points in the reactor; and (3) molecular conduction is negligible in all three phases. With these assumptions, separate energy balances were written for the gas and liquid phases taking into account the rate and heat of evaporation. These two equations were added to eliminate the evaporation term. The final energy balance in dimensionless form is:

$$\frac{1}{Pe_h}\frac{d^2\theta}{dz^2} - \frac{d\theta}{dz} + R_n^* \exp[-\gamma(1 - \theta)]v^n = 0 \qquad , \qquad (4.33)$$

where

$$Pe_h = \left[\frac{L}{D_{HL} + D_{HG}}\right] U_f \rho_f c_{p_f} \qquad (4.34a)$$

$$R_n{}^* = \frac{k_n(1 - \varepsilon_c)C_o^{\,n}(-\Delta H_r)}{T_o \rho_L c_{p_L} \alpha} = R_n \Gamma \quad ; \quad \Gamma = \frac{(-\Delta H_r)C_o}{T_o \rho_L c_{p_L} \alpha} \tag{4.34b}$$

$$U_f \rho_f c_{p_f} = U_G \rho_G c_{p_G} + U_L \rho_L c_{p_L} \quad , \tag{4.34c}$$

and α is the same as given by Eq. (4.29b), D_{HL} and D_{HG} are axial thermal dispersion coefficients in the liquid and gas phases, respectively, and all the other symbols are as given previously. Shah and Paraskos [79] further assumed that $D_{HG} << D_{HL}$, and thus D_{HG} was neglected, and $D_{HL}/\rho_L c_{pL} \simeq D_L$ for mass dispersion such that $Pe_h \simeq Pe_m \simeq Pe$. The resultant model was solved using the Danckwerts boundary conditions:

$$\left.\begin{aligned} 1 - v &= -\frac{1}{Pe}\frac{dv}{dz} \\ \theta &= \frac{1}{Pe}\frac{d\theta}{dz} \end{aligned}\right\} \quad ; \quad z = 0^+ \tag{4.35a}$$

$$\frac{dv}{dz} = \frac{d\theta}{dz} = 0 \qquad ; \quad z = 1^- . \tag{4.35b}$$

For plug flow,

$$\theta = \Gamma(1 - v) \tag{4.36}$$

and the mass balance becomes

$$-\frac{1}{Pe}\frac{d^2v}{dz^2} + \frac{dv}{dz} + R_n{}' \exp(\gamma')\exp(-\gamma' v)v^n = 0 \ , \tag{4.37}$$

where $\gamma' = \gamma\Gamma$ and $R_n{}' \Rightarrow R_n \exp(-\gamma')$, as given by Eqns. (4.29a) and (4.29c), respectively. A series solution was obtained for n = 1 and n = 2. Approximate solutions for large Pe were obtained using a perturbation technique [80], and compared with numerical solutions obtained with a Runge-Kutta technique [70]. These results are shown in Figure 15 from which it can be seen that the approximate second order perturbation solution agrees quite well with the numerical solution for Pe > 3. Also v(1) is shown to be a function of δ', and not δ and Γ separately.

Shah, Mhaskar, and Paraskos [81] developed a plug flow nonisothermal model to study the dynamics of adiabatic trickle bed reactors with aging catalyst and quench fluid injection. The adiabatic temperature rise due to heat generation by chemical reaction in hydrodesulfurization and hydrocracking trickle bed reactors is usually controlled by the addition of a quench fluid (normally hydrogen) at one or more positions along the reactor

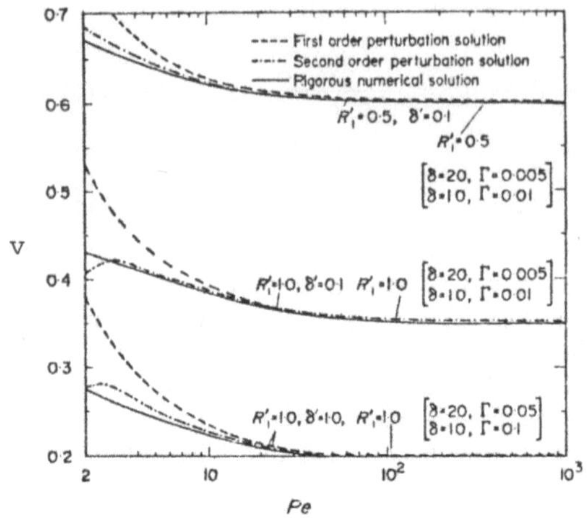

Figure 15. Comparison of numerical and approximate solutions for the nonisothermal trickle bed dispersion model (n=1) [79].

length. Also, in HDS reactors for example, catalytic activity decays nonuniformly along the reactor length due to metal and coke deposition. This is usually countered by increasing the feed temperature or by decreasing the quench, or both. Usually, this procedure continues until the attainment of a maximum allowable temperature, at which time the bed is regenerated. The time required for this procedure is known as the cycle life. The model developed by Shah, et al. [81] enables the optimization of the quench position(s) and the variation of feed temperature with time to achieve maximum cycle life. More recently, Mhaskar, Shah, and Paraskos [82] extended this model to a single liquid quench and two gas quenches. Mhaskar and Shah [83] also carried out a similar analysis for nonadiabatic operation with either constant wall temperature or constant wall heat flux. Since radial thermal gradients can be important for large-scale industrial reactors, they can be taken into account using correlations for effective radial thermal conductivity, k_e, and wall heat transfer coefficient, h_w, obtained by Specchia and Baldi [84], for example, for multiphase fixed beds operating in both the low and high interaction regimes.

A major complication in attempting to accurately model nonisothermal trickle bed reactors in which a strong exothermic reaction occurs is due to volatilization of the liquid phase, which may then react on dry catalyst at an increased rate of reaction. The primary causes for this behavior are believed to be the disappearance of the mass transport resistance offered by the liquid phase, and more effective utilization of the interior catalyst surface area due to higher gas phase diffusion coefficients. In many hydroprocessing trickle bed reactors, operated under conditions of high severity, as much as 90 to 100% of the liquid feed may be evaporated.

Sedricks and Kenney [85] studied the hydrogenation of crotonaldehyde to n-butyraldehyde on partially-wetted catalyst consisting of palladium deposited on porous alumnina pellets. These authors concluded that the difference between rates of reaction on wet and dry catalyst must be due to: (1) nonequilibrium between gas and liquid phases, i.e., interphase mass transport resistance; (2) nonequilibrated temperature, i.e., interphase heat transport resistance; or (3) differences in activation energies associated with the reaction on the catalyst in contact with the gas or the liquid. It was suggested that operation under conditions of partial wetting could be used to some advantage in cases where the reaction on the wetted catalyst is severely diffusion-limited. Reaction of the gas phase on dry catalyst would proceed at rates closer to intrinsic rates, but at liquid phase temperatures due to the rapid rate of vaporization in comparison to other transport processes. A partially wetted bed thus exhibits aspects of a gas-solid reactor and a gas-liquid contactor with the liquid phase serving mainly as *in situ* source of gas phase and a sink for products. It was further suggested that some existing reactors owe their intrinsic performance primarily to this type of mechanism [85].

On the other hand, Sedriks and Kenney [85] point out that dry areas are obvious sites for hot spots and local temperature runaway for highly exothermic reactions. Also, Hanika, et al. [86] have shown experimentally that the transition from liquid to gas phase usually creates the undesirable effect of a narrow reaction zone with steep temperature gradients. This may expose the catalyst to local overheating and subsequent deactivation, or may induce undesirable side reactions. The occurrence of hot spots has been observed during hydrogenation of α-methyl styrene [87], cyclohexene [88], and 1,5 cyclooctadiene [86]. For the hydrogenation of cyclohexene Hanika, et al. [88] observed hysteresis of the rate of reaction with respect to hydrogen flow rate, average bed temperature and liquid phase concentration of cyclohexene. These effects were attributed to transition from liquid phase reaction to gas phase reaction regimes. The transition is accompanied by a change in apparent kinetics from zero to first order in cyclohexene concentration, as well as by a change from primarily intraphase resistance to interphase resistance. Hanika, et al. [86] also studied transient axial temperature profiles in a laboratory-scale, adiabatic trickle bed reactor for the hydrogenation of 1,5 cyclooctadiene. In this work it was concluded that hot spot formation is caused by the sudden and substantial reduction of mass transport resistance accompanying the evaporation of the liquid mixture. As a result, axial heat transfer within the bed gives rise to hot spot movement resembling flame propagation during combustion. In the same paper, Hanika, et al. [86] also reported experiments with the hydrogenation of cyclohexene at approximately double the liquid flow rate for 1,5 cyclooctadiene, for which the hot spot remains stationary.

It seems clear that the nonisothermal behavior of trickle beds leading to evaporation of the liquid phase and the formation of dry spots on the catalyst which result in hysteresis and complex transient hot spot behavior could be quite important in attempting to formulate predictive and design models for trickle bed reactors. However, as concluded by Sedriks and Kenny [85],

> "Given the complexity of the hydrodynamics involved and the dependence of the extent of wetting on both the geometry of the bed and the catalyst, this is unlikely to be subject to reliable *a priori* analysis in the near future. Thus even if the intrinsic kinetics of a reaction and the heat and mass transfer characteristics of all phases are well defined, prediction of trickle bed performance can remain a hit and miss procedure. Even scale-up from pilot plant work could be unreliable because some change in overall geometry is inevitable."

4.4 Trickle Bed Cell Models

All the trickle bed models considered in the preceeding in Section 4 are of the distributed parameter-type whereby dispersion is represented by a Fickian diffusion-like formulation. As shown in Section 3, however, the staged mixing-cell models offer an alternative to this description. Mixing-cell models possess some computational advantages over dispersion models since the steady-state representation is a set of nonlinear algebraic equations, rather than a set of second order, nonlinear differential equations. Moreover, complex hydrodynamic effects such as mass interchange between fluid regions, bypassing, dead volume, backmixing, etc., are all easily incorporated with the addition of simple algebraic terms.

Ramachandran and Smith [89] formulated and analyzed a mixing-cell model for an isothermal trickle bed reactor for monotonic (m,n)th order kinetics with respect to two reactants, A and B, completely wetted particles, and incorporating all mass transport resistances. A schematic of the model is presented in Figure 16. These authors present a method for numerical solution of the resultant set of nonlinear algebraic equations for general kinetics and an analytical solution for first order kinetics in A and zeroth order in B ($m = 1$, $n = 0$). Oxidation of a liquid phase component present in excess, or hydrodesulfurization reactions first order in hydrogen are examples of the latter case. Limiting cases of very slow and rapid gas-liquid mass transfer were also considered. An example of the application of this model was presented for the oxidation of formic acid in a trickle bed reactor. The mass transfer and data given by Goto and Smith [90], and the effective intra-

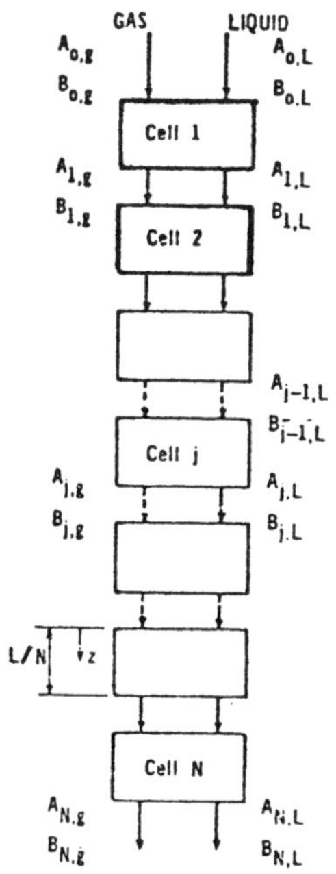

Figure 16. Mixing-cell model of an isothermal trickle bed reactor [89].

particle diffusivity and intrinsic kinetics available from the experiments of Baldi et al. [91] were used to obtain the parameters for the mixing-cell model. The experimental conversions of formic acid were found to be predicted by the mixing-cell model for values of N between 1 and 10. However, Ramachandran and Smith [89] point out that the operating conditions were such that the gas-liquid transport resistance was large, and thus the difference between observed conversions and plug flow (large N) predictions may not be due entirely to liquid phase backmixing.

Thus mixing-cell models have been shown to be just as suitable as dispersion models for the correlation and prediction of trickle bed reactor performance. The inclusion of more complex phenomena in trickle bed reactor models could probably be done most conveniently with a mixing-cell, or combined mixing-cell-dispersion model.

4.5 Regionalized Trickle Bed Model.

The preceeding trickle bed models in this section are essentially global descriptions. That is to say, they are continuum representations with characteristic semi-empirical parameters (e.g. holdup, wetting, dispersion) averaged over the entire spectrum of localized behavior (e.g., trickling films and rivulets, and stagnant liquid zones) which has been observed under trickle bed operating conditions. The relationships between these semi-empirical global parameters and the physical properties of the system are not well-defined. For example, the dependence of ln C_o/C_e on LHSV and L reported by Paraskos et al. [75] (Section 4.2) varied with temperature and the nature of the system in an unknown fashion. Also, of course, these models are incapable of describing local phenomena such as channeling and hot spots, for example.

In view of this situtation, Crine et al. [106] have suggested a more phenomenological description of the trickle bed reactor which takes into account the relative amounts and spatial distribution of localized zones of different behavior, emphasizing the discontinuous and random nature of the bed. This "regionalized" description focusses on the elementary transport cell envisioned as a sheaf of randomly directed facets representing portions of the external surface of the catalyst, intersecting at the contact point between particles. The facets may be covered with laminar films, and liquid rivulets, or may be dry. A stagnant liquid zone is located at the intersection point of the sheaf. The elementary transport cells may be coordinated in preferential zones of flow, or clusters. Three types of flow structures are identified: (i) isolated clusters or liquid blobs; (ii) liquid rivulets (anisotropic); (iii) liquid films (isotropic). For each of these structures there exists a minimum flow rate below which the structure becomes unstable. These minimum flow rates correspond to the minimization of energy dissipation associated with the creation of flow within each structure.

Crine et al. [106] applied such a description to the trickle bed hydrotreating data of Paraskos et al. [75] and Montagna et al. [76]. The global reactor efficiency, η_G, for a first order reaction in irregularly-shaped particles was given by:

$$\eta_G = \eta_E \cdot \eta_{CE} \tanh(\eta_i \phi_L/\eta_{CE})/\phi_L, \tag{4.38}$$

where η_E is the external efficiency,

$$\eta_{CE} = (a_w^2/a_t^2) + [(a_t - a_w)/a_t]\sqrt{a_w/a_t}, \tag{4.39}$$

a_t and a_w are the total and liquid film-covered catalyst surface area per unit bed volume, respectively, η_i is the wetted fraction of internal volume ($\eta_i \simeq 1$ for the present case), and ϕ_L is the Thiele modulus. The minimum flow rates for the rivulets and film structures were equated and treated as a single parameter, L_m $\sqrt{a_w/a_t}$ is the fraction of stagnant liquid zones exchanging mass with the catalyst surface. The fraction of wetted surface as a function of L_m was obtained from a numerical solution of the full mathematical description and fit to a polynomial. The parameters which best correlated the experimental data were obtained by maximizing the likelood function of a multiresponse model for conversion of sulfur, nickel, and vanadium. The different dependencies of the conversion of these three species on liquid flow rate was accounted for only by variation of L_m with catalyst size and fluid properties. L_m was shown to increase with increasing particle size (i.e., liquid-solid contacting efficiency decreases with increasing particle size) for the same feedstock, in agreement with experimental data [76]. The intrinsic rate constants computed were also in good agreement with experimental data [77]. Thus this approach can correlate trickle bed reactor behavior with a single parameter, L_m, which is independent of temperature and nature of the reaction, and varies only with fluid properties and catalyst size in a logical manner. These workers are currently developing a more detailed model along these lines to predict trickle bed performance from operating conditions.

4.6 Models for Other Flow Regimes.

As indicated in Section 2.1, trickle flow is only one of the flow regimes which can occur in multiphase fixed bed reactors. Due to the strong coupling between interphase transport phenomena and the hydrodynamics, models specific to particular flow regimes have been formulated. For example, the RTD model of Lerou et al. [29] for cocurrent pulsed downflow in fixed beds, and the various RTD models cited for trickle flow, are directed at specific flow regimes. Nevertheless, the general mathematical formulations of the dispersion and mixing-cell models presented above specifically for trickle bed reactors, are exactly the same for multiphase fixed bed reactors operating in other flow regimes. The only differences are in the magnitudes and relative importance of the various parameters representing backmixing in the fluid phases, interphase and intraparticle transport resistance, and fluid phase holdups.

For example, Pavlica and Olson [92] outlined a generalized dispersion model for the isothermal bubble column reactor with axial dispersion in both the gas and liquid phases and a pseudo-first order reaction in both phases. Shah formulated and analyzed a similar model for the same type of reactor in which a generalized (m , n)th order reaction occurs between a gaseous species A and a liquid species C in the liquid phase only, much like reaction (4.1). Thus the resultant mathematical model and the methods of solution resemble quite closely those presented in Section (4.1) for trickle bed reactors. Shah reports techniques for approximate and numerical solutions to the model for countercurrent and cocurrent flow for the cases of slow reaction (occurs in bulk fluid), fast reaction (occurs in liquid phase diffusion film), and instantaneous reaction (A and C cannot coexist in the liquid phase). Just as for the trickle bed models, these bubble column models assume constant gas phase velocity and pressure. Deckwer [93], however, has shown that gas phase variations due to the opposing effects of absorption and pressure drop could significantly affect the conversion. The resultant nonisobaric, variable gas velocity model was solved numerically.

5. SLURRY REACTORS

The major developments in the engineering and design of three-phase slurry reactors have been recently reviewed by Chaudhari and Ramachandran [21]. As for flow regimes other than trickle flow in fixed bed reactors, the multiphase models presented in Section 4 are also applicable to three-phase slurry reactors. For example, Goto and Smith [67] indicate that their three-phase packed bed model, presented in Section 4.1, with $Pe_G = \infty$ and $Pe_L = 0$ is applicable to a slurry reactor where bubbles of gas rise through the liquid creating plug flow in the gas and complete mixing in the liquid phase. Also, Ramachandran and Smith [89] state that their trickle bed mixing-cell model, discussed in Section 4.4, is directly applicable to modeling slurry reactors. However, these authors do point out that there are some practical differences between the two reactor types. They cite the use of smaller size catalyst particles in slurry reactors which reduce the importance of intraparticle diffusional intrusions and increase the specific fluid-solid interfacial area for mass transport. Although these effects tend to enhance conversion in slurry reactors, the inherent intense backmixing can reduce conversion under some conditions. It is clearly demonstrated, however, that slurry operation can yield higher conversions than in trickle flow systems with large intraparticle mass transport resistances, and vice versa [89].

5.1 Semi-Batch Slurry Reactors.

Chaudhari and Ramachandran [21] formulated and examined models for semi-batch slurry operation (batch slurry phase and continuous gas phase) in which the (m,n)th order reaction

$$A(g) + \nu B(\ell) \rightarrow \nu_p C(\ell) \tag{5.1}$$

occurs at a rate local rate of

$$\Omega = k_{(m+n)} A^m B^n . \tag{5.2}$$

For (1,1) kinetics, the analytical solution of the transient mass balance for B is:

$$\frac{B_{L,0}}{M_A} + \frac{1}{wk_2} \ln \frac{B_{L,0}}{B_L} = \frac{\nu A_{G,i}}{H_A} t \quad , \tag{5.3}$$

where $B_{L,0}$ and B_L are the initial concentration and that at batch time, t, of B in the liquid phase, w is the mass of catalyst per unit volume of slurry $A_{G,i}$ is the inlet concentration of A in the gas phase, $H_A = A_G/A_L$ is the solubility or Henry's law coefficient for A, and

$$M_A = \left[\left\{\frac{H_A\ Q}{V_L}\ (1 - \exp(-\alpha_A L))\right\}^{-1} + (K_s\ a_s)_A^{-1}\right]^{-1} \quad , \tag{5.4}$$

in which Q is the volumetric gas flow rate, V_L is the total volume of slurry, $\alpha_A = (K_L\ a_L/v_G H)_A$, and L is the total height of slurry above the gas inlet. Eq. (5.3) is applicable only for negligible intraparticle transport resistance ($\eta = 1$), which is a good approximation if the Thiele modulus, ϕ,

$$\phi = \frac{R}{3}\left[\frac{k_2 \rho_c B_{L,0}}{D_{e_A}}\right]^{1/2} < 0.2, \tag{5.5}$$

where R is the particle radius, ρ_c is the catalyst density, and D_{eA} is the effective intraparticle diffusivity. When condition (5.5) is violated, the effect of intraparticle mass transport can be taken into account by use of the appropriate isothermal effectiveness factor, η, in the transient mass balance. Upon solution, this formulation yields an integral term which is available in graphical form from Ramachandran and Chaudhari [94].

For a (1,0) reaction (e.g.,hydrogenation of nitrobenzene on Pd-based catalysts, and of styrene on 0.5% Pt-Al_2O_3 catalyst), the variation of B with batch time is:

$$(B_{L,0} - B_L)\ \frac{1}{M_A} + \frac{1}{\eta w k_1} = \frac{\nu A_{G,i}}{H_A}\, t \qquad (5.6)$$

Either analytically or numerically, similar relationships can be derived for other kinetics of interest.

Due to the batch nature of these reactors and the complex coupling between interphase and intraphase mass transport and intrinsic chemical kinetics, it is possible that the global kinetics might change with conversion or batch time. For example, the (1,0) kinetics, for which Eq. (5.6) is applicable, could shift to (1,1) kinetics when B falls below a critical concentration, $B_{L,c}$. In order to model such a situation, Eq. (5.6) would be used from $B_{L,0}$ to $B_{L,c}$ and then Eq. (5.6), or its counterpart with intraparticle resistance would be used from $B_{L,c}$ to B_L.

Chaudhari and Ramachandran [21] have also pointed out that the solubility of the gas phase reactant can be affected by concentration changes in the liquid phase. In the present context this means, for example, that the quantity $A_{G,i}/H_A$ could be a function of the concentration of B, in which case the batch time as given by Eq. (5.3) for example, would be incorrect. Taking this affect into account using an empirical correlation of the form:

$$A^* = \frac{A_{G,i}}{H_A} = (r_1 + r_2\, B_L + r_3\, C_L)\, p_{GA}\ , \qquad (5.7)$$

where r_1, r_2, r_3 are empirical constants, and p_{GA} is the partial pressure of A in the feed gas, Chaudhari and Ramachandran [21] derived the following expression for (1,1) kinetics for $\eta = 1$:

$$\frac{\ln \gamma}{M_A\, f_1} + \frac{\ln \gamma\ (B_L/B_{L,0})}{w k_2\, f_2} = \nu p_{G,A} t \qquad (5.8)$$

where

$$f_1 = r_2 - r_3 \nu_p/\nu \qquad (5.9a)$$

$$f_2 = r_1 + r_3 (C_{L,0} + \nu p B_{L,0}/\nu) \qquad (5.9b)$$

$$\gamma = \frac{f_2 + B_{L,0} f_1}{f_2 + B_L f_1}\ . \qquad (5.9c)$$

Batch time design equations similar to Eq. (5.8) could be derived for other kinetics as well. In addition to the effects of the solubility of gas phase reactant A on conversion, these authors point out that other product inhibition effects may significantly influence reactor operation and should be incorporated in modeling semi-batch slurry reactors; e.g., reaction products more strongly adsorbed on the catalyst than reactants, and catalyst deactivation by products.

Chaudhari and Ramachandran [95] have also analyzed the case of (0,0) kinetics and obtained analytical solutions for rates incorporating interphase and intraparticle transport resistance. For these kinetics, these authors have shown that for a given set of operating conditions, a critical gas phase concentration exists below which diffusional effects become important and the reaction rate becomes a function of the reactant concentration in the gas phase. Equations were developed to calculate the rate of a (0,0) reaction in a slurry reactor incorporating all the transport effects. The resultant model was demonstrated for ethynylation of formaldehyde, hydrogenation of dinitrotoluene, and the oxidation of cyclohexene in slurry systems.

Govindarao [96] formulated a generalized, nonisothermal, transient model for a slurry reactor operating continuously in a cocurrent or countercurrent mode , with appropriate boundary and initial conditions, which, in principle, could be solved numerically for the reactant concentration and the temperature in each phase. However, only the isothermal, semi-batch case with a first order reaction was actually solved. This latter model was examined for the dynamic response of the reactor to a step change in feed gas concentration. The effects of various process variables on the dynamic and steady-state performance of the reactor were evaluated.

The qualitative effects of the principal model parameters are presented below:

An increase in parameter X affects condition Y as shown.

X/Y → ↓	system dynamics	uniformity of steady-state concentration in gas phase	uniformity of steady-state concentration in liquid phase and on catalyst
gas velocity	faster	increase	increase
catalyst loading	faster	decrease	decrease
particle diameter	slower	increase	decrease
reactor diameter	faster	none	decrease
reaction rate constant	faster	decrease	decrease
gas phase Peclet number	--	decrease	minimal

Since the gas phase dispersion does not have a significant effect on the steady-state or transient behavior of the liquid phase or surface concentrations, the usual assumption of plug flow in the gas phase is satisfactory. Also, for almost the entire range of parameters examined, the liquid phase and surface concentrations at the inlet of the reactor differ significantly from those at the reactor exit. This indicates that the degree of dispersion in the slurry phase may be small, and thus the assumption of a perfectly mixed slurry phase is not generally valid.

Ramachandran and Smith [97] also studied the dynamic response of a semi-batch slurry reactor to a pulse or step input, and derived equations for the steady-state conversion and the first moment of the response in the gas phase to an impulse of reactant at the inlet. Overall irreversible and reversible first order reactions, and adsorption-reaction processes at the catalyst surface were considered from the point of view of evaluating rate coefficients from response data. Limiting cases without all the transport steps were examined. It was shown that under certain conditions it is possible to determine both the rate constant, k_1, and the effective intraparticle diffusivity, De, or the rate constant, k_1, and the liquid-particle mass transfer coefficient, K_s, from the same set of impulse response experiments. This is not possible with steady-state experiments. It was also shown how to determine the rate constant, k_1, and the gas-liquid mass transfer coefficient, K_L, or k_1 and De, provided K_s is known. In a similar fashion, the authors show that in principle, the adsorption rate and surface reaction rate constants for a two-step process can be evaluated separately, which is also not possible with steady-state measurements.

5.2 Continuous Slurry Reactors.

Goto and Smith [68] studied the performance of continuous slurry reactors for the catalytic oxidation of sulfur dioxide by molecular oxygen on activated carbon in a carbon-water slurry. These authors formulated a general model for the case of two gaseous reactants with perfect mixing of the slurry. The gas phase, however, could also have an RTD approaching that of a perfectly mixed fluid under conditions of intense agitation (i.e., the concentration of soluble reactant constant throughout all the bubbles), or could approach plug flow with less agitation. The RTD of the bubbles becomes important only for cases of intermediate solubility of the gaseous reactant in the liquid. Thus Goto and Smith considered both plug flow and perfect mixing for the gas phase.

In dimensionless form, the model with plug flow of the gas phase is given by:

$$\frac{dy_{GA}}{dz} + Da_{GA}(y_{GA} - y_{LA}) = 0 \tag{5.10}$$

$$\frac{dy_{GB}}{dz} + Da_{GB}(y_{GB} - y_{LB}) = 0 \quad (5.11)$$

$$y_{LA,0} - y_{LA} + Da_{LA}(y_{GA} - y_{LA}/\eta_{OA}) = 0 \quad (5.12)$$

$$y_{LB,0} - y_{LB} + Da_{LB}(y_{GB} - y_{LB}) - \beta_L Da_{LB} y_{LA} = 0 \quad (5.13)$$

where the reaction is first order with respect to O_2(A) and zero order with respect to SO_2(B) (i.e., (1,0) kinetics), the y values are the dimensionless concentrations (e.g., $y_{LA} = C_{LA}H_A/C_{GA,0}$ and $y_{GA} = C_{GA}/C_{GA,0}$), η_{OA} is an overall effectiveness factor, β_L is the dimensionless rate constant, and the four Da_{GL} parameters are the gas-liquid mass transfer coefficients in dimensionless form. Eqns. (5.10-5.13) were solved analytically with the boundary conditions:

$$y_{GA} = y_{GB} = 1, \text{ at } z = 0 \quad (5.14)$$

to yield expressions for the concentrations in the reactor effluent i.e., $y_{LA,e}$, $y_{LB,e}$, $y_{GA,e}$, and $y_{GB,e}$.

The model for a perfectly mixed gas phase consists of Eqns. (5.12) and (5.13) and replacement of Eqns. (5.10) and (5.11) by:

$$1 - y_{GA} - Da_{GA}(y_{GA} - y_{LA}) = 0 \quad (5.15)$$

$$1 - y_{GB} - Da_{GB}(y_{GB} - y_{LB}) = 0 \quad (5.16)$$

Of course, this model can also be solved analytically.

Goto and Smith [68] examined the effects of catalyst particle size and loading, feed concentration, and liquid flow rate on the removal of SO_2. The effect of liquid flow rate is presented in Figure 17. It should be noted that as the liquid volumetric flow rate, $Q_L \rightarrow 0$, the semi-batch case is approached. However, this case differs fundamentally from that considered by Chaudhari and Ramachandran [21] in that both reactants are introduced in the gas phase, and thus the reactor can still be operated at steady-state with no liquid flow. As can be seen from Figure 17, the amount of SO_2 removed increases with liquid flow rate and with decreasing catalyst particle size. Therefore, in this case semi-batch operation is not advantageous. It should be noted that Ramachandran and Chaudhari [98] have also developed generalized rate equations for the reaction of two gases in a catalytic slurry reactor for both (1,0) and (1,1) kinetics.

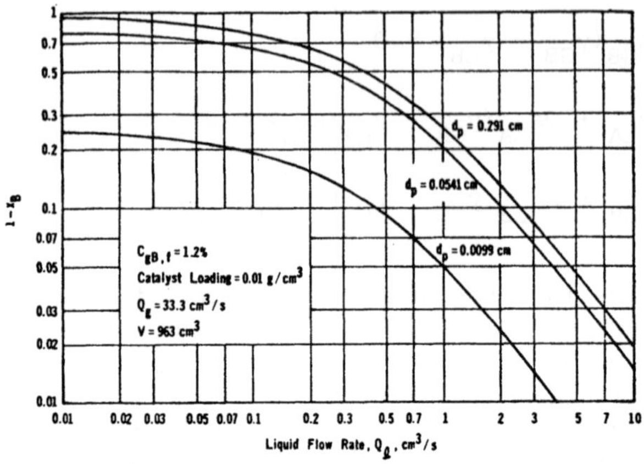

Figure 17. Effect of liquid flow rate on SO_2 removal in a continuous slurry reactor [68].

The modeling of nonagitated, three-phase, cocurrent flow, fluidized bed reactors, or more simply, three-phase fluidized beds, has not yet progressed to the point where conversions and hydrodynamic behavior, or even the holdups of the phases, can be predicted with any certainty. Many of the approaches to modeling these reactors derive from existing models of two-phase fluidized beds, due to the fundamental similarities between these two reactor types (e.g., see [99]). Even so, important differences also exist. For example, a three phase fluidized bed may either contract or expand upon the introduction of gas bubbles at the bottom of the bed [100] - unlike two-phase fluidized beds. The current status of the three-phase countercurrent fluidized or spouted bed reactor is relatively similar. However, because of the importance of these reactors in synthetic fuels processes such as Fischer-Tropsch and catalytic coal liquefaction (e.g., H-COAL), hydroprocessing and hydrocracking of heavy oils (e.g., H-OIL), the modeling of these systems is of considerable current interest and advances will be forthcoming.

6. CONCLUDING REMARKS

It is hoped that this necessarily brief attempt at reviewing the modeling aspects of multiphase chemical reactors clearly conveys the view that a considerable amount of work has been done and that, although formidable challenges still remain, much is already known in this regard; and also the impression that this area is one of considerable growing interest. In view of the complexity of the coupling between hydrodynamics, transport phenomena, and chemical reaction in these reactors, it is unlikely that universally useful, generalized reactor models will ever be developed for multiphase chemical reactors, as they have for simpler reactor types. Therefore, primarily two types of models seem most likely to enjoy the most use:

I. Simple models wherein much of the physicochemical fine structure is lumped or integrated over. This type of model will probably remain sufficient for general multiphase reactor analysis and development, where correlation and characterization of gross reactor behavior is desired.

II. Large complex industrial process models (e.g., see [10-13]) for the prediction of reactor performance for a very specific set of chemical reactions and reactor-type operating in a well-defined mode over a limited range of parameters. In view of the considerable magnitude of time and expense required for the careful collection and analysis of the large amount of data, including systematic variation of parameters and reactor size, and numerical work in model design, it is most likely that this type of model will be developed by way of a comprehensive research and development program by an industrial concern with commercial interests in the specific process. However, academic research has contributed greatly to this area in the past and will continue to do so in the future both in defining and characterizing the fundamental interrelationships among parameters in the multiphase reactor context, and in the analysis of model structure and numerical methods of solution.

In addition to the classic approaches and techniques which have proven successful in modeling simpler reactor types, most of which have been reviewed here, other methods useful in the modeling of multiphase chemical reactors are being developed and applied. For example, the volume averaging approach (e.g., see [101, 102, 103]) as by Carbonell and Whitaker [104] holds considerable promise in this regard. In addition, new numerical techniques both for the model solution and for probing interrelationships among system parameters are becoming available. An example in the latter category is sensitivity analysis of large multiparameter models. The sensitivities of model outputs and variables to model inputs such as kinetic and transport parameters, initial and boundary conditions, etc., are essentially all the gradients at the model operating point in the multidimensional parameter space. The magnitudes of these gradients can be used in model discrimination, parameter estimation, in identifying particular parameters and interrelationships which must be known to greater accuracy than others, etc. In view of the large set of sensitivities which can be generated for even a relatively simple model, the efficacy of this approach depends to a great extent on the method used to calculate the sensitivities. One particularly promising technique in this regard is the Green's function method developed by Rabitz and coworkers [105].

In view of past and recent developments, and the current high level of interest in multiphase chemical reactors, the modeling of these reaction systems seems poised for major advances in the near future.

NOTATION

A reactant concentration and reactant, or constant in Eq. (3.31).

a_L gas-liquid interfacial area

a_s liquid-solid interfacial area

a_t total catalyst surface area

a_w wetted particle surface area

B reactant/product concentration and reactant

B_o Bodenstein number

b dilution ratio of inerts to catalyst

C reactant/product or tracer concentration and reactant

c_p heat capacity

Da Damkohler number in Eqns. (4.9) and (4.24)

$Da_{G,L}$ dimensionless interphase transport coefficients in Eqns. (5.10-5.13)

D_A diffusivity of A

D_e effective diffusivity

D_H axial thermal dispersion coefficient

D_z axial dispersion coefficient

dp particle diameter

E_a activation energy

$E(t)$ residence time distribution

f $= (1 - \phi)$

f_1 constant defined in Eq. (5.9a)

f_2 constant defined in Eq. (5.9b)

Fr Froude number

G mass velocity

Ga Galileo number

g acceleration due to gravity

H_A Henry's law constant for A

$(-\Delta H_r)$ heat of reaction

h	holdup (d-dynamic, s-static)
I_1	first order Bessel function of the first kind
K, k, k_g	interphase mass transfer coefficient
k_n	nth order rate constant (per catalyst volume)
k'_n	interchange coefficient divided by phase holdup
L	reactor or bed length
LHSV	liquid hourly space velocity
m	reaction order or constant in modified bypass model
M_A	defined by Eq. (5.4)
N	number of stages or tanks
n	reaction order or number of laminar flow elements in bypass RTD model
$\bar{N}$	constant, Eq. (3.29)
p	pressure
P^*	constant in Eqns. (4.12, 4.13)
Pe	Peclet number
Q	volumetric flow rate
Q^*	constant in Eq. (4.13)
q	fraction in stagnant zone
R	universal gas constant and particle radius
Re	Reynolds number
R'_n	constant in Eq. (4.29c)
R^*_n	defined in Eq. (4.34b)
r	recycle ratio
r_1, r_2, r_3	constants in Eq. (5.7)
T	temperature
t	time
$\bar{t}$	mean residence time
t_d	delay time
t_i	breakthrough time
U	superficial fluid velocity
$\bar{u}$	mean fluid velocity
V	volume (reactor volume unsubscripted)
v	dimensionless concentration

w mass of catalyst per unit volume of slurry

We Weber number

X conversion

x bypass fraction

y mole fraction

Z axial distance

z dimensionless axial distance

Greek Letters

α constant in recycle flow model and Eq. (4.29b)

α_A constant in Eq. (5.4)

β constant in recycle flow model and Eq. (5.13)

Γ gamma function and constant in Eq. (4.36)

γ dimensionless activation energy and constant in Eq. (5.9c)

δ film thickness, delta function, and constant in Eq. (4.28c)

δ' $= \delta\Gamma$

ε void fraction

η effectiveness factor

η_L liquid-solid effectiveness factor

η_o overall effectiveness factor

η_t trickle bed reactor efficiency

μ viscosity

θ dimensionless time and temperature (eq. (4.31c)

ν stoichiometric coefficient

ρ density

σ square root of variance and surface tension

τ space time

ϕ plug flow fraction in crossflow model

χ perfectly mixed flow fraction in bypass model

Ω local reaction rate

Subscripts

A reactant A

B reactant B

C reactant C

c catalyst or critical

e reactor exit

f	flow, fluid
G,g	gas phase
h	heat
i	cell index
j	plug flow tube index
L,l	liquid phase
m	mean or mass
n	reaction order
o	initial or inlet
p	particle
s	liquid-solid interface or stagnant

REFERENCES

1. R. H. Wilhelm, Pure Appl. Chem. 4, 403 (1962).
2. G. F. Froment, Ind. Eng. Chem. 59, 18 (1967).
3. V. Hlavacek, Ind. Eng. Chem. 62, 8 (1970).
4. V. Hlavacek and J. Votruba, Chemical Reactor Theory-A Review, L. Lapidus and N. R. Amundson, eds., Prentice-Hall, Inc., Englewood Cliffs, N.J., 1977, Chap. 6.
5. J. J. Carberry, Chemical and Catalytic Reaction Engineering, McGraw-Hill, Inc., New York, 1976.
6. C. G. Hill, Jr., An Introduction to Chemical Engineering Kinetics and Reactor Design, John Wiley & Sons, New York, 1977.
7. G. F. Froment and K. B. Bischoff, Chemical Reactor Analysis and Design, John Wiley & Sons, New York, 1979.
8. J. B. Butt, Reaction Kinetics and Reactor Design, Prentice-Hall, Inc., Englewood Cliffs, N.J., 1980.
9. O. Levenspiel, Chemical Reaction Engineering, 2nd ed., John Wiley & Sons, New York, 1972.
10. V. W. Weekman, Jr., Advances in Chemistry, No. 148, ACS, Washington, D.C., 1975, p.98.
11. S. M. Jacob, B. Gross, S. E. Voltz, and V. W. Weekman, Jr., AIChE J. 22, 701 (1976).
12. K. R. Graziani and M. P. Ramage, in Chemical Reaction Engineering-Houston, V. W. Weekman, Jr. and Dan Luss, eds., ACS Symposium Series 65, American Chemical Society, Washington, D.C., 1978, p. 282.
13. M. P. Ramage, K. R. Graziani, and F. J. Krambeck, Chem. Eng. Sci. 35, 41 (1980).
14. R. Shinnar, in Chemical Reaction Engineering Reviews-Houston, D. Luss and V. W. Weekman, Jr., eds., ACS Symposium Series No. 72, American Chemical Society, Washington, D.C., 1978, p. 1.
15. C. N. Satterfield, AIChE J. 21, 209 (1975).
16. J. C. Charpentier, Chem. Eng. J. 11, 161 (1976).
17. S. Goto, J. Levec, and J. M. Smith, Catal. Rev.-Sci. Eng. 15, 187 (1977).
18. H. Hofmann, Int. Chem. Eng. 17, 19 (1977).
19. H. P. Hofmann, Catal. Rev.-Sci. Eng. 17, 71 (1978).
20. A. Gianetto, G. Baldi, V. Specchia, and S. Sicardi, AIChE J. 24, 1087 (1978).
21. R. V. Chaudhari and P. A. Ramachandran, AIChE J. 26, 177 (1980).
22. Y. T. Shah, G. J. Stiegel, and M. M. Sharma, AIChE J. 24, 369 (1978).
23. K. Østergaard, Adv. Chem. Eng. 7, 71 (1968).
24. Y. T. Shah, Gas-Liquid-Solid Reactor Design, McGraw-Hill Book Co., New York, 1979.
25. G. B. Wallis, One-Dimensional Two-Phase Flow, McGraw-Hill Book Co., New York, 1969.

26. G. W. Govier and K. Aziz, The Flow of Complex Mixtures in Pipes, Van Nostrand Reinhold Co., New York, 1972.
27. R. P. Larkin, R. R. White, and D. W. Jeffrey, AIChE J. 7, 231 (1961).
28. V. W. Weekman, Jr. and J. E. Myers, AIChE J. 10, 951 (1964).
29. J. J. Lerou, D. Glasser, and D. Luss, Ind. Eng. Chem. Fund. 19, 66 (1980).
30. N. Midoux, M. Favier, and J. C. Charpentier, J. Chem. Eng. Japan 9, 350 (1976).
31. V. Specchia and G. Baldi, Chem. Eng. Sci. 32, 515 (1977).
32. J. G. Schwartz and G. W. Roberts, Ind. Eng. Chem. Proc. Des. Dev. 12, 262 (1973).
33. J. M. Hochman and E. Effron, Ind. Eng. Chem. Fund. 8, 63 (1969).
34. W. E. Dunn, T. Vermuelen, C. R. Wilke, and R. R. Word, Ind. Eng. Chem. Fund., 16, 116 (1977).
35. I. A. Furzer and R. W. Michell, AIChE J. 16, 380 (1970).
36. R. W. Michell and I. A. Furzer, Chem. Eng. J. 4, 53 (1972).
37. J. B. Butt, AIChE J. 8, 553 (1962).
38. D. M. Himmelblau and K. B. Bischoff, Process Analysis and Simulation, John Wiley & Sons, Inc., New York, 1968.
39. R. W. Michell and I. A. Furzer, Trans. Inst. Chem. Eng. 50, 334 (1972).
40. J. F. Davidson, Trans. Inst. Chem. Eng. 37, 131 (1959).
41. J. M. Hochman and J. R. McCord, Chem. Eng. Sci. 25, 97 (1970).
42. C. J. Hoogendorn and J. Lips, Can. J. Chem. Eng. 41, 125 (1965).
43. B. A. Buffham, L. G. Gibilaro, and M. N. Rathor, AIChE J., 16, 218 (1970).
44. H. A. Deans, Soc. Petrol. Eng. J., 3, 49 (1963).
45. H. A. Deans and L. Lapidus, AIChE J., 6, 656 (1960).
46. J. Villermaux and W. P. M. Van Swaaij, Chem. Eng. Sci., 24, 1097 (1969).
47. M. Popovic and W. D. Deckwer, Chem. Eng. J., 11, 67 (1976).
48. A. Bennett and F. Goodridge, Trans. Inst. Chem. Eng., 48, T232 (1970).
49. A. Matsuura, T. Akehata, and T. Shirai, J. Chem. Eng. Japan, 9, 294 (1976).
50. W. P. M. Van Swaaij, J. C. Charpentier, and J. Villermaux, Chem. Eng. Sci., 24, 1083 (1969).
51. B. A. Buffham and L. G. Gibilaro, Chem. Eng. J., 1, 31 (1970).
52. A. J. Oorts and L. J. Hellinckx, Chem. Eng. J., 1, 147 (1979).
53. V. G. Rao and Y. B. G. Varma, AIChE J., 22, 612 (1976).
54. J. Raghuraman and Y. B. G. Varma, Chem. Eng. Sci., 28, 585 (1973).
55. R. S. H. Mah, Chem. Eng. Sci., 26, 201 (1971).
56. K. Shügerl, Proc. Int. Symp. on Fluidization, 782 (1967).
57. M. L. Michelsen and K. Østergaard, Chem. Eng. J., 1, 37 (1970).
58. K. Østergaard and M. L. Michelsen, Can. J. Chem. Eng., 47, 107 (1969).

59. L. Massimilla, A. Solimando, and E. Squillace, Brit. Chem. Eng., 6, 232 (1971).
60. Y. K. Vail, N. K. Manakov, and V. V. Manshilin, Int. Chem. Eng., 8, 293 (1968).
61. Y. Kato and A. Nishiwaki, Kogaku Kogaku, 35, 912 (1971).
62. K. Imafuku, T. Y. Wang, K. Koide, and H. Kubota, J. Chem. Eng. Japan, 1, 153 (1968).
63. S. Goto, S. Watabe, and M. Matsubara, Can. J. Chem. Eng., 54, 551 (1976).
64. Y. Sato, H. Hirose, F. Takahashi, and M. Toda, Pac. Chem. Eng. Congr., Session 8, p. 187 (1972).
65. C. N. Satterfield, A. A. Pelossof, and T. K. Sherwood, AIChE J., 15, 226 (1969).
66. V. M. H. Govindarao and K. V. R. Murthy, J. Appl. Chem. Biotechnol., 25, 169 (1975).
67. S. Goto and J. M. Smith, AIChE J., 24, 294 (1978).
68. S. Goto and J. M. Smith, AIChE J., 24, 286 (1978).
69. Y. T. Shah and J. A. Paraskos, Chem. Eng. Sci., 30, 465 (1975).
70. R. D. Mhaskar, Chem. Eng. Sci., 29, 897 (1979).
71 H. C. Henry and J. B. Gilbert, Ind. Eng. Chem. Proc. Des. Dev., 12, 329 (1973).
72. D. E. Mears, Adv. Chem. Ser., 133, 218 (1974).
73. S. S. Puranik and A. Vogelpohl, Chem. Eng. Sci., 29, 501 (1979).
74. K. Onda, H. Takeuchi, Y. Koyama, Kagaku Kogaku, 31, 126 (1967).
75. J. A. Paraskos, J. A. Frayer, and Y. T. Shah, Ind. Eng. Chem. Proc. Des. Dev., 14, 315 (1975).
76. A. A. Montagna and Y. T. Shah, Ind. Eng. Chem. Proc. Des. Dev., 14, 479 (1975).
77. A. A. Montagna, Y. T. Shah, and J. A. Paraskos, Ind. Eng. Chem. Proc. Des. Dev., 16, 152 (1977).
78. D. E. Mears, Chem. Eng. Sci., 26, 1361 (1971).
79. Y. T. Shah and J. A. Paraskos, Chem. Eng. Sci., 30, 1169 (1975).
80. A. Burghardt and T. Zaleski, Chem. Eng. Sci., 23, 575 (1968).
81. Y. T. Shah, R. D. Mhaskar, and J. A. Paraskos, Ind. Eng. Chem. Proc. Des. Dev., 15, 400 (1976).
82. R. D. Mhaskar, Y. T. Shah, and J. A. Paraskos, Ind. Eng. Chem. Proc. Des. Dev., 17, 27 (1978).
83. R. D. Mhaskar and Y. T. Shah, Ind. Eng. Chem. Proc. Des. Dev., submitted, 1977.
84. V. Specchia and G. Baldi, Chem. Eng. Comm., 3, 483 (1979).
85. W. Sedriks and C. N. Kenney, Chem. Eng. Sci., 28, 559 (1973).
86. J. Hanika, K. Sporka, V. Ruzicka, and R. Pistek, Chem. Eng. Sci., 32, 525 (1977).
87. A. H. Germain, A. G. Lefebvre, and G. A. L'Homme, Adv. in Chemistry Series, 133, 164 (1974).
88. J. Hanika, K. Sporka, V. Ruzicka, and J. Kransova, Chem. Eng. Comm., 2, 19 (1975).
89. P. A. Ramachandran and J. M. Smith, Chem. Eng. Sci., 32, 873 (1977).

90. S. Goto and J. M. Smith, AIChE J., 21, 706, 714 (1975).
91. G. Baldi, S. Goto, C. K. Chow, and J. M. Smith, Ind. Eng. Chem. Proc. Des. Dev., 13, 447 (1974).
92. R. T. Pavlica and J. H. Olson, Ind. Eng. Chem., 62, 45 (1970).
93. W. D. Deckwer, Chem. Eng. Sci., 31, 309 (1976).
94. P. A. Ramachandran and R. V. Chaudhari, Chem. Eng. J., Submitted, 1980.
95. R. V. Chaudhari and P. A. Ramachandran, Ind. Eng. Chem. Fund., 19, 201 (1980).
96. V. M. H. Govindarao, Chem. Eng. J., 9, 229 (1975).
97. P. A. Ramachandran and J. M. Smith, Chem. Eng. Sci., 32, 873 (1977).
98. P. A. Ramachandran and R. V. Chaudhari, Eng. Eng. Chem. Proc. Des. Dev., 18, 703 (1979).
99. R. C. Darton and D. Harrison, Chem. Eng. Sci., 30, 581 (1975).
100. N. Epstein, Can. J. Chem. Eng., 54, 259 (1976).
101. J. C. Slattery, AIChE J., 13, 1066 (1967).
102. S. Whitaker, AIChE J., 13, 420 (1967).
103. J. C. Slattery, Momentum, Energy and Mass Transfer in Continua, McGraw-Hill, New York, 1972.
104. R. G. Carbonell and S. Whitaker, Chem. Eng. Educ., XII, 182 (1978).
105. J.-T. Hwang, E. P. Dougherty, S. Rabitz, and H. Rabitz, J. Chem. Phys., 69, 5180 (1978).
106. M. Crine, P. Marchot, and G. A. L'Homme, Chem. Eng. Sci., 35, 51 (1980).

SCIENTIFIC BASIS FOR THE DESIGN OF TWO PHASE CATALYTIC REACTORS

Alírio E.Rodrigues

Department of Chemical Engineering
University of Porto,Portugal

This lecture will be divided into three parts.After some introductory remarks we discuss in section I the competition between reaction and transport processes in porous catalysts.Section II deals with selectivity disguises due to diffusional effects and catalyst deactivation. Finally,in Part III,modelling of two phase reactors (mainly fixed beds) is analised and correlations for the prediction of model parameters are presented.

INTRODUCTION

In the chemical industry,transformation of raw materials into "useful" products involves physical separation processes,reaction processes,energy input,as well as the control of the whole plant |1|.This corresponds to the three main branches of chemical engineering: separation engineering,reaction engineering and "information" engineering.

A simplified picture |2| is shown in Figure 1.The reaction process is often a catalytic one (80% of chemicals involve catalysis in their production |3|) and then two main problems have to be solved:

a) choice and production of the ideal catalyst for a given reaction
b) choice and design of the reactor where the reaction will be carried out

The interest of developing new catalysts is easily understandable from the economic point of view and has been stressed due to the energy crisis. Better activity, selectivity, resistance to poisoning,stability are major catalyst requirements which are directly related to the size (cost) of the reactor,size of separation unit before the reactor,size of the separation unit after the reactor and need for maintenance of product specifications,respectively.

On the other hand the choice of the type of reactor (fixed bed,fluidised bed,moving bed,transport reactor,perfectly mixed, radial flow reactor) for a given reaction is a "a priori" decision.

The design of a reactor involves |1|:

- conservation equations (mass,energy,momentum)
- equilibrium laws
- kinetic laws of transport and chemical reaction
- boundary and initial conditions
- optimization principle (a certain objective function should be defined based on different considerations : production rate,energy consuption |4|,etc)

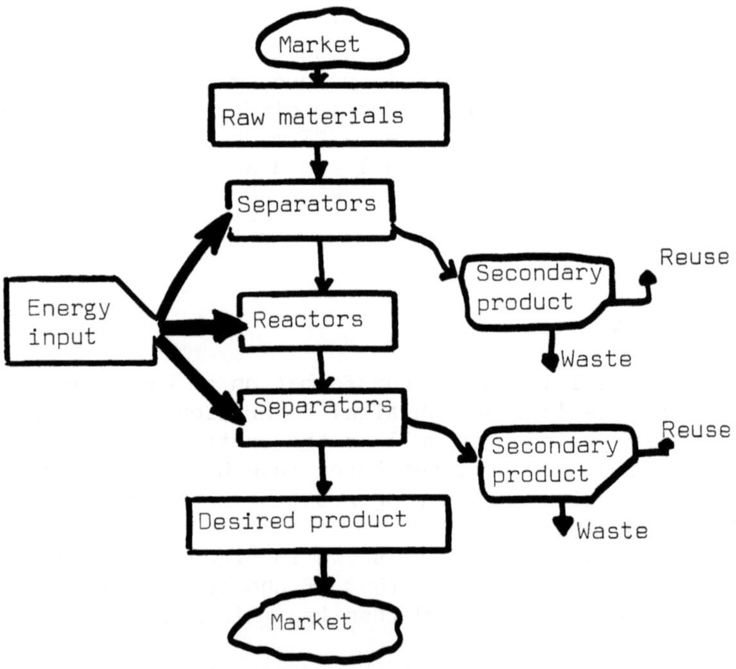

Figure 1- A simplified picture of chemical industry |2|

The simplest model for the two extreme cases of fluid/solid reactors (plug flow and perfectly mixed) assuming isothermal operation and neglecting diffusional resistances, leads to the so-called "design equations"

Plug flow reactor

$$W=Q_o c_o \int_0^{\infty} \frac{dX}{r(X)}$$

Perfectly mixed reactor

$$W=Q_o c_o X/r(X)$$

where W is the mass of catalyst to be placed in the reactor in order to obtain an extent of reaction, X when processing a feed at flowrate Q_o and total concentration (of active components), c_o and r(X) is the reaction rate referred to the mass of catalyst.
The so-called design equations can rarely be used and then we have to look first at what happens at the particle level before considering the whole reactor and calculating the correct amount of catalyst, W' for the above conditions.

I-TRANSPORT PHENOMENA AND CHEMICAL REACTION IN A CATALYST PARTICLE

1.1-Definition of catalyst effectiveness

Let us assume that a reaction $A \xrightarrow{k} B$ takes place in a porous catalyst particle. The following steps can be considered (Figure 2):

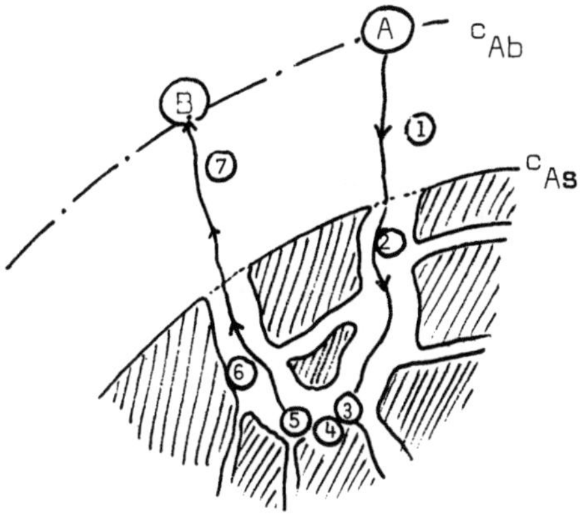

Figure 2-Transport and reaction processes in a catalyst particle.

1. diffusion of molecule A from the bulk of fluid phase (where the concentration is c_{Ab}) to the catalyst surface, through the film around the particle -external diffusion.
2. diffusion of molecule A in the pores of the catalyst
3. adsorption of molecule A on the active sites of the catalyst
4. chemical reaction at the active sites of the catalyst
5. desorption of product B
6. diffusion of molecule B through the pores to the catalyst surface
7. diffusion of B from the catalyst surface to the bulk of the fluid phase

Due to the diffusional resistances, concentration gradients can occur inside and outside catalyst particles; also if noniso - thermal reactions are taking place temperature gradients will exist. These gradients will affect the characteristics of the pellet : activity, selectivity, poisoning rate, stability. Two limiting situations can be expected |5|: in the first one the diffusion rate inside the catalyst is much higher than the reaction rate and the concentration profile inside the pellet is almost flat, i.e., $c_A(r) \simeq c_{As}$ (resistance to diffusion negligible); in the opposite situation the reaction rate is much faster than the diffusion rate and then a concentration gradient will exist within the particle (the concentration can even be zero in a region near the center of the particle; this core does not "work").

This justifies the introduction of the concept of catalyst effectiveness or catalyst utilization. The effectiveness factor, η is defined as the ratio between the actual or observed reaction rate and the reaction rate which would be observed if the whole catalyst were at the surface conditions

$$\eta = \frac{r_{obs}}{r_V} = \frac{(1/V_p)\iiint_{V_p} r(c_A, T)dV}{r(c_{As}, T_S)} \qquad |1.1|$$

It is obvious that for isothermal particles η can be greater than one only for "abnormal" reaction rates (order $n < 0$)*; for nonisothermal cases the effectiveness factor can be greater than one since the reaction rate $\underline{r}$ depends exponentially on the temperature and as a small power of concentrations.

For design purposes the amount of catlyst is then

$$W' = W/\eta$$

---

* Assuming only molecular diffusion inside the particle and no convection, besides the chemical reaction.

where W is the mass of catalyst calculated by using the simplified approach mentioned in the Introduction (design equations). This illustrates the importance of knowing the effectiveness factor.

1.2-Isothermal catalysts: chemical reaction and diffusion inside the particles

The methodology we will follow starts with the steady state mass balance over a volume element of catalyst (slab, sphere or cylinder) associated with Dirichlet boundary conditions; then normalized variables are introduced ($f=c_A/c_{As}$, $z=r/\ell$) and the dimensionless model equations obtained.

For a nth order, irreversible, reaction $A \xrightarrow{k} B$ the model equations are:

mass balance

$$\frac{d^2f}{dz^2} + \frac{\alpha}{z}\frac{df}{dz} - \phi^2 f^n = 0 \qquad |1.2|$$

boundary conditions

$$f=1 \quad , \quad z=1 \qquad |1.3a|$$

$$\frac{df}{dz}=0 \quad , \quad z=0 \quad \text{(simmetry condition)} \qquad |1.3b|$$

α=0,1 and 2 for slab, cylinder and sphere geometries, respectively.

We notice that the particle behavior is determined by the dimensionless group, ϕ (Thiele modulus)

$$\phi = \ell\sqrt{kc_{As}^{n-1}/D_{eff}} \qquad |1.4|$$

where ℓ is the characteristic dimension of the particle (radius of sphere, half thickness of a slab), c_{As} the surface concentration, k the kinetic constant and D_{eff} the effective diffusivity. We can also see that

$$\phi^2 = \frac{\text{reaction rate}}{\text{diffusion rate inside the particles}} \qquad |1.5|$$

The solution of Eq.|1.2| leads to the concentration profile inside the particle and the effectiveness factor can be calculated by

$$\eta = \frac{r_{obs}}{r(c_{As})} = \frac{\dot{N}_A}{N_A^*} = \frac{A_p D_{eff} (dc_A/dr)_{r=\ell}}{V_p \; kc_{As}} \qquad |1.6|$$

where $\dot{N}_A$ is the overall flux crossing the external surface of the catalyst, A_p and N_A^* the mass flux due to the chemical reaction taking place in the whole particle volume, V_p.
Table I summarizes the concentration profiles, effectiveness factors, and asymptotic expressions of η for the three catalyst geometries and for 1st order reactions |6|.

TABLE I-Competition between reaction and diffusion inside isothermal catalyst particles.

Geometry	slab (thickness:2ℓ)	sphere (radius:R)	cylinder (radius:R, length:L)
concentration profile inside particles	$f=\frac{\cosh(\phi z)}{\cosh\phi}$	$f=\frac{\sinh(\phi z)}{z\ \sinh\phi}$	$f=\frac{I_0(\phi z)}{I_0(\phi)}$
effectiveness factor	$\eta=\frac{\tanh\phi}{\phi}$	$\eta=\frac{3}{\phi}\left[\frac{1}{\tanh\phi}-\frac{1}{\phi}\right]$	$\eta=\frac{2}{\phi}\frac{I_1(\phi)}{I_0(\phi)}$
asymptotic expression of η	$\eta=1/\phi$, $\phi>3$	$\eta=3/\phi$, $\phi/3>3$	$\eta=2/\phi$, $\phi/2>3$

I_0 -modified Bessel functions of first kind and zero order

I_1 -modified Bessel function of first kind and first order

A generalized Thiele modulus, ϕ_p can be defined in such a way that the asymptotic expression of the effectiveness factor is the same for all geometries, $\eta=1/\phi_p$; it is sufficient to set

$$\phi_p = \frac{V_p}{A_p}\sqrt{kc_{As}^{n-1}/D_{eff}} \qquad |1.7|$$

Three working regimes can be distinguished : chemical regime ($\phi_p < 0.3$), intermediate regime ($0.3 < \phi_p < 3$) and internal diffusion regime ($\phi_p > 3$). The interest of the effectiveness factor was stated above when calculating the actual mass of catalyst to be placed in the reactor; however since $\eta = f(\phi)$ and the Thiele modulus is expressed in terms of the often unknown quantity k, one should try to obtain η in terms of <u>observable quantities</u> |7|. This can be achieved since $\eta\phi^2$ contains only measurable quantities:

$$\eta\phi^2 = \frac{\ell^2 r_{obs}}{c_{As} D_{eff}} \qquad |1.8|$$

and then one can plot $\eta = f(\eta\phi^2)$ as shown in Figure 4 ($Bi_m = \infty$).

The observed rate approaches the true value (in the absence of diffusional limitations) as particle size decreases ; this suggests a method for determining η vs d_p |8|.

The above treatment has been generalized to Langmuir-Hinshelwood kinetics (L-H) where $r = kc_A/(1+K_A c_A)$ by Rajadhyaksha et <u>al</u> |9|; they found that the $\eta - \phi$ plot for 1st order reactions <u>is</u> still valid if we define a new Thiele modulus, ϕ'

$$\phi' = \phi \frac{\sqrt{n}}{1+\omega} \sqrt{\frac{\omega}{\ln|1+\omega|}} \qquad |1.9|$$

where $\omega = Kc_{As}$. For L-H kinetics, n=1 while for n^{th} order reactions $\omega = 0$.

1.3-Isothermal catalysts: competition between external diffusion, internal diffusion and chemical reaction.

The methodology is similar to the previous one; we have just to modify the boundary conditions. Now there is resistance to mass transfer in the film and the surface concentration results from the equality of fluxes (Robin boundary conditions |10|) :

$$k_f(c_{Ab} - c_{As}) = \pm D_{eff} \left.\frac{dc}{dr}\right|_{r=\pm\ell}$$

or

$$Bi_m(1-\bar{f}) = \frac{d\bar{f}}{dz}, z=1 \qquad |1.10|$$

where $Bi_m = k_f\ell/D_{eff}$ is the mass Biot number (k_f is the film mass transfer coefficient, cm/sec and $\bar{f}$ is the reduced concentration referred to the bulk conditions, c_{Ab}).

Table II summarizes the concentration profiles and effectiveness factors $\bar{\eta}$ (referred to the bulk conditions) for 1st order irre - versible reactions.
The problem involves now two parameters: the Thiele modulus, ϕ (competition between reaction and internal diffusion) and the Biot number, Bi_m (competition between external and internal diffusion).
Since the observed reaction rate is now, for a 1st order irreversible reaction,

$$r_{obs} = \frac{\eta k \, c_{Ab}}{1 + \eta \, Da} \quad * \qquad |1.11|$$

where $Da = k\ell/k_f = \phi^2/Bi_m$ (Damkholër number), several regimes can be distinguished |11|

a) $\phi < 1$, $\eta \to 1$

For $Da \ll 1$, $r_{obs} \simeq k c_{Ab}$ (chemical reaction control)

b) $\phi > 3$, $\eta \to 1/\phi$

For $\phi \ll Bi_m$, $r_{obs} \simeq \frac{\sqrt{kD_{eff}}}{\ell} c_{Ab}$

(intermediate regime: internal diffusion + reaction control)

c) Bi_m small and high ϕ , $\phi \gg Bi_m$ ($\eta\phi^2 \simeq Bi_m$)

$r_{obs} \simeq k_f c_{Ab}/\ell$ (film mass transfer control)

Figure 3 shows a plot of $\bar{\eta} = f(\phi)$ for 1st order irreversible reactions and slab geometry of catalyst particles; for $Bi_m = \infty$ (no film mass transfer resistance) $\bar{\eta} = \eta$ and we get the classic plot for the internal diffusion+reaction problem, $\eta = f(\phi)$.

Again, for practical purposes, one should plot $\bar{\eta}$ as a function of $\bar{\eta}\phi^2$ since this quantity can be expressed in terms of known or measurable quantities (Figure 4).

Carberry |12| introduced a dimensionless group, Ca (Carberry number) which is very convenient for the analysis of external diffusion and reaction in isothermal catalysts

$$Ca = \bar{\eta} \, Da = r_{obs} \ell / k_f c_{Ab} \qquad |1.12|$$

For abnormal reactions ($n < 0$), $\bar{\eta}$ can be greater than one (Figure 5).

---

* One can see that $\bar{\eta} = \eta / (1 + \eta \, Da)$

TABLE II- Competition between reaction, internal diffusion and external diffusion in isothermal catalysts.

Geometry	slab	sphere	cylinder
Concentration profile	$\bar{f}=\dfrac{\cosh(\phi z)}{\cosh\phi+\phi\sinh\phi/Bi_m}$	$\bar{f}=\dfrac{\sinh(\phi z)}{z\{\sinh\phi+(\phi\cosh\phi-\sinh\phi)/Bi_m\}}$	$\bar{f}=\dfrac{I_o(\phi z)}{I_o(\phi)+\dfrac{\phi}{Bi_m}I_1(\phi)}$
Effectiveness factor	$\bar{\eta}=\dfrac{\tanh\phi}{\cosh\phi+\dfrac{\phi}{Bi_m}\sinh\phi}$	$\bar{\eta}=\dfrac{3}{\phi^2}\,\dfrac{\phi\coth\phi-1}{1+\dfrac{1}{Bi_m}(\phi\coth\phi-1)}$	$\bar{\eta}=\dfrac{2I_1(\phi)}{\phi I_o(\phi)+\dfrac{\phi^2}{Bi_m}I_1(\phi)}$

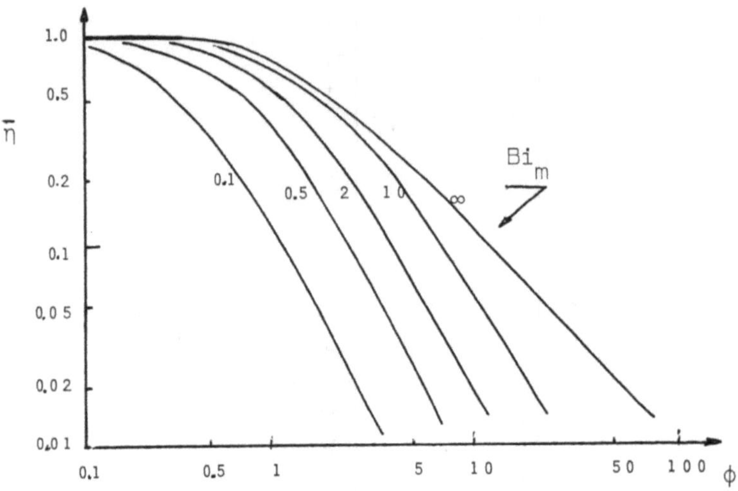

Figure 3- $\bar{\eta}$ vs ϕ for 1st order irreversible reactions and slab geometry

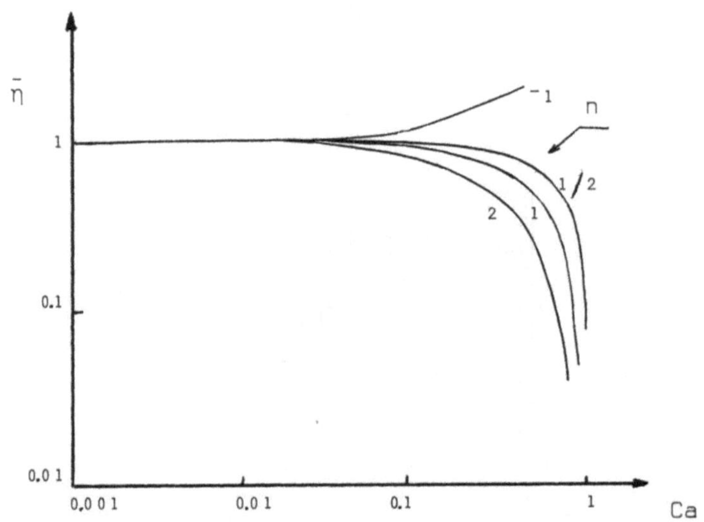

Figure 5- External diffusion and chemical reaction: $\bar{\eta}$ vs Ca.

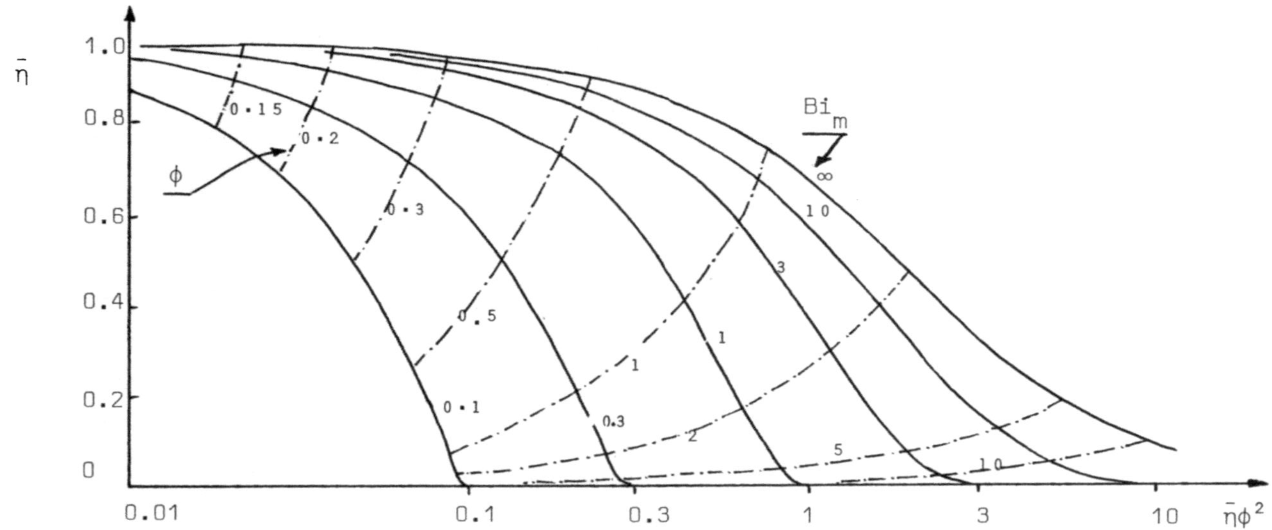

Figure 4- $\bar{\eta}$ vs $\bar{\eta}\phi^2$ with Bi_m and ϕ as parameters

1.4- Kinetic disguises due to diffusional limitations

One of the major steps in reactor design is the determination of the rate equation in the laboratory.It is then important to know to which extent diffusional limitations affect the reaction rate.

As previously defined ,the measured rate for n^{th} order irreversible reactions is

$$r_{obs} = \eta \, k(T) \, c_{As}^{n} \qquad |1.13|$$

and then the apparent order is

$$n' = \frac{d(\ln r_{obs})}{d(\ln c_{As})} = n + \frac{n-1}{2} \frac{d\ln\eta}{d\ln\phi} \qquad |1.14|$$

assuming that D_{eff} does not depend on c_{As}.For small ϕ (chemical reaction control) $n'=n$;for high values of ϕ (diffusional regime) one gets

$$n' = \frac{n+1}{2} \qquad |1.15|$$

Similarly,the apparent activation energy E' will be

$$E' = -R \frac{d(\ln r_{obs})}{d(1/T)}$$

and then

$$-E'/R = -E/R + \frac{d \ln}{d \ln} \left(- \frac{E}{2R} + \frac{mT}{2} \right)$$

In the chemical regime E'=E; in the diffusional regime

$$E' = E/2 + mRT/2 \simeq E/2 \qquad |1.16|$$

where m is due to the dependency of D_{eff} on the temperature ($D_{eff} \sim T^m$; m=1/2 for Knudsen diffusion and m=3/2 for ordinary diffusion).

Table III summarizes these findings |13|;Figure 6 shows an Arrhenius plot covering a large range of temperatures where the various transitions are shown as well as the corresponding con - centration profiles |6|.

TABLE III- Competition between external diffusion, internal diffusion and chemical reaction in isothermal catalysts -effect of operating variables on the observed kinetic parameters |13|

Controlling mechanism	Observed order	Observed activation energy	Effect of ℓ	Effect of u
Chemical reaction	n	E	—	—
Internal diffusion	$\frac{n+1}{2}$	$\simeq E/2$	$1/\ell$	—
External diffusion	1	$\simeq 0$	$(1/\ell)^a$	u^b

a,b depend on the reactor geometry

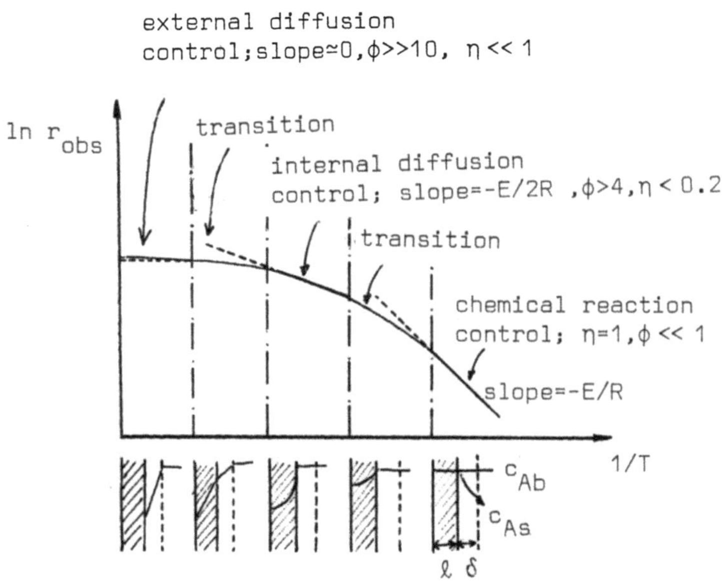

Figure 6- Different operating regimes of catalyst particles; slab geometry |6|.

1.5- Nonisothermal catalysts

Until now only concentration gradients were considered; however for nonisothermal reactions resistances to heat transfer lead to nonuniformity of temperature inside and outside the particle.

In each point of a particle the steady state temperature will be determined by the heat balance between heat generated by chemical reaction and heat removed by conduction |14|.Figure 7 shows qualitative concentration and temperature gradients for an exothermic reaction

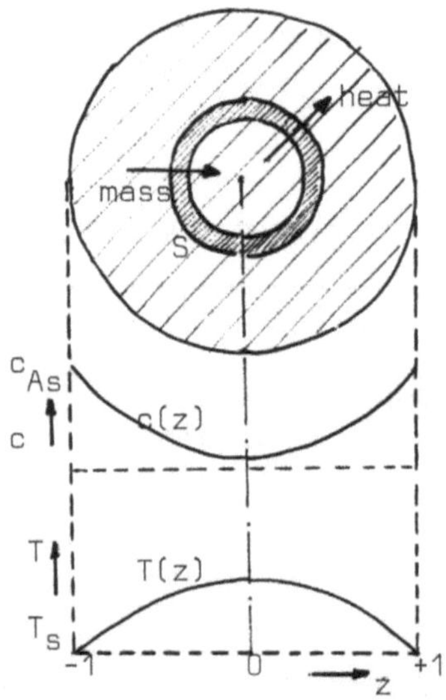

Figure 7- Sketch of concentration and temperature profiles in nonisothermal catalysts.

In steady state the global mass flux crossing any section S inside a spere is

$$\dot{N} = S\ r' = S\ D_{eff}\left(\frac{dc}{dr}\right)_S \qquad |1.17|$$

where the reaction rate r' is referred to the catalyst surface. This mass flux generates a heat flux $\dot{N}\ (-\Delta H)$ by chemical reaction which should be removed by conduction through the catalyst particle and then

$$\dot{N}\ (-\Delta H) = -\ S\lambda_{eff}\left(\frac{dT}{dr}\right)_S \qquad |1.18|$$

Combining Eqs.|1.17| and |1.18| one gets

$$T - T_s = \frac{(-\Delta H)D_{eff}}{\lambda_{eff}} (c_{As} - c) \qquad |1.19|$$

which is the Damkhöler equation,relating the temperature T and the concentration,c in a point inside the catalyst.

It is important to know a priori the maximum temperature which can be reached inside the particle;this is obtained for c=0 and then

$$T_{max} = T_s \ (1+\beta) \qquad |1.20|$$

where

$$\beta = \frac{(-\Delta H)D_{eff}c_{As}}{\lambda_{eff}T_s} \qquad |1.21|$$

is the Prater thermicity factor,based on the surface conditions. The relative temperature rise inside the particle is simply

$$\Delta T_i = (T_{max} - T_s)/T_s = \beta \qquad |1.22|$$

In general the problem of nonisothermal particles where a 1st order irreversible reaction is going on is described by the following equations:

mass balance

$$\frac{d^2f}{dz^2} + \frac{\alpha}{z}\frac{df}{dz} - \phi^2 \exp\{-\gamma(\frac{1}{\theta} - 1)\} f = 0 \qquad |1.23|$$

heat balance

$$\frac{d^2\theta}{dz^2} + \frac{\alpha}{z}\frac{d\theta}{dz} + \phi^2\beta \exp\{-\gamma(\frac{1}{\theta} - 1)\} \quad f = 0 \qquad |1.24|$$

boundary conditions

$$f = \theta = 1 \ , \ z = 1 \qquad |1.25a|$$

$$df/dz = d\theta/dz = 0 \ , \ z = 0 \qquad |1.25b|$$

where $\theta = T/T_s$ and $\gamma = E/RT_s$ (Arrhenius number).

The model parameters are now :

ϕ -Thiele modulus
β -Prater thermicity factor
γ -Arrhenius number

based on the surface conditions.
Figure 8 shows the classic η vs ϕ plot for nonisothermal particles; obviously, as we said before, the effectiveness factor can be greater than one and multiple steady state solutions can occur. For practical purposes plots of η vs $\eta\phi^2$ should be used. In Figure 9 this plot is shown for 1st order irreversible reactions; for other kinetic laws see Satterfield |7|.

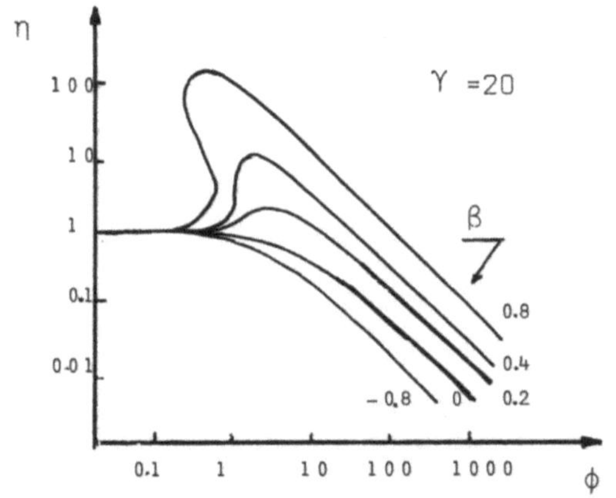

Figure 8- η -ϕ plot for nonisothermal particles and 1st order irreversible reactions (sphere).

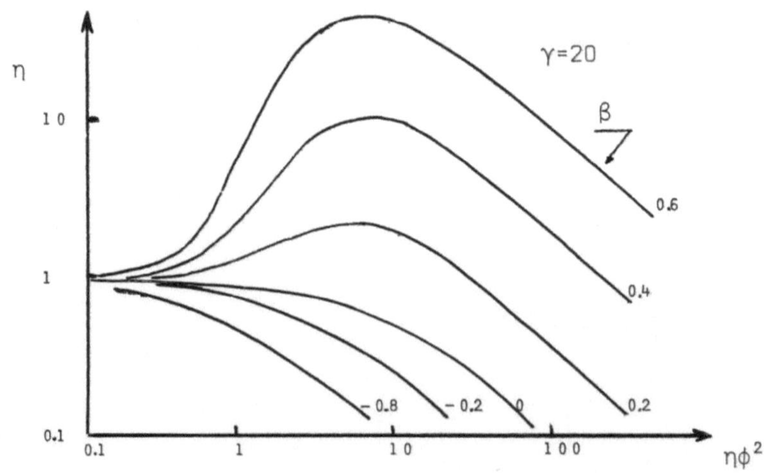

Figure 9- η vs $\eta\phi^2$ plot for nonisothermal particles and 1st order irrversible reactions.

For reactions $A + \nu B \rightarrow$ products with rate law $r=kc_A^n c_B^m$ and $M=c_{Bs}/c_{As}$ Le Goff and Zoulalian |15| give equations for calculating the effectiveness factor

-chemical reaction control ,$\eta\phi^2 << 2$

$$\eta=(1-\frac{\eta\phi^2}{4})^n \, (1-\frac{\nu}{4M}\eta\phi^2)^m \exp(\frac{\beta\gamma}{4}\eta\phi^2) \qquad |1.26|$$

-diffusion control,$\eta\phi^2 >> 2$

$$\eta=\frac{1}{2^n}\,\frac{1}{\eta\phi^2}\,(1-\frac{\nu}{2M})^m \exp(\frac{\beta\gamma}{2}) \qquad |1.27|$$

When temperature gradients in the film are considered one can write,for reactions taking place at the catalyst surface

$$\dot{N} = k_f \, a_p \, (c_{Ab}-c_{As}) = k(T_s) \, c_{As} \qquad |1.28|$$

and

$$\dot{N} \, (-\Delta H) = h \, a_p \, (T_s - T_b) \qquad |1.29|$$

where k_f(and h) is the film mass (and heat) transfer coefficient, a_p the specific area of the particle,T_s and T_b the surface and bulk fluid phase temperatures.

Thus the effectiveness factor,referred to the bulk conditions of the fluid phase is,for 1st order irreversible reactions,

$$\bar{\eta} = k(T_s)c_{As}/k(T_b)c_{Ab} \, {}^{*} \qquad |1.30|$$

leading to

$$\frac{1}{\bar{\eta}\, k(T_b)} = \frac{1}{k_f a_p} + \frac{1}{k(T_s)} \qquad |1.31|$$

and

$$T_s - T_b = \frac{(-\Delta H)k_f}{h}\,(c_{Ab}-c_{As}) \qquad |1.32|$$

The maximum temperature rise in the film is obtained for $c_{As} << c_{Ab}$ and then

$$T_{s,max} - T_b = \frac{(-\Delta H)k_f c_{Ab}}{h} \qquad |1.33|$$

---

*For reactions in nonisothermal particles we get,in general, $\bar{\eta} = r_{obs}/k(T_b)c_{Ab}$;$Ca = \bar{\eta}\,\overline{Da} = \eta Da/(1+\eta Da)$ where $\overline{Da}$ is calculated for the bulk conditions.

Finally

$$\overline{\Delta T^*_e} = \frac{T_{s,max}-T_b}{T_b} = \frac{k_f}{h}\,\frac{(-\Delta H)c_{Ab}}{T_b} = \overline{\beta}\,\frac{Bi_m}{Bi_h} \qquad |1.34|$$

where $\overline{\beta}$ is the Prater thermicity factor in the bulk conditions (c_{Ab}, T_b), $Bi_m = k_f \ell / D_{eff}$ is the mass Biot number and $Bi_h = h\ell/\lambda_{eff}$ is the heat Biot number.
From the Chilton-Colburn analogy

$$\frac{k_f}{h} = \frac{1}{\rho_f\, c_{pf}}\, Le^{-2/3} \qquad |1.35|$$

where Le=Sc/Pr is the Lewis number for the fluid phase;finally

$$b = Bi_m/Bi_h = \frac{\lambda_{eff}}{D_{eff}}\,\frac{1}{\rho_f c_{pf}}\, Le^{-2/3} = \frac{\lambda_{eff}}{D_{eff}}\,\frac{Sh}{Nu}\,\frac{D}{\lambda} = \xi_h Le_c Le^{-2/3} \qquad |1.36|$$

with $Sh = k_f\ell/D$, $Nu = h\ell/\lambda$, ξ_h is the ratio between solid and fluid heat capacities and Le_c the Lewis number for the catalyst ($Le_c = \lambda_{eff}/D_{eff}\,\rho_c c_{pc}$).

In Table IV we summarize the gradients of concentration and temperature inside and outside a nonisothermal catalyst particle with slab geometry|16|.

1.6- Practical considerations

The parameter which governs the relative importance of temperature gradients in the film and inside the particle is the ratio between the two Biot numbers (Eq.|1.36|):

$$b = Bi_m/Bi_h$$

Carberry |13| presents an estimate of values of $\overline{\gamma}$, $\overline{\beta}$ (based on bulk conditions) and b for gas/solid and liquid/solid systems. Since for gas-solid systems $\overline{\gamma}=5$ to 30, $\overline{\beta}=0.001$ to 0.3, $b=10$ to 10^4 and $\overline{\Delta T^*_e}=0.01$ to 2 one can expect that when internal resistance is important (Ca ≃0 and the ratio between temperature gradients in the particle and in the film is 1/b) thus external temperature gradients are always larger (10 to 10^4) than in the solid; as Ca increases this result still holds.
However this does not mean that the absolute temperature rise inside the particle is negligible since the upper bound is $\overline{\beta}T_b$ and unless $\overline{\beta}$ is very small the temperature rise can be considerable |17|. For liquid/solid systems $\overline{\gamma}=5$ to 30, $\overline{\beta}=10^{-3}$ to 10^{-1} ,

$b=10^{-4}$ to 10^{-1} and $\overline{\Delta T^*_e}=10^{-3}$ to 0.05 ;then the above conclusion can be misleading.One can roughly say that temperature effects inside particles are often negligible except in the case of ordinary diffusion of gases in nonmetalic solid catalysts.
In Table V values of $\overline{\beta}$, $\overline{\gamma}$, Le_c^{-1}, ϕ, Sh/Nu and B (dimensionless adiabatic temperature rise, $B=E/RT_b \cdot (-\Delta H)c_{Ab}/\rho_f c_{pf} T_b$) are shown for some industrially important reactions .

TABLE IV-Concentration and temperature gradients in porous catalysts in terms of observable quantities(slab geometry).

Model equations	
$D_{eff}\dfrac{d^2c}{dr^2}=r(c,T)$	
$\lambda_{eff}\dfrac{d^2T}{dr^2}=(\Delta H)r(c,T)$	
$D_{eff}\dfrac{dc}{dr}=k_f(c_{Ab}-c_{As})\quad ,r=\ell$	
$-\lambda_{eff}\dfrac{dT}{dr}=h(T_b-T_s)\quad ,r=\ell$	
$\dfrac{dc}{dr}=\dfrac{dT}{dr}=0\quad ,r=0$	
External concentration gradient	
$\overline{\Delta c_e}=\dfrac{c_{Ab}-c_{As}}{c_{Ab}}$	$Ca=\overline{\eta}\overline{Da}=\dfrac{r_{obs}\ell}{k_f\, c_{Ab}}$
Internal concentration gradient	
$\overline{\Delta c_i}=\dfrac{c_{As}}{c_{Ab}}$	$1-Ca$
Maximum internal temperature rise	
$\Delta T^*_i=\dfrac{T_{max}-T_s}{T_s}=\beta$	$\dfrac{\overline{\beta}(1-Ca)}{1+\overline{\beta}bCa}$
Maximum external temperature rise	
$\overline{\Delta T^*_e}=\dfrac{T_{s,max}-T_b}{T_b}$	$\overline{\beta}b$
External temperature rise	
$\overline{\Delta T_e}=\dfrac{T_s-T_b}{T_b}$	$\overline{\beta}bCa$
Maximum internal temperature rise (reduced by T_b)	
$\overline{\Delta T^*_i}=(T_{max}-T_s)/T_b$	$\overline{\beta}(1-Ca)$

Total temperature rise (reduced by T_b)	
$\overline{\Delta T_i^*} + \overline{\Delta T_e}$	$\overline{\beta}\{1 + Ca(b-1)\}$
Relative external temperature rise	
$\dfrac{\overline{\Delta T_e}}{\overline{\Delta T_i^*} + \overline{\Delta T_e}}$	$\dfrac{bCa}{1 + Ca(b-1)}$

All quantities with ($\overline{\ \ }$) are referred to bulk conditions.

TABLE V- Parameters for catalytic reactions(exothermic) |18,19,20|

Reaction	$\overline{\gamma}$	$\overline{\beta}$	$(Le_c)^{-1}$	ϕ	$\frac{Sh}{Nu}$	B	$\overline{\beta}b$
Synthesis of NH_3	29.4	6.1×10^{-5}	2.6×10^{-4}	1.2	-	0.07	0.59
Synthesis of higher alcohols from CO and H_2	28.4	8.4×10^{-4}	8.5×10^{-4}	-	-	0.30	1.2
Oxidation of CH_3OH to CH_2O	16.0	0.011	0.0045	1.1	-	7.6	17.5
Synthesis of vinylchloride from C_2H_2 and HCl	6.5	0.254	0.1	0.27	-	-	-
Hydrogenation of ethylene	23-27	0.044--0.1	0.11	0.2--2.8	580	-	-
Oxidation of H_2	6.75--7.52	0.03-0.3	0.036	0.8--2	-	-	-
Oxidation of ethylene to ethylenoxide	13.4--18	0.13	0.065	0.08	45	12.2	0.58
Hydrogenation of benzene	14-16	0.12	0.006	0.05--1.9	150	4.1	0.21
Oxidation of SO_2	14.8	0.0118	0.0415	0.9	105	-	-
Oxidation of naphtalene	22-30	0.015	-	-	2.7	34.3	0.041

Some authors suggest criteria for neglecting concentration gradients inside the particles, both for isothermal and nonisothermal cases |21|, based on Weisz and Prater arguments

Isothermal catalysts

$$\frac{(V_p/A_p)^2\, r_{obs}}{D_{eff}\, c_{As}} < 1 \qquad \text{or} \quad \eta\phi^2 < 1 \qquad |1.37|$$

Nonisothermal particles

$$\frac{(V_p/A_p)^2\, r_{obs}}{D_{eff}\, c_{As}} \exp\left(\frac{\gamma\beta}{1+\beta}\right) < 1 \quad \text{or} \quad \eta\phi^2 \exp\left(\frac{\gamma\beta}{1+\beta}\right) < 1 \qquad |1.38|$$

All we have said is based on the catalyst behavior in steady state; however in the transient regime |22| higher temperatures than the steady state temperature inside the pellet can be reached; then local overheating and sintering problems should be taken into consideration.
Multiple steady states can occur; for 1st order irreversible reactions Le Goff and Zoulalian |15|, based on simple models, give the condition for the existence of multiple steady states, i.e.,

$$\bar{\beta}b\bar{\gamma} > 4\left(1+\bar{\beta}b + \frac{Bi_m}{4}\right) \qquad |1.39|$$

which agrees with the more rigorous solution derived by Hlavacek and Kubicek |23|.
The stability of steady state solutions is deeply related with the Lewis number of the catalyst, $Le_c=\lambda_{eff}/\rho_c c_{pc} D_{eff}$ since the transient regime is described by (for first order irreversible reactions)

$$\frac{\partial f}{\partial \tau} = \frac{\partial^2 f}{\partial z^2} + \frac{\alpha}{z}\frac{\partial f}{\partial z} - \phi^2 \exp\left\{-\gamma\left(\frac{1}{\theta}-1\right)\right\} f \qquad |1.40|$$

$$(Le_c)^{-1}\frac{\partial \theta}{\partial \tau} = \frac{\partial^2 \theta}{\partial z^2} + \frac{\alpha}{z}\frac{\partial \theta}{\partial z} + \phi^2 \beta \exp\left\{-\gamma\left(\frac{1}{\theta}-1\right)\right\} f \qquad |1.41|$$

where $\tau=D_{eff}t/\ell^2$.
A great number of papers deal with the very important problem of effectiveness factor of catalysts. Calculations of η can be done using rigorous models or "simplissime" models |15|, for any geometry (slab, sphere, noncylindrical extruded catalysts |24|, cylinder..), for complex reactions |25,26| and concentration-dependent diffusion coefficients |27,28|.

1.7 -Pressure gradients in porous catalysts

This aspect is rarely taken into account in the analysis of diffusion and reaction in porous catalysts although pressure gradients affect concentrations of the reacting species and therefore the effectiveness factors.

For isothermal systems Hite and Jackson |29| made several interesting points:

a) for single reactions, $\sum_i \nu_i A_i = 0$ if there are to be no pressure gradients inside the pellet it is necessary that

$$\sum_i \nu_i \sqrt{M_i} = 0 \qquad |1.42|$$

This means that even for reactions without volume change pressure variations can occur in despite of opposite erroneous statements.

b) for a reaction $A \rightarrow \nu B$ with $p_{Bo}=0, p_{Ao}=p_o$ (surface partial pressures) one gets

$$p = \sqrt{\nu}\, p_o - (\sqrt{\nu} - 1) p_A \qquad |1.43|$$

Then, if Knudsen diffusion controls the bounds of pressure are p_o and $\sqrt{\nu}\, p_o$ (in the center of the pellet); if bulk diffusion controls the bounds are p_o and $p_o + \mu D_{eff}/B_o \cdot \ln \sqrt{\nu}$ (B_o -permeability of porous medium, μ-viscosity of the reaction mixture). Even for $B_o \rightarrow \infty$ small pressure variations can modify the fluxes (in the viscous term the product B_o grad p appears) sufficiently to change the flux ratio from its isobaric value $\sqrt{\nu}$ to the value ν required by the stoichiometry of the reaction.

c) for systems with no volume change $\sum_i \nu_i = 0$, but nonisothermal catalysts, Wong et al |30| show that the center pellet pressure in the Knudsen limit is

$$p = p_s \sqrt{1+\beta} \qquad |1.44|$$

where p_s is the pressure at the catalyst surface. The authors claim that pressure variations are due only to temperature gradients; this will be true only for binary mixtures. In the bulk diffusion limit, $p=p_s$. They also give the maximum center pellet temperature:

Knudsen diffusion control

$$T=T_s(1+\beta) \qquad |1.45|$$

Bulk diffusion control

$$T=T_s(1+\beta+\ \beta^2/4) \qquad |1.46|$$

These effects should be taken into account on the analysis of the effectiveness factor, the relative importance of diffusional regimes being indicated by $\Theta=D^e_{K,s}/D^e_s$ (ratio between effective Knudsen and bulk diffusivities at surface conditions).
For an excellent discussion of this topic we recommend the book by Jackson |31|.
It is interesting to be able to predict the effect of catalyst structure (particle size, ℓ, mean pore diameter, $\bar{d}$) and pressure on the effectiveness factor, η.
The influence of ℓ on η can be displayed by a plot of η vs ϕ^*, with Θ and $\tilde{\beta}=B_o p_o/\mu D^e_K$ as parameters, since $\phi^*=\ell\sqrt{k/D^e_K} \sim \ell/\bar{d}^{\alpha}$ (α=1/2 to 1). Similarly if we note that Θ and $\tilde{\beta}$ are proportional to $p_o\bar{d}$, the effect of $\bar{d}$ on η can be visualized from a plot of η vs Θ with $\tilde{\beta}/\Theta$ and $\phi^*\sqrt{\Theta}$ as constant parameters (assuming a constant number of active sites per pellet volume in order to get $\phi^*\sim\ell/\sqrt{\bar{d}}$); also the effect of p_o is viewed from the η vs Θ plot with $\tilde{\beta}/\Theta$ and ϕ^* as constant parameters. This results from the well established dependencies of effective Knudsen diffusivity, D^e_K, effective bulk diffusivity, D^e_s and permeability, B_o on p_o, $\bar{d}$ and ℓ, i.e.,

$$D^e_K \sim \bar{d}\ \ p^0_o T^{1/2}$$

$$D^e_s \sim \bar{d}^0\ p^{-1}_o T^{3/2}$$

$$B_o \sim \bar{d}^2\ p^0_o$$

1.8 - Intraparticle forced convection, diffusion and reaction in porous catalysts

Molecular diffusion is usually the unique intraparticle transport mechanism considered when dealing with porous catalysts. However, convective transport can have a role and one should look at the effect of this transport process on the effectiveness factor. Nir and Pismen |32| studied this problem for single reactions, while Nir |33| tackled the problem of consecutive reactions.

For slab geometry of catalyst, 1st order irreversible reactions and isothermal particles, the starting equation is:

$$\frac{d^2f}{dz^2} - \lambda \frac{df}{dz} - \phi^2 f=0 \qquad |1.47|$$

where $\lambda=v_o \ell/D_{eff}$ ($v_o = \frac{B_o}{\mu} \frac{\Delta p}{\ell}$ is the characteristic intraparticle flow velocity);for Robin boundary conditions ($f=1, z=\pm 1$)the effectiveness factor is

$$\eta= \frac{1/\alpha_1 - 1/\alpha_2}{\coth \alpha_1 - \coth \alpha_2} \qquad |1.48|$$

with

$$\alpha_1,\alpha_2= \frac{\lambda}{2} \{1 \pm \sqrt{1+ \frac{4\phi^2}{\lambda^2}}\}$$

Obviously λ is the new parameter showing the relative importance of convective and diffusional mass transfer inside the particle; for $\lambda \to 0$,$\eta \to$th ϕ/ϕ.
For the reaction controlled case,convective currents are of no importance;however in the intermediate regime convection enhances the catalyst effectiveness factor η ,i.e.,$\eta>\eta_o$ where η_o is based on a constant D_{eff} ,in absence of convection.
It is interesting to analyse the implication of neglecting the convection term in reactor design.This can be done by comparing η defined above with the effectiveness factor,$\tilde{\eta}$ calculated from a measured effective diffusivity,$\tilde{D}_{eff}$ (assuming only diffusion inside the particle).
One can show |34|that

$$\tilde{D}_{eff}= D_{eff}/f(\lambda) \qquad |1.49|$$

where

$$f(\lambda)= \frac{3}{\lambda}\{\frac{1}{\text{th } \lambda} - \frac{1}{\lambda}\}$$

Intraparticle forced convection should be considered in order to explain changes of the measured effective diffusivity,$\tilde{D}_{eff}$ with flowrate in large pore catalysts |34|.
Then $\tilde{E}=\eta/\tilde{\eta}$ can be ploted as a function of λ and ϕ (based on a true D_{eff});a summary of these results is presented in Figure 10 (iso-$\tilde{E}$ plot),which shows that in some conditions errors of 100% can be easily made in reactor design.
Wheeler |35| suggested that forced convection in catalyst pores (due to pressure drop across the reactor,reaction with volume change) should be taken into account for high pressupe reactions in large pores;he gave a criterion which should be satisfied in that case:

$$(P\,\overline{d}\,)^2 \geqslant 6\times10^{-7}\,\phi$$

where the pressure is P (atm), the pore diameter is $\overline{d}$ (cm) and ϕ the Thiele modulus.

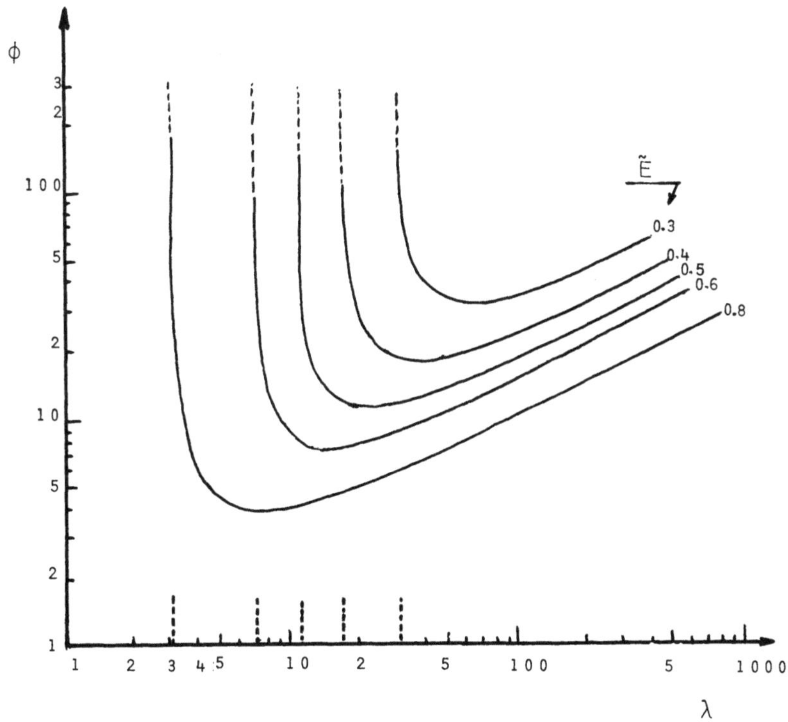

Figure 10- Iso-$\tilde{E}$ plot.

REFERENCES FOR PART I

|1| Le Goff,P.,Cours de Génie Chimique,ENSIC,Nancy(1973)
|2| Villermaux,J. and Le Goff,P.,L'Actualité Chimique,p.25, Jan.1978
|3| Richardson,J. and Luss,D.,Applications of heterogeneous catalysis,Univ.Houston (1971)
|4| Le Goff,P.,Energétique Industrielle,ed.Technique &Documentation,Paris(1979)
|5| Aris,R.,Chem.Eng.Sci.,8(1),20(1974)
|6| Emig,G. and Hofmann,H.,Int.Chem.Eng.,17(1),8(1977)
|7| Satterfield,C.,Mass transfer in heterogeneous catalysis, MIT Pres(1970)
|8| Carra,S.,"Influence of heat and mass transfer on heterogeneous catalysis" in Fundamental principles in heterogeneous catalysis,NATO ASI,Venice (1971)
|9| Rajadhyaksha,R.,Vasudeva,K. and Doraiswamy,L.,J.Catalysis, 41,61(1976)
|10| Aris,R. The mathematical theory of diffusion and reaction in permeable catalysts,vol 1.,ed.Oxford Univ.Press(1975)
|11| Carberry,J.,Cat.Rev.,3,61(1969)
|12| Cassiere,G. and Carberry,J.,Chem.Eng.Educ.,22,Winter 1973
|13| Carberry,J.,Chemical and catalytical reaction engineering, ed.Mc Graw Hill(1976)
|14| Le Goff,P. and Zoulalian,A.,Cours de Génie Chimique ,ENSIC, Nancy(1973)
|15| Le Goff,P. and Zoulalian,A.,The Chem.Eng.J.,12,33 (1976);also 12,47(1976)
|16| Bischoff,K.,Ind.Eng.Chem.Fund.,15(3),229(1976)
|17| Oh,S.,Hegedus,L. and Aris,R.,Ind.Eng.Chem.Fund.,17(4),309(1978)
|18| Hlavacek,V. and Votruba,J.,Adv.Catalysis,27,59(1978)
|19| Van den Bosch,B. and Luss,D.,Chem.Eng.Sci.,32,203(1977)
|20| Hlavacek,V. and Votruba,J.,"Steady state operations of fixed bed reactors and monolithic structures" in Chemical reactor theory,ed.Lapidus and Amundson,Prentice-Hall(1977)
|21| Petersen,E.,Chemical reaction analysis ,Prentice-Hall(1965)
|22| Carberry,J. and Butt,J.,Cat.Rev.-Sci.Eng.,10(2),221(1974)
|23| Hlavacek,V. and Kubicek,M.,Chem.Eng.Sci.,25,1537(1970)
|24| Suzuki,T. and Uchida,T.,J.Chem.Eng.Japan,12(6),425(1979)
|25| Allen,D.,Gerhard,E. and Likins,M.,AIChE J.,22(3),593(1976)
|26| Ramachandran,P.,Kam,E. and Hughes,R.,Chem.Eng.Sci.,31,244, (1976)
|27| Pereira,C. and Varma,A.,Chem.Eng.Sci.,33,396(1978)
|28| Schneider,P. Chem.Eng.Commun., 2,155(1976)
|29| Hite,R. and Jackson,R.,Chem.Eng.Sci.,32,703(1977)
|30| Wong,R.,Hubbard,G. and Denny,V.,Chem.Eng.Sci.,31,541(1976)
|31| Jackson,R.,Transport in porous catalysts ,Elsevier,1977.
|32| Nir,A. and Pismen,L.,Chem.Eng.Sci.,32,35(1977)

|33| Nir,A.,Chem.Eng.Sci.,32,925(1977)
|34| Rodrigues,A.,Ahn,B. and Zoulalian,A.,submitted to AIChEJ.
|35| Wheeler,A.,Adv.in Catalysis,3,250(1951)

II - SELECTIVITY FALSIFICATION DUE TO DIFFUSIONAL EFFECTS AND CATALYST DEACTIVATION .

In this section we will analyse diffusional falsifications of selectivity as well as catalyst deactivation in order to take these problems into account when modeling fluid/solid catalytic reactors.

2.1-The concept of selectivity

In the literature there is some indiscriminate use of the terms yield and selectivity.Let us define the following quantities for complex reaction schemes in which we want to produce P from a reactant A |1|:

i)Yield, $\rho_{P/A}$

It is the ratio between the quantity (batch reactors) or molar flow (open systems) of P actually obtained and the quantity (or molar flow) of P obtained if all A initially present were transformed in an optimal path

$$\rho_{P/A}=n_P/\nu n_{Ao} \text{ or } F_P/\nu F_{Ao} \qquad |2.1|$$

ii)Relative yield, $\Phi_{P/A}$

It is the ratio between the quantity (or molar flow) of P actually obtained and the quantity (or molar flow) of P obtained if all A effectively consumed were transformed in an optimal path

$$\Phi_{P/A}=n_P/\nu(n_{Ao}-n_A) \text{ or } F_P/\nu(F_{Ao}-F_A) \qquad |2.2|$$

iii)Selectivity, $\sigma_{P/A}$

It is the ratio between the quantity of P actually obtained and the quantity of undesirable products resulting from A

$$\sigma_{P/A} = \frac{n_P}{\nu(n_{Ao}-n_A)-n_P} \quad \text{or} \quad \frac{F_P}{\nu(F_{Ao}-F_A)-F_P} \qquad |2.3|$$

Obviously there is a relationship between relative yield and selectivity

$$1/\Phi = 1 + 1/\sigma \qquad |2.4|$$

with $0 < \Phi < 1$ and $0 < \sigma < \infty$.
These are global quantities;local relative yield and local selectivity are respectively

$$\Phi'_{P/A} = r_P/-\nu r_A \quad \text{and} \quad \sigma'_{P/A} = r_P/(-\nu r_A - r_P) \qquad |2.5|$$

where r_P is the rate of production of P and $-r_A$ the rate of disappearance of A.Convenient plots for relative yield and selectivity are shown in Figure 11.

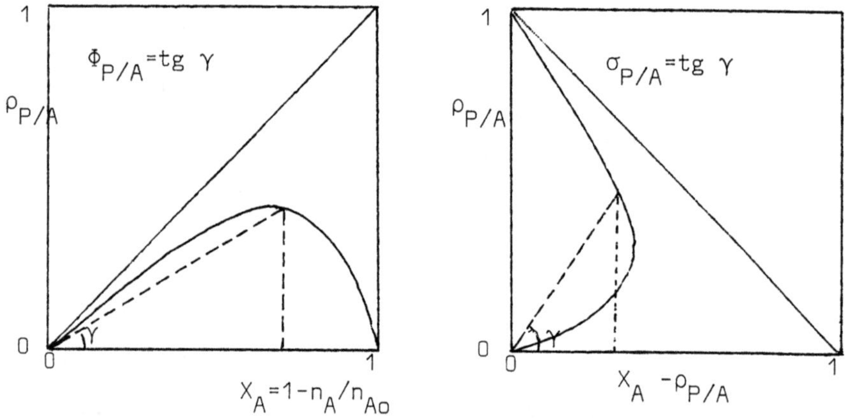

Figure 11- Relative yield and selectivity.

Before proceeding further one should explain what is the stoichiometric coefficient ν appearing in the above definitions.To do that let us assume the reaction scheme presented below;we are interested in the production of P from A.The selectivity characterizes the relative importance of the reactions leading from A to P compared with the undesirable reactions which consume A without production of P.

$$A + B \rightarrow C \qquad (1)$$
$$C \rightarrow P + S \qquad (2)$$
$$S + B \rightarrow D \qquad (3)$$
$$D \rightarrow P + R \qquad (4)$$
$$R + B \rightarrow E \qquad (5)$$

The ideal situation is such that the desirable reactions are infinitely fast compared with the secondary ones;then one can neglect the intermediates in the paths leading from A to P and adding the other reactions we get

$$A + \ldots = \nu P + \ldots. \qquad |2.6|$$

The stoichiometric coefficient ν in the definition of yield and selectivity is the maximum ν found in this way if several graphs for the reaction scheme are possible.For the above example the graph is

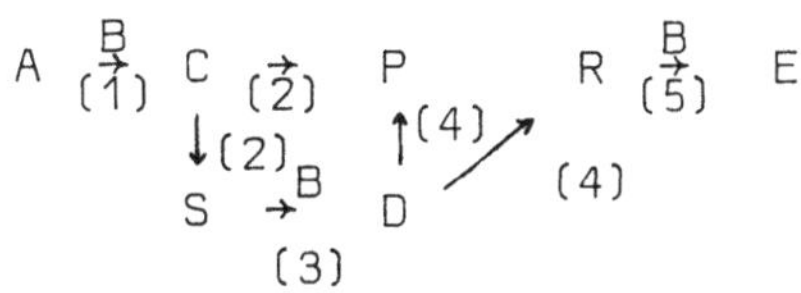

Neglecting reaction (5) which can not lead to P and adding the other reactions we obtain

$$A + \ldots \rightarrow 2P + \ldots$$

so ν=2.

2.2-Diffusional falsification of selectivity

We already analysed the disguise of kinetic parameters due to diffusional effects.Let us now review the influence of diffusion upon selectivity following the classic treatment of Wheeler |2,3|.

i) Selectivity Type I

Reaction scheme:

$A \xrightarrow{k_1} B + C$ (desirable;1st order)

$X \xrightarrow{k_2} Y + Z$ (undesirable;1st order)

In the kinetic regime the selectivity (local) is:

$$\sigma'_o = (r_A/r_X)_o = k_1 c_{Ao}/k_2 c_{Ao}$$

In the diffusional regime the selectivity is now

$$\sigma' = \eta_1 k_1 c_{Ao}/\eta_2 k_2 c_{Ao} = \sigma'_o \, \eta_1/\eta_2$$

For large Thiele moduli ,ϕ_1 and ϕ_2 one gets

$$\eta_1/\eta_2 = \phi_2/\phi_1 = \sqrt{k_2/k_1} = \sqrt{1/s}$$

with $s = k_1/k_2$;finally

$$\sigma' = \sigma'_o \sqrt{1/s} \qquad |2.7|$$

The intrinsec selectivity is reduced by diffusional intrusions; remedies for this can be either decreasing particle diameter or increasing pore size.

ii) Selectivity Type II

Reaction scheme:

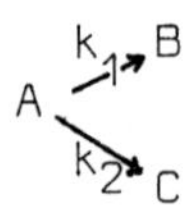

(desirable;order n)

(undesirable;order m)

The intrinsec selectivity is $\sigma'_o = (r_B/r_C)_o = s\, c_{Ao}^{n-m}$ .In the diffusional regime $\sigma' = r_B/r_C = s\, c_A^{n-m}$ and then

$$\sigma' = \sigma'_o \, (c_A/c_{Ao})^{n-m} \qquad |2.8|$$

Three cases can occur:

a) n=m

Diffusional effects do not affect selectivity although they reduce activity.

b) n> m

Diffusional resistances reduce selectivity.

c) n< m

Diffusional limitations enhance selectivity.

iii) "Selectivity" Type III

Reaction scheme:

$$A \underset{k_1}{\rightarrow} B \underset{k_2}{\rightarrow} C$$

The intrinsec yield is $\Phi'_o = (-r_B/r_A)_o = 1 - \frac{1}{s}\frac{c_{Bo}}{c_{Ao}}$;if initially no B is present,then $\Phi'_o=1$.
In the diffusional regime

$$\Phi' = \frac{s}{s-1}\left\{1 - \frac{\tanh(\phi_1/\sqrt{s})}{\sqrt{s}\ \tanh\phi_1}\right\} \qquad |2.9|$$

For $\phi_1 > 3$,

$$\Phi' = \sqrt{s}/(\sqrt{s}+1) \qquad |2.10|$$

The mathematical development can be illustrated for this last reaction scheme.Assuming first order reactions with no volume changes and constant effective diffusivities for A and B species, $D_{A,eff}$ and $D_{B,eff}$ respectively, we get the steady state mass balance equations for spherical particles:

component A

$$-\frac{1}{r^2}\frac{d}{dr}\left(r^2 D_{A,eff}\frac{dc_A}{dr}\right) = -k_1 c_A \qquad |2.11a|$$

component B

$$-\frac{1}{r^2}\frac{d}{dr}\left(r^2 D_{B,eff}\frac{dc_B}{dr}\right) = k_1 c_A - k_2 c_B \qquad |2.11b|$$

The boundary conditions are:

$$dc_A/dr = dc_B/dr = 0 \ , \ r=0 \qquad |2.11c|$$

$$c_A = c_{Ao}(1-X) \ , \ c_B = c_{Ao}Y \quad \text{at } r=R_p \qquad |2.11d|$$

Using normalized concentrations $f_A=c_A/c_{Ao}$, $f_B=c_B/c_{Ao}$, reduced radial coordinate,$z=r/R_p$ and defining

$$\phi_1 = R_p\sqrt{k_1/D_{A,eff}} \qquad |2.12a|$$

$$\phi_2 = R_p\sqrt{k_2/D_{A,eff}} \qquad |2.12b|$$

$$\phi_3 = R_p\sqrt{k_3/D_{B,eff}} \qquad |2.12c|$$

one gets, after a new change of variables $q_A=f_Az, q_B=f_Bz$,

$$\frac{d^2q_A}{dz^2} - \phi_1^2 q_A = 0 \qquad |2.13a|$$

$$\frac{d^2q_B}{dz^2} + \phi_2^2 q_A - \phi_3^2 q_B = 0 \qquad |2.13b|$$

$$q_A = 1-X \ , \ q_B = Y \ , \ z=1 \qquad |2.13c|$$

$$q_A = q_B = 0 \ , z=0 \qquad |2.13d|$$

Then the solution of Eqs |2.13a| and |2.13b| with the boundary conditions, Eqs. |2.13c| and |2.13d| leads to

$$q_A = (1-X) \frac{sh(\phi_1 z)}{sh\phi_1} \qquad |2.14|$$

or

$$c_A/c_{As} = \frac{sh(\phi_1 z)}{sh\ \phi_1} \qquad |2.15|$$

Similarly

$$c_B/c_{As} = \{\frac{c_{Bs}}{c_{As}} - \frac{\phi_2^2}{\phi_3^2 - \phi_1^2}\}\frac{sh(\phi_3 z)}{z\ sh\phi_3} + \frac{\phi_2^2}{\phi_3^2 - \phi_1^2}\ \frac{sh(\phi_1 z)}{z\ sh\phi_1} \qquad |2.16|$$

The effectiveness factors for the reactions : disappearance of A and net production of B , are respectively

$$\eta_1 = \frac{1}{\phi}\{\frac{1}{th(3\phi)} - \frac{3}{\phi}\} \qquad |2.17|$$

and

$$\eta_2 = \frac{\alpha}{\phi(1-c_{Bs}/sc_{As})}\{\frac{c_{Bs}}{c_{As}}(\frac{1}{\phi} - \frac{1}{\sqrt{\alpha s}\ th\frac{3\phi}{\sqrt{\alpha s}}}) + \frac{1}{(\frac{1}{s}-\alpha)}(\frac{1}{\sqrt{\alpha s}\ th\frac{3\phi}{\sqrt{\alpha s}}} - \frac{1}{th\ 3\phi})\} \qquad |2.18|$$

with $\phi = \phi_1/3$, $s=k_1/k_2$ and $\alpha=D_{B,eff}/D_{A,eff}$.

The local yield is then

$$\Phi' = \alpha \left\{ \frac{\frac{c_{Bs}}{c_{As}}\left[\frac{1}{\phi} - \frac{1}{\sqrt{\alpha s}\ th(3\phi/\sqrt{\alpha s})}\right] + \frac{s}{1-\alpha s}\left[\frac{1}{\sqrt{\alpha s}\ th(3\phi/\sqrt{\alpha s})} - \frac{1}{th 3\phi}\right]}{\frac{1}{th(3\phi)} - \frac{3}{\phi}} \right\}$$

In the diffusional regime

$$\Phi' = \frac{\sqrt{\alpha s}}{\sqrt{\alpha s}+1} - \frac{c_{Bs}}{c_{As}}\sqrt{\alpha/s} \qquad |2.19|$$

If no B is initially present, then

$$\Phi' = \sqrt{\alpha s}\ /(\sqrt{\alpha s}+1) \qquad |2.20|$$

which in the case of $\alpha=1$ gives the result presented by Wheeler, Eq. |2.10| (Figure 13). Obviously in this case $\sigma' = \sqrt{s}$.

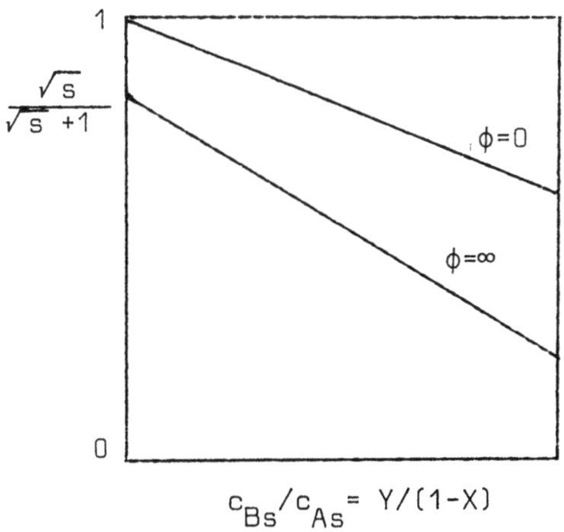

Figure 13 - Φ' vs c_{Bs}/c_{As}

$\phi \to 0 \quad \Phi' = 1 - c_{Bs}/sc_{As}$

$\phi \to \infty \quad \Phi' = \frac{\sqrt{s}}{\sqrt{s}+1} - \frac{1}{\sqrt{s}}\frac{c_{Bs}}{c_{As}}$

2.3- Catalyst deactivation

The main processes leading to catalyst deactivation are poisoning by chemisorption of poisons on active sites, fouling origi-

nated by physical blocking of the surface,usually by coke deposition,and sintering due to heat effects and leading to a loss of surface area.
This can be expressed in terms of an observed kinetic constant, k_{obs} which is related to the intrinsec kinetic constant,k_{int} by

$$k_{obs} = k_{int} S_v \eta \qquad |2.21|$$

Sintering and poisoning lead to a decrease on S_v (specific area, cm^2/gr) by loss of surface area and active sites,respectively ; fouling leads to a decrease of η |4|.
Regeneration of the catalyst is possible when it is deactivated by fouling,sometimes possible when deactivation is due to poisoning and never possible if deactivation is due to sintering.
This change of the kinetic constant with time or,as Froment suggests,the change of k with the reaction conditions and local composition of the reaction mixture should be incorporated in the design equations |5|.
Two limiting cases of poisoning can be considered :

a) uniform poisoning

b) pore-mouth or shell-progressive poisoning

Let us analyse these situations:

a) Uniform poisoning

If α is the poisoned fraction of the catalytic surface, then for first order irreversible reactions,the reaction term which appears in the mass balance equation will include $k(1-\alpha)$; the concentration profile for the poisoned pellet (slab geometry) will be:

$$f = \frac{ch(\phi\sqrt{1-\alpha}\, z)}{ch(\phi\sqrt{1-\alpha})} \qquad |2.22|$$

Let us define a poisoning factor

$$P = \frac{\text{rate with poisoning}}{\text{rate of the unpoisoned catalyst}}$$

Then we obtain

$$P = \frac{(\phi\sqrt{1-\alpha})\tanh(\phi\sqrt{1-\alpha})}{\phi \tanh \phi} \qquad |2.23|$$

For small values of ϕ (surface completely available)

$$P = 1-\alpha \quad \text{(non-selective poisoning)} \qquad |2.24|$$

which is represented by curve A in Figure 14.
For high values of ϕ ,then

$$P = \sqrt{1-\alpha} \quad \text{(anti-selective poisoning)} \qquad |2.25|$$

corresponding to curve B in Figure 14;i.e.,activity is lost at a rate slower than that of poisoning.

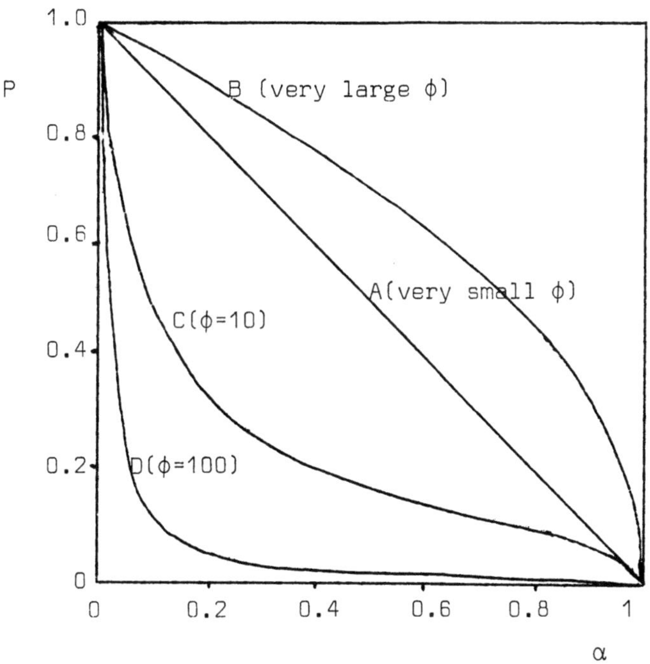

Figure 14-Types of poisoning curves for porous catalysts |2| .

b) Pore-mouth poisoning

This is sketched for a pore or slab catalyst in Figure 15. It is equivalent to a shell progressive poisoning.The mathematical development for a slab catalyst and first order irreversible reaction leads to

$$\frac{d^2c}{dz^2} - \phi^2 c = 0 \qquad |2.26|$$

where $\phi=\ell\sqrt{k/D_{eff}}$ and $z=r/\ell$. The boundary conditions are now

$$dc/dz=0, z=0 \quad |2.27|$$

$$c=c_1, z=1-\alpha \quad |2.28|$$

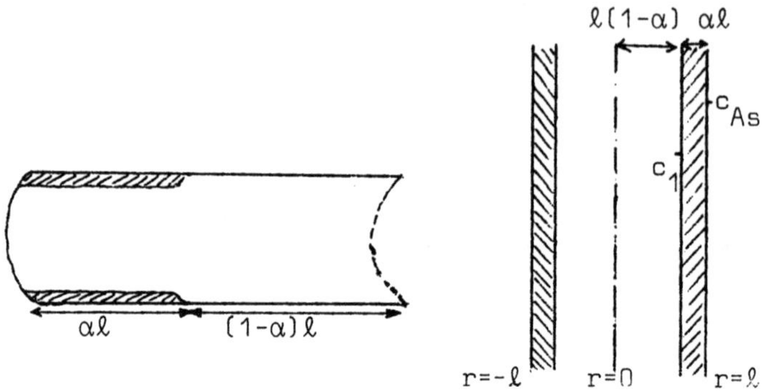

a) pore b) catalyst slab

Figure 15-Pore mouth poisoning

The solution is then

$$c=c_1\frac{ch(\phi z)}{ch\{\phi(1-\alpha)\}} \quad |2.29|$$

The concentration c_1 is determined by the equality of fluxes

$$\frac{c_{As}-c_1}{\alpha\ell}=\frac{1}{\ell}\left.\frac{dc}{dz}\right|_{z=1-\alpha} \quad |2.30|$$

and so

$$c_1=\frac{c_{As}}{1+\alpha\phi\ th\{\phi(1-\alpha)\}} \quad |2.31|$$

Finally

$$P=\frac{th\{\phi(1-\alpha)\}}{(th\ \phi)(1+\alpha\phi\ th\{\phi(1-\alpha)\}} \quad |2.32|$$

For small values of ϕ, again $P=1-\alpha$; however for large ϕ

$$P = \frac{1}{1 + \alpha\phi} \qquad |2.33|$$

which shows that a small degree of poisoning leads to a large decrease of activity (selective poisoning;curves C and D in Figure 14).

These considerations indicate that the rates of poisoning and deactivation should be considered while choosing a catalyst.Some poisoning studies as well as regeneration of coked catalysts involve propagation of heat ant concentration waves through fixed bed reactors.This general background of percolation processes|6| finds here an interesting field of application.Complete reviews on catalyst deactivation have been published by Butt|7,8| .

REFERENCES FOR PART II

|1| Villermaux,J.,Introduction à la theorie des reacteurs chimiques, ENSIC,Nancy(1973)

|2| Wheeler,A.,Adv.Cat.,3,250(1951)

|3| Wheeler,A.,"Reaction rates and selectivity in catalyst pores" in Catalysis,Chapter 2,105(1955),ed.Emmett.

|4| Richardson,J. and Luss,D.,Applications of heterogeneous catalysis,Univ.Houston(1971)

|5| Froment,G.,"A quantitative approach of catalyst deactivation by coke formation",in Catalyst deactivation ,pg1,ed.B.Delmon and G.Froment,Elsevier(1980)

|6| Rodrigues,A.E. and Tondeur,D.,Percolation processes:theory and applications , Sijthoff&Noordhoff(1980)

|7| Butt,J.,Adv.Chem.Ser.,109,259(1972)

|8| Butt,J.,ACS Symp.Series,72,288 (1978),ed.Luss and Weekman.

III - MODELS FOR TWO-PHASE FIXED BED CATALYTIC REACTORS

In this section we will review models for two-phase fixed bed catalytic reactors as well as the correlations used in the prediction of the various parameters involved in design.

3.1-Generalized model

Let us consider an element of a cylindrical fixed bed reactor. In transient operation,mass and energy balances for the fluid phase are respectively

$$u_i \frac{\partial c}{\partial z'} + \frac{\partial c}{\partial t} - \mathcal{R} = D_{ax} \frac{\partial^2 c}{\partial z'^2} + D_r \left(\frac{\partial^2 c}{\partial R^2} + \frac{1}{R} \frac{\partial c}{\partial R} \right) \qquad |3.1|$$

(each term has the dimensions of (amount of substance)xvolume^{-1} xtime^{-1})

and

$$u_i \rho_f c_{pf} \frac{\partial T}{\partial z'} + \rho_f c_{pf} \frac{\partial T}{\partial t} - \mathcal{Q} = \rho_f c_{pf} D_{h,ax} \frac{\partial^2 T}{\partial z'^2} + \rho_f c_{pf} D_{h,r} \left(\frac{\partial^2 T}{\partial R^2} + \frac{1}{R} \frac{\partial T}{\partial R} \right) \qquad |3.2|$$

(each term with dimensions cal/vol. time)

When these equations are put in dimensionless form some <u>reactor parameters</u> |1,2,3| can be identified:

$Pe_{ax} = u_i d_p / D_{ax}$ axial mass Peclet number

$Pe_r = u_i d_p / D_r$ radial mass Peclet number

$A = L/R_o$, $n = L/d_p$,$m = R_o/d_p$ aspect ratios

$Pe_{h,ax} = u_i d_p / D_{h,ax}$ axial heat Peclet number

$Pe_{h,r} = u_i d_p / D_{h,r}$ radial heat Peclet number

$\tau = L/u_i$ space time

General boundary conditions can be expressed as follows:

Inlet $z'=0$ $\quad u_i c_o = u_i c - D_{ax} \partial c / \partial z'$ |3.3a|

$$u_i \rho_f c_{pf} T_o = u_i \rho_f c_{pf} T - \rho_f c_{pf} D_{h,ax} \partial T / \partial z' \qquad |3.3b|$$

Outlet $z'=L$ $\quad \partial c / \partial z' = \partial T / \partial z' = 0$ |3.3c|

Center line of the reactor

$R=0 \quad \partial T / \partial R = \partial c / \partial R = 0$ |3.3d|

Reactor wall, $R=R_o$

$$\partial c / \partial R = 0 \qquad |3.3e|$$

$$\rho_f c_{pf} D_{r,h} \partial T / \partial R = U(T - T_c) \qquad |3.3f|$$

Thus a new dimensionless number will arise

$Bi_{h,w} = UR_o / \rho_f c_{pf} D_{h,r}$ heat Biot number at the wall conditions

Besides the fluid phase equations one should write the conservation equations for the particle, Eqs |1.40| and |1.41|. The external (interphase) resistances are introduced as boundary conditions for the particle equations. Thus the particle parameters are

Intraparticle
- ϕ -Thiele modulus
- β -thermicity factor
- λ -intraparticle Peclet number

Interphase
- Bi_m -mass Biot number
- Bi_h -heat Biot number

The reaction parameters are

- γ - Arrhenius number
- N_r - number of reaction units $(k\tau c^{n-1})$

When dealing with the a priori design of a fixed bed reactor one should be able to estimate the various parameters involved in these physical and chemical dimensionless numbers, i.e.

heat and mass axial dispersion, $D_{h,ax}$ and D_{ax}

heat and mass radial dispersion, $D_{h,r}$ and D_r

mass (particle) effective diffusivity, D_{eff}

heat (particle) effective conductivity, λ_{eff}

film heat (mass) fransfer coefficient, h and k_f

fluid-wall heat transfer coefficient, U (unidimensional models) and h_w (bidimensional models)

In the following we will present Froment's classification of fixed bed reactors derived from the above generalized model and correlations available for the estimation of the necessary design parameters.

3.2-Classification of Froment for fixed bed reactor models

A classification of models for fixed bed reactors has been already suggested by Froment |4,5,6|.He divided models into two classes: pseudo-homogeneous, in which external and internal resistances due to diffusional limitations are not taken into account and heterogeneous models when these resistances are considered.
In each class of models two groups are considered: unidimensional, in which only changes in the axial direction are included and bidimensional when both axial and radial variations are present.

Table VI summarizes the classification proposed by Froment. At this point one should mention several reviews on modeling of catalytic reactors |7,8,9,10,11|.

TABLE VI- Froment's classification of fixed bed reactor models

Models	Pseudo-homogeneous (PH) $T_s=T_b=T$; $c_s=c_b=c$	Heterogeneous(HT) $T_s \neq T_b \neq T$; $c_b \neq c_s \neq c$
Unidimensional	PH1- Ideal model	HT1-Ideal model+external resistance
	PH2- Ideal model+axial dispersion	HT2-Ideal model+external resistance +internal resistance
Bidimensional	PH3- Ideal model +radial dispersion	HT3-Ideal model+ internal and external resistances +radial dispersion

Let us now present the model equations for each category of models according to the classification of Froment.

Model equations according to Froment's classification

- Pseudo-homogeneous models (PH)

  - Unidimensional models

PH1 - Ideal model

Assumptions: plug flow of the fluid phase; no diffusional resistances

Mass balance

$$u \frac{dc_b}{dz'} + \rho_b r = 0 \qquad |3.4|$$

Heat balance

$$u\rho_f c_{pf} \frac{dT_b}{dz'} - (-\Delta H)\rho_b r + \frac{2U}{R_o}(T_b - T_w) = 0 \qquad |3.5|$$

Momentum balance

$$-\frac{dp_t}{dz'} = \frac{2f\rho_f u}{gd_p} \qquad |3.6|$$

Boundary conditions

$$z' = 0 \; , T_b = T_o \; , \; c_b = c_o \qquad |3.7|$$

In the above equations $\rho_b r$ is the rate of chemical reaction relatively to the bed volume, u is the superficial velocity, f is the friction factor and the overall heat transfer coefficient U is defined by

$$\frac{1}{U} = \frac{1}{h_i} + \frac{\ell}{\lambda_w} \frac{A_b}{A_m} + \frac{1}{h_o} \frac{A_b}{A_o} \qquad |3.8|$$

with λ_w -thermal conductivity of the wall (kcal/m.hr.^oC), h_i, h_o -heat transfer coefficients, bed side and heat transfer medium side, respectively (kcal/m^2hr ^oC), ℓ-wall thickness(m), and A_b, A_o, A_m -heat exchanging surfaces, bed side, heat transfer medium side and log-mean of A_b and A_o (m^2). Bulk concentrations, temperature and total pressure are respectively, c_b, T_b and p_t while the wall temperature is T_w. Fluid properties are density, ρ_f and heat capacity, c_{pf}.

PH2 -Ideal model + axial dispersion

Model equations:

Mass balance

$$u\frac{dc_b}{dz'} + \rho_b r = \varepsilon D_{ax}\frac{d^2c_b}{dz'^2} \qquad |3.9|$$

Heat balance

$$u\rho_f c_{pf}\frac{dT_b}{dz'} - (-\Delta H)\rho_b r + \frac{2U}{R_o}(T_b - T_w) = \lambda_{ax}\frac{d^2T_b}{dz'^2} \qquad |3.10|$$

Boundary conditions

$$z'=0 \qquad uc_o = uc_b - \varepsilon D_{ax}\frac{dc_b}{dz'} \qquad |3.11a|$$

$$u\rho_f c_{pf} T_o = u\rho_f c_{pf} T_b - \lambda_{ax}\frac{dT_b}{dz'} \qquad |3.11b|$$

$$z'=L \qquad \frac{dc_b}{dz'} = \frac{dT_b}{dz'} = 0 \qquad |3.11c|$$

In the above equations $(-\Delta H)$ is the heat of reaction, D_{ax} is the mass axial dispersion and $D_{h,ax}=\lambda_{ax}/\rho_f c_{pf}$ is the heat axial dispersion coefficient.

-Bidimensional model

PH3-Ideal model +radial dispersion

Model equations:

Mass balance

$$u\frac{\partial c_b}{\partial z'} + \rho_b r = \varepsilon D_r\left(\frac{\partial^2 c_b}{\partial R^2} + \frac{1}{R}\frac{\partial c_b}{\partial R}\right) \qquad |3.12|$$

Heat balance

$$u\rho_f c_{pf}\frac{\partial T_b}{\partial z'} - (-\Delta H)\rho_b r = \lambda_r\left(\frac{\partial^2 T_b}{\partial R^2} + \frac{1}{R}\frac{\partial T_b}{\partial R}\right) \qquad |3.13|$$

Boundary conditions

$$z'=0,\ 0 < R < R_o,\quad c_b=c_o,\ T_b=T_o \qquad |3.14a|$$

$$R=0, \forall z',\ \partial T_b/\partial R=0 \qquad |3.14b|$$

$$R=R_o, \forall z',\ \partial c_b/\partial R=0 \qquad |3.14c|$$

$$\partial T_b/\partial R = -(h_w/\lambda_r)(T_b - T_w) \qquad |3.14d|$$

In the above equations D_r is the mass radial dispersion coefficient, $D_{h,r}=\lambda_r/\rho_f c_{pf}$ is the heat radial dispersion coefficient and h_w is the fluid-wall heat transfer coefficient. When comparing models PH3 and PH1 the relationship $1/h_i=1/h_w + R_o/2\lambda_r$ should be used.

- Heterogeneous models (HT)

  -Unidimensional models

HT1-Ideal model + external resistances

Model equations:

Fluid phase

mass balance

$$u\frac{dc_b}{dz'}+k_f a_v(c_b-c_s)=0 \qquad |3.15|$$

heat balance

$$u\rho_f c_{pf}\frac{dT_b}{dz'}-ha_v(T_s-T_b)+\frac{2U}{R_o}(T_b-T_w)=0 \qquad |3.16|$$

Catalyst particle

mass balance

$$\rho_b r_s=k_f a_v(c_b-c_s) \qquad |3.17|$$

heat balance

$$(-\Delta H)\rho_b r_s=ha_v(T_s-T_b) \qquad |3.18|$$

Boundary conditions

$$z'=0,\ T_b=T_o \text{ and } c_b=c_o \qquad |3.19|$$

In these equations r_s is the reaction rate at the surface conditions, k_f and h are the film mass transfer and film heat transfer coefficients and $a_v=6(1-\varepsilon)/d_p$.

HT2-Ideal model+external resistances+internal resistances

Fluid phase

mass balance

$$u\frac{dc_b}{dz'}+k_f a_v(c_b-c_s)=0 \qquad |3.20|$$

heat balance

$$u\rho_f c_{pf} \frac{dT_b}{dz'} - ha_v(T_s - T_b) + \frac{2U}{R_o}(T_b - T_w) = 0 \qquad |3.21|$$

Catalyst (slab)

mass balance

$$D_{eff} \frac{d^2c}{dx^2} - \rho_b r = 0 \qquad |3.22|$$

heat balance

$$\lambda_{eff} \frac{d^2T}{dx^2} + \rho_b r(-\Delta H) = 0 \qquad |3.23|$$

Boundary conditions

$$x = \ell, \quad dc/dx = dT/dx = 0 \qquad |3.24|$$

$$x = 0, \quad k_f(c_b - c_s) = D_{eff} \frac{dc}{dx} \qquad |3.25a|$$

$$h(T_s - T_b) = \lambda_{eff} \frac{dT}{dx} \qquad |3.25b|$$

$$z = 0, \quad T_b = T_o \text{ and } c_b = c_o \qquad |3.26|$$

In the model equations D_{eff} is the mass effective diffusivity in the particle, λ_{eff} is the effective thermal conductivity of the catalyst, x is the catalyst coordinate, c and T the concentration and temperature inside the particle and c_b, T_b the concentration and temperature in the bulk of the fluid phase.

-Bidimensional model

HT3-Ideal model+external resistances+internal resistances+ +radial dispersion

Model equations:

Fluid phase

mass balance

$$u \frac{\partial c_b}{\partial z'} + k_f a_v(c_b - c_s) = \varepsilon D_r\left(\frac{\partial^2 c_b}{\partial R^2} + \frac{1}{R}\frac{\partial c_b}{\partial R}\right) \qquad |3.27|$$

heat balance

$$u\rho_f c_{pf} \frac{\partial T_b}{\partial z'} - ha_v(T_s - T_b) = \lambda_r\left(\frac{\partial^2 T_b}{\partial R^2} + \frac{1}{R}\frac{\partial T_b}{\partial R}\right) \qquad |3.28|$$

Catalyst particle

mass balance

$$k_f a_v (c_b - c_s) = \bar{\eta}\, \rho_b r_b \qquad |3.29|$$

heat balance

$$h a_v (T_s - T_b) = \bar{\eta} \rho_b r_b (-\Delta H) + \lambda_{eff} \left(\frac{\partial^2 T}{\partial R^2} + \frac{1}{R} \frac{\partial T}{\partial R} \right) \qquad |3.30|$$

Boundary conditions

$$z' = 0 \;, c_b = c_o \text{ and } T_b = T_o \qquad |3.31a|$$

$$R = 0, \; \forall z' \quad \partial c_b / \partial R = \partial T_b / \partial R = \partial T_s / \partial R = 0 \qquad |3.31b|$$

$$R = R_o, \forall z' \quad \partial c_b / \partial R = 0 \qquad |3.31c|$$

$$R = R_o, \forall z' \quad h_w^f \, (T_w - T_b) = \lambda_r \, \partial T_b / \partial R \qquad |3.31d|$$

$$h_w^s \, (T_w - T\;) = \lambda_{eff} \partial T \,/ \partial R \qquad |3.31e|$$

Model equations were written for steady-state operation.

Runaway criteria

On the basis of the PH1 model a thermal sensivity analysis can be made showing that hot spots can exist in the reactor and sometimes runaway occur as a result of small variations in inlet or cooling conditions.However original profiles are recovered if these perturbations are also removed.

Several criteria have been suggested to find conditions to avoid runaway;they are based on different observations |6,10| :

a) limiting condition $dQ_r/dT = dQ_g/dT$ with Q_r-heat removed and Q_g-heat generated by reaction; if $dQ_r/dT > dQ_g/dT$ no runaway occur (Hlavacek et al)
b) limiting condition $d^2T/dz'^2 = d^3T/dz'^3 = 0$
If the temperature profile is concave ,$d^2T/dz'^2 < 0$ no runaway occur (Dente and Collina).
c) Barkelew's criterion is based on the envelope of a family of curves relating the hot spot temperature with B(dimensionless adiabatic temperature rise) and $a = 2\ L\ St/B\ R_o Da$.
d) Van Welsenaere and Froment present two criteria based on the phase diagram (c,T).In this plot the trajectory (c_m, T_m) corresponding to the hot spots is drawn.
The first criterion states that trajectories leading to runaway will intercept the "c_m -curve" beyond its maximum.

A second criterion is based in the fact that temperature profiles close to the runaway develop inflection points before the maximum. In the (c,T) plane this is shown by the loci of inflection points, "c_i-curves".
To avoid runaway a trajectory should not intercept "c_i-curve" (Figure 16).

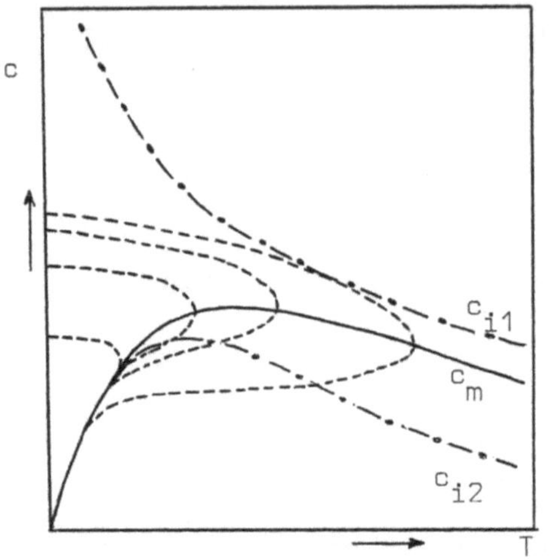

Figure 16-Trajectories in the c-T plane

These criteria are compiled in Figure 17 |6| as a plot of N/S vs S; N/S is the heat removed per unit volume at $E/RT_r^2\ (T-T_r)=1$ and $S=B\gamma$ where B is the dimensionless adiabatic temperature rise.
A line can be defined which separates the insensitive zone from the runaway region.

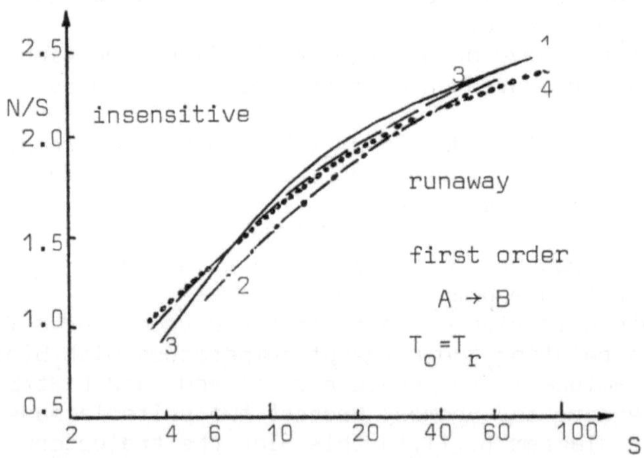

Figure 17-Runaway criteria : 1-Barkelew; 2-Dente and Collina; 3-Hlavacek et al; 4-Van Welsenaere and Froment. |6,12,13,14|

3.3-Estimation of parameters involved in reactor design

Let us now review some theoretical concepts and correlations available for the prediction of the parameters needed for fixed bed reactor design.

3.3.1- Pressure drop

The pressure drop through beds can be estimated by a number of correlations some of them leading to very different results.A classic correlation is the Ergun equation:

$$\frac{\Delta P}{L} = 150 \frac{(1-\varepsilon)^2}{\varepsilon^3} \frac{\mu}{d_p^2} u + 1.75 \frac{1-\varepsilon}{\varepsilon^3} \frac{\rho}{d_p} u^2 \qquad |3.32|$$

The contributions of laminar flow ($\Delta P/L \sim u$) and turbulent flow ($\Delta P/L \sim u^2$) are evident ,the transition being around Re≃10. Jerónimo |15| gave an unified treatment of hydrodynamics in porous media,particularly useful for fluidised beds with high porosities. His idea is that the pressure drop by unit of bed length results from the drag force exerted by the fluid on the particles contained in the unit volume

$$\frac{\Delta P}{L} = \underbrace{C_D \frac{\pi d_p^2}{4} \frac{\rho u^2}{2}}_{\text{force on an isolated particle}} \underbrace{\frac{1-\varepsilon}{\pi d_p^3/6}}_{\text{number of particles in the bed volume}} \underbrace{f(\varepsilon)}_{\text{correction for the presence of the other particles}} \qquad |3.33|$$

In the above equation C_D is the drag coefficient and u the superficial velocity;for a cubic packing

$$f(\varepsilon) = \{ 1-1.21(1-\varepsilon)^{2/3} \}^{-2.21} \qquad |3.34|$$

and

$$C_D = \frac{24}{Re}(1+0.22\, Re^{1/2} + 0.005\, Re^{1.12}) \qquad Re < 2\times10^5 \qquad |3.35|$$

In the laminar region,Eq.|3.33|deviates by 0.1% of predicted values by Ergun equation.However in the intermediate region values predicted by Ergun,Jerónimo and Brownell et al|16| are quite different as shown in Figure 18 (friction factor ,$\overline{f}$ vs Re).

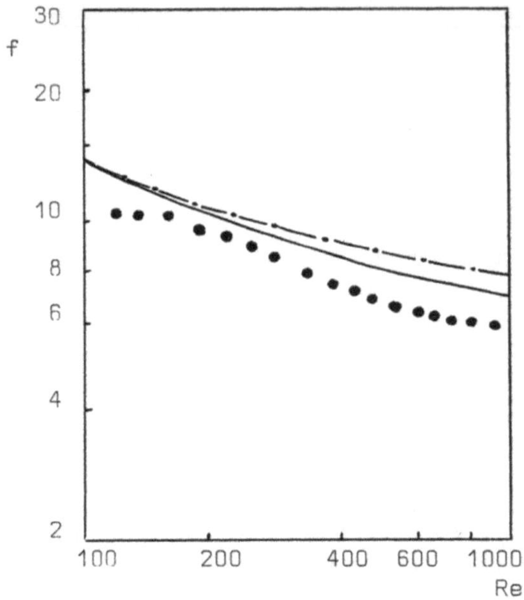

Figure 18- Friction factor, f vs Re |15|

...... Brownell et al

______ Jerónimo $f=\overline{3/8}\ C_D(1-\varepsilon)f(\varepsilon)$

-.- $f(\text{Ergun})=150(1-\varepsilon)^2/2\varepsilon^3 Re \quad +1.75(1-\varepsilon)/2\varepsilon^3$

3.3.2-Axial and radial mass dispersion

The statistical theory of diffusion-random walk wodel-was the basis for the Einstein relationship

$$D = \ell^2/2\tau_D \qquad |3.36|$$

where ℓ is a mixing or diffusion length and τ_D the diffusion time.

For axial dispersion , $\ell \simeq d_p$ and $\tau_D \simeq d_p/u_i$ so $D_{ax}=u_i d_p/2$ and then

$$Pe_{ax} = u_i d_p/D_{ax}=2 \qquad |3.37|$$

For radial dispersion, $\ell \simeq d_p/2$ and $\tau_D \simeq d_p/u_i$ so $D_r=u_i d_p/8$ and then

$$Pe_r=u_i d_p/D_r=8 \qquad |3.38|$$

Experimental data from several authors are reported in Figures 19, 20 and 21, showing that limiting values of axial and radial Peclet numbers for mass dispersion are |17,18|

$$Pe_{ax}=2 \qquad |3.39|$$

$$Pe_{r}=10 \qquad |3.40|$$

in good agreement with Einstein predictions.

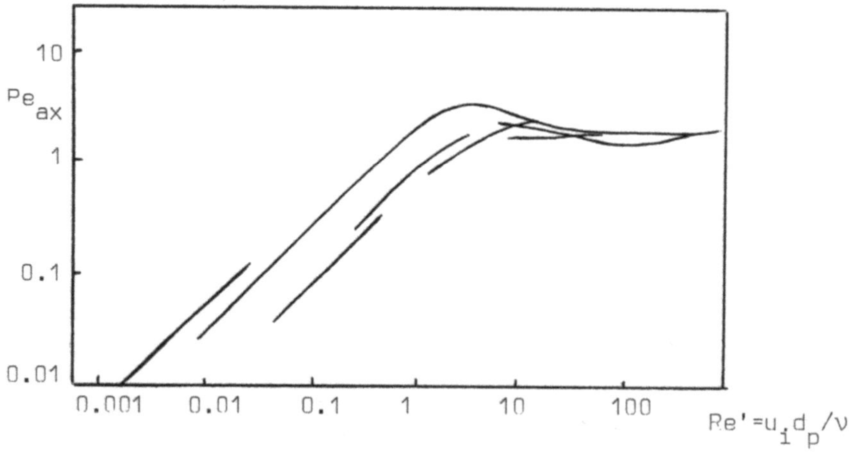

Figure 19- Axial mass dispersion in packed beds: gas/solid systems

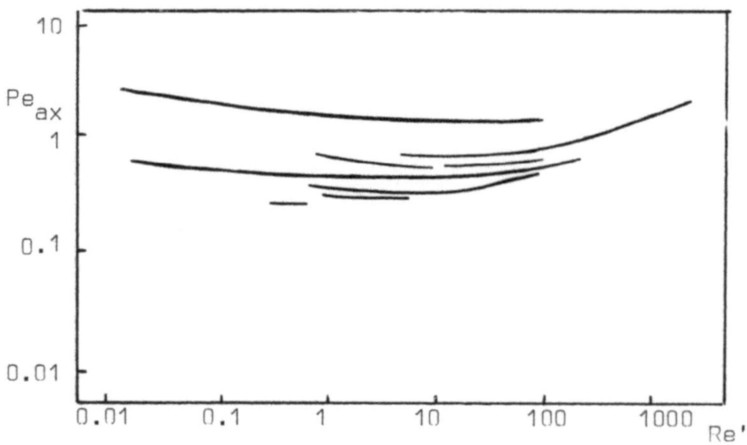

Figure 20- Axial mass dispersion in packed beds: liquid/solid systems

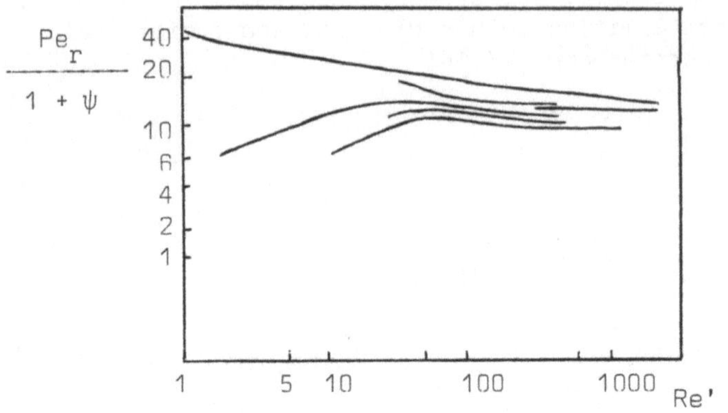

Figure 21- Radial mass dispersion in packed beds; $\psi = 19.4(d_p/d_t)^2$ |6,11|

As Wicke |17| pointed out, flow dispersion effects in packed beds are due to a number of elementary processes:

a-molecular diffusion in the voids between the particles
b-branching effect of the solid packing
c-eddy diffusivity (the voids act as mixing cells)
d-channeling (due to irregularities of packing and wall effects)

The radial dispersion coefficient is then

$$D_r = \frac{D_m}{\tau} + 0.1\ u_i d_p \qquad |3.41|$$

where D_m is the molecular diffusion and τ is the tortuosity factor. In the direction of flow the dispersive processes lead to a residence time distribution from which an axial dispersion coefficient is calculated

$$D_{ax} = \frac{D_m}{\tau} + 0.5\ u_i d_p \qquad |3.42|$$

Axial and radial dispersion are in fact interconnected by the mechanism of Taylor dispersion and so it can be shown that

$$D_{ax} = \frac{D_m}{\tau} + \frac{0.5\ u_i d_p}{1 + \frac{10\ D_m}{u_i d_p\ \tau}} \qquad |3.43|$$

which accounts for the maximum of the curve Pe_{ax} vs Re' in the transition region.

3.3.3-Axial and radial heat dispersion

Axial heal Peclet numbers in the turbulent region approach values similar to axial mass Peclelt numbers, i.e., 1 to 2 |19,20|. For design purposes, the correlation proposed by Votruba et al |21| can be used:

$$\frac{1}{Pe_{h,ax}} = \frac{\lambda_e^0/\lambda_f}{Re\ Pr} + \frac{14.5}{d_p(1+\frac{A}{Re\ Pr})} \qquad |3.44|$$

where $Pe_{h,ax}$ is based in the superficial velocity, λ_e^0 is the effective thermal conductivity of the stagnant bed, λ_f is the fluid thermal conductivity, $Re=ud_p/\nu$, $Pr=\mu c_{pf}/\lambda_f$ and A is a constant which takes values from 0 to 5. This correlation should be used together with

$$\frac{\lambda_e^0}{\lambda_f} = \left(\frac{\lambda_s}{\lambda_f}\right)^{0.28 - 0.757\log\varepsilon - 0.057\log\frac{\lambda_s}{\lambda_f}} \qquad |3.45|$$

where λ_s is the solid thermal conductivity.

Radial heat Peclet numbers are somewhat smaller than the corresponding numbers for mass dispersion, i.e., $Pe_{h,r}=5$ to 10. The effective radial thermal conductivity, λ_r can be estimated by |11|:

$$\lambda_r=\varepsilon\left(\lambda_f+\frac{ud_p\rho_f c_{pf}}{Pe_r}+\frac{4e\sigma T^3 d_p}{2-e}+(1-\varepsilon)\frac{h'd_p}{2+\frac{h'd_p}{\lambda_s}}\right) \qquad |3.46|$$

where e is the emissivity of the particle, σ is the Stephan-Boltzman constant and h' includes all heat transfer paths from one particle to the next; each term represents the contribution of one mechanism: molecular conduction, eddy conduction, radiation through the void space and conduction through the solid. After some rearrangements

$$\frac{1}{Pe_{h,r}} = \frac{1}{Pe_r} + \frac{0.3\ St}{1+0.35\ Bi_{h,s}} + \frac{2e\sigma T^3}{u\rho_f c_{pf}} \qquad |3.47|$$

where $St=h/u\rho_f c_{pf}$ and $Bi_{h,s}=hd_p/\lambda_s$.

Experimental data from various authors in terms of λ_r vs Re are shown in Figure 22.

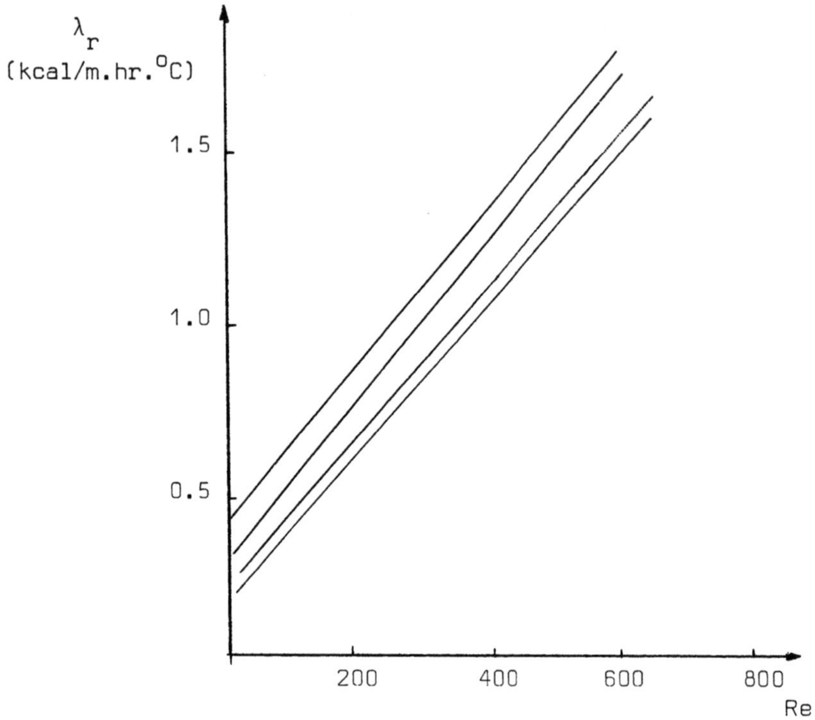

Figure 22- Radial thermal conductivity vs Re

De Wasch and Froment |22| proposed

$$\frac{\lambda_r}{\lambda_f} = \frac{\lambda_r^0}{\lambda_f} + \frac{0.14 \text{ Re Pr}}{1+46(d_p/d_t)^2} \qquad |3.48|$$

where λ_r^0 is the static contribution for the thermal conductivity of the bed.

3.3.4-Overall heat transfer coefficient,U and fluid-wall heat transfer,h_w

When dealing with unidimensional models one uses an overall heat transfer coefficient,U as defined in Eq.|3.8|.In these calculations we need to know how to estimate h_i;De Wasch and Froment |22| suggested the relationship

$$\frac{h_i d_p}{\lambda_f} = \frac{h_i^o d_p}{\lambda_f} + 0.033 \text{ Pr Re} \qquad |3.49|$$

or

$$\text{Nu} = \text{Nu}^o + 0.033 \text{ Pr Re}$$

where $Pr=\mu_f c_{pf}/\lambda_f$ is the Prandtl number, $Re=u\, d_p/\nu_f$ is the Reynolds number, d_t is the reactor diameter, λ_e^0 is the static contribution of the thermal conductivity of the bed and $h_i^o=2.44\ \lambda_e^0/d_t^{4/3}$ |23|.
On the other hand, bidimensional models require the use of heat transfer coefficients between tube walls and packed beds. Beek |11| suggests the following correlations

-cylinders

$$Nu_w = \frac{h_w d_p}{\lambda_f} = 2.58\ Re^{1/3} Pr^{1/3} + 0.094 Re^{0.8} Pr^{0.4} \qquad |3.50a|$$

-spheres

$$Nu_w = 0.203\ Re^{1/3} Pr^{1/3} + 0.220 Re^{0.8} Pr^{0.4} \qquad |3.50b|$$

De Wasch and Froment |22| suggest the use of the following relationship for air flow through packed beds

$$\frac{h_w d_p}{\lambda_f} = \frac{h_w^o d_p}{\lambda_f} + 0.631\ d_t \text{ Pr Re} \qquad |3.51|$$

with d_t expressed in meters, $h_w^o = 8.4\lambda_r^0/d_t^{4/3}$ in kcal/m^{2}hr^{o}K and λ_r^0 in kcal/m.hr.^{o}K (Figure 23). The static contribution has been evaluated for a number of catalysts and d_t=0.157m:

	λ_r^0 (kcal/m hr ^{o}K) for catalysts
SO_2 oxydation	0.176
Phtalic anhydride synthesis	0.241
NH_3 synthesis	0.360

For more details on heat transfer in fixed beds we recommend the review by Schlunder |23| and papers by Balakrishan and Pei |24,25,26|.

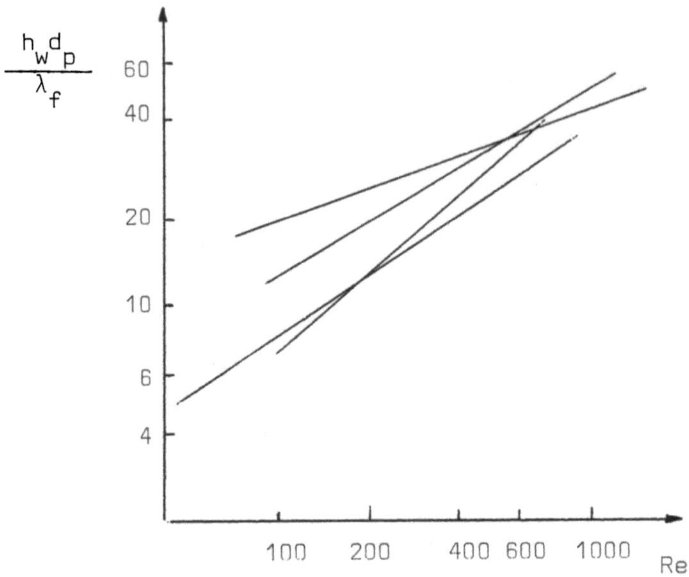

Figure 23-Fluid-wall heat transfer data

3.3.5-Film mass (and heat) transfer coefficients

It is well known that the minimum Sherwood number for mass transfer between a sphere and a stagnant infinite medium is

$$Sh_{min}=k_f d_p/D_m=2$$

Similarly for heat transfer

$$Nu_m=hd_p/\lambda_f=2$$

(The thermal diffusivity will be $\lambda_f/\rho_f c_{pf}$ $\{L^2t^{-1}\}$; k_f has dimensions of Lt^{-1} and the corresponding film heat transfer coefficient -with the same dimensions- is $h/\rho_f c_{pf}$).

Any fluid motion will contribute to increase k_f; then Froessling equation applies

$$Sh=2 +0.6\ Sc^{1/3}Re^{1/2}$$

or

$$Nu=2 +0.6\ Sc^{1/3}Re^{1/2}$$

However, it can be shown that

$$Sh = \frac{2D_s}{D_s - d_p}$$

if the stagnant medium is in a shell of diameter D_s.
If $D_s \to \infty$,then $Sh_{min}=2$;otherwise,if the sphere is surrounded by other spheres one gets $Sh_{min} < 2$ (if two spheres are touching, the distance between centers is $\ell=2d_p$ and then $Sh_{min}=1.386$).
In fixed beds,experimental data for mass and heat transfer are usually reported in terms of "j" factors

$$j_m = \frac{k_f}{u} Sc^{2/3} \quad \text{and} \quad j_h = \frac{(h/\rho_f c_{pf})}{u} Pr^{2/3}$$

and the Chilton-Colburn analogy enables us to write

$$j_m = j_h$$

Several correlations have been suggested,for example

-Carberry's correlation ($0.1 < Re < 1000$)

$$j_m = \frac{1.15}{\sqrt{\varepsilon}} Re^{-1/2} \qquad |3.52|$$

-Wilson and Geankoplis

$$j_m = \begin{cases} 0.250\ Re^{-1/3} & (55 < Re < 1500) \\ 1.09\ Re^{-2/3} & (0.001 < Re < 55) \end{cases} \qquad |3.53|$$

Complete reviews of available correlations are those of Dwivedl et <u>al</u>|27|,Karabelas et <u>al</u>|28|, Pandey et <u>al</u>|29|.
Figure 24 shows a rough picture of the experimental data available. One should notice that in fixed beds of fine particles Sh_{min} is lower than 2 by several orders of magnitude.This was nicely explained by several authors |23,30,31,32|.
Schlunder|23,30| and Martin |32| were able to explain why the observed Sh(or Nu) are so low in the low Peclet number zone. They start from the fact that,in a packed bed,the porosity depends on the radial position as shown in Figure 25.Then,the fluid velocity near the wall is higher than in the center region;if the section is divided into a region of "low porosity",ε_1 with area $(1-\phi)\Omega$ and a "high porosity" region near the wall with porosity ε_2 and area $\phi\Omega$ we get, for low Re,

$$\omega = u_2/u_1 = (1-\varepsilon_1)^2\, \varepsilon_2^2 \,/ (1-\varepsilon_2)^2\, \varepsilon_1^2$$

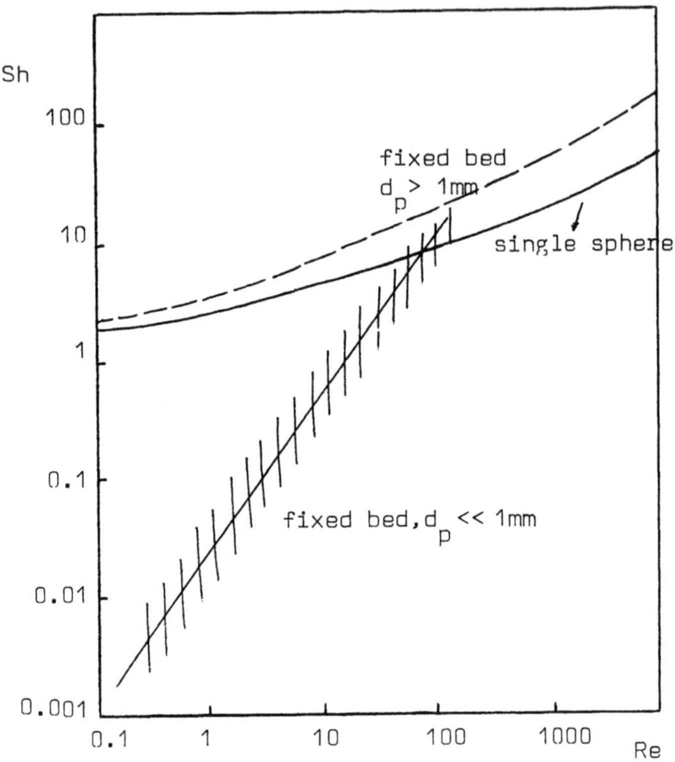

Figure 24-Experimental data for film mass (heat)transfer in fixed beds.

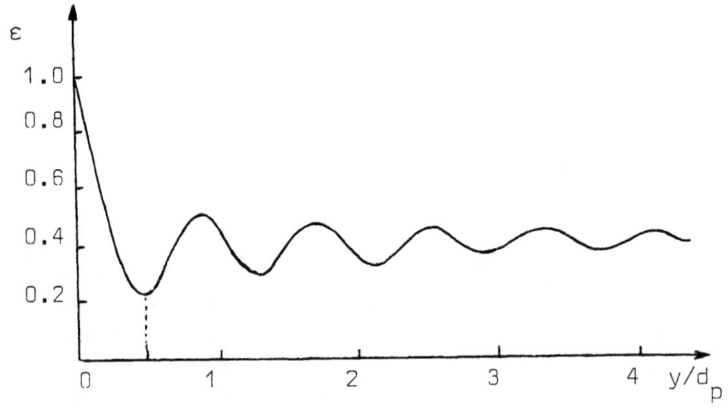

Figure 25- Bed porosity vs radial position (y is the distance from the wall)

Typically $\varepsilon_1=0.4$ and $\varepsilon_2=0.5$, so $\omega=2.81$. The flowrate fraction near the wall is then

$$\alpha_2=Q_2/(Q_1+Q_2)=\cfrac{1}{1+\cfrac{1-\phi}{\phi\omega}}$$

and $\phi\simeq d_p/R_o$.
For a single, isolated, sphere, Sh_o based in the superficial velocity u is

$$Sh_o=2+aSc^{1/3}\,Re^{1/2}$$

and for a fixed bed Sh_{fb} is

$$Sh_{fb}=f(\varepsilon)Sh \qquad |3.54|$$

with Sh based in the intersticial velocity and $f(\varepsilon)=1+1.5(1-\varepsilon)$.
For low Peclet numbers ($Pe=Re\ Sc<100$), low Sh values can be explained by the nonuniform distribution of flow in the two zones of porosity. One can show that,

$$\lim_{Pe\to 0} Sh = \frac{1}{6(1-\varepsilon)(1-\frac{\phi_1}{\alpha_2})\frac{2\alpha_2}{1-\alpha_2}}\frac{d_p}{L}\,Pe \qquad |3.55|$$

$$\phi_1=\phi\,\frac{1-\varepsilon_2}{1-\varepsilon_1}$$

and, in the intermediate Peclet region

$$Sh=-\frac{\ln\alpha_2}{6(1-\varepsilon)}\frac{d_p}{L}\,Pe \qquad |3.56|$$

so $Sh\sim Pe$ (hydraulically controlled regime). This is in agreement with observed results as shown in Figure 26.
The conclusion is that the average mass (heat) transfer coefficient in the hydraulically controlled region has no physical meaning and "one can not compare these coefficients with the rate of chemical reaction in order to decide which is the controlling process. One has to compare local transfer rates with local reaction rates".
This idea is close to that developed by Fedkiw and Newman |31|; they define an effective mass transfer coefficient, k_m for the overall reactor by

$$u(c_{in}-c_{out})=a\,k_m\,L\,\Delta c_{ln}$$

or in terms of bed Peclet number($u/aD_m=Pe_b$) and bed Sherwood number($\varepsilon k_m/aD_m=Sh_b$),

$$c_{out}/c_{in} = \exp(-\frac{aL}{\varepsilon}\frac{Sh_b}{Pe_b})$$

This k_m is related with the film mass transfer coefficient k_f based in a <u>local</u> concentration driving force and then it is easily shown that k_m becomes linearly proportional to the Peclet number as $Pe\to 0$ and also that k_f is a constant in the limit Pe=0.

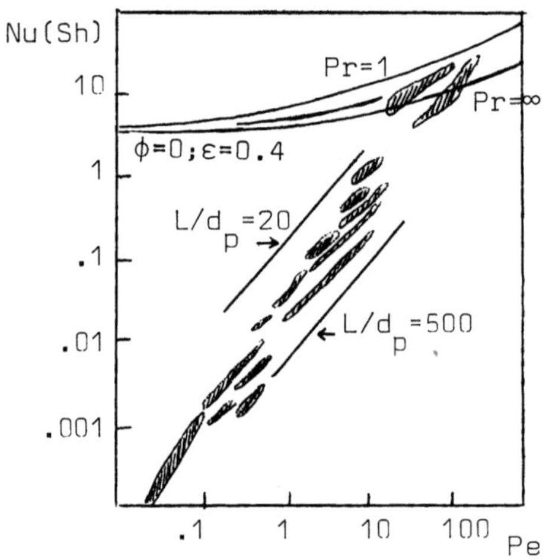

Figure 26- Nu(Sh) vs Pe;experimental and theoretical (——) predictions. ($Pr=0.7, \varepsilon=0.4, \varepsilon_2=0.5, \phi=0.06$)

3.3.6-Effective diffusivity (mass and thermal) of catalyst particles

Before providing methods for the estimation of D_{eff} and λ_{eff}(effective thermal conductivity of particles) let us briefly explain how to characterize physically a porous particle.This can be done by measuring the specific surface area (S_{BET}),the mean pore size,$\bar{r}$,the pore size distribution,f(r) and the particle porosity,ε_p.

a)Particle porosity,ε_p

Let us define first the following quantities:

Bulk density, ρ_b

ρ_b=mass of catalyst/bed volume=m/v_b

Apparent density of the particle, ρ_a

$\rho_a = m/v_{particle}$

Solid density, ρ_v

$\rho_v = m/v_{solid}$

Since $v_{bed} = v_{interstices} + v_{particles}$ and $v_{particle} = v_{pores} + v_{solid}$ one can determine v_{solid} by helium displacement and $v_{particle}$ by mercury displacement at 1 atm, and then

$$\varepsilon_p = \frac{v_{pores}}{v_{particle}} = 1 - \frac{v_{solid}}{v_{particle}} = 1 - \frac{\rho_a}{\rho_v} \qquad |3.57|$$

b) Specific surface area, S_{BET}

This is obtained from the adsorption equilibrium isotherm of N_2 at 77 K on the catalyst. The BET model assumes:

- multilayer adsorption
- fixed number of active sites
- no interaction between adsorbed molecules
- constant heat of adsorption in each layer equal to E_1 in the first and E_L (heat of liquefaction) for the other layers.

This leads to the BET equation, in a linearized plot of $p/v(p_o-p)$ vs p/p_o

$$\frac{p}{v(p_o-p)} = \frac{1}{v_m\,c} + \frac{c-1}{v_m\,c}\,\frac{p}{p_o} \qquad |3.58|$$

where v is the adsorbed volume/gram of catalyst, v_m the monolayer volume/gram of catalyst obtained from the slope and intercept of Eq|3.58|, p/p_o the relative pressure and c the BET parameter. The specific surface area is then

$$S_{BET}(cm^2/gr) = v_m(cm^3/gr)\,\frac{1}{\bar{V}}(gmole/cm^3)\;N(molecule/gmole)\sigma$$

|3.59|

where

$\sigma' = 16.2\text{x}\ 10^{-16}\ \text{cm}^2$/molecule (area occupied by the nitrogen molecule)

N=Avogadro number

$\bar{v}$=molar volume of nitrogen

In general

$$\sigma'(\text{cm}^2)=1.091\ (M/N\rho)^{2/3} \qquad |3.60|$$

where ρ is the density of the liquid adsorbate and M its molecular weight.

c)Pore size distribution

c1 -From the knowledge of the particle specific pore volume and S_{BET} one can get the average pore size (if we assume cylindrical pores and a monodisperse size distribution)

$$\bar{r} = 2\ v_{pores}/S_{BET} \qquad |3.61|$$

c2 -The macropore distribution can be obtained from a mercury penetration experiment.Taking into account that the surface tension of Hg is 473 dynes/cm,the wetting angle θ=140° we get

$$\bar{d}(\mu m)=175/P(psia) \qquad |3.62|$$

where $\bar{d}$ is the pore diameter and P the applied pressure.One easily observes that only the macropores ($\bar{d}$> 400 Å) are available to this technique(to get $\bar{d}$=120 Å,P=15000 psia).
Most authors neglect the desorption branche in mercury porosimetry; however important information is contained there as shown in Figure 27 |33|

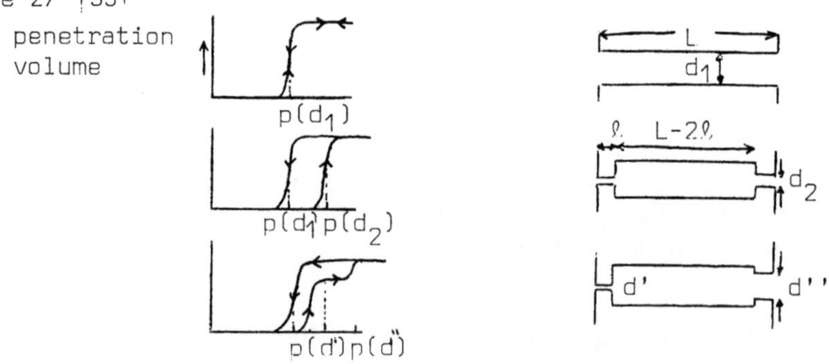

Figure 27-Hysteresis in Hg porosimetry and pore shape.

c3 -The pore size distribution for $d < 400$ Å should be obtained from the desorption curve in a BET experiment. The main idea is that pores fill by radial capillary condensation while vertical emptying of pores occurs during desorption; thus Kelvin equation for desorption is

$$p_{des} = p_o \exp(-2\sigma\bar{v}/rRT)$$

or

$$r = \frac{2\sigma\bar{v}}{RT \ln(p_o/p_{des})} \qquad |3.63|$$

where $\bar{v}$ is the molar volume of the adsorbate and r is the pore radius. One should take care, when calculating the pore size distribution $f(r)dr$, that an adsorbed layer of thickness t stays on the pore walls

$$t(\text{Å}) = 4.3\left\{\frac{5}{\ln(p_{des}/p_o)}\right\} \qquad |3.64|$$

so the pore radius is given by Eq. |3.63| plus t. Again hysteresis is related to the pore shape |34| as shown in Figure 28.

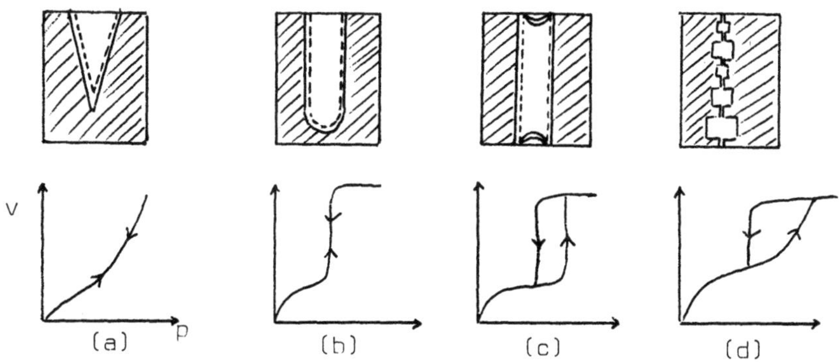

a,b - open-ended pores; no hysteresis
c,d - open pores

Figure 28- Hysteresis and pore shape.

d) Effective mass diffusivity, D_{eff} and effective thermal conductivity, λ_{eff}

In general the mass conservation equation for a component

$\underline{i}$ is

$$\frac{\partial c_i}{\partial t} + \nabla . \underline{N}_i = r_i$$

where $\underline{N}_i$ is the molar flow of $\underline{i}$ relatively to a fixed reference.
Also

$$\underline{N}_i = c_i \underline{v}^* + \underline{J}_i^*$$

where

$$\underline{v}^* = \Sigma x_i \underline{v}_i$$

is the mean molar velocity and x_i the molar fraction; $\underline{J}_i^*$ is the molar diffusion flux relatively to $\underline{v}^*$.
The diffusional mechanisms occuring inside the catalysts are: molecular diffusion (if the mean free path, $\ell <<$ pore radius, r), Knudsen diffusion ($\ell >> r$) and surface diffusion (not important at high temperatures).
For multicomponent mixtures and unidirectional diffusion one can write

$$\underline{J}_i^* = -c \, \mathcal{D}_{im} \nabla x_i = \underline{N}_i - x_i \sum_{j=1}^{n} \underline{N}_j$$

and then

$$\mathcal{D}_{im} = \frac{1 - x_i \dfrac{\sum_{j=1}^{n} \underline{N}_j}{\underline{N}_i}}{\sum_{\substack{j=1 \\ j \neq i}}^{n} \dfrac{1}{\mathcal{D}_{ij}} \left(x_j - x_i \dfrac{\underline{N}_j}{\underline{N}_i} \right)} \qquad |3.65|$$

Particular cases are:

a) $\mathcal{D}_{im} = \mathcal{D}_{i1}$, all components in diluted solution in component 1

b) $\mathcal{D}_{im} = \mathcal{D}$, since $\mathcal{D}_{ij} = \mathcal{D}$ for all components

c)

$$\mathcal{D}_{1m} = \frac{1 - x_1}{\sum_{j=2}^{n} \dfrac{x_j}{\mathcal{D}_{ij}}}$$, if components 2 to n have the same velocity.

$\mathcal{D}_{ij}$ can be estimated for binary mixtures of gases, liquids |35|.
For Knudsen diffusion:

$$D_{Ki}=9700\ r\sqrt{T/M_i} \qquad |3.66|$$

with the pore radius in cm.

For binary mixtures of A and B, neglecting pressure variations, a global diffusion coefficient can be defined

$$\mathcal{D}=\frac{1}{\dfrac{1}{D_{KA}}+(1-\alpha x_A)\dfrac{1}{D_{AB}}} \qquad |3.67|$$

and $\alpha=1+(N_B/N_A)$. For equimolecular counterdiffusion, $N_A=-N_B$ and then

$$\mathcal{D}=\frac{1}{\dfrac{1}{D_{KA}}+\dfrac{1}{D_{AB}}} \qquad |3.68|$$

For porous catalysts an effective diffusivity is defined as

$$D_{A,eff}=\varepsilon_p\mathcal{D}/\tau \qquad |3.69|$$

where τ is a tortuosity factor (usually takes values between 2 and 10, but greater values have been reported |36,37|).

The prediction of the effective thermal conductivity for porous catalysts is more difficult. Harriot |38| compares three models for the estimation of λ_{eff}: packed bed, consolidated particle and porous particle models.

In the first one the effective thermal conductivity for $\varepsilon=0.4$ depends more on λ_f than on λ_s. For high λ_s/λ_f the controlling resistance is the film of fluid near the point of contact between the particles. The conductivity of the fluid in the pores is

$$\lambda_{f,p}=\frac{\lambda_f}{1+(2\beta\ell/d)}$$

with $\beta\simeq1$, ℓ is the normal mean free path and d is the size of the pores between particles ($d\simeq0.1\ d_p=\bar{d}$ for a bed of porous catalysts).

The porous solid model leads to

$$\frac{\lambda_{eff}}{\lambda_s}=\frac{h^2}{L^2}+\chi\left(1-\frac{h}{L}\right)^2+2\left(1+\frac{h}{L}+\frac{L}{\chi h}\right)^{-1}$$

with $h/L=\bar{d}/(1+\bar{d})$ and $\chi=\lambda_{f,p}/\lambda_s$;results are shown in Figure 29. Some authors based on results for sintered metals proposed $\lambda_{eff}=(1-\varepsilon)\lambda_s/\tau$.The prediction of the effective thermal conductivity is an actual problem in chemical engineering,although some progress in the estimation of packed bed conductiviy has been made|38|.

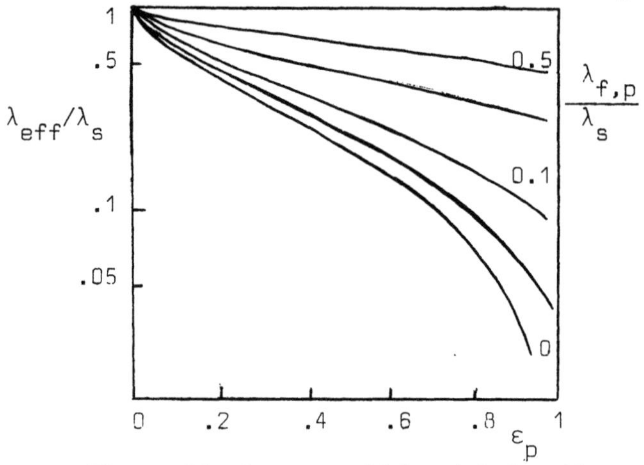

Figure 29- Porous solid model for the prediction of the effective thermal conductivity of catalysts.

3.4-Technological aspects of fluid/solid reactors

Let us classify the fluid/solid catalytic reactors into several classes:

1-Fixed beds

The catalyst particles of d_p=1 to 10 mm are packed in tubes;if one tube is used the system is often adiabatic(oxydation of SO_2,methanol synthesis).In practice the bed is divided into sections and the temperature adjusted at the inlet of each section by cooling (or heating) or by quench.The fluid flow can be towards the bottom or towards the top.

When the reaction is highly exothermic exchange surfaces are inserted in the bed or a multitubular reactor is used;a fluid around the tubes ensures thermal exchange (VCM production from HCl and acethylene,phtalic anhydride synthesis)

2-Moving bed

The catalyst moves towards the bottom by gravity;the fluid flow can be cocurrent,countercurrent or horizontal-crossflow.

The interest of the moving bed lies on the possibility of continuous catalyst regeneration.

3-Suspended catalyst systems

In this case catalyst particles have d_p = 50μ to 1mm;the suspension can be achieved by mechanical stirring,fluidisation or pneumatic transport.

The first technique,only interesting for liquid/solid systems (hydrotreating of vegetable oils),can be used either in batch or continuous processes (single perfectly mixed reactor or tanks in cascade);in this situation the problem is the separation of the catalyst from the fluid and its recycle to the reactor.

Alternatively the agitation can be provided by circulation of the liquid phase;the heat exchanger is inserted in the circulation loop.

The fluidised bed reactor has some advantages:in the case of highly exothermic reactions,temperature control is easier than in fixed beds and continuous regeneration of the catalyst is possible (catalytic cracking).However,lower conversion and selectivity than in the fixed bed are expected for "normal" reactions in gas/solid systems.

Finally the transport reactor studied in depth at Nancy |39,40,41| can find applications when the frequency of regeneration is high,diffusional limitations and thermal exchange problems are encountered ("riser" for catalytic cracking).

In Tables VII and VIII we summarize the characteristics of the different technologies for two-phase reactors (fluid/solid).

TABLE VII-COMPARISON OF TECHNOLOGIES FOR FLUID/SOLID CATALYTIC REACTORS [42]

Technology characteristics: nature of catalyst packing and thermal exchange / Criteria of comparison	FIXED BED		MOVING BED	SUSPENDED CATALYST SYSTEMS			
	Adiabatic G/S	Multitubular G/S	Adiabatic G/S	Mechanical stirring (batch) L/S	Mechanical stirring (continuous) L/S	Fluidised bed; G/S or L/S	Transport reactor; G/S
Characteristics related to the <u>particle</u>							
activity	+	+	+	++(G+L/S)	++	++	++
selectiviy	+	+	+	++	++	++	++
stability	+	+	o	+	+	+	+
regenerability	+	o	++	o	-	++	++
cost of catalyst	++	++	-	-	o	-	-
Characteristics related to the <u>technology</u>							
concentration gradient	++	++	++	++	-	-	++
temperature control	-	o	-	++	++	+	+
capacity	++	+	+	-	o	+	+
flexibility	+	o	o	++	+	-	-
simplicity	++	-	-	++	+	o	o
separation catalyst/products	++	++	+	o	-	o	-
replacement of catalyst	-	-	+	++	o	+	+
pressure drop	+	+	+	o	o	+	+
fiability	++	+	o	++	++	+	o

++ very good ; + good ; o reasonable ; - bad

TABLE VIII- UTILIZATION OF DIFFERENT TECHNOLOGIES |42|

	BATCH	CONTINUOUS			THERMAL EXCHANGE		FLUID PHASE
		TUBULAR	PERFECTLY MIXED	STAGED	ADIABATIC	with thermal exchange	G or L
FIXED BED	+	+++	+	+	+++	++	G +++ L +
MOVING BED	-	++	-	-	++	+	G ++ L +
SUSPENDED CATALYST :							
-MECHANICAL STIRRING	++	-	++	+	+	++	G + L ++
-FLUIDISED BED	-	-	++	+	++	++	G ++ L +
-TRANSPORT REACTOR	-	++	-	-	++	+	G ++ L +

+++ often used
++ sometimes used
+ possible but rare
- not realized

REFERENCES FOR PART III

|1| Carberry,J. and Butt,J.,Cat.Rev.Sci.Eng.,10(2),221(1974)
|2| Hulburt,H.,P.C.Reilly Lectures,Uni.Notre Dame (1967)
|3| Carberry,J.,Chemical and catalytic reaction engineering ,ed. Mc Graw Hill (1976)
|4| Froment,G.,Adv.Chem.Series,109,1(1972)
|5| Froment,G.,Ind.Eng.Chem.,59(2),18(1967)
|6| Froment,G. and Bischoff,K.,Chemical reactor analysis and design, ed.J.Wiley (1979)
|7| Karanth,N. and Hughes,R.,Cat.Rev.Sci.Eng.,9(2),169(1974)
|8| Gilles,E.,"Reactor models" in 4th Int.Symp.Chem.React.Eng., Heidelberg(1976)
|9| Gros,J. and Bugarel,R.,The Chem.Eng J.,13,165(1977)
|10| Hlavacek,V.,Ind.Eng.Chem.,62(7),9(1970)
|11| Beek,J.,Adv.Chem.Eng.,3,203(1962)
|12| Van Welsenaere,R. and Froment,G.,Chem.Eng.Sci.,25,1508(1970)
|13| Barkelew,C.,Chem.Eng.Prog.,Symp.Series,55(25),38(1959)
|14| Dente,M. and Collina,A.,Chim.et Industrie,46,752(1964)
|15| Jerónimo,M.,"Unified treatment of fluid/solid relative movement" Internal Report,Dept.Chem.Eng.,Univ.Porto (1980)
|16| Brownell,L.,Dombrowsky,H. and Dickey,C.,Chem.Eng.Prog.,46,415 (1950)
|17| Wicke,E.,Adv.Chem.Series,148,75(1975)
|18| Levenspiel,O. and Bischoff,K.,Adv.Chem.Eng.,4,95(1963)
|19| Asbjornsen,O. and Wang,B.,Chem.Eng.Sci.,26,585(1971)
|20| Smith,J.M.,The Chem.Eng.J.,5,109(1973)
|21| Votruba,J.,Hlavacek,V. and Marek,M.,Chem.Eng.Sci.,27,1845(1972)
|22| De Wasch,A.and Froment,G.,Chem.Eng.Sci.,26,629(1971)
|23| Schlunder,E.,ACS Symp.Series,72,110(1978)
|24| Balakrishnan,A. and Pei,D.,Ind.Eng.Chem.Proc.Des.Dev.,18(1),31 (1979)
|25| Balakrishnan,A. and Pei,D.,Ind.Eng.Chem.Proc.Des.Dev.,18(1),40 (1979)
|26| Balakrishnan,A.and Pei,D.,Ind.Eng.Chem.Proc.Des.Dev.,18(1),47 (1979)
|27| Dwivedl,P. and Upadhyay,S.,Ind.Eng.Chem.Proc.Des.Dev.,16(2),157 (1977)
|28| Karabelas,A.,Wegner,I. and Hanratty,T.,Chem.Eng.Sci.,26,1581(1971)
|29| Pandey,D.,Upadhyay,S.,Gupta,S. and Misra,P.,J.Scient.&Ind.Res., 37(5),224(1978)
|30| Schlunder,E.,Chem.Eng.Sci.,32,845(1977)
|31| Fedkiw,P. and Newman,J.,Chem.Eng.Sci.,33,1043(1978)
|32| Martin,H.,Chem.Eng.Sci.,33,913(1978)
|33| Androutsopoulos,G. and Mann,R.,Chem.Eng.Sci.,31,1131(1976)
|34| Voyutsky,S.,Colloid chemistry ,ed.Mir Publ.(1978)
|35| Reid,R.,Sherwood,T. and Prausnitz,J.,The properties of gases and liquids,ed.Mc Graw Hill(1977)
|36| Luss,D.,"Interactions between transport phenomena and chemical rate processes",4th Int.Symp.Chem.React.Eng.,Heidelberg(1976)

|37| Satterfield,C.Mass transfer in heterogeneous catalysis,MIT Press (1970)

|38| Harriot,P.,The Chem.Eng.J.,10,65(1975)

|39| Baeza,M.S.,"Etude du transport pneumatique vertical a forte concentration du solide",Thesis,Univ.Nancy(1970)

|40| De Lasa,H.,"Etude des phenomenes diffusionels dans un reacteur a transport pneumatique",Thesis,Univ.Nancy(1971)

|41| Yousfi,Y.,"Contribution a l'etude d'un reacteur a transport pneumatique:hydrodynamique et transfert de chaleur",Thesis, Univ.Nancy(1973)

|42| Trambouze,P. and Van Landeghem,H.,IFP-Internal Report(1976)

INTRAPARTICLE AND INTERPHASE TRANSPORT RESISTANCES IN GAS-LIQUID-SOLID CATALYTIC REACTORS

NORMAN H. SWEED

OXIRANE INTERNATIONAL
PRINCETON, NJ 08540

INTRODUCTION

Transport resistances can have a profound influence on the performance of chemical reactors which contain solid catalyst, a gas, and a liquid. This review examines the effects of intraparticle, liquid-solid, and gas-liquid mass transfer resistances on several types of three-phase reactors including trickle beds and slurries.

The review has four sections. The first is an overview of reaction with mass transfer in three-phase systems. The second section looks at intraparticle mass transport in both completely wet and partially wet porous catalyst pellets. The third section deals with the effectiveness of liquid-solid contacting and correlations for liquid-solid mass transfer. The final section treats gas-liquid mass transfer.

OVERVIEW OF REACTION WITH MASS TRANSFER

Consider the path a gaseous reactant molecule must follow to get from the bulk gas phase to an active catalytic site within a porous catalyst particle. The molecule enters the reactor with the bulk gas, either in discrete bubbles (as in a slurry or bubble column) or in a continuous phase (as in a trickle bed). It then crosses the gas film which separates the bulk gas from the gas-liquid interface. See Figure 1. At the interface the molecule dissolves and then moves across the liquid film resistance to the bulk liquid. Once in the bulk liquid the molecule "swims" along until it comes into the vicinity of a catalyst

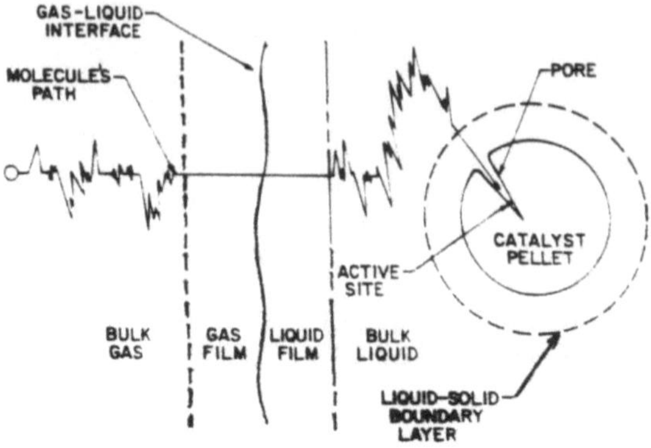

FIGURE 1. GAS MOLECULE'S PATH BETWEEN THE BULK GAS AND AN ACTIVE SITE.

particle. It then has to cross the liquid boundary layer surrounding the particle, and finally diffuse through the pores of the catalyst to an active site where (hopefully!) reaction will occur.

Not all these resistances play an important role in every reactor even though they are always present. For example, the gas film resistance is negligible in most three-phase reactors because the gas phase reactant concentration is usually kept high. In slurries, the intraparticle mass transport resistance is often negligible because slurries usually contain very small catalyst particles.

If we are to interpret laboratory data correctly and apply them successfully to the design of commercial reactors, we have to understand all these transport steps. The resistance that controls the rate in a small laboratory reactor can be very different from what controls in a large industrial unit.

To quantify the relative importance of the various resistances, and to understand how the physical and chemical parameters of the reactor system influence performance, we shall make use of a simple mathematical model of a three-phase reactor. At steady state the rate of each of the individual transport steps is equal to the rate of chemical reaction which in turn is equal to the overall reaction rate, R. Consider the case where the reaction is first order in the gaseous component. The rates of the individual steps are given in Equations 1-4.

$R = k_G a_L(C_G - C_{iG})$ mass transfer rate from bulk gas to gas-liquid interface (1)

$= k_L a_L(C_{iL} - C_L)$ mass transfer rate from gas-liquid interface to bulk liquid (2)

$= K_S a_S(C_L - C_S)$ mass transfer rate from bulk liquid to solid pellet interface (3)

$= kC_S\eta$ reaction rate for pellet = reaction rate based on surface concentration C_S x effectiveness factor to correct for intraparticle resistance (4)

where a_L = gas-liquid interfacial area/unit reactor volume

a_S = liquid-solid interfacial area/unit reactor volume

C_G = gas phase concentration of reactant

C_L = liquid phase concentration of reactant

C_S = liquid phase concentration of reactant at the catalyst pellet's external surface

k_G = gas side mass transfer coefficient at gas-liquid interface

k_L = liquid side mass transfer coefficient at gas-liquid interface

K_S = liquid side mass transfer coefficient at pellet surface

k = first order reaction rate constant

η = catalyst effectiveness factor

and subscript i refers to interface.

At the gas-liquid interface we assume that equilibrium exists and that Henry's law applies:

$$C_{iG} = H\ C_{iL} \quad (5)$$

Combining (1) through (5) gives:

$$R = \left(\frac{a_S}{a_L k_G} + \frac{a_S H}{a_L k_L} + \frac{H}{K_S} + \frac{H}{k\eta} \right)^{-1} C_G \qquad (6)$$

The four terms in parentheses are, from left to right, the contributions from the gas side resistance, the liquid side resistance, the liquid resistance at the particle surface, and the kinetic rate and intraparticle diffusion.

Equation 6 shows that even with four transport resistances, the overall rate is still first order in reactant concentration, though the overall rate constant (i.e., the reciprocal of the quantity in parentheses) is a complicated function of transport and chemical rate constants, two interfacial areas, and Henry's Law constant. Which terms in the overall rate constant predominate and hence control the rate depend on the type of reactor (e.g., trickle bed, slurry, etc.) and how it is operated.

The liquid-solid interfacial area, a_S, is equal to the external surface area of the particles, a_p, only if the particles are fully immersed in the liquid. For particles of equal size:

$$a_p = \frac{6\ m}{\rho_p d_p} \qquad (7)$$

where d_p = particle diameter

m = catalyst loading, mass of catalyst per volume of reactor

ρ_p = catalyst particle density

In a trickle bed, a_p cannot be varied over a very wide range. Catalyst loading is highly constrained since the volume of solids in a packed bed is usually close to 60% of the bed volume. The particle diameter can vary somewhat, but pressure drop becomes unacceptably large in commercial reactors if d_p gets too small. In a slurry, on the other hand, a_p can vary over several orders of magnitude. Catalyst loading can range from very little for an active catalyst (<< 1%) to nearly fixed bed levels in fluid bed applications. Particle sizes, too, can vary widely, the lower limit being the difficulty of recovering the catalyst from the slurry.

The gas-liquid interfacial area, a_L, is a much more complex function of reactor type, gas and liquid flow rates, agitation level, interfacial tension, tendency of the liquid to foam, and other parameters. This will be discussed below.

The Henry's Law constant, H, is usually a physical chemical parameter for the reaction system that the reactor designer has to live with. However there are processes in which the reactor liquid is not just reactants and products, but is specially tailored to increase the gas solubility. Hydrogen donor solvents used in the liquification of coal can carry much more hydrogen than can be dissolved in ordinary liquids. (Of course in this case the hydrogen capacity is due to chemical reaction and not just solubility.)

The chemical rate constant, k, is a function of temperature and also the loading of active catalyst (e.g., Pt) on the support (e.g., Al_2O_3).

If we are to be able to design these multiphase reactors a priori, we need correlations for the remaining parameters--a_L, k_G, k_L, K_S, and η. The remainder of this review will focus on such correlations.

INTRAPARTICLE RESISTANCES

The coupling of diffusion and reaction in porous catalyst pellets was first analyzed by Jüttner in 1909; but it was not until 30 years later when Thiele (1939), Zeldowitsch (1939), and Damkohler (1936) independently developed the concept of the effectiveness factor, that the idea was disseminated widely throughout the scientific community. In the last 40 years a vast number of papers have been published on the subject. Two monographs are particularly useful. Satterfield's book on "Mass Transfer in Heterogeneous Catalysis" is an excellent treatment of the subject from the practical, applied viewpoint. Aris's two-volume set on "The Mathematical Theory of Diffusion and Reaction in Permeable Catalysts" is an encyclopedic treatment of the theory. Interestingly, Volume I of Aris's work lists over 900 literature references, yet such is the history of the field that very few of these treat the case where the catalyst pores contain two phases--both liquid and gas.

In multiphase reactors, catalysts can be either completely wet (i.e., have their pores filled entirely with liquid), or partially wet (i.e., have two phases in their pores). When they are completely wet, the theory used by Thiele is applicable. This occurs in slurry reactors for example. If they are only partially wet, the situation is much more complex; and no theory fully explains the behavior.

Completely Wet Catalyst

When catalyst pores are filled only with liquid, transport of reactant to catalytic sites is by diffusion. As molecules diffuse into the pellet some react away leaving the concentration of reactant in the interior of the particle lower than at the surface. For reactions of positive order this means that catalyst in the interior of the pellet is less effective than the same amount of catalyst near the particle's surface. The effectiveness factor, η, defined in Equation 8, is an overall measure of how much intraparticle mass transport influences the overall reaction rate.

$$\eta = \frac{\text{actual reaction rate for a catalyst pellet}}{\text{reaction rate if all catalytic surface in the pellet were exposed to the external surface concentration}} \tag{8}$$

Equation 8 assumes that the pellet is isothermal; this is reasonable for catalyst immersed in liquid.

For a completely wet, isothermal, spherical catalyst pellet in which a first order reaction occurs, the material balance is given by Equation 9:

$$\frac{1}{r^2}\frac{d(r^2\, dC/dr)}{dr} = \phi^2 C \tag{9}$$

where ϕ = Thiele modulus = $a\sqrt{k/D}$

a = pellet radius

C = reactant concentration in the pellet

D = effective diffusivity

k = first order rate constant per unit volume of pellet

r = radial coordinate

The usual boundary conditions for Equation 9 are:

$$C = C_S \quad \text{at} \quad r = a \tag{10}$$

$$\frac{dC}{dr} = 0 \quad \text{at} \quad r = 0 \tag{11}$$

Solving Equation 9 with these boundary conditions gives the reactant concentration profile in the pellet:

$$C(r) = C_S \frac{a \sinh(\phi r/a)}{r \sinh(\phi)} \tag{12}$$

Integrating the reaction rate, kC(r), over the pellet volume gives the numerator for Equation 8. The effectiveness factor for this case is:

$$\eta = \frac{3}{\phi}\left(\frac{1}{\tanh \phi} - \frac{1}{\phi}\right) \tag{13}$$

Figure 2 shows the effectiveness factor plotted vs the Thiele modulus. Small particles, such as those used in slurries, usually have small values of ϕ and high effectiveness. Larger particles have higher ϕ and are more likely to be less effective.

Satterfield, et al (1968) studied the effectiveness factor of 1/8" x 1/8" alumina pellets impregnated with palladium for the hydrogenation of α-methylstyrene to cumene at 70° C to 115° C. Effectiveness factors for these liquid-filled pellets ranged from 0.07 to 0.13. From Figure 2 we can estimate the Thiele modulus for this case to be about $\phi = 30$ (for $\eta = 0.1$). To make this catalyst highly effective, we have to reduce ϕ to about 1.0 by reducing particle diameter by a factor of 30, or from about 3 mm to 0.1 mm diameter. With this Pd/Al_2O_3 catalyst we have the choice of using a trickle bed reactor and settling for low effectiveness (hence higher palladium cost) or a slurry reactor with high effectiveness but the cost of separating catalyst from liquid. The 0.1 mm particles would give too high a pressure drop in most commercial trickle bed applications.

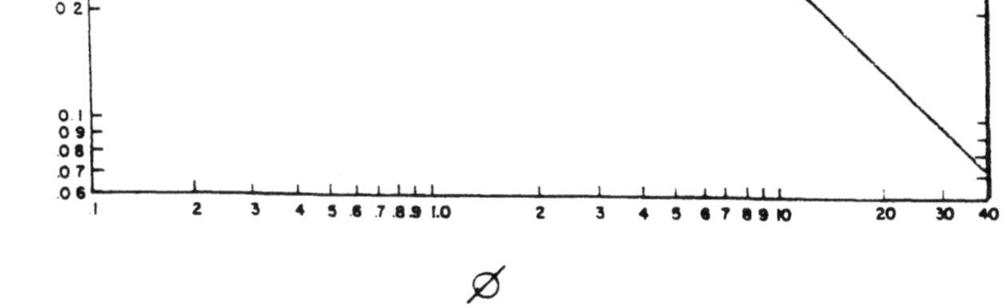

FIGURE 2. EFFECTIVENESS FACTOR, η, VS THIELE MODULUS, ϕ, FOR A SPHERE.

While the mathematical approach for determining catalyst effectiveness is elegant, the usual laboratory method is to vary particle size and observe the reaction rate. Of course, external mass transfer resistances must be eliminated in such experiments. If reducing the particle size has no effect, then the original large particle is fully effective. On the other hand, if smaller particles give a higher rate, then the large particle is not fully effective.

This technique for finding effectiveness factors works well in two-phase systems (gas-solid, liquid-solid). However it can cause problems in three-phase systems, particularly trickle beds, even if the pores are completely filled with liquid.

Van Klinken and van Dongen (1980) found that by filling the interstitial spaces in a laboratory trickle bed with fine particles of 0.2 mm inert silicon carbide, the conversion increased. The 1.5 mm spherical Ni-Mo/Al_2O_3 catalyst was being used to hydrodenitrogenate a Kuwait vacuum distillate. While some of the increase comes from better wetting of the pellet's external surface and some from reduced axial dispersion, the authors found that most of the increase was due to larger liquid holdup, which is another way of saying longer liquid residence time.

If the experiment had been a test of catalyst effectiveness and if whole pellets had been used in one case and crushed in another--but without inerts in either--the higher conversion with crushed could easily have been attributed to higher effectiveness factor even though the real cause was increased holdup. Satterfield (1970, pp 93-94) cites two papers (LeNobel and Choufoer, 1959, and van Zoonen and Douwes, 1963) which report higher rates for smaller particles in trickle beds. Though these authors attribute the improved performance to a reduction of intraparticle mass transfer resistance, van Klinken and van Dongen's recent work suggests a different interpretation. Clearly, measuring catalyst effectiveness in a trickle bed can lead to problems.

The use of inerts to improve laboratory trickle bed performance did not originate with van Klinken and van Dongen. It was mentioned by many earlier workers including Mears (1971), Sedricks and Kenney (1973), DeBruijn (1976), and Koros (1976). There is no reason to expect that the use of fine inert solids in a commercial reactor will have any effect on conversion since these units already have high holdup and good catalyst wetting. The inerts would, however, cause much higher pressure drops.

Fully wet pellets can give lower effectiveness than Figure 2 suggests if they are incompletely irrigated by liquid as can happen at low liquid mass velocities. If a pellet is wet on one side only, for example, Equations 9 and 10 are no longer valid, the former because concentration becomes a function of more than just radial position, and the latter for obvious reasons. To get a feeling for the effect of partial irrigation, consider the simple example of a catalyst slab. The Thiele modulus for a slab of thickness 2a is the same as defined earlier if both faces of the slab are irrigated. If only one face is irrigated, the diffusion distance (i.e., the length term in the Thiele modulus) increases from a to 2a, and so the value of the Thiele modulus doubles. From Figure 2 we see that the higher Thiele modulus gives lower effectiveness. Thus, low mass velocity can cause partial irrigation which reduces catalyst effectiveness factor. In commercial reactors with high mass velocity, irrigation is usually complete.

Partially Wet Catalyst Pellets

When a reactor loaded with dry catalyst is started up, it can take many hours until the pores fill with liquid. During this period of incomplete wetting, the effectiveness factor can either be greater or less than with liquid-filled pellets.

Sedricks and Kenney (1973) studied the reduction of crotonaldehyde in a trickle bed of Pd/Al_2O_3 with well wetted and poorly wetted catalyst. The well wetted case used a preflooded reactor; the poorly wetted used an initially dry reactor. The dry catalyst gave significantly higher reaction rates. Apparently liquid in the pores severely limits the transport of hydrogen, and so this resistance controls the reaction rate. In dry pores, on the other hand, hydrogen diffuses readily. Since it can react with crotonaldehyde vapor on the catalyst, the reaction proceeds quickly. Preflooded catalyst gives lower rates. As the initially dry pores fill with liquid over time (see Figure 3), the rate declines to that of the preflooded catalyst.

Satterfield and Ozel (1973) also found dry catalyst more effective, in this case for benzene hydrogenation. Again hydrogen transport is limiting in liquid pores, and benzene is sufficiently volatile that it can react rapidly with hydrogen in dry pores.

If substantial heat is released on reaction, a hot spot can develop which keeps a dry catalyst dry by vaporizing incoming feed. The wet or dry phenomena can lead to multiple reactor steady states.

Wet catalysts can dry out. If liquid flow to a portion of

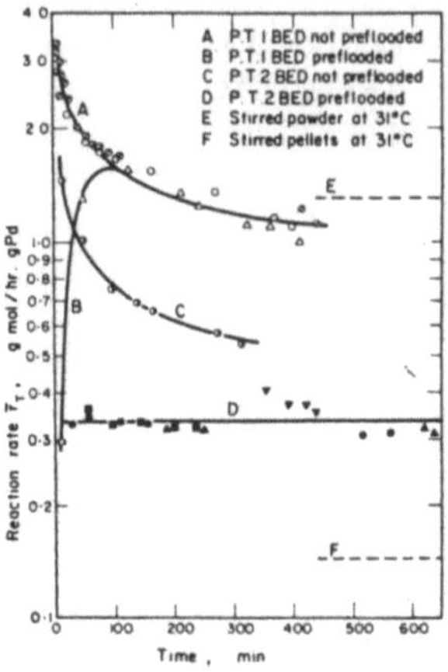

FIGURE 3. REACTION RATES FOR INITIAL DRY AND PREFLOODED Pd/Al_2O_3 CATALYST FOR THE REDUCTION OF CROTONALDEHYDE IN A TRICKLE BED. (AFTER SEDRICKS AND KENNEY (1973)).

a trickle bed becomes poor and some catalyst is not irrigated, heat generated by the reaction, which normally would be removed by the liquid flow, can raise catalyst temperature and vaporize liquid from the pores. When flow returns the dry pores may not refill.

The discussion above showed that if the reactants are volatile, dry catalyst can be more effective than wet. However if one of the reactants is nonvolatile, no reaction occurs on the dry portion of the catalyst, and partially wet catalyst is less effective than fully wet. Since particles are usually well wet in slurry reactors, the above discussion applies primarily to trickle beds.

One situation that has not been investigated extensively is the case where a gas is evolved from a liquid. The decomposition of hydrogen peroxide to water and O_2, and the decomposition of hydrazine (N_2H_4) to N_2 and H_2 are typical of this case. The latter reaction was studied by Sangiovanni and Kestin (1971) in connection with the use of hydrazine as a rocket monopropellant. When liquid hydrazine first contacts a dry catalyst pellet, the pores begin to fill by capillary action. At the same time, decomposition of both hydrazine liquid and vapor begins to generate gas in the pores. Since the pore mouths are sealed with liquid, the pressure inside the pellets rises as gas is formed until suddenly the pellets explode. This increases the external surface of the pellets and the overall reaction rate (i.e., effectiveness) increases. Of course such behavior is not usually desirable.

LIQUID-SOLID MASS TRANSFER

The rate of liquid-solid mass transfer depends on the liquid-solid interfacial area, a_S, and the mass transfer coefficient, K_S. As stated above, a_S is related to the particle surface area, a_p, but may be less than a_p if the surface is not fully wet. Since both a_S and K_S are so dependent on reactor type, we shall treat one type at a time.

Trickle Beds

Liquid-solid mass transfer coefficients have been correlated by many workers for a wide variety of particle sizes, flow rates, and flow regimes. Some correlations are based on the rate of dissolution of a slightly soluble substance into liquid (e.g., benzoic acid, naphthalene, or β-naphthol into water) and some on actual reactions. Hirose, et al (1974) studied the dissolution of benzoic acid spheres (0.28 to 1.27 cm diameter) for trickle flow, pulsed flow, and bubble flow regimes. Their data are plotted in Figure 4. Over the entire range of flows K_S is approximately twice as large as in liquid-only flow. The reason

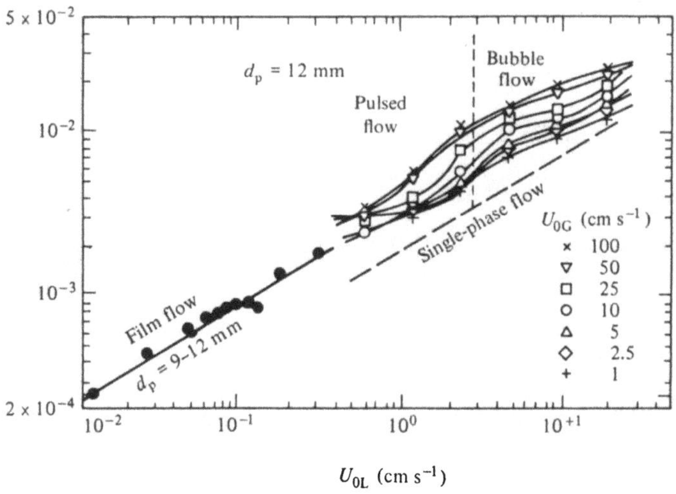

FIGURE 4. LIQUID-SOLID MASS TRANSFER COEFFICIENT K_S IN A TRICKLE BED. (AFTER HIROSE, ET AL 1974)

why K_S is larger in three-phase flow than two-phase flow is that the actual linear liquid velocity is higher in the three-phase case. In three-phase flow, gas occupies part of the so-called bed void volume making the liquid fraction less than in the liquid-filled case. Hirose, et at used these correlations for <u>liquid filled</u> reactors:

$$\varepsilon J_D = 0.8\ Re_L^{-0.5} \qquad 20 < Re_L < 200 \tag{14}$$

$$\varepsilon J_D = 0.53\ Re_L^{-0.42} \qquad 200 \leq Re_L < 5000 \tag{15}$$

$$\text{where } J_D = \frac{K_S a_S}{V_{OL} a_p} \left(\frac{\mu_L}{\rho_L D} \right)^{2/3}$$

$$Re_L = \frac{d_p V_{OL} \rho_L}{\mu_L}$$

a_p = surface area of particles/volume of reactor

a_S = liquid-solid interfacial area/volume of reactor

D = diffusivity

d_p = pellet diameter

K_S = liquid-solid mass transfer coefficient

V_{OL} = superficial liquid velocity

ε = bed void fraction

μ_L = liquid viscosity

ρ_L = liquid density

Dharwadkar and Sylvester (1977) proposed the following correlation for all flow regimes after surveying much literature data:

$$\frac{K_S}{V_{OL}} \left(\frac{\mu_L}{\rho_L D} \right)^{2/3} = 1.637\ Re_L^{-0.331} \tag{16}$$

The gas velocity is not included in this equation because the authors claim its effect on K_S is small.

None of the above three equations considers the influence of surface tension, and none of them includes the effect of gas velocity. Speccia, et al (1976) presented the correlation in Equation 17 which includes these parameters, the latter by way of the liquid holdup.

$$\ln\left[\frac{K_s a_s d_p}{a_p D}\left(\frac{\mu_L}{\rho_L D}\right)^{-1/3}\right] = 7.82\sqrt{\ln(We')} - 1.29\ln(We') - 7.61 \qquad (17)$$

where We' = modified Weber number

$= 1000\ V_{OL}^{2}\rho_L d_p/H_t^{2}\sigma_L$

H_t = liquid holdup per unit volume of empty reactor

σ_L = surface tension

Most of the correlations above contain the ratio of a_s/a_p, the fraction of the external pellet surface actually wet. The reason for this is that much of the transport resistance between liquid and solid in trickle beds is due to the less than 100% effective wetting of the surface. This ratio is a measure of the "contacting efficiency" and depends on not only the wet and dry areas, but also on semistagnant or slow-moving liquid regions in the bed.

There has been some controversy in the literature about contacting efficiency. Weekman (1976), in an excellent review, points out that while most workers agree that higher liquid and gas mass velocities give higher reaction rates, they disagree on the reasons. Many explanations have been proposed that deal with liquid holdup effects, mass transfer effects, axial dispersion, catalyst wetting, and flow maldistribution. We will treat only part of this subject here.

In 1976, Schwartz, et al, presented data that showed contacting efficiency was independent of flow rate. Koros (1976), on the other hand, showed that catalyst efficiency increases with flow rate up to a linear mass velocity of about 3 kg/m^2s. Above this mass velocity (which is not a universal constant) efficiency is high and unchanging. The data of Schwartz, et al (1976) may have been taken under conditions where high efficiency prevailed at even low flow rates so that higher flows could not improve the situation.

Koros's study (1976) is a particularly interesting one. To eliminate gas-liquid mass transfer effects, he used the irreversible, first order, solid catalyzed decomposition of hydrogen peroxide on charcoal. This reaction is unaffected by the gas phase except as it influences fluid flow. At a first look, Koros's finding that higher liquid mass velocity increased catalyst utilization might be attributed to better solid-liquid mass transfer. However when Koros tested this hypothesis by more than doubling the intrinsic reaction rate, he found catalyst utilization unchanged. Had mass transfer been controlling, the utilization efficiency should have dropped. Thus mass transfer could not be limiting. At very low mass velocities, mass transfer limitations were observed. (Note: Koros's paper does state how the intrinsic rate was increased.)

Another hypothesis for the improved catalyst utilization at higher flow rates is that flow uniformity across the bed increases with mass velocity. However, Koros found that even poor uniformity still gave good utilization. This finding confirms other work that shows a high degree of radial mixing in trickle beds even with flow maldistribution. Koros also showed that the geometry of the bed--i.e., the details of how it is packed--has an influence on contacting efficiency.

A good physical picture of why catalyst utilization increased with flow rate even though mass transfer is not limiting has not emerged. We are left with a useful rule of thumb that at liquid mass velocities above 4 - 5 kg/m^2s, the catalyst utilization is likely to be good.

Slurries and Three-Phase Fluidized Beds

For very small particles suspended in a liquid, the mass transfer coefficient is correlated by the Sherwood number:

$$Sh = \frac{K_S d_p}{D} = 2.0 \tag{18}$$

This is strictly valid only for a stationary particle in an infinite quiescent medium. To explain and correlate data when flows are present, two theories have been used. One uses the slip velocity, i.e., the relative velocity between a particle and the liquid, to correlate <u>vs</u> Sherwood number. A typical correlation is the following due to Ranz and Marshall (1952):

$$Sh = 2.0 + 0.6\ Re_p^{1/2} Sc^{1/3} \tag{19}$$

where Re_p is the Reynolds number in which the velocity is the slip velocity.

The problem with theories of this type is that the slip velocity is not usually known and is not usually a single value.

The other approach for getting mass transfer data is based on Kolmogoroff's theory of isotropic turbulence. Instead of a slip velocity it used the power input per unit mass of slurry. The energy is dissipated into turbulent eddies which promote mass transfer. This approach suffers from the fact that in practice equal power inputs don't necessarily give equal mass transfer coefficients. The type and location of the impeller makes a difference.

Nevertheless this theory is useful. The theory (Brian, et al (1969)) says that the mass transfer coefficient, K_S, is a function of Reynolds number where the Reynods number takes on one of two forms depending on whether the eddies are larger or smaller than the particles. Thus:

$$Re_V = \left(\frac{e d_p^4 \rho_L^3}{\mu_L^3}\right)^{1/2} \qquad \text{for } l_e \gg d_p \tag{20}$$

$$Re_I = \left(\frac{e d_p^4 \rho_L^3}{\mu_L^3}\right)^{1/3} \qquad \text{for } l_e \ll d_p \tag{21}$$

where e = energy supplied to the liquid per unit mass

l_e = eddy size

$$= \left(\frac{\mu_L^3}{\rho_L^3 e}\right)^{1/4} \tag{22}$$

Levins and Glastonbury (1972) developed the following correlation which includes the ratio of impeller to vessel diameter (d_I/d_V):

$$\frac{K_s d_p}{D} = 2.0 + 0.47\, Re_I^{0.62} \left(\frac{d_I}{d_v}\right)^{0.17} \left(\frac{\mu_L}{\rho_L D}\right)^{0.36} \tag{23}$$

In bubble columns, the energy input to the liquid can be estimated from Equation 24:

$$e = V_{OG}\, g \tag{24}$$

where g = gravitational acceleration

V_{OG} = superficial gas velocity

GAS-LIQUID MASS TRANSFER

The rate of mass transfer from gas to liquid is strongly dependent on the interfacial area, a_L, as well as the mass transfer coefficients. Since a_L is very reactor dependent, this section is organized by reactor type.

Trickle Beds

In trickle beds the liquid side mass transfer coefficient per unit volume of reactor, $k_L a_L$, is greatly influenced by momentum transfer between the gas and liquid phases. It is therefore not surprising that the rate of energy dissipation per volume of reactor, E_L, has been used to correlate $k_L a_L$. E_L is usually given by Equation 25:

$$E_L = (\Delta P/\Delta Z)_{LG}\, V_{OL} \tag{25}$$

where E_L = rate of energy dissipation, W/m^3

$(\Delta P/\Delta Z)_{LG}$ = pressure drop per unit length for liquid-gas flow in a trickle bed, N/m^3

V_{OL} = superficial liquid velocity, m/s

For low value of E_L (i.e., for the trickle flow regime), Charpentier (1976) recommends the following correlation:

$$k_L a_L = 0.0011\, E_L \frac{D}{2.4 \times 10^{-9}} \tag{26}$$

where D = molecular diffusivity in the liquid, m^2/s

$5 < E_L < 100\ W/m^3$

For very low flow rates (i.e., $E_L < 5$) Charpentier recommends using $k_L a_L = 0.008 s^{-1}$ for particles with $d_p > 2$ mm.

In the pulse and spray flow regimes, Reiss (1967) measured $k_L a_L$ for Rashig rings, Intalox saddles, and spheres using water and ammonia or oxygen. He found:

$$k_L a_L = 0.0173\ E_L^{0.5} \tag{27}$$

for $E_L > 60 - 100\ W/m^3$

Satterfield (1975) reviewed Reiss's work and suggested that Equation 27 be modified to account for different diffusivities and viscosities. Satterfield suggests using

$$k_L a_L = 0.0173\ E_L^{0.5} \left(\frac{D}{2.4 \times 10^{-9}} \right)^{0.5} \tag{28}$$

where D_A = molecular diffusivity, m^2/s

if the liquid viscosity were near that of water. If not, then use

$$D \mu_L^{0.8} = \text{constant} \tag{29}$$

to account for viscosity differences.

For the gas side mass transfer coefficient, Charpentier (1976) suggests using Reiss's correlation (1967) for the pulse and spray regimes:

$$k_G a_L = 2.0 + 0.1\ E_G^{0.66} \tag{30}$$

where $E_G = (\Delta P/\Delta L)_{LG}\ V_{OG}$, W/m^3.

For the trickle regime, Charpentier (1976) suggests use of countercurrent flow data though this gives a conservative estimate of $k_G a_L$.

The gas-liquid interfacial area, a_L, in trickle beds (i.e., cocurrent down flow packed reactors) is larger than for countercurrent flow and is usually much larger than a_S. Charpentier (1976) recommends correlations for a_L/a_s for the various flow regimes. For pulsed and spray flow regimes

$$\frac{a_L}{a_S} = 0.25 \left[\left(\frac{\Delta P}{\Delta Z} \right)_{LG} \left(\frac{\varepsilon}{a_S} \right) \right]^{0.5} \tag{31}$$

where ε = void fraction of packed bed

For trickle flow

$$\frac{a_L}{a_S} = 0.05\left[\left(\frac{\Delta P}{\Delta Z}\right)_{LG}\left(\frac{\varepsilon}{a_S}\right)\right]^{1.2} \tag{32}$$

Slurry Reactors

In slurry reactors, the liquid side mass transfer coefficient, k_La_L, is usually measured using physical absorption rates in experiments where gas side resistance is negligible. This can be done using either a pure gas or a slightly soluble one. Most k_La_L data are based on experiments where no solids are present. Tamhankar and Chaudbari (1979) and Joosten, et al (1977) found that the presence of particles has little or no effect on k_La_L. Slesser, et al (1968) found otherwise. It is suggested that solid particles influence viscosity and bubble coalescence.

Calderbank and Moo-Young (1961) correlated k_L for swarms of bubbles rising in a nonagitated reactor. However correlations for a_L are needed to use their equation. The area, a_L, has been studied extensively (see Chaudhari and Ramachandran, 1980) as a function of the power input per unit reactor volume. Joosten, et al (1977) showed that the power input needs to include both the power from the mechanical agitator and from the gas, though most correlations ignore the latter.

Calderbank and Moo-Young (1961) suggested

$$k_L = 0.42\left(\frac{(\rho_L-\rho_G)\ \mu_L g}{\rho_L^2}\right)^{1/3}\left(\frac{D\rho_L}{\mu_L}\right)^{1/2} \tag{33}$$

When no agitation is used (Froment and Bischoff, 1979):

$$a_L = 2\ V_{OG}\left(\frac{\rho_L g}{\sigma_L}\right)^{1/2}\left(\frac{\sigma_L g}{\rho_L}\right)^{-1/4} \tag{34}$$

When agitation is used, Yagi and Yoshida (1975) propose using:

$$\frac{k_La_L}{D}\ d_I^{\,2} = 0.060\ a^{1.5}b^{0.19}c^{0.5}d^{0.6}e^{0.32} \tag{35}$$

where $a = \dfrac{d_I^{\,2}\ N\ \rho_L}{\mu_L}$

$$b = \frac{d_I N^2}{g}$$

$$c = \frac{\mu_L}{\rho_L D}$$

$$d = \frac{\mu_L V_{OG}}{\sigma}$$

$$e = \frac{N d_I}{V_{OG}}$$

and d_I = impeller diameter, cm

D = diffusivity, cm^2/s

g - gravitational acceleration, cm/s^2

N = speed of agitation, s^{-1}

V_{OG} = superficial gas velocity, cm/s

σ = liquid surface tension, dynes/cm

The reader is referred to Professor Shah's book (1979) and the paper by Chaudhari and Ramachandran (1980) for additional discussion.

ADDITIONAL LITERATURE

There are a number of review papers and monographs on the subject of this paper. Several that the author recommends for additional information on mass transfer in three-phase reactors include Professor Shah's excellent book on "Gas-Liquid-Solid Reactor Design" (1979), Hofmann's review of "Multiphase Catalytic Packed Bed Reactors" (1978), Goto, et al's review of "Trickle Bed Oxidation Reactors" (1977), and Weekman, V.W., "Hydroprocessing Reactor Engineering" (1976).

REFERENCES

Aris, R.,"The Mathematical Theory of Diffusion and Reaction in Permeable Catalysts" Volumes I and II, Clarendon Press, Oxford (1975)

Brian, P.L.T., Hales, H.B., and Sherwood, T.K., AIChEJ 15, 727 (1969)

Calderbank, P.H. and Moo-Young, M.B., CES 16, 39 (1961)

Charpentier, J.C., Chem Eng J 11, 161 (1976)

Chaudhari, R.V. and Ramachandran, P.A., AIChEJ 26, 177 (1980)

Damkohler, G., Z. Electrochem 42, 846 (1936)

DeBruijn, A., 6th Intl. Cong. Catal., paper B34 (1976)

Dharwadkar, A. and Sylvester, N.D., AIChEJ 23, 376 (1977)

Froment, G.F. and Bischoff, K.B., Chemical Reactor Analysis and Design, Wiley, N.Y. (1979)

Goto, S., Levec, J., and Smith, J.M., Catal. Rev.-Sci. Eng. 15, 187 (1977)

Hirose, T., Toda, M. and Sato, Y., J Chem Eng (Japan) 7, 187 (1974)

Hoffman, H.P.. Catal. Rev.-Sci. Eng 17, 71 (1978)

Jooseten, G.E.H., Schilder, J.G.M., and Janssen, J.J., CES 32, 563 (1977)

Juttner, F., Z Phys. Chem. 65, 595 (1909)

Koros, R.M., 4th Intern. Symp. Reaction Eng., p 372, Heidelberg (1976)

Lemay, Y., Pineault, G., and Reuther, J.A., IEC Proc. Design & Dev. 14, 280 (1975)

LeNobel, J.W. and Choufoer, J.H., Fifth World Petroleum Congress Proceedings Section III, Paper 8, Fifth World Petroleum Congress, Inc., New York, 1959

Levins, D.M. and Glastonbury, J.R., CES 27, 531 (1972)

Mears, D.E., CES 26, 1361 (1971)

Ranz, W.E. and Marshall, W.R., CEP 48, 141 (1952)

Reiss, L.P., IEC Process Design & Devel 6, 486 (1967)

Sangiovanni, J. and Kesten, A.S., CES 26, 533 (1971)

Sano, Y., Yamaguchi, N., and Adachi, T., J. Chem Eng Japan 7 255 (1974)

Satterfield, C.N., Ma, Y.H., and Sherwood, T.K., CEP Symp Series 28, 22 (1968)

Satterfield, C.N., "Mass Transfer in Heterogeneous Catalysis", M.I.T. Press, Cambridge, MA, 1970

Satterfield, C.N., and Ozel, F., AIChEJ 19, 1259 (1973)

Satterfield, C.N., AIChEJ 21, 209 (1975)

Schwartz, J.G., Weger, E., and Dudukovic, M.P., AIChEJ 22, 894 (1976)

Sedricks, W. and Kenney, C.N., CES 28,559 (1973)

Shah, Y.T., "Gas-Liquid-Solid Reactor Design", McGraw-Hill, 1979

Slesser, C.G.M., Allen, W.T., Cumming, A.R., Pawlowsky, U., and Shields, J., Chem Reaction Eng - Proceedings of Fourth European Symp, Brussels, 41 (1968)

Speccia, V., Baldi, G., and Gianetto, A., Proceedings of the Fourth Intl Symp on Chem Reaction Eng, p 390, Heidelberg, 1976

Tamhankar, S.S. and Chaudbari, R.V., IEC Fundam 18, 406 (1979)

Thiele, E.W., Ind Eng Chem 31, 916 (1939)

Van Klinken, J. and van Dongen, R.H., CES 35, 59 (1980)

Van Zoonen, D. and Douwes, C.Th., J Inst Petrol 49, 383 (1963)

Weekman, V., Proceedings of the Fourth Intl Symp on Chem Reaction Eng, p 615, Heidelberg, 1976

Yagi, H. and Yoshida, F., IEC Proc Design and Devel 14, 488 (1975)

Zeldowitsch, Y.B., Acta Physicochim U.R.S.S. 10, 583 (1939)

GENERAL CHARACTERISTICS OF MULTIPHASE GAS-LIQUID REACTORS : HYDRODYNAMICS AND MASS TRANSFER

J.C. Charpentier

Laboratoire des Sciences du Génie Chimique
CNRS-ENSIC, 1, rue Grandville, 54042 NANCY (France)

ABSTRACT. Design equations of multiphase reactors show that hydrodynamics, physical kinetics, chemical thermodynamics and kinetics are strongly connected. The lectures include the following sections : 1- Presentation and industrial use of gas-liquid reactors, 2- Design equation of gas-liquid reactors, 3- Some considerations on solubility and diffusivity of gases in liquids, 4- Gas-liquid mass transfer parameters : technique measurement and some data, 5- Prediction of the performances of industrial reactors by laboratory-scale models : simulation. Application to absorption packed columns and venturi-jet reactors.

1. INTRODUCTION

Gas-liquid reactions and absorptions are widely used in five main fields of the chemical industry that are |1|

- the liquid phase processes : oxidation, hydrogenation, sulfonation, nitration, halogenation, hydro-halogenation, alkylation, sulfation, polycondensation, polymerization...
- the gas scrubbing : CO_2, H_2S, CO, SO_2, NO, NO_2, N_xO_y, HF, SiF_4, Cl_2, P_2O_5, Phosgene, Hydrocarbons... (very often to fight against the air pollution).
- the manufacturing of pure products : H_2SO_4, HNO_3, $BaCO_3$, $BaCl_2$, adipic acid, Nitrates, Phosphates...
- Biology : aerobic fermentation, oxidation of sludges, manufacturing of proteins from hydrocarbons, biological waste treatment...
- Multi and two-phase flow in petroleum and nuclear engineering.

The heart of these processes is an absorber or a reactor, of particular configuration best suited to the chemical absorption or reaction being carried out (fig. 1).

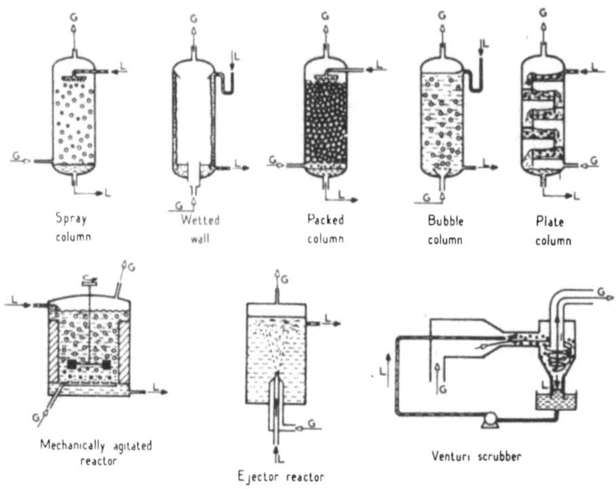

Figure 1 - Main types of gas-liquid reactors.

Indeed gas-liquid contactors present different shapes such as tubular spray columns, wetted falling film columns, packed columns, bubble columns, plate columns, mechanically agitated reactors, jet and venturi reactors, gas lift loop reactors... These different configurations are mainly due to the competition between the various phenomena involved with the chemical thermodynamics and the physical and chemical kinetics that intervene simultaneously and that determine the choice of the equipment which will have to work in the optimal energetic and hydrodynamic conditions. *We want to show here the problems specific to gas-liquid reactors used in chemical and parachemical industries and to present the theoretical bases and the necessary experimental data to choice one of the reactors of figure 1 or at least to simulate it by laboratory mock-ups.*

So we are going to propose in the following text :

- some design equations of multiphase reactors which will show that hydrodynamics, physical kinetics, chemical hydrodynamics and kinetics are strongly connected,
- some data on physico-chemical parameters (gas solubility and diffusivity in liquids) and gas-liquid mass transfer parameters (interfacial area and mass transfer coefficients) with some emphasis on trickle-bed and mechanically stirred reactors,

- some considerations on simulation of industrial gas-liquid reactors by laboratory scale models : application to packed columns and venturi-jet reactors.

Complementary hydrodynamics, heat transfer and design considerations on the system geometry will be presented by Baldi in the same NATO Summer School Program.

2. HYDRODYNAMICS MASS TRANSFER AND CHEMICAL THERMODYNAMICS AND KINETICS ARE STRONGLY CONNECTED IN GAS-LIQUID REACTORS

To study the gas-liquid mass transfer phenomena, it is convenient to consider first of all steady-state situation in which the composition of the gas and the liquid are statistically constant when *averaged* with respect to any temporal fluctuation and *over a specified region such as a short vertical or horizontal slice of a tubular column or the volume of an open agitated vessel*. Then, although the situation in gas-agitated liquid reactor is a rather complicated one as often diffusion, convection and reaction proceed simultaneously and the nature of the convective movements of both gas and liquid are ill-defined, however several useful predictions have been performed to describe the behavior of such high complicated system in using highly-simplified models which simulate the situation sufficiently well for practical purposes without introducing a large number of parameters : the Film Model and the Surface Renewal Models.

The type of chemical systems which have received most attention is the one in which the dissolved gas solute (component A) undergoes an irreversible second-order reaction with a reactant (component B) dissolved in the liquid to give a product (component Prod.). The stoichiometry of the reaction is formulated by

$$A + zB \xrightarrow{k_2} Prod$$

with the rate equations,

$$r_A = k_2 C_A C_B, \qquad r_B = z r_A$$

This could represent the oxidation of an organic liquid by air as an example. z is the number of moles of B reacting with one mole of A in the liquid phase holdup β ($\alpha = 1-\beta$ is the gas holdup) and $k_2 = k_{AB}$ is the rate constant.

Let us consider now that this reaction is carried out first in a tubular bubble column and then in a mechanically agitated gas-liquid reactor and let us write the design equations in both cases.

In the tubular reactor of figure 2 where it will be assumed that the gas phase is in piston flow and the liquid phase is in axial dispersed piston flow, the mass balance (in gmole.s^{-1}) and the heat balance (in cal.s^{-1}) within a small horizontal slice of volume Ωdx will lead to

$$-q_G \frac{P}{RT} dy_A - k_G aP(y_A - y_A^*)\Omega dx = -q_G \frac{P}{RT} dy_A - Ek_L a(C_A^* - C_A)\Omega dx = 0 \qquad (1a)$$

(convection) (gas-liquid mass transfer)

$$D_L \frac{d^2C_A}{dx^2} \beta\Omega dx - q_L dC_A + Ek_L a(C_A^* - C_A)\Omega dx - \beta r_A \Omega dx = 0 \qquad (2a)$$

(axial dispersion) (convection) (gas-liquid mass transfer) (chemical reaction)

$$D_L \frac{d^2C_B}{dx^2} \beta\Omega dx - q_L dC_B - \beta z r_A \Omega dx = 0 \qquad (3a)$$

(axial dispersion) (convection) (chemical reaction)

$$D_L \rho_m C_{pm} \frac{d^2T}{dx^2} \Omega dx - \rho_m C_{pm} q_L dT - \beta r_A \Delta H_A - UA(T - T_c) = 0 \qquad (4a)$$

(axial dispersion) (convection) (chemical reaction) (heating or cooling exchange with outside)

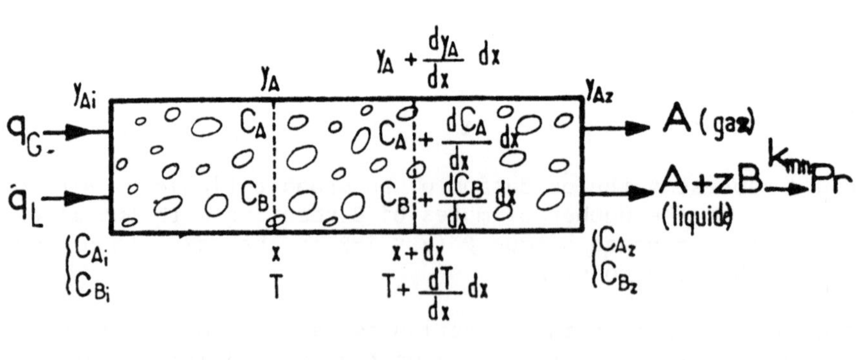

Figure 2 - Tubular reactor.

Similarly in the mechanically stirred reactor of figure 3 where it will be assumed that both phases are hydrodynamically perfectly macromixed, the mass and the heat balances for the entire volume V_R of the open vessel will lead to

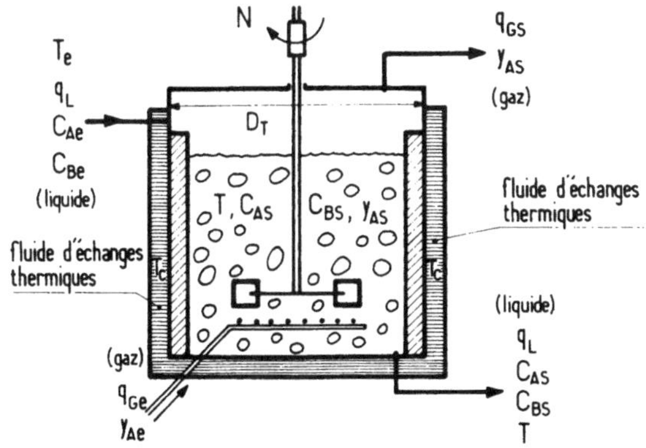

Figure 3 - Mechanically agitated reactor.

$$q_{Ge}y_{Ae}-q_{Gs}y_{As}-k_G aP(y_{As}-y_A^*)\beta V_R = q_{Ge}y_{Ae}-q_{Gs}y_{As}-Ek_L a(C_A^*-C_{As})\beta V_R=0 \tag{1b}$$

(convection) ↑, (gas-liquid mass transfer) ↑

$$q_L(C_{Ae}-C_{As}) + Ek_L a(C_A^*-C_{As})\beta V_R - \beta r_A V_R = 0 \tag{2b}$$

(convection) ↑, (gas-liquid mass transfer) ↑, (chemical reaction) ↑

$$q_L(C_{Be}-C_{Bs}) - \beta z r_A V_R = 0 \tag{3b}$$

(convection) ↑, (chemical reaction) ↑

$$q_L\rho_m c_{pm}(T_e-T) - \beta r_A V_R \Delta H_A - UA(T-T_c) = 0 \tag{4b}$$

(convection) ↑, (chemical reaction) ↑, (heating or cooling exchange with outside) ↑

The first equations (1a, 1b) mean that while the gas phase flows through the considered volume of reactor, the decrease of the gas phase mole fraction y_A is due to the rate of component A transferred into the liquid phase across the interfacial area per unit volume of reactor or of liquid respectively by means of the gas phase mass transfer coefficient k_G and the liquid-film mass transfer coefficient k_L. So component A can diffuse on each side of the interface under the driving force terms $(y_A-y_A^*)$ and $(C_A^*-C_A)$. In this expression of the rate of absorption, the physical kinetics is represented by the Film Model. E is the enhancement factor due to the chemical reaction and that characterizes the effect of the chemical reaction on the physical kinetics of the mass transfer (E = 1 for an absorption or a desorption in absence of chemical reaction).

The second equations (2a, 2b) mean that while the liquid phase flows through the considered volume of reactor either in axial dispersed piston flow or perfectly macromixed, the gas solute, transferred from the gas phase and then dissolved into the liquid phase, reacts with the component B at the kinetic rate of reaction r_A which is a strong function of temperature and concentration in most cases.

The third equations (3a, 3b) concern the component B and involve the same terms as the previous equations except the mass transfer term as it is assumed that this component is not volatile.

The fourth equations (4a, 4b) concern the heat transfer inside the mixture of density ρ_m and heat capacity C_{pm}. They account for energy input in the feed, for heat generation by reaction and for heat exchanged through cooling jacket or coils at temperature T_c, of exchange surface A and with an overall heat transfer coefficient U.

The equations 1 to 4 show how strongly connected are the phenomena concerning

- the hydrodynamics of the two-phase flow, characterized here by the liquid holdup β and the liquid axial dispersion coefficient D_L
- the chemical thermodynamics, characterized here by the solubility $C_A^* = p_i/He = y_A^*P/He$, i.e., the concentration of dissolved gas A at interface, in equilibrium with gas at interface, by the diffusivities D_A, D_B of reactants in liquid phase and the heat of reaction ΔH_A
- the physical kinetics of the mass and heat transfer characterized here by the transfer parameters k_Ga, a, k_La, U
- the chemical kinetics characterized by the rate constant and the partial orders of reaction.

Indeed it is only when the experimental data of the different previous parameters, specific to multiphase flow reactors, are estimated that the system of equations may be solved with the ad hoc boundary conditions. Thus for a tubular reactor the temperature profile T(x) and/or the concentration profiles $C_A(x)$, $C_B(x)$ and $y_A(x) = p_A(x)/P$ as function of the length of the reactor x can be determined. Moreover this design procedure rates the capacity of a reactor for a given production and conversion, and searches for the size which meets a predetermined plant capacity.

Therefore it is seen that the choice of a gas-liquid reactor, then its sizing and its working for given conditions requires knowledge of a great number of parameters, mainly those involved in the global rate of mass transfer between gas and liquid per unit volume of reactor,

$$\Phi = \phi \cdot a = k_G a P(y_A - y_A^*) = k_G a(p - p_i) = E k_L a(C_A^* - C_A) \qquad (5)$$

and their variations with the operating conditions (flow rates of gas and liquid phases q_G, q_L, stirring rate N...) during the scaling up or the scaling down for a reactor of a known type or for the various types of reactors of figure 1.

Indeed the competition between the physical and the chemical kinetics at the level of the mass transfer between gas and liquid (the mass transfer reaction regime where the reaction belongs) may allow for the choice of the type of gas-liquid reactor |2|. This is clearly shown on figure 4 that represents schematically the concentration profiles for A and B on each side of the interface, first in the phase bulks where the partial pressure p and the concentrations $C_A = C_{Ao}$ and $C_B = C_{Bo}$ are uniform, then in the laminar films situated close to the interface (the thickness are $\delta_G = D_G/k_G$, with $D_G = 0$ for a pure gas, and $\delta_L = D_A/k_L$) where the resistance to mass transfer and/or the chemical reaction are located. Eight distinct kinetic regimes are observed. The formulation of the specific rate of absorption per unit interfacial area ϕ shows that for instantaneous reaction (cases A and B) or for rapid chemical reaction (cases C and D), the chemical reaction is occuring only in the liquid-film during the transportation of component A. The concentration of A in the bulk of liquid is virtually zero (r_A=0 in eqs. 2 and 3). For such reactions (i.e. absorption of acid gases) the liquid holdup is not important and equation (5) shows that the global rate of absorption will be large if the interfacial area a and the mass transfer coefficients k_L, k_G are large. This is called the physical kinetic regime characterized by the Hatta number $Ha = \sqrt{D_A k_2 C_{Bo}}/k_L > 3$. Use of a packed column, plate column or a venturi scrubber would be suitable. On the contrary, at the opposite for the very slow chemical processes in liquid bulk (case H), no reaction occurs in the film and mass transfer is used to keep the bulk concentration of component A close to the saturation value ($C_{Ao} \simeq C_A^*$). This is called the chemical kinetic regime characterized by Ha << 1. For such reactions (oxidation,

Regime	Interface concentration profiles	Rate equations	
Instantaneous	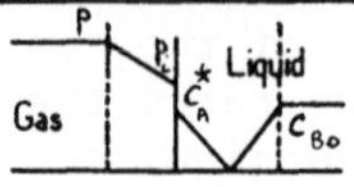	$\varphi = \dfrac{\frac{p}{He} + \frac{D_B}{D_A}\frac{C_{Bo}}{Z}}{\frac{1}{k_L} + \frac{1}{He\,k_G}}$	A
Instantaneous and surface	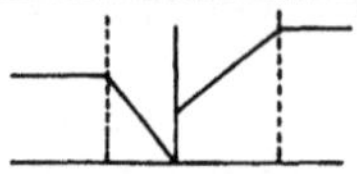	$\varphi = k_G\ p$	B
Rapid	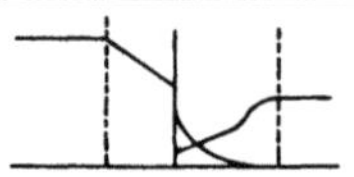	$\varphi = \dfrac{p}{\frac{1}{k_G} + \frac{He}{E\,k_L}}$	C
Rapid pseudo Ist or mth order	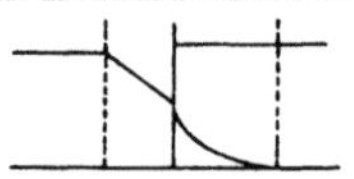	$\varphi = \dfrac{p}{\frac{1}{k_G} + \frac{He}{\sqrt{D_A\,k_2\,C_{Bo}}}}$	D
Intermediate	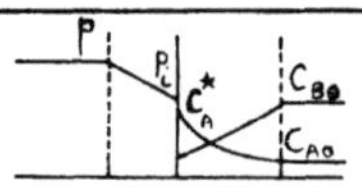	No exact general expression developed	E
Intermediate	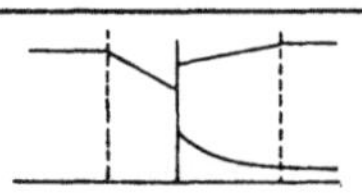	No exact general expression developed	F
Slow diffusional process	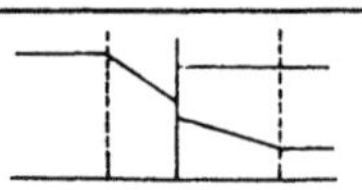	$\varphi = \dfrac{p}{\frac{1}{k_G} + \frac{He}{k_L} + \frac{He\ a}{\beta\,k_2\,C_{Bo}}}$	G
Very slow chemical process in the bulk of the liquid	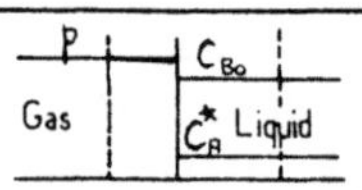	$R_A = k_2\ C_A^*\ C_{Bo}\,\beta$	H

Figure 4
"Interface concentration profiles for the height destinct kinetic regimes for mass transfer with chemical reaction".

hydrogenation, chlorination in organic liquid phase, some fermentations...), sufficient interfacial area is necessary but a high liquid holdup is the more important requirement. Indeed the rate of transformation of component A in liquid-bulk is

$$R_A = k_A C_A^* C_{Bo} \beta$$

and it is not necessary to spend a great amount of energy to obtain high values of k_L, k_G and a. Bubble column, packed column and agitated reactors would be suitable.

For the intermediary regimes (cases E, F and G) where the reaction may occur more or less rapidly in both liquid-film and liquid-bulk, chemical kinetics and physical kinetics may have a comparable influence in the global rate of absorption and both interfacial and hydrodynamic parameters such as interfacial area and liquid holdup should be high. Use of mechanically agitated tank, ejector reactors or plate columns would be suitable depending on the type of chemical reaction encountered.

To help for the application of the previous considerations, the use of Table I will give an idea of the order of magnitude of the liquid holdup and the mass transfer parameters in most of gas-liquid reactors represented in figure 1.

Type of reactor	β (% gas-liquid volume)	k_G $(gmol/cm^2.s.atm)\times10^4$	k_L $(cm/s)\times10^2$	a cm^2/cm^3 reactor	k_La $(s^{-1})\times10^2$
Packed columns					
- counter-current	2-25	0.03-2	0.4-2	0.1-3.5	0.04-7
- co-current	2-95	0.1-3	0.4-6	0.1-17	0.04-102
Plate columns					
- Bubble cap	10-95	0.5-2	1-5	1-4	1-20
- Sieve plates	10-95	0.5-6	1-20	1-2	1-40
Bubble columns	60-98	0.5-2	1-4	0.5-6	0.5-24
Packed bubble columns	60-98	0.5-2	1-4	0.5-3	0.5-12
Tube reactors					
- horizontal and coiled	5-95	0.5-4	1-10	0.5-7	0.5-70
- vertical	5-95	0.5-8	2-5	1-10	2-100
Spray columns	2-20	0.5-2	0.7-1.5	0.1-1	0.07-1.5
Mechanically agitated bubble reactors	20-95	-	0.3-4	1-20	0.3-80
Submerged and plunging jet	94-99	-	0.15-0.5	0.2-1.2	0.03-0.6
- Hydrocyclone	70-93	-	10-30	0.2-0.5	2-15
- Ejector reactor	-	-	-	1-20	-
- Venturi	5-30	2-10	5-10	1.6-25	8-25

Table I - Liquid holdup, mass transfer coefficient and effective interfacial area in gas-liquid reactors.

Thus, in using the film model, to tell whether the reaction is slow, intermediate or fast, it has been necessary to focus on a unit surface of gas-liquid interface and to define a film-conversion parameter or a diffusion-reaction parameter by the ratio

$$\frac{\text{Maximum possible conversion in liquid-film}}{\text{Maximum diffusion transport through the liquid-film}} = \frac{k_2 C_A^* C_{Bo} \delta_L}{D_A \left(\dfrac{C_A^* - 0}{\delta_L}\right)} = \frac{k_2 C_{Bo} \delta_L^2}{D_A} = \frac{D_A k_2 C_{Bo}}{k_L^2} = Ha^2$$

So appears the significance of the Hatta number : roughly if $Ha^2 >> 1$, all reaction occurs within the film and surface area is the controlling factor. On the contrary, if $Ha^2 << 1$, no reaction occurs within the film, and bulk volume controls the rate with no benefit from an increase of interfacial area. *Therefore the Hatta number is actually the criterium for whether the reaction occurs completely in the bulk of liquid necessitating a large volume of liquid, or completely in the boundary layer necessitating contacting devices which create large interfacial area, or in both the bulk and the film necessitating both large volume of liquid and large interfacial area.* This parameter defines both the reaction type and the regime. Recently Barona has succinctly summarized the different reaction and regime situations of figure 4 with

- Bulk reactions : $Ha^2 \leq 0.01$
  . very slow chemical process (case H) :

$$Ha^2 \leq 0.05 \left(\frac{a}{\beta} \frac{D_A}{k_L}\right)$$

  . Slow diffusional process (case G) :

$$0.05 \left(\frac{a}{\beta} \frac{D_A}{k_L}\right) < Ha^2 \leq 0.01$$

- Split reactions (cases E and F) : $0.01 < Ha^2 < 16(E_i - E)/(E_i - 1)$

  with $E_i = 1 + \dfrac{D_B}{D_A} \dfrac{C_{Bo}}{zC_A^*}$ being the instantaneous enhancement factor.

- Film reactions : $Ha^2 \geq 16(E_i - E)/(E_i - 1)$
  . film regime (cases C and D) :
  $16(E_i - E)/(E_i - 1) \leq Ha^2 < (4E_i - 3)^2$

. instantaneous regime (cases A and B) :

$$Ha^2 \geq (4E_i-3)^2$$

The specific rate equations may have an explicite expression (cases A, B, D, G and H) or not (cases C, E and F). In this last case, the rate will be calculated by eq. (5) where the enhancement factor E is evaluated from the diagram of fig. 5 in which E is plotted as a function of the Hatta number, the concentration-diffusion parameter $Z_D = E_i-1$ and the film-total volume ratio parameter for an irreversible second-order reaction, $(a/\beta)(D_A/k_L)$. Different regions have been represented on this diagram which could be summarized in region D where the contactor works as an absorber and in regions A, B, C where it works as a reactor.

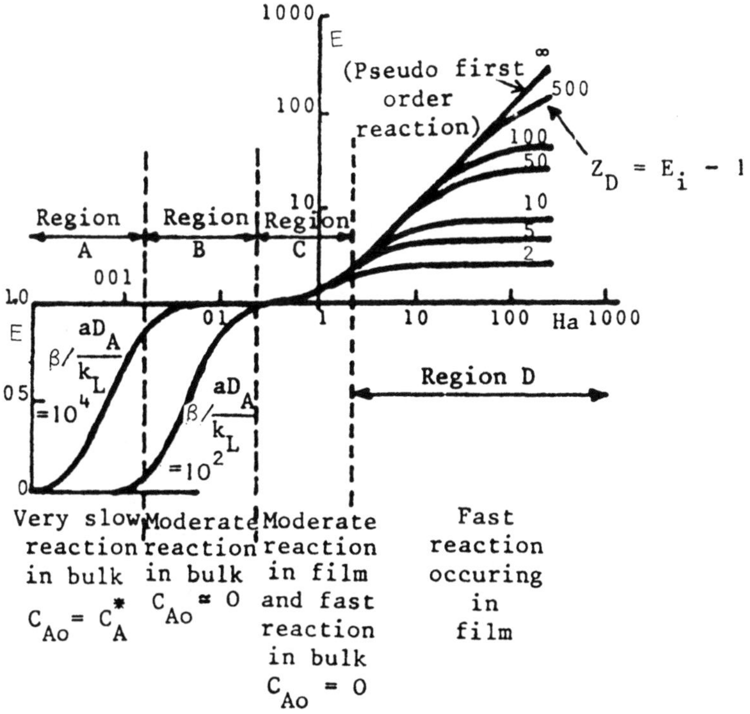

Figure 5 - Enhancement factor for second order reaction.

There exists also several other implicit or explicit approximate correlations between E, Ha and E_i |2|. Furthermore such a diagram or approximate application have been developed for irreversible m,nth-order reaction or reversible, consecutive and parallel reactions |4-7|. In such cases, the Hatta number is defined by

$$Ha^2 = \frac{2}{m+1} \frac{D_A k_{mn} C_A^{*m-1} C_{Bo}^n}{k_L^2}$$

Moreover a third phase may complicate the situation, like in a hydrogenation, where both the dissolved gas and the liquid must reach the surface of a solid catalyst suspended in the liquid or packed in the reactor, where the reaction takes place. Then, mass transfer between the liquid and the particle catalyst surface becomes important and must be taken into account together with the intrinsic kinetics occuring at the catalyst surface sites in the design equations (2, 3). So, since many practical catalytic reactions involving a gaseous reactant A can be characterized as

$$A(gas) + B(liq) \xrightarrow{Catalyst} Prod,$$
with intrinsic rate = $kC_{AC}C_{BC}$

where C_{AC} and C_{BC} are the concentrations on the surface catalyst, the relevant balances for design must be written with E = 1 (no reaction in the liquid) and with the chemical reaction term as

$$\eta k C_{AC} C_{BC} = k_s a_s (C_{Ao} - C_{AC}) = k_s a_s (C_{Bo} - C_{BC}) \qquad (6)$$

k_s is the liquid-solid mass transfer coefficient, a_s is the liquid-solid interfacial area per unit reactor volume, k is the intrinsic surface reaction rate constant per unit catalyst volume and η is the catalyst effectiveness factor for completely wetted particle calculated from the Thiele Modulus defined by $(V_p/S_p)\sqrt{k_V/D_i}$ where D_i is the solute diffusivity inside the catalyst pores and V_p and S_p are the catalyst particle volume and external surface. In his lecture Baldi will show that because of the hydrodynamic behaviour of gas-liquid-solid reactors, the particles may be partially wetted leading to use for the reactor modeling an apparent contacting effectiveness or an apparent kinetic rate constant.

Finally the choice of a gas-liquid reactor, its rating for production capacity or its scaling-up require the simultaneous knowledge of the parameters involved in the *chemical thermodynamics and kinetics* together with the *mass transfer reaction regime* and the variation of the *mass transfer* parameters with the working *hydrodynamic and energetic* operating conditions. It has to be noted also that the hydrodynamic conditions must be also taken into consideration for the sizing in diameter for tubular reactor which depends on such considerations as flow configuration, flooding, entrainment, foaming effects...

For more details on the theory of gas-liquid and gas-liquid-solid reactors, it should be recommended the three admirably clear expositions in the books of Astarita |6|, Danckwerts |7| and Shah |8|.

3. SOME CONSIDERATIONS ON SOLUBILITY AND DIFFUSIVITY OF GASES IN LIQUIDS

The design of equipment for the diffusional separation of gas-liquid mixture is determined by two major considerations, namely, the distribution of the components between phases in a state of thermodynamic equilibrium (solubility of gases in liquid), and the rate at which mass transfer occurs under conditions prevailing in the equipment (liquid diffusivity and chemical reaction). Therefore solubilities and diffusivities of gases in liquids are practically always required for design of gas-liquid process and obtaining solubility and diffusivity data for the gas-liquid system under consideration can very often be a chalenging problem so wide is the range of solutes and solvents the chemical engineer or researcher may encounter.

Although the theory on solubility and diffusivity has not yet progressed to the point where purely theoretical predictions are possible, many strides have been made in that direction in the last decade and our purpose is to present here only the most representative empirical or semi-empirical correlations or references to consult for practical but, unfortunately, very often specific use. *So the reader should always keep in mind that relationships presented below are of course to be used only in the absence of experimental data and after a careful study of the conditions of validity.*

3.1 Solubility of gases in liquids

Different approaches have been made depending whether the solutions are electrolytic or nonelectrolytic.

The main sources for gas solubility data are Linke and Seidell's handbook |9| (inorganic compounds), Seidell and Linke's handbook |10| (inorganic and organic compounds) which do not contain thermodynamic data, Appendix 3 in Hildebrand, Scott and Prausnitz's excellent book |11|, Wilhem and Battino's paper |12| that surveyed the available literature through the end of 1970 and more recent papers by Prausnitz et al. |13-18|, Battino et al. |19-22|, Hildebrand and Lamoreaux |23| and Fleury and Hayduck |24|. Battino and Clever |25| have also given a comprehensive review of the physico-chemical aspects of the solubilities of gases in liquids, including molten salts and liquid metals, together with a great deal of references to solubility determinations.

3.11 Nonelectrolytes.

There are many ways to express the solubility of a gas in a liquid the most popular being the Ostwald coefficient, defined as

$$Os = \frac{V_A^*}{V_B} = \frac{\text{Volume of gas absorbed at temperature T}}{\text{Volume of absorbing liquid at temperature T}} \quad (7)$$

or the Henry's law constant (K_{Ho} or He),

$$p_A = K_{Ho}X_A^* = HeC_A^* \quad (8)$$

p_A is the partial pressure of the gas over the solution which is in equilibrium with the mole fraction X_A^* (also called mole fraction gas solubility) or with the concentration C_A^* of the solute in solution.

If ideal gas behavior is assumed, the mole fraction of the dissolved gas is,

$$X_A^* = \left| (RT/Os.p_A V_B^\circ)+1 \right|^{-1} \quad (9)$$

where R is the gas constant, T is the Kelvin temperature and V_B° is the molar volume of the pure solvent.

Equation (8) is a simplified approximation at low pressures of the equation for ideal solutions,

$$f_A = K_H X_A^* \quad (10)$$

$$\text{with, } K_{Ho} = \lim_{X_A^* \to 0} K_H \quad (11)$$

where f_A is the fugacity of the gas solute.

In ideal solutions K_{Ho} is only temperature dependent. It generally increases with increase in temperature while the solubility of the gas decreases.

However, if in the gas phase, mole fraction y_A is used instead of partial pressure p_A, the equilibrium constant is K_{Ho}/P and therefore the equilibrium term depends also on total pressure P.

When temperature and pressure are constant, the dependence of the equilibrium values are expressed graphically by a straight line (equilibrium line) passing through the origin with a slope of K_{Ho} or He.

The Henry's law constants, particularly He, can be satisfactorily used to expressed solubility, but it must be always kept in mind from thermodynamics that the Henry's law, which is valid for ideal solutions, is applicable only over a restricted range for dilute solutions which means in practise that it is just a limiting law for sparingly soluble gases and for readily soluble gases at low concentrations. In such cases values of K_{Ho} or He, for aqueous solution of different gases can be found for instance in the classical handbooks |26-28|. For higher concentrations, the solubility is generally lower than that given by Henry's law. In such cases K_{Ho} or He depends on the composition of the liquid, and the equilibrium line becomes an equilibrium curve.

So when the component behaves in a nonideal manner in both the gaseous phase and solution, solubility can be obtained as follows

$$f_A = \gamma_A P_{SA} X_A^*$$

where P_{SA} is the extrapolated vapor pressure of the component A.

In accordance with the theory of regular solution |11|, the activity coefficient can be calculated with,

$$\ln\gamma_A = \ln \frac{\left(X_A^*\right)_{ideal}}{\left(X_A^*\right)_{real}} = \frac{V_A^\circ}{RT} (\delta_A - \delta_B)^2 \qquad (12)$$

V_A° is the partial molar volume of the solute gas, δ_A and δ_B are the parameters of the solubility, sometimes referred to as the cohesive energy density (a measure of the forces between the molecules),

$$\delta = \left(\frac{\Delta E}{V^\circ}\right)^{1/2}$$

ΔE is here the molar energy of vaporization and V° is the molar volume.

V_A° is often assumed to be simply the molar liquid at 25°C but it may be considered linearly variable with T/T_c |28|. If water is the solvent, V_A° is determined approximatively as the molar volume of the component at its boiling point in the liquid state at the given temperature if the temperature is above the critical.

Values of δ_A, δ_B and V_B° can be obtained either in Reid, Prausnitz and Sherwood's book |28| or for 59 compounds in the article by Wilhem and Battino |12| in which solubility data for 16 gases in 39 solvents at 25°C are presented (also included in this last paper are some data for several fluorine containing gases). A detailed analysis of the cohesive energies of 31 hydrocarbons of different families enabled Maffiolo et al. |29| to propose a simple correlation to predict the cohesive energy of any liquid hydrocarbon from its molar volume and the polarizabilities and Van der Waals volume of the group forming it. Always in the case of hydrocarbons, it is also important to mention the set of equations derived by Meisner |30| to calculate molar volumes from carbon number and temperature for homologous unbranched hydrocarbons in the liquid state.

If the size of the molecules of the gas solute differs greatly from that of the molecules of the solvent, the following equation may be used,

$$\ell n\gamma_A = \frac{V_A^\circ}{RT}(\delta_A - \delta_B)^2 + \ell n\frac{V_A^\circ}{V_B^\circ} + \left(1 - \frac{V_A^\circ}{V_B^\circ}\right) \qquad (13)$$

Equations (12) and (13) have proved to be remarkably satisfactory for solutions where both solvent and solute are non-polar and for slightly polar solvents. A more thorough discussion of the use of these equations and references to gas-liquid systems is given by Battino and Clever |25|. Moreover Hildebrand and Lamoreaux |23| have presented an interesting empirical linear relationship and a graphical plot allowing for the calculation of the solubility of any inert gas except H_2 in any non-polar solvent except fluorochemicals, from the energy of vaporization of the gas at its boiling point and the solvent cohesive energy density. Complementary and alternate results were then afford by Fleury and Hayduck |24| that still apply for perfluorinated gases and solvents.

For gas solute in large-molecule solvents, Chappelow and Prausnitz |15| proposed a modified equation for γ_A in which both size and shape of the molecules are taken into consideration and provide accurate gas-solubility data for 26 hydrocarbon gas-liquid systems in the temperature range 25° to 200°C in the vicinity of 1 atm pressure. Complementary informations on solubility of organic solutes in polymers have been recently proposed by Maloney and Prausnitz |31| and Stiel and Harnish |32|.

The temperature dependence of solubility has been very often used to get the molar heat of solution ΔH (the difference between the partial molar enthalpy of solute at infinite dilution in solvent and the enthalpy of ideal gas at the same temperature). In this case, plots of ℓnX_A^* against 1/T are usually linear to the accuracy with which gas solubility is commonly measured assuming a temperature independent ΔH and calculating it from the slope. However it is important to note that for several binary systems a plot of gas solubility versus temperatures goes through a minimum. For instance, in measuring the solubilities of hydrogen, methane, ethane in hexadecane, bicyclohexyl and diphenylmethane, in the temperature range 25-200°C, Cukor and Prausnitz |14| have obtained that the solubility of hydrogen rises and that of ethane falls with increasing temperature while the solubility of methane goes through a flat at minimum in the region near 150°C. As suggested by these authors, it is likely this minimum may be observed in the low pressure solubilities for any gas in any non-polar solvent provided the range of temperature is chosen adequat. The same trend has been observed by Preston and al. |13| and Chappelow and Prausnitz |15| which reenforces the very interesting theoretical work of these last authors.

Precedent laws are apparently obeyed to only moderate pressures and give considerable deviations from the experimental results if the pressure is high even if X_A^* is small. In such case and for solutions of pure gases, the following equation can be used,

$$\ln \frac{f_A}{X_A^*} = \ln K_H + \frac{V_A^\circ (P-P_{SB})}{RT} \qquad (14)$$

where P_{SB} is the solvent saturation vapor pressure, P the total pressure and K_H the Henry's constant related to the partial molar fugacity of the solute at the solvent saturation pressure. This equation has been successfully used to fit high pressure gas solubility in water, methanol and hydrocarbon and hydrogen solubility in binary and ternary mixed hydrocarbon solvents |25|. Also, in addition to the effect of total pressure Orentlicher and Prausnitz |33| have presented an equation taking into account the effect of composition,

$$\ln \frac{f_A}{X_A^*} = \ln K_H + \frac{V_A^\circ (P-P_{SB})}{RT} + \frac{A(X_B^2-1)}{RT} \qquad (15)$$

where A is an empirical coefficient depending only on temperature for a given hydrogen-solvent system and this equation make it possible quickly to calculate the solubility of hydrogen at low temperatures and high pressures in pure and mixed non-polar solvents. It is important to note that these authors also summarize the manner to make good estimates of solubility of hydrogen in mixed solvent.

A correlation essentially identical to Eq. (15) has been proposed by Krichevsky and Ilniskaya |34| to fit the solubility of mixture of gases.

For gas solubilities in mixed non electrolyte solvents, Battino and Clever |25| have presented numerous references for different systems and O'Connell and Prausnitz |35| have proposed an equation for a gas dissolved in any number of miscible solvents.

Solubilities in water of many gases range in solute mole fraction from about 0.000007 for helium to about 0.3 for ammonia at 25°C and previous correlations expressed or cited to predict gas solubilities in non-polar or regular solutions are not successful in polar, associating solvents such as water or alcohols.

Using solubilities of 20 gases in water and other solvents (methanol, ethanol, butanol, acetone, methyl acetate, acetic acid, ethylene glycol and chlorobenzene) selected from literature (when available), Hayduck and Laudie |36| developed a systematic approach to relate the effect of association or H-bonding on gas solubilities. A large reduction in solubility from the ideal solubility is attributed to strong H-bonding which appears to have the effect of excluding solute molecules. H-bonding factors were defined by Hayduck and Laudie for each gas, based on ratio of its actual solubility to its ideal solubility. The conclusions of these authors are that H-bonding factors for gases in water are related to those in the primary, normal alcohols but are essentially independent of H-bonding factors in chlorobenzene, a slighty polar solvent. Similarly, H-bonding factors in acetone are related to those in methyl acetate, acetic acid and ethyleneglycol. Except in those cases where the dissolved gas reacts chemically with the solvent (i.e. C_2H_3Cl, C_2H_2, C_2H_4, N_2O, CO_2 and SO_2 in acetone), gas solubilities can be approximately predicted from relation between H-bonding factors in the solvents, provided the solubility is known in at least one of them. Hayduck and Laudie |36| have also found that, like for non-polar solvents, solubilities of all gases in water tends towards a constant molar concentration at the solvent critical temperature (fig. 6a). More recently Beutier and Renon |37| have shown that this is true for all gases and solvents and they derived an exact value of the reference solubility from thermodynamic considerations (fig. 6b). They have also shown that enthalpies of solution of all gases are infinite at the solvent critical temperature.

Cysewski and Prausnitz |16| have presented a semi-empirical correlation that is useful for estimating gas solubilities at modest pressure and not restricted to non-polar systems nor to temperature near 25°C. It does not include liquid metals or molten salt. This correlation for the reduced $K_H V_B^o/RT$ is proposed in terms of parameters determined empirically from 76 binary systems which characterize the interaction of a single solute molecule with a matrix of solvent molecules. It may be used in either of two ways over a wide range of temperature and for a variety of gases in typical non-polar and polar solvents, including water : first, if a

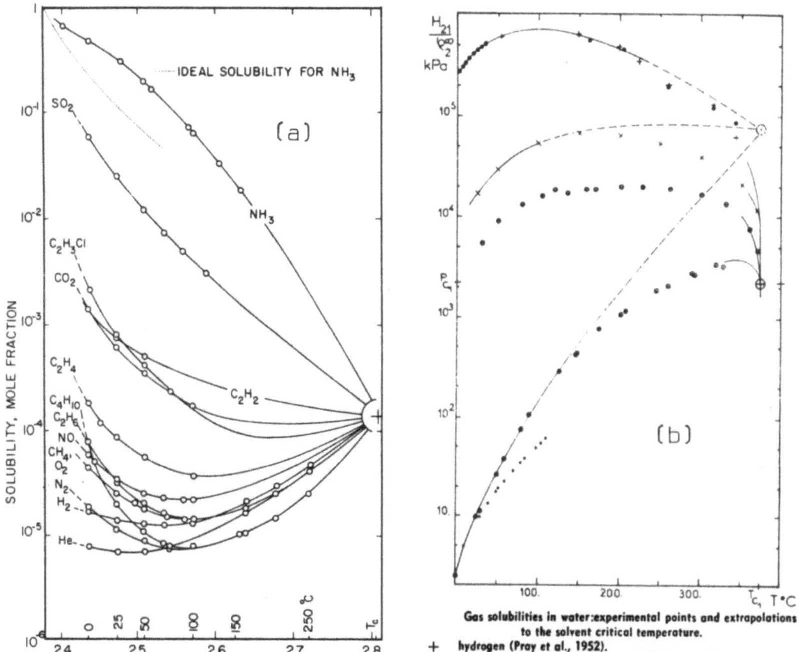

Solubilities in water.

Gas solubilities in water:experimental points and extrapolations to the solvent critical temperature.

+ hydrogen (Pray et al., 1952).
● oxygen (Pray et al., 1952; Benson and Krause, 1976).
× carbon dioxide (Houghton et al., 1957; Ellis and Golding, 1963).
⊖ hydrogen sulfide (Clarke and Glew, 1971; Selleck et al., 1952; Kozintseva, 1964).
⊙ ammonia (Jones, 1963).
• sulfur dioxide (Beuschlein and Simenson, 1940).
⊕ reference solubility from this work.
- - - Hayduk's graphical extrapolations.
— real behavior suggested by experimental points and this work.

Figure 6 : Solubilities in water

solubility is known at one temperature, the correlation predicts, often with a good accuracy, the solubility of another temperature ; second, if no solubility data are available, the correlation may be used to obtain a reasonable estimate (within a factor of 2). Simultaneously 289 literature data on the standard entropy of solution and 408 literature data on Henry's constant were correlated by De Ligny et al. |38| by regression analysis. The 20 gases molecules range in size from He to C_3H_8 and the 39 solvent molecules from CS_2 to polyethylene of molecular weight of 10^5 and the solvents range in polarity from hexane to methanol. Moreover for associating solutions it should be noted that a comprehensive and promising model has been presented by Tamir and Wisniak |39| that is suitable for the case of binary mixtures containing two associating species and that takes into account hetero-dimerization in the vapour phase (between two fatty acids for example).

Furthermore an interesting empirical correlation for non-polar gas solubilities in mixed aqueous alcohol solutions (such as methanol, ethanol, isopropanol) has been also presented by Tokunaga |40| in terms of an excess quantity of gas solubility defined as,

$$\ell nK = \ell n(Os)_m - \Phi \ell n(Os)_{alc} - (1-\Phi)\ell n(Os)_{wat} = \Phi(1-\Phi)(A+B\Phi+C\Phi^2+D\Phi^3)$$

where Φ is the alcohol volume fraction in liquid phase and $(Os)_m$, $(Os)_{alc}$, $(Os)_{wat}$ are the Ostwald coefficients for the solute gas in aqueous alcohol solution, alcohol and water. The empirical coefficients seem to depend only on the nature of the alcohol and temperature, regardless the kind of the solute gas (O_2, CO_2, N_2 and even CH_4).

It must be noted also the practical data on the industrial important solubility of mixtures of acidic gases in amines reported in references |97-99|. At last it is important to note that for several systems, there exist a linear relationship between log Os and solvent surface tension σ for a gas in a serie of solvents at a constant temperature |25|.

3.12 Electrolytes

The solubility of electrolyte solutions can be estimed by the empirical method of Van Krevelen and Hoftijzer |41| that relates the solubility in solution to that in pure solvent at the same temperature by means of the Sechenov expression,

$$\log_{10} \frac{He}{He_B} = hI \tag{16}$$

where He_B is the value in pure solvent and I is the ionic strenght of the solution,

$$I = \frac{1}{2}\sum c_i z_i^2$$

C_i is the concentration of ions of valency Z_i.

The salting coefficient h is the sum of contributions referring to the species of gas h_G and to the species of positive and negative ions present,

$$h = h_G + h_+ + h_-$$

Van Krevelen and Hoftijzer |41|, Barrett |42| and Onda et al. |43| have evaluated h for various species. The values are presented in Tables II and III.

It is assumed that the range in h with temperature is confined to the change in h_G and it seems that effect of pressure on h_G values is small (up to 200 atm |43|).

Moreover Tiepel and Gubbins |44| presented a theory for salt effects based on perturbation theory for mixtures and applied to the prediction of thermodynamic properties of gases dissolved in any electrolyte solutions. This theory satisfactorily predicts the heat of solution and the concentration and pressure dependence of activity coefficients of non-polar solutes when salting-out occurs i.e. when the molecules and ions are not very large, provided that chemical association between ions and solute does not occur. Unfortunately when the ions are large or when there are association forces, salting-in occurs and the theory gives poorest results. Anyway Tiepel and Gubbins have presented comprehensive and complementary results of those of Tables II and III for salting coefficients for CH_4, C_2H_6, SF_6 and C_6H_6 in LiCl, NaCl and KI.

In mixed electrolytes it may be assumed that the value of He is given by an expression of the form |7|

$$\log_{10} \frac{He}{He_B} = h_1 I_1 + h_2 I_2 + \dots \qquad (17)$$

where I_1 is the ionic strenght attributable to species n° 1 of electrolyte and h_1 has the value characteristic of that electrolyte. This additive rule for salting-out parameters has been confirmed by Onda et al. |43| for the absorption of CO_2, C_2H_4 and C_2H_2 in three mixed salts.

It is also important to note that the results of Tables II and III are valuable for industrial ionic strenght that are often much higher than 2gmoles/l.

An important application is the knowledge of the solubility of oxygen in electrolyte solutions, in aqueous solutions of organic compounds and in fermentation broths containing a variety of nutrients and metabolic products. For this purpose, Schumpe et al.

Table II : Values of h_+ and h_-

h_+ ℓ/g.mole				h_- ℓ/g.mole	
H^+	- 0.1110	Fe^{++}	- 0.0602	Cl^-	0.3416
Na^+	- 0.0183	Co^{++}	- 0.0534	Br^-	0.3310
K^+	- 0.0362	Ni^{++}	- 0.0520	I^-	0.3124
NH_4^+	- 0.0737	Cd^{++}	- 0.0062	NO_3^-	0.3230
Sr^+	- 0.0445	Cr^{+++}	- 0.0986	OH^-	0.3875
Li^+	- 0.0416			CNS^-	0.2612
Cs^+	- 0.0584			HSO_3^-	0.3869
Rb^+	- 0.0449			HS^-	0.3718
Mg^{++}	- 0.0568			HCO_3^-	0.4286
Zn^{++}	- 0.0590			CO_3^{--}	0.3754
Ca^{++}	- 0.0547			SO_4^{--}	0.3446
Ba^{++}	- 0.0473			SO_3^{--}	0.3275
Mn^{++}	- 0.0624			PO_4^{---}	0.3265
				$C_6H_5O^-$	0.4084
				MnO_4^-	0.2600

Table III : Values of h_G in ℓ/g.mole

T°C	H_2	O_2	CO_2	N_2O	H_2S	NH_3	C_2H_2	C_2H_4	SO_2
0	—	- 0.1653	- 0.2110	—	—	—	—	—	—
10	- 0.2170	—	—	- 0.2156	—	—	—	—	—
15	- 0.2197	- 0.1786	- 0.2222	- 0.2118	—	—	- 0.2124	- 0.2003	—
20	- 0.2132	- 0.1771	—	- 0.2128	—	—	—	—	—
25	- 0.2115	- 0.1892	- 0.2277	- 0.2141	- 0.2551	- 0.2394	- 0.2240	- 0.1951	- 0.3154
35	—	—	—	—	—	—	—	—	- 0.3122
40	—	—	- 0.2327	- 0.2179	—	—	—	—	—

T°C	N_2	He	Ne	Ar	Kr	NO
25	- 0.1904	- 0.2222	- 0.2240	- 0.1866	- 0.1762	-0.1825

|45| have regrouped all the literature data, including NH_4Cl, $CaCl_2$, K_2SO_4, $KHSO_4$, $MnSO_4$, Na_2HPO_4 and several mixtures of these salts in the temperature range 10 to 40°C, in equations (16) and (17) where h_G is not taken into account, i.e.,

$$\log_{10} \frac{He}{He_B} = \sum_i H_i I_i \tag{18}$$

with the H_i values at 25°C in the following Table. This formulation has given excellent agreement with the experimental data and should be recommended for solubility of oxygen in electrolyte solutions.

Ion Specific Constants for Salting-Out Oxygen at 25°C |45|

Cation	H_i (liter/g-ion)	Anion	H_i (liter/g-ion)
H^+	−0.774	F^-	0.982
Li^+	−0.664	Cl^-	0.844
Na^+	−0.550	Br^-	0.820
K^+	−0.596	I^-	0.821
Rb^+	−0.662	OH^-	0.941
Cs^+	−0.666	NO_3^-	0.802
NH_4^+	−0.720*	ClO_4^-	0.888
NEt_4^+	−0.912*	SO_4^{2-}	0.453
Mg^{2+}	−0.314*	CO_3^{2-}	0.485
Ca^{2+}	−0.303*	PO_4^{3-}	0.320
Ba^{2+}	−0.297*		
Mn^{2+}	−0.311*		

* Calculated from measured Sechenov constants with optimized H_i values of anions.

To end this abstract on solubility, it is important to remember that the properties of non ideal and ideal solutions are appreciably different when the gas reacts in solution. In such case precedent equations (i.e. Henry's law) may be still applied to the concentration of the unreacted solute, determined from the equilibria of the reactions taking place in the solution.

3.2 Diffusivity in liquids

In contrast with the situation for gases where the correlations are well established |28|, there are no satisfactory theoretical methods of predicting diffusivities in liquid systems. Different approaches have been made depending whether the solutions are electrolytic or nonelectrolytic. Most studies have been devoted to the estimation of diffusion coefficients in very dilute solutions although some papers have published about the substantial variations that happen with increasing concentration of the diffusing solute.

The theories and the experimental methods available for estimating diffusivities in liquid are well reviewed by Kamal and Canjar |46| Nienow |47|, Bretsznajder |48|, Dullien et al. |49, 50| and Simons and Ponter |51|.

3.21 Nonelectrolytes

Several correlations for dilute solutions are presented in Table IV with the average errors given by Reid, Prausnitz and Sherwood |28|, Skelland |52| and the cited authors.

In this table, A is the solute (dissolved gas or reactant), B is the solvent. D_A is the diffusion coefficient in cm^2/s, μ_{AB} and μ_B are the solution and the solvent viscosities in cP., T is the absolute temperature in K, M_B is the molecular weight of the solvent, V_{mA} and V_{mB} are the solute and solvent molecular volumes at the normal boiling point in cm^3/g.mole. The molecular volumes of some simple substances are given in Table V. Molecular volumes of complex substances are estimated by adding up the contributions of the atoms in the molecule according to Table VI, and substracting according to Table VII in the case of ring-compounds.

For the estimation of diffusivity, one of the best known and most convenient relationship to use is that of Wilke and Chang |53| which is based on the Stokes Einstein equation. In this formula the associated parameter "x" allows for differences in solvent behaviour : x = 2.6 for water, x = 1.9 for methanol, x = 1.5 for ethanol, x = 1.0 for benzene, ether, heptane and other unassociated solvents. The average error of the prediction, for the systems surveyed by the authors, was about 10 % and the relationship cannot be used when a complex is formed between solute and solvent. It is interesting to note that Akgerman and Gainer |59| have reported that for amylalcohol, isobutyl alcohol, ethyleneglycol and glycerol, the associated parameter can be approximated using,

$$x = \left(\frac{\Delta E_{vap} \text{ of hydrogen-bonded substance}}{\Delta E_{vap} \text{ of homolog}} \right)^{0.6}$$

The homolog is defined by substituting a "$-CH_3$" group for the "$-OH$" group.

The uncertainty involved in assigning values to x for new solvents not covered in the original work has resulted in efforts to eliminate this factor from the formula. Scheibel |54| and Lusis and Ratcliff |57| proposed equations in which the associated parameter was eliminated by the introduction of the ratio of the molar volumes of the solvent and the solute. Sirataman and al. |55| and Reddy and Doraiswamy |56| replaced it by introducing functions of the latent heats of vaporization at their normal boiling temperatures (in cal/gram) or of the molar volumes. In the meantime, recommendations on the basis of Table IV would be to use, especially for the liquid reactant diffusivity, Eqs. (2) or (4) (when applicable) or (6) for diffusion in organic solvents |52| and Eq. (7) of this table when water is the solvent. Note also for a

Table IV : Relationships for diffusivity in very dilute binary solutions of non-electrolytes

Equation Number	Equation	Average error % Organic solvents	water as solvent	water as solute	Overall (by authors)	Reference
1	$\frac{D_A \mu_{AB}}{T} = 7.4 \times 10^{-8} \frac{(x M_B)^{1/2}}{V_{mA}^{0.6}}$	27	11	up to 250	10	Wilke and Chang [53]
2	$\frac{D_A \mu_{AB}}{T} = 8.2 \times 10^{-8} \left[\frac{1 + \left(\frac{3 V_{mB}}{V_{mA}}\right)^{2/3}}{V_{mA}^{1/3}}\right]$	25	11	up to 250	-	Scheibel [54]
3	$\frac{D_A \mu_B}{T} = 5.4 \times 10^{-8} \left(\frac{M_B^{1/2} \Delta E_B^{1/3}}{V_{mA}^{0.5} \Delta E_A^{0.3}}\right)^{0.93} \times \left(\frac{T}{\mu_B}\right)^{0.07}$	26	12	12	13	Sirataman and al.[55]
4	$\frac{D_A \mu_B}{T} = 10 \times 10^{-8} \frac{M_B^{1/2}}{(V_{mA} \cdot V_{mB})^{1/3}}$ for $\frac{V_{mB}}{V_{mA}} < 1.5$	15	9	-	13.5	Reddy and Doraiswamy [56]
5	$\frac{D_A \mu_B}{T} = 8.5 \times 10^{-8} \frac{M_B^{1/2}}{(V_{mA} \cdot V_{mB})^{1/3}}$ for $\frac{V_{mA}}{V_{mB}} > 1.5$	18	-	26	18	Reddy and Doraiswamy [56].
6	$\frac{D_A \mu_B}{T} = 8.5 \times 10^{-8} \frac{\left[1.40 \left(\frac{V_{mB}}{V_{mA}}\right)^{1/3} + \frac{V_{mB}}{V_{mA}}\right]}{V_{mB}^{1/3}}$	16	-	-	-	Lusis and Ratcliff [57]
7	$D_A = 14 \times 10^{-5} \mu_{wT}^{-1.1} \times V_{mA}^{-0.6}$	-	-	-	5	Othmer and Thakar [58]

Table V : Molecular volume V_m of simple substances, in cm^3/g.mole [26]

Air	29.9	COS	51.5	N_2	31.2
Br_2	53.2	H_2	14.3	NH_3	25.8
Cl_2	48.4	H_2O	18.9	NO	23.6
CO	30.7	H_2S	32.9	N_2O	36.4
CO_2	34	He	16	O_2	25.6
		I_2	71.5	SO_2	44.8

Table VI: Contributions to molecular volumes of complex substances, in cm^3/g.mole [26]

As	30.5	F	8.7	P	27.0	Sn	42.3
Bi	48.0	Ge	34.5	Pb	48.3	Ti	35.7
Br	27.0	H	3.7	S	25.6	V	32.0
[illegible]	14.8	Hg	19.0	Sb	34.2	Zn	20.4
Cr	27.4	I	37.0	Si	32.0		

[illegible], terminal	21.6	N tertiary amine	10.8
Cl, medial	24.6	O	7.4
N double bonded	15.6	O methyl esters and ethers	9.1
N triple bonded	16.2	O ethyl esters and ethers	9.9
N primary amine	10.5	O higher esters and ethers	11.0
N secondary amine	12	O acids	12.0
		O joined to S, P, N	8.3

Table VII Deduction from molecular volumes for rings, in cm^3/g.mole [26]

3-membered ring	6.0
4-membered ring	8.5
5-membered ring	11.5
6-membered ring	15.
Naphtalene ring	30.
Anthracene ring	47.5

Moreover the diffusivity D_A^m for dilute solute A in mixed solvents B and C may be determined by the equation of Leffler and Cullinan |61|,

$$D_A^m \mu_{ABC} = \left(D_{AB}\mu_B\right)^{X_B} \times \left(D_{AC}\mu_C\right)^{X_C}$$

where μ_{ABC} is the viscosity of the solvent mixture B and C, D_{AB} and D_{AC} are the binary diffusivities in very dilute solutions estimated by one of the correlations presented in Table IV and X_B, X_C are the mole fractions of B and C.

In thermodynamically non-ideal solutions, the effect of concentration on diffusion coefficient $(D_A)_{conc}$ is estimated as a first approximation in terms of the activity coefficient of the solute γ_A, the binary diffusivities in very dilute solutions and the mole fractions X_A, X_B of A and B |138|

$$(D_A)_{conc} = \left(D_A\right)^{X_B} \left(D_B\right)^{X_A} \left(1 + \frac{d\ln\gamma_A}{d\ln X_A}\right)$$

D_B is here the molecular diffusivity in very dilute solution of species B in A. This relationship has been modified by Leffler and Cullinan |61| to take into consideration the viscosity of the solution and the pure components and its application will be suggested in the rare opportunity where the dilute gas in the liquid cannot be considered as a very dilute phase.

It is important to note also that the effect of pressure may be usually neglected.

As last before to use one of the above relationships, including or not the effect of concentration, it will be judicious to check first whether the studied gas-liquid system is included in the tabulation of experimental results of diffusivity prepared by Johnson and Babb |62| or in scattered tables presented by Ponter et al. |63, 64|, Tyn |65|, Akgerman and Gainer |59|, Perry |26| and De Kee and Laudie |66|.

3.22 Electrolytes

Diffusion of a dissolved ionized electrolyte involves the diffusion of both cations and anions which, because of their smaller size, diffuse more rapidly than the undissociates molecules. Both types of ions diffuse at the same rate so that the electrical neutrality of any solution is preserved.

On the assumption of complete dissociation in a solution containing two species of ions, the diffusion coefficients of electrolytes can be predicted very accurately at infinite dilution using the equation,

$$D_A = 8.931\times10^{-10}\ T\left(\frac{\lambda_+^\circ\lambda_-^\circ}{\Lambda^\circ}\right)\left(\frac{Z_+ + Z_-}{Z_+Z_-}\right)$$

where D_A = combined diffusivity of electrolyte, cm^2/s,
Z_-, Z_+ = absolute values of anion and cation valence,
$\Lambda^\circ = \lambda_+^\circ + \lambda_-^\circ$ = electrolyte conductance at infinite dilution, mho×cm²/g×equivalent,
λ_-°, λ_+° = anionic and cationic conductances at infinite dilution, mho×cm²/g×equivalent,
T = temperature, K.

Useful tabulations of ionic conductances at infinite dilution in water are given in Perry |26| at 25°C and in Bretsznajder |48| at three other temperatures (0 ; 18 and 100 °C).

Diffusion coefficients at higher concentrations (up to 2N) may be evaluated from the relationship proposed by Gordon |67|.

The general case where the solution contains more than two species of ions has been considered by Vinograd and Mc Bain |68|. Equations have been proposed for both ions, which have been used judiciously by several authors for irreversible second-order reaction, acetic acid vapour - NaOH system |69|, for irreversible instantaneous reaction, HCl - NaOH system |70, 7| and CO_2 - solutions of KOH, NaOH and LiOH in which the reaction between CO_2 and OH^- is either irreversible rapid pseudo first-order or instantaneous depending on the gas-liquid contactor used (laminar jet or falling film) |71|. It is seen that the enhancement factors calculated with molecule diffusion are much lower than those calculated on the more realistic basis of ion diffusion.

Nevertheless in most investigations where a gas absorption takes place into a liquid containing several species of ion, the investigators have ignored the complications introduced by differences in ionic diffusion coefficient and, as it has just been outlined, the results may depart significantly from those obtained in using equations concerning unioniezed molecules. A comprehensive tabulation of experimentally measured diffusivities for electrolytes has been presented by Robinson and Stokes |72|. Relationships and reserves presented above are of course to be used only in the absence of experimental data.

But in fact, for quick and practical calculations, it is convenient to use the formula of Wilke and Chang |53| which is sufficiently simple and which has been verified on many systems either electrolyte or nonelectrolyte. For instance, comparing with experimental results of Wise and Houghton |73| obtained for the diffusion coefficients of ten slightly soluble gases (H_2, O_2, N_2, Air, He, A, CH_4, C_2H_6, C_3H_8, n-C_4H_{10}) in water at 10-60°C, Shrier |74| has shown that the use of the Wilke and Chang's formula gives predicted values (with the exception of H_2 and He) which are adequate for many practical applications. More generally an examination of the literature data for diffusivity in aqueous solutions at 25°C led Hayduk and Laudie |75| to suggest that a revised association parameter for water x = 2.26 instead of 2.6, still improves the accuracy of the equation of Wilke and Chang. So it is only for small solute gas molecule such as H_2 or He or for viscous (μ_B > 3 cP) or highly non-ideal system (such as gas into aqueous alcohol mixture) that the theory developed by Akgerman and Gainer |59| taking into account molecular interaction as well as solvent viscosity is recommended. Recently Mazarei and Sandall |76| have measured then reviewed the literature data on diffusion coefficients for helium and hydrogen in water at 25°C and the following mean values are respectively suggested, $D_A = 4.8\times10^{-5}$ cm^2/s for H_2 and $D_A = 6.5\times10^{-5}$ cm^2/s for He.

In the case of the absorption of CO_2 which is often used to determine interfacial parameters in gas-liquid reactors, it is interesting to note that Ratcliff and Holdcroft |77| found that the diffusivities in aqueous solutions of chlorides, nitrates and sulphates of sodium and magnesium vary roughly as $\mu_B^{-0.64}$ rather than as μ^{-1}, as in the Wilke and Chang formula, but these results were obtained for small variations of liquid viscosities ($0.9<\mu_B<1.7$ cP). The same conclusions apply to results of Nijsing and al. |71| and Danckwerts and Alper |78| that found the diffusivity of CO_2 in sulphates of sodium and magnesium or in sodium carbonate to vary as μ_B at the power (-0.85) to (-0.82). More specifically, Yasunishi and Hoshida |79| measured molecular diffusivities of CO_2 in aqueous solutions of 15 electrolytes and correlate the data in terms of $D_A\mu_B^n$ = constant. All the monovalent electrolytes studied gave a value of n = 1 and with electrolytes containing one or two divalent ions, a reasonable correlation was obtained with n = 0.5 but only with measurements at 25°C and liquid viscosities up to 7 cP.

All these difficulties to know either the solubility or the diffusivity of the gas with a good accuracy are avoided with the use of the laboratory models such as laminar jet or wetted wall column presented in next section for the determination of the interfacial parameters or the chemical kinetics. Indeed only a mathematical combination of solubility and diffusivity such as $C_A^*\sqrt{D_A}$ is necessitated with a high precision and not the separate values of C_A^* and D_A. But for industrial design, the simultaneous knowledge of these two parameters is often necessary, which requires the use of above correlations or literature data that are unfortunately very often determined in separate operating conditions or from different experimental studies.

4. GAS-LIQUID MASS TRANSFER PARAMETERS. MEASUREMENT AND SOME DATA

The gas holdup, the interfacial area and the mass transfer coefficient are the main variables determining the mass transfer rates in gas-liquid contacting device. The methods used to measure these parameters can be classified into two categories : local measurements with physical techniques such as light scattering and reflection techniques, photographical and electrical and electrochemical techniques, and global measurements with chemical techniques. Each method has its advantages and its drawbacks. Let us describe and comment them and give some data with a major emphasis on trickle-bed reactors and mechanically stirred gas-liquid reactors.

4.1 Physical techniques

These methods are mainly devoted to the measurement of gas holdup, bubble size and specific surface area in gas-liquid dispersions usually encountered in bubble columns, plate columns, mechanically agitated tank and spray towers. Any two of these interfacial parameters are sufficient to define all three since they are interrelated as,

$$a' = \frac{6\alpha}{d_{SM}} \quad \text{and} \quad d_{SM} = \frac{\sum_i n_i d_b^3}{\sum_i n_i d_b^2} \tag{18}$$

where α is the gas holdup and d_{SM} is the volume-surface mean diameter or Sauter mean diameter. d_b is the diameter of a single bubble or drop and n_i is the number of bubbles or drops of diameter d_b.

4.11 Gas holdup.

The gas holdup α is determined by directly measuring the height of aerated liquid Z_a and that of the clear liquid without aeration, Z. The average gas holdup is then evaluated from the relation,

$$\alpha = \frac{Z_a - Z}{Z_a} \tag{19}$$

This method, most often used for plate columns, vertical bubble column and for mechanically agitated tank |2| is not time consuming but is not very accurate (15-20 % accuracy) especially when waves or foams are occuring on the top of the dispersion.

An alternate and more accurate manometric technique has been used by Reith et al. |80| and Burgess and Calderbank |81| where the gas holdup in the dispersion is computed from measurements of the clear liquid height in the dispersion at successive manometer tappings on the side of the froth container. Linek and Mayrhoferova |82| used an electrical technique to measure the dispersion height. The method is based on measuring the surface elevation at certain selected points by means of an electrically conductive tip. The height is determined by the vertical position of the tip at which the sum of contact times equals one half of the measurement period. The accuracy of the measured value of the total surface elevation is claimed by the authors to be ± 0.2 mm, and the gas holdup is then calculated from the total surface elevation and the cross section of the reactor.

The gamma-ray transmission technique previously employed by Vermeulen et al. |83| has been used to determine the point gas holdup in mechanically agitated tank by Calderbank, in plate columns and in packed columns |2|. The principles of the application of gamma-ray absorption to holdup measurements depend on the use of the relationship,

$$\mathrm{Ln}\,\frac{I_o}{I} = \lambda \rho s \qquad (20)$$

I_o/I is the intensity ratio between incident beam of radiation (I_o) and trasmitted beam (I), λ is the mass absorption coefficient, values of which for most atoms have been published |84|, ρ is the density to be measured (simply related to gas holdup) and s is the thickness of the absorbing medium.

Thus Calderbank |85| used a cesium 137 source in conjunction with a scintillation counter and scaler. Traverses of the reactor were made, and readings taken of the gamma-radiations transmitted through the empty reactor, the reactor filled with liquid and the reactor containing the dispersion. The point gas holdup was then calculated by,

$$\alpha = \frac{\log t_2/t_1}{\log t_o/t_1} \qquad (21)$$

where t_o, t_1 and t_2 are the times for a fixed number of counts respectively for the reactor empty, fulled with liquid and containing the dispersion.

4.12 Bubble size.

The Sauter mean particle size of dispersion, Eq. (18) is evaluated directly by a statistical analysis of photomicrographs or high speed flash photographs when the dispersion is dynamically maintained. Photographs are taken through the wall of the transparent reactor or in the interior of the reactor with the aid of an intrascope.

To avoid any wall effect or perturbation effect that may occur with these methods, a sampling apparatus for the photographic technique has been proposed by Kawecki et al. |86|. Bubbles are extracted from the tank containing the dispersion by means of a tube connected to a small square-section column through which a continuous flow of liquid and bubbles rises. The flowrate is choosen high enough so that the small differences in the free rise velocity of the bubbles do not affect the mean residence time of the bubbles in the column. The bubbles in the column are then photographed and their diameters are determined from enlargements of the negatives projected on a screen. An alternative sampling method has been used by Todtenhauft |87| where bubbles extracted from the dispersion are photographed through a calibrated capillary tube.

It should be noted that the photographic technique for measurement of bubble size most often gives local values, does not take into account large bubbles and is applicable only in the case of small gas holdup, that is usually a low gas velocity.

4.13 Interfacial area.

The local interfacial contact area is determined directly by the light transmission and reflection techniques.

In the light transmission technique, a parallel beam of light is passed through the dispersion and a photocell is placed at a large distance from it. Light scattered by the bubbles passes outside the photocell and is lost while that part of the incident parallel beam which passes through the dispersion without meeting any obstacle is recorded by the photocell placed at the extremity of an internally blackened tube. Calderbank |85| showed that for scattering bubbles, which are large in comparison with the wavelength of light, the scattering cross-section is equal to its projected area. Furthermore, the total interfacial area per unit volume of the dispersion equal four times the projected area per unit volume, thus giving the following equation,

$$\ln \frac{I_o}{I} = \frac{a'L}{4} = \ln \frac{t}{t_o} \qquad (22)$$

where L is the optical path length. By connecting the photocell to a light quantity meter and electric timer, it is possible to measure the time for a given quantity of light to be received by the photocell when the light passes through the liquid (t_o) and through the dispersion (t). This technique holds for values of a'L < 25 when multiple scattering is negligible and for bubble diameter larger than 50 μ. Landau et al. |88| extended this technique to conditions when the light source and photocell are placed outside the column and multiple scattering is taken into account. An interesting empirical correlation based on anisotropic scattering in all six mutually perpendicular directions is given,

$$\frac{a'L}{4} = \frac{\ln(I_o/I)}{I-\Phi}$$

with $\Phi = 1 - 6.59 \left(\frac{1-\exp(-0.233\tau)}{\tau}\right)$

which is valid for a'L up to 100. This represents a fourfold increase of the range of the applicability of the light attenuation technique which is not time consuming. The range of interfacial area is up to 8 cm^{-1} with fractions of light transmitted less than 0.02 and values of a' are approximating by 5 % those obtained simultaneously by the photographic method.

With optically dense dispersion such as those occuring in plate columns, the transmission method may fail because of intense multiple scattering. In such cases a reflectivity probe may be used |89| where the optical reflectivity or backward scattered light from dispersion is measured. The specific interfacial area is calculated by,

$$\frac{R_\infty}{R} - 1 = \frac{46.5}{a'} \tag{23}$$

where R is the intensity of light reflected back by the dispersion, and R_∞ is the intensity of light reflected back when a' is infinite (as obtained by extrapolation of a plot of 1/R against 1/a' to 1/a' = 0). It was found that for a given dispersion R_∞ is a function of the refractive index ratio.

4.14 Comments on physical methods. Need of chemical methods as a complementary and not an alternative tool.

The specific contact surface area necessitated for mass transfer in a gas-liquid dispersion or more generally in any type of gas-liquid reactor is defined as the interfacial contact area of all the bubbles or drops or phase element such as films or rivulets within a volume element divided by the volume of that element.

It is necessary to distinguish between the over-all specific contact area S for the whole reactor with volume V_R and the local specific contact area S_i for a small volume element ΔV_i. In practise ΔV_i is directly analysed by the physical methods. The main difficulty in the determination of the over-all specific contact area from the local specific contact area is that S_i depends strongly on the position of ΔV_i in the reactor which is a consequence of a variation in the local gas holdup and/or the local Sauter mean diameter as seen in Eq. (18). So there is a need for a direct determination of the over-all interfacial contact area over the whole reactor which is possible with the use of the chemical technique. But it can already be stated that this technique that allows for over-all values cannot be used without certain restrictions that arise from the results observed with the physical methods. For example the chemical method can hardly be used with certain restrictions for fast-coalescing systems since the presence of a chemical compound will probably reduce the coalescence rates and in fast-coalescing systems, as it is observed with the physical methods, the value of the specific contact area may depend strongly on the position in the reactor, which reduces the meaning of an average value obtained with the chemical methods. In fact both physical and chemical techniques should be simultaneously necessary to qualify the phenomenae that occur in gas-liquid reactor. While the chemical methods offer over-all values of interfacial area that are practical and immediatly available for design, a complementary information is however desirable to follow the variations of the local interfacial parameters such as α, d_{SM} within the reactor, which is very important for the knowledge of the trends to follow in the scaling-up design. But this complementary information that is only possible with the physical methods, should be obtained from local simultaneous measurements of two of the three interfacial parameters and not only from one single local measurement of d_{SM} or α. In order to gather these complementary informations simultaneously, the electroresistivity probe technique, proposed by Burgess and Calderbank |81| for the measurement of bubble properties in bubble dispersion is very promising.

A three dimensional resistivity probe with five channels was designed in order to sence the bubble local interface approache angle as well as measure bubble size and velocities in sieve tray froths. Moreover the probe system accepts only those bubbles whose central axes are coincident with the vertical probe axis, this discrimination function being achieved with the aid of an on-line computer, receiving signals from the five channels communicating with the probe array. Gas holdup, gas flow specific interfacial area and even gas and liquid-side mass transfer efficiencies have been calculated directly from the local measured distributions of bubbles size and velocity. The derived values of the dispersion parameter for air-water system have been found in excellent agreement with independently observed and previously published data. This promising method has revealed an interesting result : the magnitude of the interfacial areas reported compare very favourably with those computed from the chemical technique but are lower than those measured using photography. Whereas photography through the container wall appears to truncate data above equivalent diameters of about 15 mm, the probe used by Burgess and Calderbank only deletes data below a diameter of 4 mm. So the obvious difference between the interfacial values measured by chemical technique and photographic technique may be due to the presence of the small number fraction of large bubbles that dominate the interfacial area parameters of an assembly of small and large bubbles such as that met in sieve tray, and to their apparent inadvertent omission by photography owing to observational effects, as confirmed later by Kurten and Zehner in mechanically agitated reactor |90|.

This is a good example of the limitations of the physical methods when they provide one single information within a local volume (d_{SM}) while the other information α, is obtained for the global volume of the reactor.

So before to present now the chemical methods that are time consuming, it is important to insist and to keep in mind that a perfect knowledge of the mass transfer phenomenae that occur inside a gas liquid reactor necessitates in fact simultaneous local physical and global chemical techniques whereas the chemical methods offer ex abrupto global values without any indication on the in situ spatial variations. Ignoring it may be misleading for reactors that present important inhomogeneity |91|.

4.2 Chemical techniques

The chemical techniques to determine gas-liquid interfacial area and mass transfer coefficient have been intensively developed for the 15 last years. The principles of these techniques are the measurement of the absorption rates where an absorbed gas undergoes chemical reaction with precisely known kinetics. A gas A is absorbed into a liquid and there undergoes a reaction with a dissolved reactant B,

$$A + zB \xrightarrow{k} \text{Products}$$

By suitably choosing the solubility, the concentration of the reactant and the rate of reaction, either the mass transfer coefficients, or the interfacial area or both groups of parameters can be deduced from the over-all rate of absorption |92|. Generally, but not always, a steady flow of each phase through the reactor is assumed.

4.21 Determination of k_La.
If the resistance to transfer of component A is entirely in the liquid phase ($k_La << He k_G a$), the global rate of absorption, per unit volume of a gas-liquid system, is, in the absence of chemical reaction,

$$\Phi = \phi \cdot a = k_L a(C_A^* - C_{Ao}) \quad (24)$$

where C_A^* and C_{Ao} are the concentration respectively at the interface and in the bulk of liquid.

The value of k_La can in certain circumstances be determined by purely physical experiments in the reactor. For instance, k_La may be evaluated from the observed total rate of absorption in the case of a piston-like countercurrent flow of the two-phases where C_A^* is a known function of C_{Ao} or when the gas and the liquid are well stirred so that C_A^* and C_{Ao} are the same at all points. However there are two possible difficulties about the determination of k_La by the physical absorption method. One is that the flow-patterns and residence time distribution of the phases may be undetermined (C_{Ao} is not a known function of C_A^*) and hence the value of k_La cannot be deduced from Φ. The other is that in efficient contacting devices the gas and liquid may approach equilibrium quite closely. As the determination of k_La depends on the difference between the actual and the equilibrium extent and this necessitates an extremely accurate measurement of the flowrates, this method may become impracticable.

4.22 Determination of k_La with an irreversible slow reaction.
The problem raised by the saturation of the liquid can be avoided by using a solution which reacts with the dissolved gas in the slow reaction regime i.e. the reaction is too slow to affect the rate of absorption directly, but, on the other hand, the reaction is fast enough to reduce the bulk concentration of dissolved gas effectively to zero. If the considered reaction is irreversible and second-order (first order with respect to both components A and B) this leads to,

$$\Phi = \phi \cdot a = k_L a C_A^* = R_A = k_2 C_A^* C_{Bo} \beta \quad (24)$$

The conditions to be satisfied are,

$$k_L a << \beta k_2 C_{Bo} \quad (25a)$$

$$\text{and} \quad \frac{aD_A}{\beta k_L} << 1 \quad \text{and} \quad \frac{D_A k_2 C_{Bo}}{k_L^2} = Ha^2 << 1 \quad (26a)$$

Thus the rate of absorption Φ is the same at all points in the reactor and the residence time distribution of the liquid is irrelevant.

It is possible similarly to make use of reaction which is m,nth-order, the local rate being thus $k_{mn}C_A^m C_B^n$. The condition for $C_{Ao}=0$ to be zero is,

$$k_L a << \beta k_{mn} C_A^{*m-1} C_{Bo}^n \tag{25b}$$

and for no reaction in the film

$$Ha^2 = \frac{2}{m+1} \frac{D_A k_{mn} C_A^{*m-1} C_{Bo}^n}{k_L^2} << 1 \tag{26b}$$

It is not always easy to satisfy both conditions (25a or 25b) and (26a or 26b) simultaneously ; thus, if $k_{mn}C_{Bo}^n$ is made large enough to satisfy condition (25b) (i.e. $C_{Ao} = 0$) it may become too large for condition (26b) to be satisfied (no reaction in the film). When condition (26b) is satisfied but condition (25b) is not, Eq. (24) may be written,

$$\Phi = \phi \cdot a = k_R a C_A^*$$

with $\dfrac{1}{k_R a} = \dfrac{1}{k_L a} + \dfrac{1}{k_{mn} C_{Ao}^{m-1} C_{Bo} \beta}$

If $k_{mn}C_{Bo}^m$ is varied, keeping $k_L a$ constant, a plot of $1/k_R a$ against $1/k_{mn}C_{Bo}^n$ will be a straight line of slope $1/C_{Ao}^{m-1}\beta$ with intercept $1/k_L a$. This offers a method of determining $k_L a$ when it is not possible to satisfy condition (25b).

Some suitable chemical systems for the determination of $k_L a$ in the slow reaction regime are presented in Table VIIIa |2|.

4.23 Determination of $k_L a$ with an irreversible instantaneous chemical reaction.

There is another and quite different method whereby a chemically reacting system can be used to determine $k_L a$. This consists in the use of an irreversible instantaneous reaction where the rate of absorption is

$$\Phi = \phi \cdot a = k_L a C_A^* E_i = k_L a C_A^* \left(1 + \frac{D_B}{zD_A} \frac{C_{Bo}}{C_A^*}\right) \tag{27}$$

The condition to be satisfied is Ha > 10E_i. If in addition $C_{Bo} << C_A^*$, then the rate of absorption is,

$$\Phi = \phi \cdot a = k_L a \frac{C_{Bo}}{z} \frac{D_B}{D_A}$$

The rate of absorption is independent of the concentration of component A in the gas phase. It is also independent of the residence time distribution in the gas phase. In practise it is found that the use of Eq. (27) is available when E_i > 4. Some suitable chemical systems for the determination of $k_L a$ in the instantaneous regime are presented in Table VIIIb.

4.24 Determination of a.

When the reaction between the components A and B in the liquid phase is mth-order in A and nth-order in B, under certain circumstances the concentration of component B is the same everywhere as it is in the bulk of the solution ($k_{mn} C_{Bo}^n$ is constant). The reaction is said rapid pseudo-mth order in A (Fig. 4, D). The conditions for this situation are 3 < Ha << E_i and the rate of absorption is expressed by,

$$\Phi = \phi \cdot a = E k_L a C_A^* = Ha k_L a C_A^* = a \left(\frac{2}{m+1} k_{mn} D_A C_A^{*m+1} C_{Bo}^n \right)^{1/2} \qquad (28)$$

Thus, the rate of absorption is independent of k_L, that is, of the hydrodynamic conditions. Thus, provided that the average specific rate of absorption,

$$\phi = \left(\frac{2}{m+1} k_{mn} D_A C_A^{*m+1} C_{Bo}^n \right)^{1/2} \qquad (29)$$

is known and C_A^* and C_{Bo} have effectively the same value in all parts of the system, the specific interfacial area is determined directly from the measurement of the rate of absorption Φ.

It is not always necessary to know the kinetics of the reaction in order to determine ϕ . Indeed the values of ϕ may be measured by absorbing the component A into the same solution in some laboratory apparatus with a known interfacial area, such as laminar jet, wetted wall column, stirred cell... The agitation speed or the flowrate should be varied to confirm that ϕ is really independent of k_L and β. In reactors when the residence time distribution of the gas phase is unknown, it is sometimes possible to choose conditions such that there is practically no change in the partial pressure of the gas by using a high gas flowrate and keeping the value of ϕ low. Moreover if component A is diluted by a carrier-gas it must be confirmed that the gas-side resistance is negligible. Some suitable aqueous, organic and viscous chemical systems for

Table VIIIa Chemical systems used for the determination of k_La in slow reaction regime. |2|

Solute gas A	Reactant B	Catalyst in the absorbent
CO_2	K_2CO_3+$HKCO_3$	NaClO
O_2 diluted with air	CuCl	
O_2 diluted with air	Na_2SO_3	$CuSO_4$
O_2 diluted with air	Na_2SO_3	$CoSO_4$
O_2	Glucose	enzyme glucose oxidase

Table VIIIb: Chemical systems used for the determination of k_La in instantaneous reaction regime. |2|

Solute gas A	Reactant B
NH_3	H_2SO_4
SO_2, Cl_2, HCl	NaOH
H_2S HCl CO_2	Amines
O_2 diluted with air	$Na_2S_2O_4$

Table IX : Chemical systems used for the determination of a in rapid pseudo-mth regime. |2|

Solute gas A	Reactant B	Catalyst
CO_2 diluted with air	Na_2CO_3+$HNaCO_3$ K_2CO_3+$HKCO_3$	As $(OH)_2$ O^- ClO^-
	Aqueous or organic solutions of amine	
	LiOH - NaOH KOH - $Ba(OH)_2$	
	Na_2S	
COS diluted with air	Aqueous solutions of amine	
O_2 in air	$Na_2S_2O_3$	
O_2 diluted with air	Na_2SO_3	Co SO_4 Cu SO_4
Isobutylene in C_4 Fraction or air	H_2SO_4	
CO_2 diluted with air	Cyclohexylamine in toluene or xylene containing 10 % Isopropanolamine	
	Cyclohexylamine in cyclohexanol	
	Monoethanolamine in aqueous di or polyethylene glycol	
CO_2 and O_2 diluted with air	Aqueous cuprousamine complex solution	
O_2 diluted with air or N_2	Propionaldehyde	Manganese propionate
O_2 in air	C_{10} trialkylaluminium dissolved in organic solvents	
Cl_2	p-cresol dissolved in dichlorobenzene	
H_2	Edible oil	Ziegler-Natta
Desorption of Isoamylene into N_2	n-Heptane Toluene	

Table X : Chemical systems used for the determination of k_Ga in instantaneous and surface regime |2|

Solute gas A	Insoluble gas diluent	Reactant B
SO_2 or Cl_2	Air Freon 12 Freon 22 Freon 114	NaOH
NH_3	Air Freon 12 Freon 22 Freon 114	H_2SO_4
Triethylamine	Air Freon 12 Freon 22 Freon 114	H_2SO_4
I_2	Air	NaOH
SO_2	Air	Na_2SO_3
Propylene CO	air	Cuprousamine complex solution

Table XI : Chemical systems for the determination of k_L and a with the use of Danckwerts plot. |2|

Solute gas A	Reactant B	Catalyst
CO_2 diluted with air	$HNaCO_3+Na_2CO_3$	Arsenite
CO_2 diluted with air	$HNaCO_3+Na_2CO_3$	Hypochlorite
O_2 in air	CuCl	
O_2 in air	Na_2SO_3	$CoSO_4$

the determination of a in the pseudo-mth regime are presented in Table IX |2|.

4.25 Determination of k_Ga.
If the resistance to transfer of component A is entirely in the gas phase ($Hek_Ga << k_La$), the rate of absorption, per unit volume of a gas-liquid system is, in the absence of a chemical reaction,

$$\Phi = \phi \cdot a = k_G a(p-p_i) \tag{30}$$

where p and p_i are the pressure respectively in the bulk of gas and at the interface.

The value of k_Ga can in certain circumstances be determined by purely physical experiments in the reactor, for instance in the case of piston-like countercurrent gas-liquid flow or when both phases are well-stirred. However, the rate of absorption depends on the residence time distribution in both phases that may be undetermined and in addition there is normally an appreciable resistance on the liquid side that must be taken into account. Thus the liquid side resistance can be eliminated and the rate of absorption can be made independent of the liquid side residence time distribution by using a solution which reacts instantaneously and irreversibly with the dissolved gas so that there is no back-pressure. Therefore

$$\Phi = \phi \cdot a = k_G a(p-p_i) = k_L a C_A^* \left(1 + \frac{D_B}{zD_A}\frac{C_{Bo}}{C_A^*}\right) = k_L a C_A^* E_i \tag{31}$$

The condition to satisfy is $Ha > 10E_i$ and in this case, the dissolved gas A reacts instantaneously with the component B in a plane beneath the interface where the concentration of both components is zero (Fig. 4, A). If the reaction plane is now at the interface (surface reaction, $p_i = C_A^* = 0$), it may be written,

$$p_i = HeC_A^* = \frac{He\left(k_G ap - k_L a \frac{D_B}{zD_A} C_{Bo}\right)}{Hek_G + k_L} = 0 \tag{32}$$

where $\Phi = k_G ap = k_L a \dfrac{D_B}{zD_A} C_{Bo}$

This is the situation where the dissolved component B reaches the interface by diffusion through the liquid as fast as gaseous component A reaches it by diffusion through the gas and the transfer process is controlled by diffusion in both phases. Moreover if at all points in the reactor the following condition is satisfied,

$$k_G ap - k_L a \frac{D_B}{zD_A} C_{Bo} < 0 \tag{33}$$

the rate of absorption remains equal to $k_G p$ but the interfacial concentration of component B is greater than zero (Fig. 4, B). Thus provided Eq. (33), the absorption process is entirely controlled by the transport of component A across the gas-film and,

$$\Phi = k_G ap \tag{34}$$

This forms the basis of methods for measuring $k_G a$. Indeed if the entrance & exit low partial pressure of the soluble gas p_e and p_s are measured for the reactor with a lenght H, $k_G a$ is calculated by

$$k_G a = \frac{G_m}{PH} \ln \frac{p_e}{p_s}$$

where G_m is the superficial molar mass flowrate of the insoluble gas (in gmole/cm^2 s) and P is the total pressure (in atm).

It is important to note that in all cases the calculation of $k_G a$ should be based on analysis of the gas stream as a small error in the analysis of the liquid stream can lead to large errors in the calculated value of $k_G a$. Some suitable chemical systems for the determination of $k_G a$ in the instantaneous regime are presented in Table X |2|.

4.26 Simultaneous measurement of mass transfer coefficients and interfacial area.

In certain circumstances, it may occur mass transfer resistance in both phases which necessitates the knowledge of both $k_G a$ and $k_L a$. Moreover, for the purpose of calculating the effect of a chemical reaction on the rate of absorption of a gas, it is generally necessary to know the parameters k_L and a separately. In using not only the two-film model but also the surface-renewal model, the separate mass transfer parameters can be determined by the chemical methods with pseudo-1st order regime or instantaneous regime.

Thus the condition to satisfy the pseudo-1st order in component A, in using the Danckwerts model |7|,

$$\sqrt{1+Ha^2} < \sqrt{\frac{D_A}{D_B}} + \sqrt{\frac{D_B}{D_A}} \frac{C_{Bo}}{zC_A^*} = E_i \sqrt{\frac{D_A}{D_B}} \tag{35}$$

whence, if there exist a mass transfer resistance in both phases,

$$\Phi = \phi \cdot a = p \left[\frac{1}{k_G a} + \frac{He}{k_L a \sqrt{1+Ha^2}} \right]^{-1} \tag{36}$$

Two important cases are now presented whether k_L and a or k_G and a are necessitated.

— Simultaneous determination of k_L and a. When the gas phase resistance is negligible, the rate of absorption is given by Eq. (36) with $k_G a = 0$, that is,

$$\Phi = \phi \cdot a = \frac{p}{He} k_L a \sqrt{1+Ha^2} = a C_A^* \sqrt{D_A k_2 C_{Bo} + k_L^2} \tag{37}$$

If the rate of absorption is measured with different values of $k_2 C_{Bo}$ and the hydrodynamic conditions remain constant, a plot of Φ^2 against $k_2 C_{Bo}$ gives a straight line with slope $D_A a^2 C_A^{*2}$ and intercept $(k_L a C_A^*)^2$. This is the so-called Danckwerts plot. If C_A^* and D_A are known both $k_L a$ and a can be determined.

Note that a similar result is obtained using the double-film model to interpret an irreversible (m,n)-th order when the condition $Ha \ll E_i$ can be satisfied but the condition $Ha > 3$ cannot be satisfied |2|.

In using the Danckwerts plot, it is necessary to ensure that the physical properties of the system do not alter as $k_2 C_{Bo}$ is changed. For this reason it is more convenient to make use of a catalytic reaction and to change k_2 by adding small amounts of catalyst. If the catalyst is sufficiently powerful, the reaction rate can be varied over a wide range without substantially altering the concentration of the solution. The rate of absorption is thus,

$$\Phi = \phi \cdot a = a C_A^* \sqrt{D_A k_c C_{Cat} + k_L^2} \tag{38}$$

The concentration C_{Cat} of the catalyst can be varied and a and k_L determined plotting Φ^2 against C_{Cat}. Some suitable chemical systems for the determination of k_L and a with the use of Danckwerts plot are presented in Table XI.

It is interesting to know that the hydrodynamic conditions (i.e. k_L) may be influenced by the chemical reaction, causing a change in a and k_L with a change in the reaction rate. *Thus complementary informations of those given by the Danckwerts plots can be offered when the rates of simultaneous chemisorption of one gas (giving a* *with a pseudo-mth order reaction) and of physical absorption (or desorption) of another (giving k_L or k_La) are determined experimentally |93-96, 127|.*

— Simultaneous determination of k_G and a. If it is not possible to keep the gas-side resistance negligible, it is still possible to determine k_G and a by use of a fast irreversible pseudo-mth order reaction (Ha >> 1 and $\sqrt{1+Ha^2} \approx Ha$). Thus the rate of absorption expressed by Eq. (36) becomes,

$$\Phi = \phi \cdot a = p \left(\frac{1}{k_G a} + \frac{He}{k_L a Ha} \right)^{-1} \tag{39}$$

$$\text{whence, } \frac{p}{\Phi} = \frac{1}{a} \left(\frac{1}{k_G} + \frac{He}{k_L . Ha} \right) = \frac{1}{a} \left[\frac{1}{k_G} + \frac{He}{\left(\frac{2k_{mn}}{m+1} D_A C_A^{*m-1} C_{Bo}^n \right)^{0.5}} \right] \tag{40}$$

Thus, if k_2C_{Bo} or $k_{mn}C_{Bo}^n$ is varied a plot of p/Φ against $He/\sqrt{D_A k_A C_{Bo}}$ or against

$$\frac{He}{\left(\frac{2k_{mn}}{m+1} D_A C_A^{*m-1} C_{Bo}^n \right)^{0.5}}$$

will give a straight line of intercept $(1/k_G a)$ and of slope (1/a), so that a and k_G can be calculated simultaneously. Convenient systems for this method are the absorption of dilute CO_2 into aqueous solutions of NaOH or amines.

4.27 Laboratory equipment to determine the physico-chemical parameters necessitated by the chemical methods.

It has just been seen that the chemical determination of the interfacial mass transfer parameters necessitates the knowledge of

- kinetics of the choosen chemical gas-liquid system
- diffusivities of the soluble gas or gases and the reactant or reactants in the liquid solution
- solubility of the gas or the gases
- practically a mathematic grouping of these parameters

$$C_A^* \sqrt{D_A} \text{ or } \left(\frac{2D_A}{m+1} k_{mn} C_{Bo}^n C_A^{*m+1} \right)^{1/2} \text{ or } \ldots$$

These physico-chemical parameters are determined in laboratory equipments in which the contact time between the gas and the liquid and the interfacial area are carefully controlled. The principles of the technique, the laboratory equipments used, and the application to the oxidation of aqueous sodium sulphite solution will be now presented.

4.271 Determination of the kinetic regime of the chemical gas-liquid system. If again we consider the irreversible second order reaction

$$A + zB \xrightarrow{k_2} \text{Products}$$

height kinetic regimes are identified in terms of the two-film theory, as seen in Figure 4. The knowledge of the variables whose variations led to the identification of one of these regimes is due to the noticeable distinct difference in the form of the rate equations. Thus, as explained by Levenspiel and Godfrey |100| Table XII shows which factors affect the rate in each regime. For a given regime, a "+" sign indicates that a change in that particular variable affects the rate, a "-" sign indicates that it does not and a "?" sign means a probable effect, but that the defining rate equations are not available.

Moreover the schematic organigram of Table XIII shows how to identify the kinetic regime for a given gas-liquid reaction, from a series of systematically planned experiments where the variables are varied independently and leads to the determination of the kinetic regime of the chemical gas-liquid system used |100, 101|. Very often the corresponding rate is a function of the physico-chemical parameters that must therefore be determined.

Determination of the physico-chemical parameters. The physico-chemical parameters are determined in the laboratory equipment presented in Fig. 7 by dynamic methods, in which a jet or a film of liquid moves continuously through the gas, to which it is exposed for a known length of time θ (this contact-time being from 10^{-4} s to a few seconds). Moreover in these equipments, the interfacial area A_m is known and the movement of the liquid is carefully controlled so that it can be considered as quiescent during its passage through the gas. Thus, during the contact-time θ each element of liquid absorbs the same amount $Q(\theta)$ of gas per unit area as through it were stagnant and infinitely deep (as in the Higbie model). In each laboratory equipment, the different values of the contact-time θ are generally obtained in varying the liquid volumetric flow rate Q_L and the length of the liquid film or jet h that is the geometrical parameters. Indeed the contact-time is defined as,

TABLE XII : FACTORS AFFECTING THE OVERALL RATE OF ABSORPTION-REACTION

VARIABLE		KINETIC REGIME							
		A	B	C	D	E	F	G	H
concentration of reactant B in the bulk of the liquid	C_{Bo}	+	−	+	+	?	?	+	+
partial pressure of component A in the bulk of the gas	p	+	+	+	+	?	?	+	+
Interfacial area	a	+	+	+	+	+	+	+	−
Liquid holdup	β	−	−	−	−	+	+	+	+
Liquid side mass transfer coefficient	k_L	+	−	+	−	?	?	+	−
Gas side mass transfer coefficient	k_G	+	+	+	+	?	?	+	−
rate constant for the second order chemical reaction	k_2	−	−	+	+	?	?	+	+

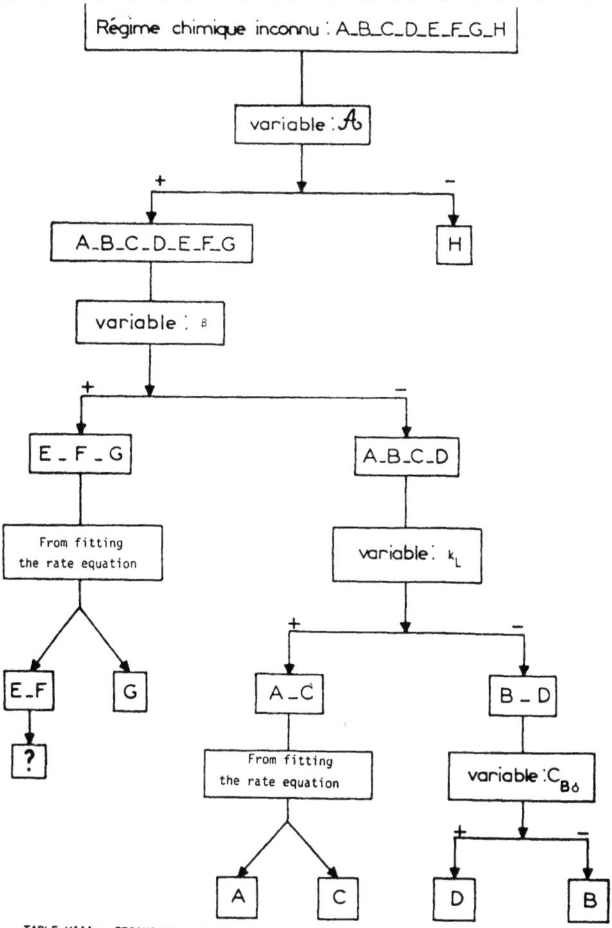

TABLE XIII : ORGANIGRAM FOR THE IDENTIFICATION OF THE KINETIC REGIMES

$$\theta = \frac{h}{u_s} \tag{41}$$

where u_s is the velocity at the surface that is a function of Q_L and the geometry of the equipment (diameter of the jet or film, angle of the cone...). Expressions of θ for different laboratory equipments are presented in Table XIV.

If $Q(\theta)$ is the amount of gas absorbed by unit interfacial area during the contact-time θ, the average rate of absorption during this time is $Q(\theta)/\theta$. Since the total area exposed in the laboratory equipment is A_m, the measured rate of absorption $\Phi(\theta)$ into the film is related to $Q(\theta)$ by,

$$\frac{\Phi(\theta)}{A_m} = \frac{Q(\theta)}{\theta} = \phi \tag{42}$$

The absorption rate $\Phi(\theta)$ (in gmole/s or cm^3/s of component A), is measured experimentally, and $Q(\theta)/\theta$ is calculated for different kinetic regimes from the Higbie theory |7|. The contact-time θ is calculated from Eq. (41) and can be altered by altering Q_L and the geometrical parameters of the considered laboratory equipment. Thus, in carrying out experiments with the same chemical systems and with the same kinetic regime as those used to determine the mass transfer parameters in the industrial gas-liquid reactor, $\Phi(\theta)$ can be determined as a function of θ in the laboratory equipment and the variations of $Q(\theta) = [\theta\Phi(\theta)/A_m]$ for the different values of θ allow for the determination of the necessary physicochemical parameters. This is done as follows whether the absorption is accompanied by a chemical reaction or not.

For the case of purely physical absorption into liquid initially gas-free ($C_{Ao} = 0$), from the Higbie theory, the amount of component A absorbed per unit area during the time θ is |7|,

$$Q(\theta) = 2C_A^* \sqrt{\frac{D_A\theta}{\pi}} \tag{43}$$

It follows from Eq. (42) that,

$$\Phi(\theta) = \frac{Q(\theta)}{\theta} A_m = 2A_m C_A^* \sqrt{\frac{D_A}{\pi\theta}} = \gamma(\theta) C_A^* \sqrt{D_A} \tag{44}$$

where $\gamma(\theta)$ is a characteristic parameter of each laboratory apparatus presented in Figure 7. A plot of measured $\Phi(\theta)$ against $\gamma(\theta)$ at various liquid flowrates Q_L and film or jet lengths h should give a straight line through the origin of slope $C_A^*\sqrt{D_A}$. If the so-

TYPE	LAMINAR JET	CYLINDRICAL WETTED WALL	CONIC WETTED WALL	SPHERICAL WETTED WALL	STRING OF DISCS	ROTATING DRUM	STIRRED VESSEL	STIRRED VESSEL
SCHEME	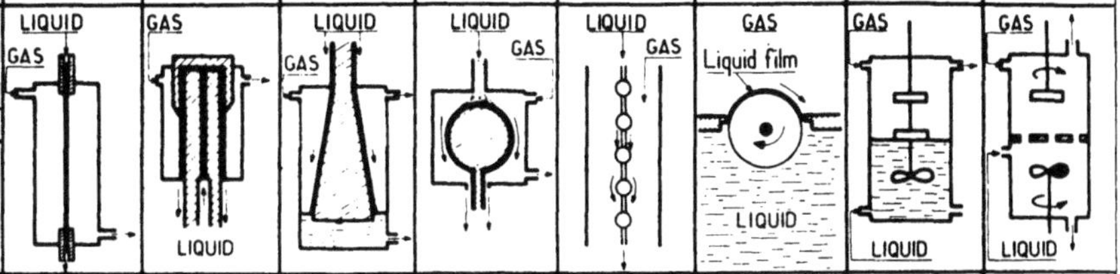							
k_L cm/s	0.016 _ 0.16	$3.6.10^{-3}$ _ 0.016	5.10^{-3} _ 0.011	5.10^{-3} _ 0.016	$3.6.10^{-3}$ _ 0.016	0.016 _ 0.36	$1.6.10^{-3}$ _ 0.02	2.10^{-3} _ 0.02
CONTACT TIMES	$10^{-3}.10^{-1}$ s	10^{-1}. 2 s	0.2 . 1 s	0.1 . 1 s	10^{-1} . 2 s	2.10^{-4}. 10 s	0.06 . 10 s	0.08 . 10 s
INTERFACIAL AREA	0.3 . 10 cm^2 high precision	10 . 100 cm^2 high precision	80 cm^2 high precision	10 . 40 cm^2 high precision	30 . 360 cm^2 moderate precision	diameter 10 cm lenght 12 cm high precision	80 cm^2 good precision	diameter 10 cm lenght 15 cm 2 _ 30% open

Figure 7 - Principal types of laboratory equipment.

Table XIV : Caracteristic parameters of laboratory equipments |2|

	θ	$\gamma(\theta)$
Laminar jet	$\frac{\pi d^2 h}{4 Q_L}$	$4\sqrt{Q_L h}$
Cylindrical wetted wall	$\frac{2h}{3}\left(\frac{3\mu_L}{\rho_L g}\right)^{1/3}\left(\frac{\pi d}{Q_L}\right)^{2/3}$	$d\sqrt{6h}\left(\frac{Q_L}{d}\right)^{1/3}\left(\frac{\pi\rho_L g}{3\mu_L}\right)^{1/6}$
Conic wetted wall	$\frac{4\pi R^2}{5\sin\alpha}\left(\frac{3\mu_L}{2\pi\rho g R\cos\alpha}\right)^{1/3}\frac{1}{Q_L^{2/3}}\left[\left(1+\frac{\sin\alpha}{R}\ell\right)^{5/3}-1\right]$	$\frac{2(\pi\ell^2\sin\alpha+2\pi R\ell)}{\sqrt{\pi}}\frac{Q_L^{1/3}}{\sqrt{\delta}}$ $\delta=\frac{4\pi R^2}{5\sin\alpha}\left(\frac{3\mu_L}{2\pi\rho g R\cos\alpha}\right)^{1/3}\left[\left(1+\frac{\sin\alpha}{R}\ell\right)^{5/3}-1\right]$
Spherical wetted wall	$\left(\frac{8}{4.5}\right)^2\pi R_s^{5/6}\left(\frac{2\pi\rho_L g}{3\mu_L}\right)^{-1/6}Q_L^{-1/3}$	$4.5\left(\frac{2\pi\rho_L g}{3\mu_L}\right)^{1/6}Q_L^{1/3}R_s^{7/6}$
Rotating drum	$\frac{Z_r}{u}$	$4\,Q_L^{1/2}\sqrt{\frac{W_r Z_r}{2\pi\delta_r}}$

α angle of the conic wetted wall
ℓ generating line of the cone
R radius of the basis of the cone
R_s radius of the spherical wetted wall

W_r belt width of the rotating drum
Z_r exposed belt lenght
δ_r film thickness

lubility of gas is known or determined separately, its diffusivity may be determined. Values of $\gamma(\theta)$ for several equipments are presented in Table XIV.

For the case of absorption accompanied by a chemical reaction, from the Higbie theory, $Q(\theta)$ is deduced from the solution of the equation,

$$D_A \frac{\partial^2 C_A}{\partial x^2} - r(x,\theta) = \frac{\partial C_A}{\partial \theta}$$

where $r(x,\theta)$ is the rate per unit volume of liquid at which the reaction is destroying the solute gas at time θ and at distance x from the interface. Analytical or numerical solutions of the diffusion-reaction equations are available for a number of kinetic regimes. For an irreversible first or pseudo-first order reaction, with the Hatta number called Ha',

$$Ha' = \sqrt{\frac{\pi}{4} k_2 C_{Bo} \theta} << \sqrt{\frac{D_A}{D_B}} + \frac{C_{Bo}}{z C_A^*} \sqrt{\frac{D_B}{D_A}} = E_i \tag{45}$$

Danckwerts |7| has proposed the following approximative solutions,

$$Q(\theta) = C_A^* \sqrt{D_A k_2 C_{Bo}} \left(\theta + \frac{1}{2 k_2 C_{Bo}}\right)$$

to within 3 % when $k_2 C_{Bo} \theta > 2$ and,

$$Q(\theta) = 2 C_A^* \sqrt{\frac{D_A \theta}{\pi}} \left(1 + \frac{k_2 C_{Bo} \theta}{3}\right)$$

to within 5 % when $k_2 C_{Bo} \theta < 1/2$. Thus, the experimental determination of the amount $\Phi(\theta)$ of the gas absorbed in function of the time θ leads to the following results :

- for long contact-time ($\theta > 2/k_2 C_{Bo}$) a plot of $Q(\theta) = [\theta \Phi(\theta)/A_m]$ against θ will give a straight line of slope $C_A^* \sqrt{D_A k_2 C_{Bo}}$ and of intercept $(C_A^*/2)\sqrt{D_A/k_2 C_{Bo}}$ and the ratio slope to intercept will give $k_2 C_{Bo}$.

- for short contact-times ($\theta < (1/2) k_2 C_{Bo}$) a plot of $Q(\theta)/\sqrt{\theta}$ against θ will give a straight line of slope $2/3 [(k_2 C_{Bo})/\sqrt{\pi}] C_A^* \sqrt{D_A}$ and of intercept $(2/\sqrt{\pi}) C_A^* \sqrt{D_A}$.

In principle, therefore, both k_2 and $C_A^* \sqrt{D_A}$ can be estimated from each set of experiments but in practice it is found that the measurements a long contact-time give more accurate values of $C_A^* \sqrt{D_A k_2}$

while those at short contact-times give more accurate values of $C_A^* \sqrt{D_A}$ |7|.

If in addition of Eq. (45), the condition,

$$Ha' > 3 \quad (\text{or } k_2 C_{Bo} \theta > 12)$$

is satisfied, then, within 5 %,

$$Q(\theta) = \theta C_A^* \sqrt{D_A k_2 C_{Bo}} \tag{46a}$$

This is the case of the fast pseudo-first condition where the rate of absorption $Q(\theta)/\theta$ is the same at all points on the surface and therefore independent of the hydrodynamics.

Similar considerations apply to fast pseudo-mth order reaction of component A, in which case |7|,

$$Q(\theta) = \theta \sqrt{\frac{2}{m+1} D_A k_{mn} C_A^{*m+1} C_{Bo}^n} \tag{46b}$$

Under these circumstances the measurements of $Q(\theta)$ from experimental $\Phi(\theta)$ in Eq. (42) do not lead to separate values of $C_A^* \sqrt{D_A}$ and k_2 or of $C_A^{*(m+1)/2} \sqrt{D_A}$ and k_{mn}. Nevertheless they lead to the value of the grouping of the parameters $C_A^* \sqrt{D_A k_2 C_{Bo}}$ or $\sqrt{\frac{2}{m+1} D_A k_{mn} C_A^{*m+1} C_{Bo}^n}$ that are those necessitated for the determination of the interfacial area in the gas-liquid reactor (Eq. 28) without a detailed knowledge of each physico-chemical parameter.

As last for an irreversible instantaneous reaction, that is when $Ha' \gg E_i$ the solution of the diffusion-reaction equations governing this case have been given by Danckwerts |7|. This leads to,

$$Q(\theta) = 2E_i C_A^* \sqrt{\frac{D_A \theta}{\pi}} = 2C_A^* \sqrt{\frac{D_A \theta}{\pi}} \left[\sqrt{\frac{D_A}{D_B}} + \frac{C_{Bo}}{zC_A^*} \sqrt{\frac{D_B}{D_A}} \right]$$

In such case, in addition to the quantities C_A^*, D_A and C_{Bo} (which enter into the pseudo-first order case), the diffusivity D_B of the reactant B is also involved. It is possible to infer the value of one of these quantities from the measurements of $Q(\theta)$ from experimental $\Phi(\theta)$, if the values of the others are known and can be estimated. For example if C_{Bo} and z are known and if C_{Bo} is sufficiently small for C_A^* and D_A to have substantially the same value as in pure solvent, then the diffusivity D_B can be determined. An interesting possibility is the reduction of C_A^* to a value much less than C_{Bo}, in which circumstances (D_B/D_A) (C_{Bo}/zC_A^*) may be much more than one (in practise very often $C_{Bo}/zC_A^* \gg 1$) and thus the quantity of component A absorbed per unit area of the laboratory equipment in time θ is,

$$Q(\theta) = 2C_A^* \sqrt{\frac{D_A\theta}{\pi}} \sqrt{\frac{D_A}{D_B}} \left(1 + \frac{D_B}{D_A}\frac{C_{Bo}}{zC_A^*}\right) = \frac{2C_{Bo}}{z}\sqrt{\frac{D_B\theta}{\pi}}$$

Thus the diffusivity D_B can be determined without the uncertainties involved in the estimation of C_A^* and D_A.

4272 Application to the kinetics of the oxidation of aqueous sodium sulphite solutions. The absorption of oxygen into sulphite solutions is very often used to measure interfacial areas in pilot plant or industrial reactor, due its lack of toxicity, fire safety and its cheap cost for large-scale equipment. Though general review articles have been published on that topic |2, 101-103| there still exist some controversy on the kinetics of this reaction. To illustrate the strategy presented in the previous sections, an investigation of this reaction in laboratory equipment experimented by Laurent |101|, will be presented now for the experimental conditions that are usually considered as suitable to determine interfacial area in gas-liquid reactors. The reaction is,

$$O_2 + 2Na_2SO_3 \xrightarrow[k_c]{CoSO_4} 2Na_2SO_4$$

with the rate, $r = k_c C_A^m C_B^n C_C^q = k_c C_{O_2}^m C_{Na_2SO_3}^n C_{CoSO_4}^p$

In using the technique proposed just before, the kinetic regime, the partial order in components and the kinetic parameters are investigated in the experimental conditions,

$$\begin{cases} 0.2 \le p_{O_2} \le 0.8\text{-}1 \text{ Atm} \quad ; \quad 0.4 < C_{Na_2SO_3} \le 0.8 \text{ gmole/l} \\ 10^{-4} \le C_{CoSO_4} \le 1\text{-}2\times10^{-3} \text{ gmole/l} \\ C_{Na_2SO_3} + C_{Na_2SO_4} = \text{const} \quad ; \quad 20 \le t \le 50\ °C \ ; \ 7.5 \le pH \le 8.5 \end{cases} \qquad (47)$$

Three laboratory equipments have been used (Figure 7) : a cylindrical wetted wall (d = 1.6 cm, h = 10 cm), a conic wetted wall (α = 14°, ℓ = 14 cm, $2R_S$ = 3.8 cm) and a stirred vessel (D = 10.1 cm, V_R = 1 l, V_L = 0.5 l).

Table XV : Determination of the kinetic regime for the oxidation of aqueous sodium sulphite.

A - Variation of the variable, interfacial area A_m

Laboratory apparatus	Interfacial area $A_m(cm^2)$	Φ'_{O_2} (cm^3/s)	
Cylindrical wetted wall	49.8	0.195	t = 20°C P_{O_2} = 1 Atm pH = 8.5 $C_{Na_2SO_3}$ = 0.8 g/ℓ C_{CoSO_4} = 0.1 g/ℓ
Conic wetted wall	74	0.296	
Stirred cell	78	0.308	

B - Variation of the variable, liquid holdup $\beta = V_L/V_R$

Laboratory apparatus	$V_L = \beta V_R$ (cm^3)	φ'_{O_2} $(cm^3/cm^2.s)$	
Cylindrical wetted wall	1.5	2.10^{-3}	t = 20°C P_{O_2} = 1 Atm pH = 7.5 $C_{Na_2SO_3}$ = 0.8 g/ℓ C_{CoSO_4} = 0.1 g/ℓ
Conic wetted wall	1	2.10^{-3}	
Stirred cell	500	2.10^{-3}	
	1000	2.10^{-3}	

$C_{Na_2SO_3}$ gmole/ℓ	P_{O_2} = He $C^*_{O_2}$ atm	C_{CoSO_4} gmole/ℓ	Température	pH	n	m	q	References
0 to 0.08	0.2 and 1	0 to 7.10^{-7}	20°C	8.0 to 9.2	1	1	1	|106|
0.00375 to 1	0.1 to 1	10^{-3} to 10^{-4}			1 if $[Na_2SO_3] < 0.06$ 0 if $[Na_2SO_3] \simeq 0.25$ <0 if $0.25 < [Na_2SO_3] < 1$	0 1 2		|107|
25 to 100 $g.\ell^{-1}$	1	10^{-6} to 10^{-3}	15 to 33°C	7.5 to 8.5	0 if $50 < [Na_2SO_3] < 100$	1	1	|108|
0.4 to 0.8	0.1 to 1	4.10^{-4}	30°C	7.5 to 8.5	-1	2 decreasing order if $P_{O_2} \geq$ 1atm if $[Na_2SO_3]$=0.4 gmole/ℓ	1	|109|
100 $g.\ell^{-1}$	0.2	10^{-3}	30°C					|110|
0.25 to 0.795	0.2 to 1	0 to 10^{-3}	15 to 35°C	7.8 to 9.2	0	1 if p_{O_2} > 0.9 atm 2 if p_{O_2} < 0.9 atm	1	|111|
0.3 to 1	0.2 to 1	5.10^{-6} to 10^{-4}	30°C	8.5	variable order	continuous variation from 2 to 1		|112|
50 to 100 $g.\ell^{-1}$	0.2 to 1	2.10^{-5} to 2.10^{-3}	15 to 60°C	7.5 to 8.0	0	2	1	|7 113 114|
0.4 to 0.8	0.2 to 1	3.10^{-6} to 3.10^{-3}	15 to 60°C	7.5 to 8.5	0	2	1	|103 102 115|
0.4 to 0.8	0.1 to 0.5	$\leq 10^{-4}$	20°C	7.5 to 8.0	0	2		|116|
0.8	0.3 to 7	10^{-4} to 10^{-3}	30°C	8.3	0	2 if p_{O_2} < 0.8 atm 1 if p_{O_2} > 1.1 atm	1	|117|

Table XVI : Kinetic orders published for the oxidation of aqueous sulphite solutions. |101|

— Determination of the kinetic regime (use of the organigram of Table XIII). The influence of the interfacial area A_m is presented in Table (XV, A). For the same operating conditions, the volumetric absorption rate of oxygen Φ_{O_2} (in cm^3/s) varies proportionally to the interfacial area of the three equipment. Therefore this reaction is not a very slow chemical process (regime H, Table XIII and Fig. 4).

The influence of the liquid holdup β is presented in Table (XV, B). For the same operating conditions, the average volumetric rate of absorption of oxygen ϕ'_{O_2} (in cm^3/cm^2 s) is independent of the liquid holdup. So the kinetic regimes E, F and G (Table XIII and Fig. 4) can be eliminated.

The influence of the liquid side mass transfer parameter k_L which is a function of the hydrodynamics of the apparatus, is obtained in changing the geometry of the apparatuses (3 apparatuses are used here and the stirring speed N_L is varied in the stirred cell) and in varying the liquid flow rate Q_L in each apparatus. The variations of $\phi' = \phi'_{O_2}$ with Q_L for the wetted films equipments and with N_L for the stirred cell are presented in Fig. 8. In a certain range of these latter variables, (that is in a defined range of contact-times for the three equipment), values of ϕ' are independent of k_L. Therefore in this range of variables, the kinetic regimes A and C (Table XIII and Fig. 4) can be eliminated.

The influence of the concentration of the reactant C_{Bo}, the sodium sulphite concentration, is also presented as an example in Fig. 8 for the cylindrical wetted wall. As ϕ' is influenced by the concentration of the reactant, thus the kinetic regime B (Table XIII and Fig. 4) can be eliminated and the reaction can be considered as a rapid pseudo-mth order type (regime D, Table XIII and Fig. 4) (of course this deductive sequence has to be repeated for the different operating conditions previously given).

— Determination of the partial orders of the reactants. If the gas phase resistance is negligible, the rate equation corresponding to the kinetic regime D is, within 5 %,

$$\Phi = \phi' a = a \sqrt{\frac{2}{m+1} D_A k_{mn} C_A^{*m+1} C_{Bo}^n C_{Cat}^q} \qquad (48)$$

that is, for any laboratory equipment, from Eqs. (42) and (46b).

$$\Phi(\theta) = A_m \frac{Q(\theta)}{\theta} = \phi' A_m = A_m \sqrt{\frac{2}{m+1} D_{O_2} k_{mn} C_{O_2}^{*m+1} C_{Na_2SO_3}^n C_{CoSO_4}^q} \qquad (49)$$

Remind that the validity of these equations is verified if $3 < Ha \ll E_i$. Then it is necessary to know the partial orders m,

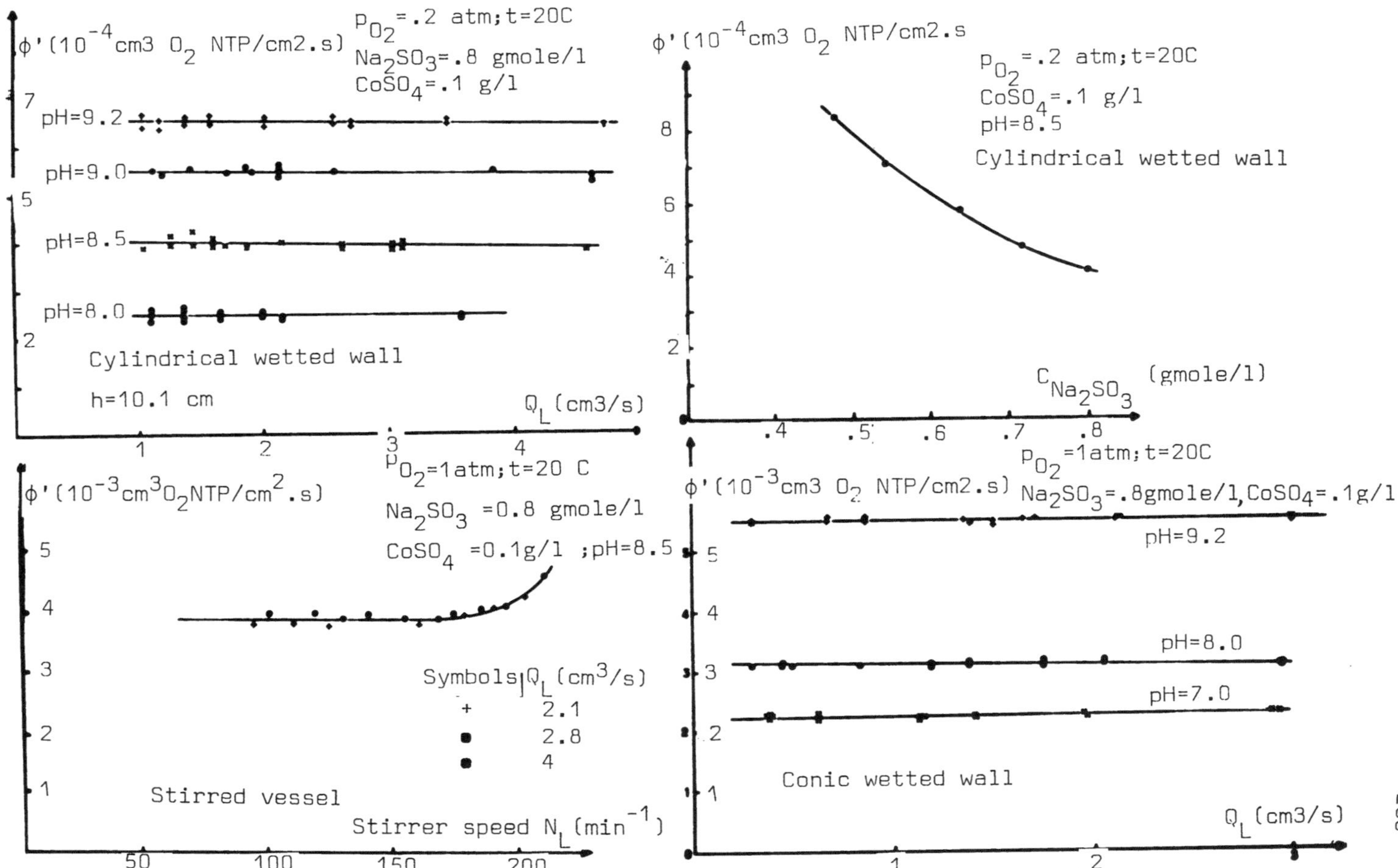

Figure 8 - Influence of hydrodynamic conditions on average volumetric absorption rate.

n and q to determine the interfacial area in a gas-liquid reactor and this in using Eq. (48) in the range of the operating conditions. The values of these orders in the published literature are presented in Table XVI. The results show large discrepancies, especially for the values of the order in oxygen and in sodium sulphite though the experimental conditions are often practically identical. For example the value of n is 1 or 2 or varies continuously from 1 to 2.

However in the kinetic regime D, the variations of the average rate of absorption with the concentration of each component should be expressed by Eq. (49) and the experimental values measured in the laboratory equipment where the interfacial area A_m is known, should lead to the values of the partial orders. Examples of such variations in the three laboratory equipment are presented in Fig. 9a.

A typical variation of the average volumetric rate of absorption of oxygen ϕ' (here in $cm^3 O_2/cm^2$ s) in function of $p_{O_2} = He C^*_{O_2}$. The reaction is second order in oxygen as the slope of the straight lines $[\log\phi'$ vs $\log p_{O_2}$ is $(m+1)/2 = 3/2]$ for p_{O_2} in the range 0.2 to 0.8 atm. Indeed in some experiments, it has been observed that, in the range 0.8 to 1 atm, second order may be not fulfilled any longer as for this range and for the higher values of p_{O_2}, it has been observed that ϕ' was less increasing with p_{O_2} |101, 117|. Any way in practise, most of the experiments in pilot plants or in industrial reactors are carried out with oxygen from air at atmospheric pressure. So it must be concluded that in such case the order in oxygen appears to be m = 2.

The influence of the sulphite concentration on ϕ' is presented in Fig. 9b. As previously explained, the specific absorption rate decreases as $C_{Na_2SO_3}$ increases (curve a). However the experiments carried out in keeping the initial sulphite and sulphate concentration constant by addition of sodium sulphate show a less marked decrease (curve b). Finally it is observed that the specific rate of absorption is independent of the sulphite concentration when the experiments are carried out in keeping a constant total sulphite and sulphate concentration ($C_{SO_3^{--}} + C_{SO_4^{--}}$ = constant).

This is performed in recycling the solution partially oxidized in sulphate into the laboratory equipment (curve c). The simultaneous presentation of these three experimental types of measurements may explain the different values of n published in the literature.

The decrease of curve (b) may be explained by the variation of the ionic strength I and undesirable influence of impurities whose influence increases with the concentration of sulphite |111| or by the considerable reduction of the solubility of oxygen at high concentrations of sulphite |109|. In the experiments corresponding

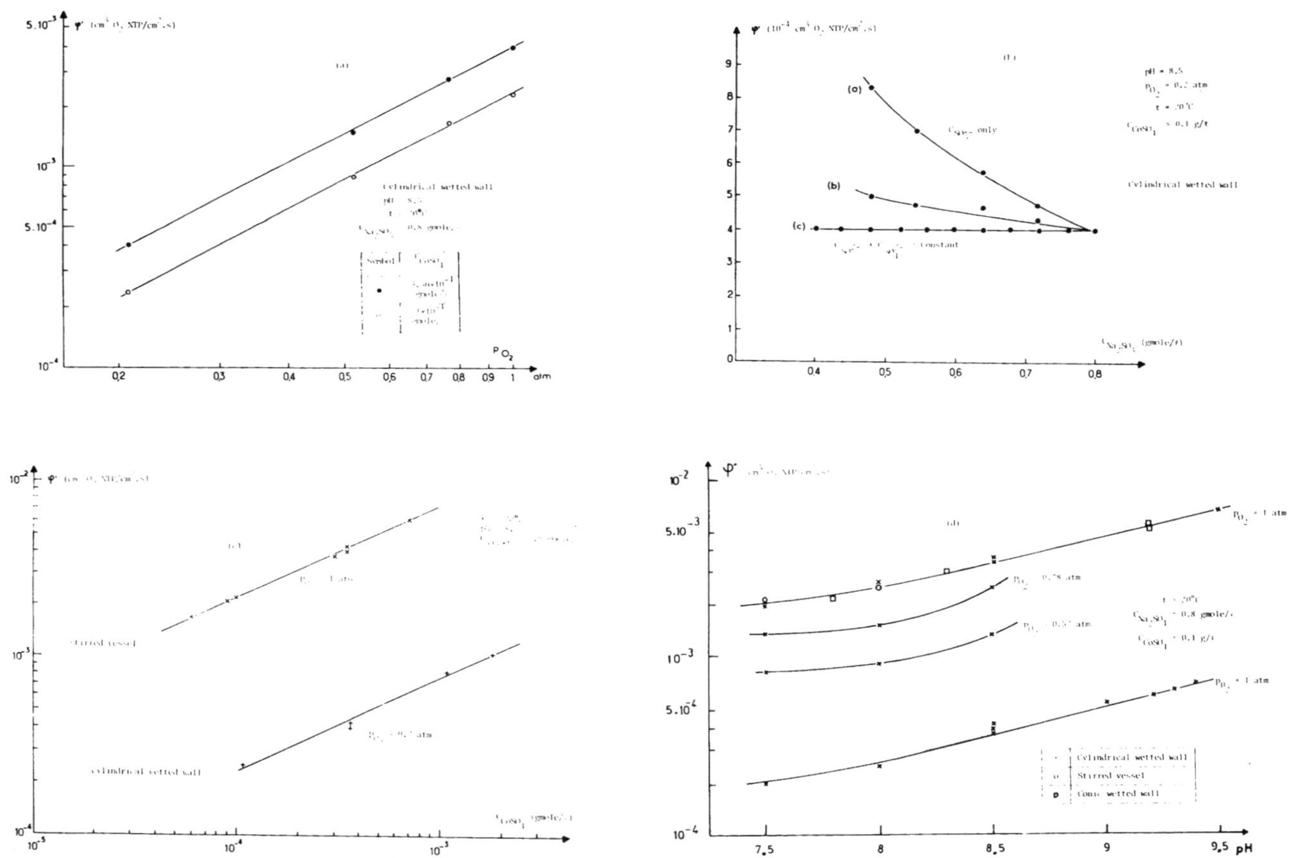

Figure 9 - Influence of component concentration and pH on average volumetric absorption rate.

to the curve (b), the ionic strength I is constant and the difference with the curve (a) may be due to a different solubility and diffusivity of O_2 in the solution. At last it is important to note that in fact the curve (c) represents the actual variations of the sulphite oxidation system in the working experimental conditions of an industrial reactor. Indeed, for a given initial sulphite concentration, the total amount of sulphite concentration, produced sulphate concentrations and impurities is maintained constant during the oxidation within the reactor. So, in such experimental conditions, the average rate of absorption is independent of the sulphite concentration in the range 0.4 to 0.8 gmole/l and it can be assumed that the order in sodium sulphite is zero.

The influence of the catalyst concentration on ϕ' is presented in Fig. 9c. The order in catalyst is 1 as the slope of the straight lines log ϕ'vs log C_{CoSO_4} is (q/2) = 1/2, for cobalt concentration in the range 10^{-4} to $(1-2)\times10^{-3}$ gmole/l. For higher cobalt concentrations, a solid precipitate is formed when the diluted cobalt sulphate is added to the sodium sulphite solution.

An example of the variations of the volumetric average rate of absorption with the pH is also prescrited in Fig. 9d, where it is shown that ϕ' increases with increasing pH.

The considerations just presented on the partial orders for the previously defined operating conditions are sufficient to use this reaction for the industrial practise because only the knowledge of the rate of absorption per unit interfacial area is necessitated to determine the interfacial area of an industrial contactor without a detailed knowledge of the kinetic rate constant and activation energy. However it may be interesting to compare the values of these parameters obtained in similar conditions by the different authors to show the errors that could be made when different qualities of products are used.

Indeed when m, n and q as well as the diffusivity and the solubility of O_2 in the solution are known, the second order reaction rate constant $k_{mn} = k_2$ and its variations with the pH and with the temperature can be determined. The variations of k_2 with the cobalt concentration and the pH are presented in Fig. 10 together with the values of k_2 derived from the well known work of Reith and Beek |103|. The agreement is good though the values obtained by Laurent |101| are systematically higher. A comparison with work of Linek and Meyrhoferova |111| is not possible because in the range of variables of Fig. 10 the authors measured a second order reaction rate in oxygen, but they did not interprete their data in second order rate constants.

Furthermore the Arrhenius equation gives a reasonable description of the dependence of k_2 on the temperature. A plot of $\log k_2$ vs 1/T for different values of pH gives straight lines and the activation energies E_a (Fig. 10) are calculated from their slope. Values of E_a reported by different authors are presented in Table XVII and again an important discrepancy of the results can be observed which, as expected, shows that the rate data of the reaction should never taken from the literature because the quoted constant depends strongly on the purity of the solution.

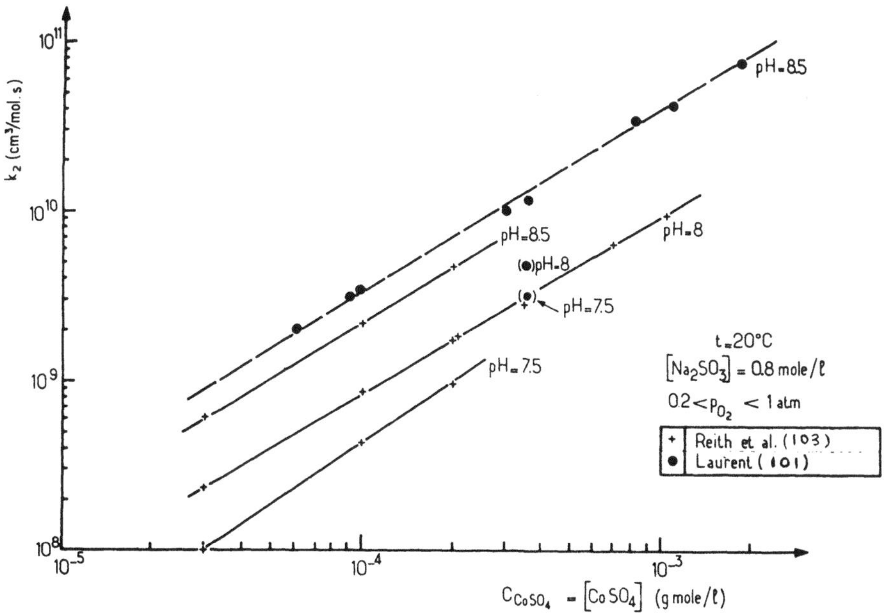

Figure 10 - Second order reaction rate constant k_2 as a function of cobalt concentration and pH.

Table XVII Published activation energies for the reaction of oxidation of sodium sulphite (E_a in cal/gmole) [2]

pH / t (°C)	7.5	8	8.5	Référence
15 - 60	13,421	13,875	12,918	(R 7)
15 - 60		18,800		(R 9)
15 - 33	12,000	12,000	12,000	(D 13)
15 - 35	12,000 to 13,000			(L 17)
30		10,526		(W 3)
20 - 50	12,250	12,000	12,250	(L 2)

It can be concluded that the oxidation of aqueous sodium sulphite solutions with cobaltous sulphate as a catalyst proves to be a convenient model reaction for the determination of interfacial area in gas-liquid reactors such as packed column, plate column, mechanically agitated reactor . Of course the kinetics of the reaction are not simple as many variables which are not independent influence the reaction rate and many of the discrepancies in measurements published can be logically explained from an insufficient knowledge of the reaction kinetics. But provided the range of cobalt concentrations, sulphite concentrations, pH and temperatures previously indicated are satisfied, the reaction is zero order in sulphite, first order in cobalt and second order in oxygen. So the specific rate of absorption is,

$$\phi = C_A^{*\,3/2} \sqrt{\frac{2}{3} D_{O_2} k_2}$$

as the following conditions are verified,

$$3 < \frac{\sqrt{D_{O_2} k_2 C_A^*}}{k_L} << 1 + \frac{D_{SO_3Na_2}}{D_{O_2}} \frac{C_{SO_3Na_2}}{2C^*_{O_2}}$$

Finally, provided these conditions are met, the interfacial area in a given reactor of volume V_R can be found by measuring the rate of absorption of oxygen $\Phi V_R = \phi \cdot a V_R$. *The quantity* $\phi = |\Phi(\theta)/A_m|$ *must be found with the same partial pressure of oxygen, using the same absorbent at the same temperature and pH, in a laboratory equipment such as a wetted falling film, laminar jet or a stirred vessel, in which the value of the interfacial area* A_m *is known. It must be insisted on the fact that the same products in both types of equipments must be used so that the influence of the catalyst, the impurities... will be the same*

Impurities such as oil traces in the air, or contact of the solution with metallic pieces should be avoided. From literature it also appears that the use of copper as a catalyst instead of cobalt leads to hasardeous results and the rate of absorption may even decreased.

It should be pointed out that at oxygen pressure higher than about 0.8-1 Atm, the order in oxygen is not 2 any more and Alper |117| has even shown that the reaction becomes first-order in oxygen in the range 1 to 7 atm. This important result should be kept in mind when this model reaction is used to determine the interfacial area of a reactor working at high pressure.

At last, compared with the common model reaction using CO_2 absorption, oxidation has the advantage of the low solubility of oxygen. So it can be used when conversion of the gas phase is very fast as in a stirred tank but the phenomenae such as foaming that sometimes occur, depending on the quality of the products used, should be carefully controlled and avoided. *Also the results are typical of those in ionic solutions, whose properties may differ from those of other process liquids.*

4.273 Limits of the chemical technique when applied to gas-liquid systems inhibiting bubble coalescence in such equipment as bubble columns, mechanically agitated tanks and plate columns.

Interfacial parameters in these reactors (especially stirred tanks) are generally determined in assuming the liquid phase perfectly mixed, which is realistic, and the gas phase either in piston flow or perfectly mixed. When the gas is dispersed into clean liquids such as water, the coalescence and the redispersion due to bubble interaction, either inside the dispersion, or with the cavities behind the stirrer blades assure a perfect mixing of the gas phase. On the contrary for liquids involving the presence of dissolved electrolytes or surface active agents that diminish or totally inhibit the coalescence, neither the perfectly nor the piston flow model may represent conveniently the behaviour of the gas, even though the bubbles may be considered independent in a first approximation.

To define the limits of the chemical technique to measure interfacial parameters in such inhibiting coalescence or foaming dispersions, Midoux et al. |131| propose a flow model for shrinking and not shrinking bubbles that seems more realistic than the piston flow. Then criteria are deduced to insure, when the real behaviour of the gas phase is ignored, the independence within 10 % relatively to this behaviour of the mass transfer parameter data experimentally measured in a laboratory-scale well stirred tank by the chemical technique.

Assumptions for not shrinking bubbles are (1) the bubbles originated by the distributor do not loose their identity up to the moment they leave the reactor. The distribution function $f(\delta_o)$ of the size of the bubbles $\delta_o = d_b/d_{SM}$ is represented by the Bayens distribution

$$f(\delta_o) = K\delta_o^2 \exp(-k^2\delta_o^2)$$

with $k = 8/3\sqrt{\pi}$ and $K = 4k^3/\sqrt{\pi}$

(2) Each bubble is perfectly mixed except for instantaneous chemi-

cal regime. (3) Mass transfer is located in the liquid phase and the true liquid-side mass transfer coefficient only depends on the square root of the bubble diameter. (4) For each bubble, the liquid phase composition is invariable, i.e. the liquid evolution time is sufficiently slow to be neglected when compared to the residence time of a bubble inside the dispersion. Moreover the residence time of each bubble reduced by the gas residence time in dispersion, depends on the magnitude of its diameters as,

$$\theta_o = \frac{t_{\delta_o}}{\tau_G} = \frac{\lambda}{\delta_o^{\nu}} \quad \text{with} \qquad \lambda = \frac{\int_0^\infty \delta_o^{\nu+3} f(\delta_o) d\delta_o}{\int_0^\infty \delta_o^{3} f(\delta_o) d\delta_o}$$

The probability for the bubble to leave or to stay inside the dispersion depends on the competition between the drag force and the buoyancy force acting on it, this normally leads to $1 \leq \nu \leq 2$ |132| with the limiting values $\nu \to 0$ corresponding to a piston flow of identical bubbles and $\nu \to \infty$ corresponding to a short circuit behaviour.

The previous assumptions together with the classical relationships of the different specific absorption rates, depending on the chemical reaction regime used to determine interfacial parameters, lead to the theoretical mean reduced values of the gaseous reactant in the gas exit stream and to absorption efficiency E_A for a gas-liquid dispersion, where a_o is the geometrical specific interfacial area and $\overline{k_L}$ is the mean true liquid-side mass transfer coefficient defined for a bubble of diameter d_{SM}. Thus, with this value of E_A, apparent values a_{app} and $(k_L a)_{app}$ may be calculated in assuming that the gas phase is perfectly micromixed and then compared with a_o and $\overline{k_L} a_o$. Then the curve representing the variation of the ratio between the apparent values and the geometrical values, with the characteristic depletion time of a bubble reduced by the gas residence time, $\overline{\theta_o} = d_{SM}\gamma_o/6\overline{\Phi}_o\tau_G$ will approximate the unity if certain conditions are fulfilled (γ_o is here the inlet solute gas concentration in gas-bulk and $\overline{\Phi}_o$ is the corresponding absorption rate for a bubble of diameter d_{SM}). More precisely the estimated mass transfer interfacial parameters will approximate the true value within $\pm$ 10 % if certain conditions for $\overline{\theta_o}$ and E_A are fulfilled. These conditions are a minimum value of the depletion time to exceed or a maximum value of the efficiency to do not exceed. These extrema are presented on Table XVIII where it is seen that the limiting values depend on the order m with respect to the gaseous reactant and on the reduced characteristic diffusion time $\overline{\theta_D} = d_{SM}^2/4D_G\tau_G$.

Though these results may be considered only like sufficient conditions, they state precisely the limits of the chemical technique. So in the slow chemical regime (used to determine $k_L a$), it is

TABLE XVIII LIMITING VALUES OF $\overline{\theta}_0$ AND E_A FOR THE VARIOUS CHEMICAL REGIMES USED TO MEASURE THE MASS TRANSFER PARAMETERS

Regime	Parameter to determine	Order m	Minimum value of $\overline{\theta}_0$	Maximum value of E_A
Physical absorption or slow chemical	k_La		3.3	0.25
Intermediary pseudo 1-n^{th}-order chemical	a and k_La	1	4.0	0.20
Rapid pseudo m-n^{th}-order chemical	a	0	1.35	0.48
		½	1.80	0.41
		1	2.60	0.30
		2	3.70	0.21
		2	4.80	0.16
Instantaneous chemical	k_La		3.3	0.27
Instantaneous chemical at the interface	k_Ga		$\overline{\theta}_D > 20$	0.55

easy to satisfy the condition $\overline{\theta}_0 > 3.3$. Let us consider, for example, the sodium sulfite system in the presence of Co^{++} ions as a catalyst with the currently used experimental conditions $C_{Bo} = 0.2$ kgmole/m^3, $\overline{\Phi}_0 = 8.27\times10^{-9}$ kgmole/m^2.s, $\gamma_0 = 8.32\times10^{-3}$ kgmole/m^3, $[Co^{++}] = 10^{-6}$ kgmole/m^3, and $d_{SM} = 10^{-3}$ m. This gives $\tau_0 = 168$ s and then $\tau_G < 51$ s to satisfy $\theta_0 = \tau_0/\tau_G > 3.3$. Such a gas residence time has been used in practice in industrial size equipment.

On the contrary, the same gas-liquid reaction with the conditions of rapid pseudo second-order regime in oxygen (to determine a), that is, $C_{Bo} = 0.8$ kgmole/m^3, $|Co^{++}| = 3.56\times10^{-4}$ kgmole/m^3, $\overline{\Phi}_0 = 1.56\times10^{-7}$ kgmole/m^2.s, cannot be used easily. Indeed, it is calculated from the condition $\overline{\theta}_0 > 3.7$ proposed in Table XVIII, that $\overline{\tau}_0 = 8.9$ s and $\tau_G < 2.4$ s. Such a small gas residence time limits the applicability of this chemical method to the small scale units. This may explain why interfacial areas have been measured in tanks the volume of which are not bigger than 1 m^3, with gas superficial velocity as high as 0.047 m^3/m^2.s |114|. Complementary information for the case of shrinking bubbles may be found in ref. |131|.

Moreover for the cases where the solute gas concentration changes considerably from inlet to outlet, a representative driving force must be defined in order to evaluate accurately the mass transfer parameters. Hassan and Robinson |133| have described a model of gas bubble coalescence-redispersion interactions that proposes that the effective driving force for the mass transfer is the most significantly affected by coalescences between bubbles of minimum solute gas partial pressure, i.e. those having exit gas composition, and bubbles of near maximum solute composition (closest to inlet gas composition) which are about to undergo their first coalescence subsequent to having been freshly sparged into the dispersion. This model leads to evaluation of a mass transfer effective coalescence frequency F (F = 0 for bubble bursting at the free liquid surface, F = 100 for fairly well mixed) and to an integral average mass transfer driving force correction factor $f = \overline{p}/p_s$ when $\overline{p}$ is the gas residence time average partial pressure and p_s is the exit stream partial pressure. This factor is used to correct the apparent driving force based on exit gas composition for the imperfect mixing of the dispersed gas phase resulting from the restricted coalescence and redispersion interaction that is characteristic of electrolyte solutions in which the chemical technique is applied. Using the equation of the specific absorption rate to determine interfacial areas, the correction factor is applied to the experimentally-measurable exit gas composition in order to determine the relevant residence time average driving force. Thus conditions for the efficiency E_A and the coalescence frequency F are researched so that the gas dispersion can be considered as perfectly mixed (f = 1). It is obtained that for all practical purposes the gas dispersion can be considered to be perfectly mixed ($f = 1 \pm 0.1$) as long as $E_A < 0.8$ and $F > 40$ are satisfied simultaneously. Both these criteria generally will be met in the case of sparingly-soluble gas absorption without chemical reaction in turbulent liquids such as water which do not appreciably hinder bubble coalescence. However, for $E_A > 0.8$ as can often be the case in chemically-reactive systems and for $F < 20$ as has been found to be the case in coalescence-inhibiting liquids such as aqueous electrolyte solutions, f is significantly greater than 1.0 particularly for $E_A > 0.90$ (typical of the absorption of CO_2 by aqueous solutions of strong alkalis in mechanically agitated tanks with a relatively high specific interfacial area). In such later cases, the specific area cannot be computed accurately from specific absorption rate equations merely using $p_i = HeC_A^* = p_s$; rather one must substitute $p_i = \overline{p} = fp_s$ in such cases which will yield a lower value of the specific area than that calculated using p_s alone. Finally application of the correction factor to the chemical technique leads to an improved estimation of the mass transfer effective specific interfacial area in mechanically agitated tanks compared to merely assuming that the gas phase is perfectly mixed, where it may be overestimated by a factor 1.4 to 3.5.

In conclusion, before using for scaling up the literature reported data for aqueous ionic liquids, it is suggested to verify if the complementary conditions proposed by the previous authors |131, 133| are verified.

4.274 Application of the chemical technique to the case of an organic liquid phase.

The gas-liquid reaction systems that are used to determine the interfacial area and the mass transfer coefficients in gas-liquid reactors are almost exclusively constituted of reactants in an aqueous solution. In fact, many industrial gas-liquid reactors work with organic liquids, and it would be desirable to have some organic reactant systems in order to be able to measure the mass transfer parameters or at least to estimate the difference in their values when applying the data obtained with non-organic systems to reactors operating with organic liquids. There are a few investigations on such organic reactions on their application to gas-liquid reactors |118-124|. We want to present here some succinct considerations on the reaction most often suggested these last years |118, 120, 121| : the reaction between carbon dioxyde and cyclohexylamine (CHA) in a toluene (TOL) plus 10 % isopropanol (IPA) solution. The addition of small amount of IPA does not alter the physical properties of the solution appreciably and enables the carbonated product to be kept in solution |121|. The apparent rate of reaction is expressed by

$$r = k_{mn} C^{m}_{CO_2} C^{n}_{CHA}$$

with an overall stoichiometry

$$CO_2 + 2R_2NH \rightarrow R_2NCOO^- + R_2NH_2^+$$

The results of the chemical thermodynamic and kinetic study by Alvarez |122| will be briefly commented here as an application of the previous considerations on the laboratory apparatuses. All the experimental work was carried out in a laboratory cylindrical wetted-wall which was also used to determine the diffusion coefficient of CO_2. The height h was modified between 6 and 14 cm and the cylinder diameter d is 1.56 cm. The liquid flowrate Q_L was varied in the range 1-4.6 $cm^3.s^{-1}$. Except for the solubility and diffusivity, all the measurements were performed at 20°C. The experimental conditions are

$$0.01 < p = p_{CO_2} < 0.16 \text{ Atm}$$

$$0.09 < C_{Bo} = C_{CHA} < 2.7 \text{ kgmoles.m}^{-3}$$

The use of the contact-time analysis together with the value of the contact-time θ presented in Table XIV gives,

$$\theta = \frac{h}{u_s} = \frac{2h}{3} \left(\frac{3\mu_L}{\rho_L g}\right)^{1/3} \left(\frac{\pi d}{Q_L}\right)^{2/3} \qquad (50)$$

Then the knowledge of the interfacial area $A = \pi dh$ and the measurement of the rate of physical absorption with initial free gas-liquid Φ leads to the following expression (including Eq. 50)

$$\Phi = \phi \cdot A = 2AC_A^* \sqrt{\frac{D_A}{\pi\theta}} = Kh^{0.5} Q_L^{0.33} C_A^* \sqrt{D_A} \qquad (51)$$

that allows for the determination of $C_A^* \sqrt{D_A}$ and then for the CO_2 solubility when the diffusivity is known. Some data are presented in Figure 11a and 11b. Moreover it has been observed that the solubility is practically independent of the amine concentration C_{Bo} ($C_A^* = 0.104$ kgmole.m^{-3}.atm^{-1} at 20°C) while the diffusivity could be represented by $D_A = 4.1\sqrt{1-1.2x}$ where x is the molar fraction of amine |120|.

Now when the gas phase resistance is considered, the specific rate of absorption for a rapid m,nth-order reaction ($3 < Ha << E_i$) may be expressed as |7|

$$\frac{Ap}{\Phi} = \frac{p}{\phi} = \frac{1}{k_G} + \frac{He}{\left(\frac{2}{m+1} D_A k_{mn} C_A^{*m-1} C_{Bo}^n\right)^{0.5}} \qquad (52)$$

Though equation (52) is not explicit ($p_i = HeC_A^*$), if the conditions for a rapid reaction are verified, the representation p/ϕ versus $C_A^{*(1-m)/2} \times C_{Bo}^{-n/2}$ will give, for the right values of m and n, a straight line and its slope will lead to the calculation of the kinetic constant k_{mn} if C_A and D_A are known |120|. Such a graphical plot is presented in Figure 11 which shows that the order with respect to CO_2 is m = 1 (Fig. 11c), then the order with respect to CHA is n = 2 and the gas phase resistance is not significant (Fig. 11d), ($1/k_G = 0.93 \times 10^{+8}$ m^2.s.Pa.kgmole^{-1}). So the values of the specific rate of absorption are such that this reaction may be used to determine interfacial area in reactors working with organic liquids except if the equipment is very efficient.

Indeed in well agitated reactors or in too long packed columns, the efficiency may be very high and the conditions of a rapid chemical might not be respected everywhere in the contactor. In fact the solubility and the specific rate are usually much higher with purely organic systems than with purely aqueous or aqueous organic systems. This has led Alvarez |122| to study several systems taking into consideration the foaming effect or the viscous effect encountered with organic liquids and which are not taken into consideration when the mass transfer parameters are measured with aqueous ionic systems such as sulfite oxidation and CO_2-aqueous NaOH. These systems are the rapid pseudo m,nth-order reaction between CO_2 and monoethanolamine (MEA), or diethanolamine

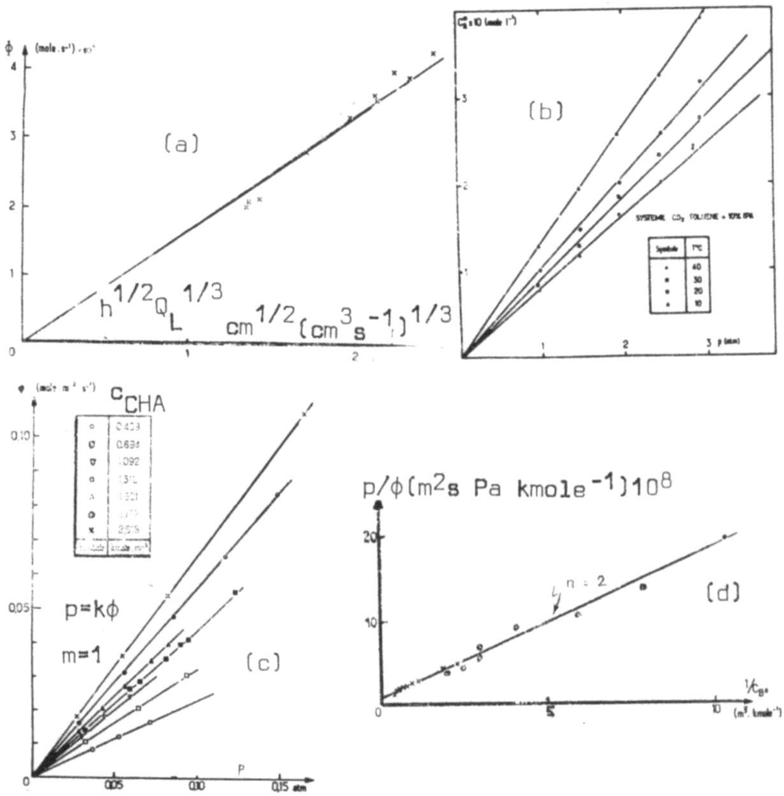

Figure 11 - Solubility and kinetic data of the system CO_2-cyclohexylamine in toluene + 10 % Isopropanol

(DEA) or cyclohexylamine (CHA) contained in different organic solvent solutions. These are presented in the following Table.

SYSTEM	10^{+5} $kmole.m^{-2}.s^{-1}$	C_{Bo} $kmole.m^{-3}$	μ_L centipoise	foaming effect
CO_2-MEA in water	$76.9\ pC_{Bo}^{0.5}D_A^{0.5}$	0.2-2.0	$1+0.32C_{Bo}$	-
CO_2-DEA in water	$1.1\ pC_{Bo}D_A^{0.5}$	0.2-0.8	$1+0.54C_{Bo}$	+
CO_2-CHA in toluene	$5.8\ pC_{Bo}D_A^{0.5}$	0.1-2.0	0.65	-
CO_2-MEA in ethanol	$6.3\ pC_{Bo}D_A^{0.5}$	0.2-1	1.64	-
CO_2-DEA in ethanol	$1.4\ pC_{Bo}D_A^{0.5}$	0.2-2.2	2.23	+
CO_2-MEA in ethyleneglycol	$4.1\ pC_{Bo}D_A^{0.5}$	0.2-1.1	$21+2C_{Bo}$	-
CO_2-DEA in ethyleneglycol	$0.8\ pC_{Bo}D_A^{0.5}$	0.2-2.1	21-40	-
CO_2-CHA in ethyleneglycol	$122\ pC_{Bo}^{0.5}D_A^{0.5}$	0.2-2.0	26-31	-
CO_2-NaOH in water	$2.8\ pC_{Bo}^{0.5}D_A^{0.5}$	0.2	1.0	-
O_2-Na_2SO_3 in water	$1.2\ p^{1.5}D_A^{0.5}\times10^{-3}$	0.4-0.8	1.5	+

Each time the carbone dioxide partial pressure $p_{CO_2} = p$ has been varied between 10^{-2} and 0.2 atm. except for the system CO_2-DEA in polyethyleneglycol where it has been varied up to 1 Atm. The chemical systems giving the smallest specific rate are thus recommended and seem more appropriate than the chemical system CO_2-CHA in toluene for very efficient gas-liquid reactor.

4.3 Mass transfer data with chemical and purely physical processes

In order to characterize the gas-liquid mass transfer performances of a reactor, k_La, k_Ga and a are determined by the above methods. But the question may arise whether the values of these parameters so obtained are valid only for similar mass transfer conditions and how they can be extrapolated to other operating conditions (for example from chemical absorption to physical absorption or vaporization). This means in other words, is the value of k_L or a, corresponding to the hydrodynamic conditions of chemical absorption, the same as the value of k_L or a corresponding to the hydrodynamic conditions of physical absorption ?

In fact in the case of a packed column some zones of liquid in the packing are almost motionless and they probably will be saturated by the absorbing gas during a physical absorption and hence almost ineffective to mass transfer. When the absorbing capacity of the liquid is increased by a chemical reactant or in evaporation operations, these zones will be still effective and the measurement of the amount of gas absorbed will give the feeling that k_L and a have greater values. Also in a mechanically agitated reactor the effective interfacial area in the case of a reacting system may be not equal to the effective interfacial area in the case of a physical absorption or desorption where the absorbing capacity of the liquid is not increased. Moreover in the case of the absorption system with a fast chemical reaction the mass transfer coefficient is independent of the hydrodynamics and equal at every point in the vessel and all the interfacial area distributed in all parts of the agitated vessel contributes equally to the mass transfer. But in the case of physical desorption or absorption, the mass transfer coefficient cannot be assumed to be equal at all points in the vessel (it can have quite different values around the agitator and far away from the agitator or in the region towards the top surface) and the interfacial region in different parts of the reactor do not contribute equally to the mass transfer |91|. The same considerations apply for any gas-liquid reactor that offer hydrodynamic inhomogeneity.

For the above reasons it can be seen for example why the assumption of the same value of the interfacial area in physical and chemical absorption can lead to some incertainty, especially if the mass transfer coefficient is obtained from the ratio k_La/a, where k_La is measured by physical absorption or desorption and a from chemical absorption in two different series of experiments. The effective interfacial area in the case of the fast reaction system where the absorbing capacity is increased by a chemical reactant is probably much larger than the effective interfacial area for physical absorption or desorption and in fact the semistagnant liquid zones in a packing or in the bulk of an agitated tank can be more and less effective to mass transfer depending on the ratio between the absorbing capacity and the rate of absorption, as pointed out by Joosten and Danckwerts |126|. These authors introduced a parameter γ,

$$\gamma = \frac{1 + \dfrac{C_{Bo}}{zC_A^*}}{E}$$

that is the ratio between the increase of liquid absorption capacity and the increase of mass transfer due to chemical reaction. The experimental results showed that for a physical absorption ($\gamma = 1$) and absorption with instantaneous chemical reaction ($\gamma \simeq 1$) the effective areas are the same while for $\gamma >> 1$ the interfacial area increased. The different effective areas found by Joosten and Danckwerts |126| or by Laurent |101| using equal packing and similar absorbing solutions, can be explained by the different γ values in the different cases.

However the different values of k_La, which depends on the type of mass transfer process (vaporization, chemical and physical absorption) are not only due to variations of liquid areas involved in mass transfer operations, but they are also due to variations of local mass transfer coefficients inherent to these zones |96, 128, 130| or variations of the mass transfer coefficient by the existence of interfacial turbulence caused by chemical absorption enhancing the physical absorption |193|.

So the technique of conducting simultaneous absorption with fast pseudo-mth order reaction and physical absorption or desorption concurrently is certainly a promising attempt to understand the whole complex problem of transfer in gas-liquid reactor |93-95, 127|. This leads to the simultaneous measurement of k_La and a whence k_L. But it may let some doubt on the value of k_L that, for example, can be changed by the presence of the simultaneous chemical reaction. As explained by Prasher |96|, it will be still more promising to lead these simultaneous experimental measurements in using a chemical reaction in a regime where both hydrodynamics and reaction have comparable effects (a Danckwerts plot where k_L and a are determined by the chemical reaction simultaneously with k_La determined by the physical absorption or desorption).

In conclusion it is important to know that the effective interfacial area and the liquid mass transfer coefficient determined by chemical methods using fast reactions are probably larger than the effective interfacial area for physical desorption or absorption. A correction relative to the interfacial area may be realized with the parameter γ previously defined but it must be kept in mind that also the values of k_L may vary simultaneously and should also necessitate a correction.

So finally, a perfect knowledge of gas-liquid mass transfer data theoretically necessitates simultaneous measurements of mass transfer parameters by two ways : local physical measurement of a, k_L and k_La and global chemical or physical measurement of a, k_L and k_La to follow the variations in situ of the mass transfer parameter. It seems important to remark that the techniques exist now and that the next years will provide fruitful data on that topic including nuclear engineering |129|.

4.4 Some data on mass transfer coefficients and interfacial areas. Use for scaling-up.

The choice of a suitable gas-liquid reactor is very often a question of matching the reaction kinetics with the characteristics of the proposed reactors. The specific interfacial area a, liquid holdup β and mass transfer coefficients k_La, k_Ga are the more important characteristics of gas-liquids reactors. Table I gives the typical values of these parameters that are provided by some typical contactors presented in Fig. 1, for fluids with properties not very different of those of air and liquids of viscosity μ_L smaller than 5 cP and for nonfoaming liquids.

A synthesis of the published values in 1976 has been proposed by Charpentier |139|. Since then excellent reviews have been published on each type of equipment by Zlokarnik |140|, for sparged stirred tanks and surface aerators, by Schügerl et al. |141| and Botton et al. |134| for bubble columns, by Hofmann |142 | for trickle-beds, by Shah |8, 137| and Chaudhariand Ramachandran |143| for three phase slurry reactors, by Blenke |144| for loop reactors, by Stichlmaier and Mersmann |145| for plate columns and by Brauer |146| for less conventional reactors such as multi-stage rotative discs, fluidized reactors, parallel tube packings...

We shall only present here some considerations on trickle-bed reactors and on gas-liquid stirred reactors.

4.41 Trickle-bed reactors.

Cocurrent gas-liquid flow in packed beds, packing being either catalytic or inert, is advantageously employed in the petroleum or chemical industries. One of the necessary conditions for the successfull modeling of mass transfer in packed bed reactors is a careful study of the hydrodynamics of this three-phase system (fluid flow patterns, pressure losses and liquid holdup).

In terms of absolute and relative phase velocity, concurrent flow that is not bounded by the flooding phenomena, offers a greater range of hydrodynamic patterns, that must be determined before presenting any mass transfer results.

In the case of cocurrent downflow, the packed bed reactor works in two main regimes :

. the "*trickle flow regime*" in which initially for a zero gas rate, the liquid phase trickles over the packing in the discontinuous shape of films, rivulets and drops near a stagnant continuous gas phase.

. the "*single phase liquid regime*" in which initially for a zero gas rate, the liquid phase fills the void of the packing.

Then, when keeping the mass liquid flowrate L constant and introducing and increasing the mass gas flowrate G the following visualized flow patterns are observed :

. for the initially trickle flow regime ($L < 20\ kg.m^{-2}.s^{-1}$) for spheres, beads and pellets, it appears a trickling flow of the liquid phase and a continuous flow of the gas phase ("trickling flow"), then alternate gas-rich and liquid-rich parts passing through the column ("pulse flow") and finally a poorly defined quiver that may be assumed to be a continuous gas phase with a fraction of the liquid phase suspended as a mist in the gas stream and with the other fraction covering the packing surface in the shape of a film ("spray flow").

It has to be noted that for small values of L ($L < 2\text{-}5\ kg.m^{-2}.s^{-1}$) there is not enough liquid to wet the whole packing surface and the pulse flow which is due to the obstruction of the liquid phase to the gas phase passage in the void fraction of the packing, is not encountered.

When the liquid phase is foaming in presence of the gas flowrate, there exist two new flow regimes, i.e. foaming flow and foaming pulsing flow that appear between trickling flow and pulsing flow.

Flow patterns and the transitions from one form to another as a function of the phase flowrates have been described by several authors. For example, the relative location of each of the encountered flow patterns in terms of the superficial mass velocities is presented on diagrams of Fig. 12 for various gases and water and different foaming and non foaming hydrocarbons flowing through beds of 3 mm glass and alumina spheres and 1.8×6 mm and 1.5×5 mm alumina pellets |147|. Other diagrams may be found in literature for other shape and size of packings but unfortunately most of them are confined to aqueous liquid-air system |148, 155|.

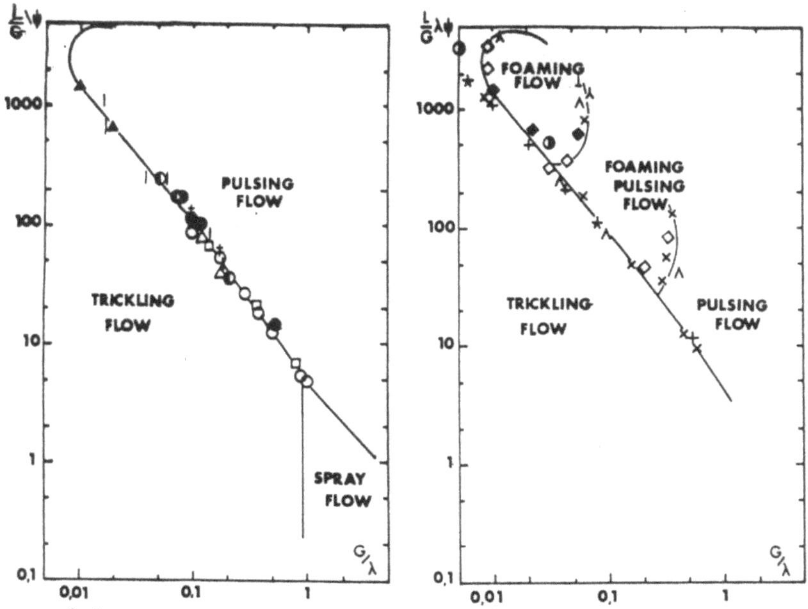

(a). Flow Pattern Diagram for Nonfoaming Liquids.

$$\lambda = \left[\frac{\rho G}{\rho_{wat}} \cdot \frac{\rho L}{\rho_{air}}\right]^{0.5} \qquad \psi = \frac{\sigma_{wat}}{\sigma_L}\left[\frac{\mu_L}{\mu_{wat}}\left(\frac{\rho_{wat}}{\rho_L}\right)^2\right]^{0.33}$$

System	Packing	Key
Water-air	Spherical catalyst	○
Cyclohexane-air	Spherical catalyst	◐
Water-air	Cylindrical catalyst 1	□
Cyclohexane-nitrogen	Cylindrical catalyst 2	△
Gasoline-carbon dioxide	Cylindrical catalyst 2	●
Gasoline-nitrogen	Cylindrical catalyst 2	■
Gasoline-helium	Cylindrical catalyst 2	▲
Petroleum ether-nitrogen	Cylindrical catalyst 2	|
Petroleum ether carbon dioxide	Cylindrical catalyst 2	‡

(b) Flow Pattern Diagram for Foaming Liquids.

System	Packing	Key
Kerosene-air	Spherical catalyst	◑
Desulfurized gas oil-carbon dioxide	Cylindrical catalyst 2	+
Desulfurized gas oil-air	Cylindrical catalyst 2	×
Desulfurized gas oil-helium	Cylindrical catalyst 2	⋏
Nondesulfurized gas oil-carbon dioxide	Cylindrical catalyst 2	★
Nondesulfurized gas oil-air	Cylindrical catalyst 2	∧
Nondesulfurized gas oil-helium	Cylindrical catalyst 2	⊥
Kerosene-air	Cylindrical catalyst 2	◇
Kerosene-nitrogen	Cylindrical catalyst 2	◆

Figure 12 - Flow regime diagram in trickle-bed reactors.

. Besides when the reactor operates initially with the single phase of liquid ($L > 20$ kg.m^{-2}.s^{-1}) either upward or downward, the following patterns are observed : after the introduction of gas, small bubbles appear unbroken in the continuous liquid phase ("bubble flow"), then these bubbles begin to coalesce and to surround several packings elements ("distorted bubble flow"), then a pulse flow and a spray flow are encountered for higher gas flow rates.

The relative location of each of these flow patterns may be found in literature for various shapes of packing but mainly again for air-water systems |149|.

Therefore, as already explained, it is easy to understand that the mass transfer data will depend on these numerous flow patterns as they offer many gas-liquid geometric configurations. So obviously mass transfer properties in cocurrent flow are related to gas and liquid energy dissipation and correlations for pressure loss and liquid holdup will be most often necessary.

4.411 Pressure drop and liquid holdup correlations.

There are several correlations to predict these parameters, the most famous being proposed by Larkins et al. |150| for co-current bubble packed reactors (± 20 %)

$$\log \frac{\delta_{LG}}{\delta_L+\delta_G} = \frac{0.416}{(\log X)^2+0.666}, \qquad \delta_{LG} = \left(\frac{\Delta H}{Z}\right)_{LG} + \frac{\beta\rho_L+(1-\beta)\rho_G}{\rho_m}$$

$$\log\beta = -0.774 + 0.525(\log X) - 0.109(\log X)^2$$

$$0.05 < X = \sqrt{\frac{\delta_L}{\delta_G}} < 30$$

δ_L and δ_G are the friction pressure losses that would exist if the liquid and the gas were assumed to flow alone separately in single phase flow with the same rates as those in two phase flow. For design, values for δ_L and δ_G may be obtained by an Ergun's type relationship as a first approximation,

$$\frac{f}{2} = \frac{\Delta H}{Z}\,\frac{\varepsilon^3}{\rho u^2 a_g(1-\varepsilon)} = h_K\,\frac{\mu a_g(1-\varepsilon)}{\rho u} + h_B$$

with different values of h_K and h_B for different catalyst packings |152-154| (Fig. 13), but a quick and easy experimental measurement of h_K and h_B for the packing to be used, will lead to more accurate values. Note that Larkins and al's relationship is not valuable when liquid is trickling but several other relationships have been proposed including this regime. For air-water flows through beds of spheres, Sato et al. |155| proposed a similar expression which is symetric about X = 1.2 instead of X = 1 :

$$\log \frac{\delta_{LG}}{\delta_L+\delta_G} = \frac{0.70}{(\log X/1.2)^2+1.00}$$

An alternate relationship has also been proposed by the same authors,

$$\left(\frac{\delta_{LG}}{\delta_L}\right)^{0.5} = 1.30 + 1.85X^{0.85} \qquad \text{for } 0.1 < X < 20$$

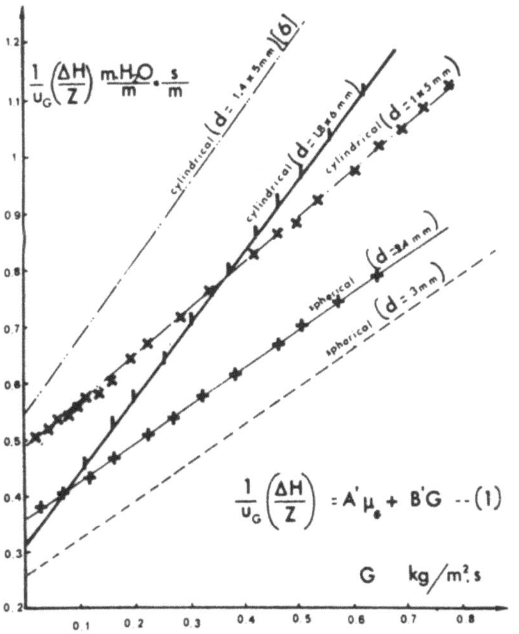

Figure 13. Influence of particle shape and diameter on pressure loss for single flow of the gas through the dry packing. |152|

Besides, it is also possible to state that for a quick and first approximation in the case of trickling flow, pressure loss may be expressed by Ergun's type relationships in which the porosity ε is replaced by $\varepsilon(1-\beta)$, the total liquid extraparticle holdup β being calculated by correlations proposed in Ref. |152-155|.

In the previous equations, u is the superficial velocity, ε the intergranular porosity of the packing, a_g the specific area of the packing, μ and ρ the viscosity and density of the considered phase, ρ_m is the density of the manometric fluid.

Instead of using the Larkins et al's two phase parameter δ_{LG} proportional to the resultant of the friction forces, Charpentier et al. |151-154| have suggested for the high gas-liquid interaction systems to use the two-phase parameters ξ_{LG}, ξ_L and ξ_G proportional to the frictional power,

$$\xi_{LG} = \frac{1}{\varepsilon}\left(\frac{L}{\rho_L} + \frac{G}{\rho_G}\right)\left(\frac{\Delta H}{Z}\right)_{LG} + \frac{L+G}{\varepsilon\rho_m}$$

with the limiting cases whether the gas is flowing alone in single phase flow through the packing or the liquid phase is flowing alone as a trickling flow over the packing. Thus a representation was adopted between these parameters with ψ_L or β versus χ',

$$\psi_L = \sqrt{\frac{\xi_{LG}}{\xi_L}}, \qquad \chi' = \sqrt{\frac{\xi_L}{\xi_G}}$$

L and G are the liquid and gas superficial mass flowrates. Finally for the case of foaming and non foaming hydrocarbons and viscous liquids together with the packing diameter (d_p = 1-3 mm) of the catalyst encountered in the trickle-bed reactors, the following correlations are suggested to determine the pressure drop and the liquid holdup :

. for nonfoaming and non viscous liquids (cyclohexane, petroleum ether, gazoline..) or for foaming liquids in trickle flow, (within ± 20 %)

$$\Phi_L = \left(\frac{\delta_{LG}}{\delta_L}\right)^{0.5} = 1 + \frac{1}{X} + \frac{1.14}{X^{0.54}}, \qquad \beta = \frac{0.66X^{0.81}}{1+0.66X^{0.81}}$$

with $0.1 < X < 80$

. for foaming hydrocarbons (gas oils, kerosene...), except at trickle flow (within ± 20 %)

$$\psi_L = \left(\frac{\xi_{LG}}{\xi_L}\right)^{0.5} = 1 + \frac{1}{X'} + \frac{6.55}{X'^{0.43}}, \qquad \beta = \frac{0.92X'^{0.30}}{1+0.92X'^{0.30}}$$

for $0.05 < X' < 100$.

. for viscous organic liquids (ethylene- and polyethyleneglycol...) within ± 20 %,

$$\psi_L = 1 + \frac{1}{X'} + \frac{7.11}{X'^{0.55}}, \qquad \beta = \frac{4.83X'^{0.58}}{1+4.83X'^{0.58}}$$

for $0.05 < X' < 100$.

It is important to note that the data usually obtained with air and water are not very representative, especially for foaming fluids. Thus it is suggested to use complementary specific correlations depending on the aqueous or the non aqueous nature of the liquid |156-158|. *Moreover it should be also emphasized that most of the data concern small diameter columns (D < 10 cm) with a good initial distribution. For reactors up to as much as 3 m in diameter, it may be that some segregation in the flow of the phases leads to less important gas-liquid interaction and therefore to smaller pressure loss especially for the non foaming liquids* |135|.

4.412 Mass transfer correlations.

Most of the published studies are reported in Table XVIII |151|.

Liquid phase mass transfer coefficient k_La is affected both by the gas rate and the liquid rate. The following correlation is recommended by Reiss |159| and Satterfield |160|

TABLE XVIII

Experimental data available in the literature for cocurrent gas–liquid downflow in packed beds |151|

Reference	Packing d (mm)	Packing ϵ	Column D (cm)	Column Z (m)	Gas	Solute	Liquid	Superficial velocity (10^{-2} m s^{-1}) u_L	u_G	$\frac{\Delta P}{Z}$	β	k_La (s^{-1})	Relationships (MKS units)
Reiss (1967)	12.5 (b)	0.72–0.78	7.5–10	0.5	Air	NH_3	H_2O	4–30	60–450	+	+	0.1–3	$k_Ga = 2.0 + 0.1E_G^{0.66}$
	25 (b)	0.79–0.88											$k_La = 0.03E_L^{0.5}$
	76 (b)	0.79			Air	O_2	H_2O						
	25 (c)	0.68	41										$E_G = \left(\frac{\Delta P}{Z}\right)_{LG} u_G E_L = \left(\frac{\Delta P}{Z}\right)_{LG} u_L$
Gianetto *et al.* (1970, 1973)	6 (a)	0.41	8	0.2	Air	CO_2	NaOH (0.5–2N)	0.08–4.4	40–230			0.011–1.6	$\frac{a}{a_g(1-\epsilon)} = 0.27\left\{\left(\frac{\Delta P}{Z}\right)_{LG}\frac{\epsilon}{a_g(1-\epsilon)}\right\}^{0.5}$
	6 (c)	0.59		0.4	Air	NH_3	Na_2SO_4 (1N)	0.25–4.3	40–220	+	–	0.009–0.9	$\frac{k_G\epsilon}{u_G} = 0.035\frac{(\Delta P/Z)_{LG}}{\Psi(\rho_G u_G^2 + \rho_L u_L^2)}$
	6 (b)	0.50		0.6								0.012–1.2	
	6 (b)	0.71			Air	O_2	NaOH (2N)	0.25–4.3	40–220			0.007–0.6	$\frac{k_L\epsilon}{u_L} = 0.03\left[\left\{\frac{\epsilon(\Delta P/Z)_{LG}}{a_g(1-\epsilon)\rho_L u_L^2}\right\}^{0.07} - 1\right]$
Nagel *et al.* (1972)	5 (a)	0.41	11	3.2	Air	CO_2	Na_2SO_3	0.3–3	1.5–33	–	–	–	Experimental curves
	5 (b)	0.64						0.3–3	1.5–60	+	–		$\frac{a}{a_g(1-\epsilon)}$ vs. u_G for u_L = const.
Ufford and Perona (1973)	6.3 (b)	–		0.25	Air	CO_2	H_2O	0.4–8.3	4.3–12	–	–	0.006–0.136	$k_La = 13.5u_L u_G^{0.75}$
	12.7 (b)	–	10	0.30									$k_La = 2.96\, u_L^{0.93} u_G^{0.42}$
	19 (c)	–											$k_La = 2.73\, u_L^{0.82} u_G^{0.46}$
Sato *et al.* (1974)	2.6 (a)	0.37	6.6	0.25	N_2	O_2	H_2O	1–20	5–100	+	+	0.02–3	$\frac{k_La}{\epsilon} = 0.074\left(\frac{u_G}{\epsilon}\right)^{0.6} E_L^{0.31}$
	5.6 (a)	0.39											
	8 (a)	0.41			N_2	CO_2	NaOH (0.1–1N)	1–10	1–100				$\frac{a}{\epsilon} = (1-\beta)E_L^{0.4}$
	12 (a)												$E_L = \left(\frac{\Delta P}{Z}\right)_{LG}\left(\frac{u_L}{\epsilon\beta}\right)$
Shende and Sharma (1974)	5 (a)	–	10		Air	CO_2	NaOH						Experimental curves
	10 (b)	–		0.84									a vs. u_G and a vs. u_L
	12.5 (b)	–			Air	O_2	$S_2O_4Na_2$	0.1–3.5	50–300	–	–	–	$k_Ga = Au_G^m u_L^n$
	16 (b)	–	15	0.90	N_2	O_2	$S_2O_4Na_2$						k_Ga in g mol cm^{-3} s^{-1} atm^{-1}
	16 (d)	0.93		1.21									10^5A / m / n
													25.4 (c) 1.64 0.64 0.38
	25.4 (b)	0.74	20		Air	SO_2	NaOH						25.4 (d) 1.57 0.65 0.40
	25.4 (c)	0.70–0.92											16 (d) 0.61 0.87 0.34
	25.4 (d)	0.90–0.94											
Goto and Smith (1975)	4.1 (a)	0.37	2.6	0.15	O_2	O_2	H_2O	0.05–0.5	0–0.8	–	+	0.002–0.007	$\frac{k_La}{D_A} = 4440\left(\frac{\rho_L u_L}{\mu_L}\right)^{0.40}\left(\frac{\mu_L}{\rho_L D_L}\right)^{0.5}$
	2.9 (e)	0.44			N_2	O_2	H_2O					0.005–0.014	$\frac{k_La}{D_A} = 9080\left(\frac{\rho_L u_L}{\mu_L}\right)^{0.41}\left(\frac{\mu_L}{\rho_L D_L}\right)^{0.5}$
Lemay *et al.* (1975)	6.3 (a)	0.45	7.6	0.76	Air	–	H_2O	1.2–4	27–55	–	–	–	$k_La\left(\frac{\mu_L}{\rho_L D_L}\right)^{0.5}$ vs. E_L for pulsing flow only

(a) Spheres; (b) Raschig rings; (c) saddles; (d) Pall rings; (e) pellet

$$k_L a = 0.0173 E_L^{0.5} \left(\frac{D_A}{2.4\times10^{-9}}\right)^{0.5}$$

where E_L is an energy dissipation term (in W/m^3) for liquid flow evaluated as $E_L = |(\Delta H/Z)_{LG} \times u_L|$; u_L is the superficial liquid velocity (in m/s) and D_A is the diffusivity of the gas in the liquid (in m^2/s). If the viscosity of the liquid differs much from that of water, a complementary correction should be applied. However it is important to note that most of the results tested to obtain Reiss's relationship are concerned with pulse flow and spray flow ($E_L > 60\text{-}100\ W/m^3$). For lower gas and liquid rates corresponding to a trickling flow of liquid over the packing, for which reported values of k_La vary from 0.01-0.1 s^{-1}, reported experimental values of k_La are smaller than those predicted by the above equation. A comparison of the experimental data published by the different authors has led Charpentier to recommend, as a first approximation, for ionic liquids,

$$k_L a = 0.0011 E_L \left(\frac{D_A}{2.4\times10^{-9}}\right)$$

for $5 < E_L < 100\ W\times m^{-3}$ with eventualy a complementary correction for viscosity. For smaller values of u_L and u_G, $\Delta P/Z$ is equal to a few cm H_2O/m of packing in trickle beds and the use of an energy dissipation term is irrelevant in trickle flow. In this case, mean value of 0.008 s^{-1} will be a good approximation for packing with $d_p > 2$ mm. For more precise results, see the experimental data of Goto et al. |161, 162|. *Finally, it is important to note that at high flowrate of gas and liquid, values of k_La may exceed 1 sec^{-1}, which can hardly be attained in other types of gas-liquid contactors such as a bubble column or an agitated vessel. On the contrary, when the liquid is trickling over the packing, k_La values are of the same order of magnitude as those obtained in countercurrent under the same working conditions.*

Gas phase mass transfer coefficient k_Ga is also affected both by the gas rate and the liquid rate. Most extensive results have been proposed by Reiss |159| for 12.5, 25 and 76 mm polyethylene Raschig rings and 25 mm intalox saddles, by Gianetto et al. |163| for 6 mm spheres, Berl saddles and glass and ceramic Raschig rings and by Shende and Sharma |164| for 25 mm ceramic - intalox saddles and propylene Pall ring and 16 mm stainless steel Pall ring. In these different studies, the values of k_Ga vary between 2 and 70 s^{-1} for liquid rates ranging from 0.5 to 30 cm/s and gas rates from 40-450 cm/s but there exists some discrepancy in the way of representating the results. So for packings not studied by the previous worker, use of the Reiss's relationship will be a first approximation for any flow pattern except when liquid is trickling,

$$k_G a = 2.0 + 0.1 E_G^{0.66}$$

where E_G is an energy dissipation term (in W/m^3) for gas flow evaluated as $E_G = [(\Delta P/Z)_{LG} \times u_G]$. *As for the trickling flow regime, the $k_G a$ values for concurrent flow are higher than those for countercurrent flow, when comparable, so the use of the correlations valuable for countercurrent will give conservative results.*

Effective interfacial area a increases with an increase in the gas as well as the liquid superficial velocity, unlike the countercurrent operation where the effective interfacial area is independent of the superficial gas velocity up to the loading point. The values of effective interfacial area offered by various packings differ substantially. For example, the extensive work of Shende and Sharma |164| leads to the conclusion that polypropylene Pall rings show better performance than polypropylene Intalox saddles and that the highest values of interfacial area are offered by stainless steel Pall rings for each size packing. In fact interfacial area are much larger than those obtained with countercurrent flows and very much greater than the geometric packing area $a_c = a_g(1-\varepsilon)$ itself for high liquid and gas flowrates (it may exceed 100 % for spray flow pattern). Following Gianetto et al. |163| it is likely that for pulse and spray flow, where a great fraction of the liquid phase is in the shape of droplets, there exists a relationship between a and the energy supplied per feed fluid volume unit and per column unit, that is the pressure gradient if gravity is ignored. The correlation proposed by the authors is,

$$\frac{a}{a_g(1-\varepsilon)} = 0.25 \left[\left(\frac{\Delta P}{Z}\right)_{LG} \frac{\varepsilon}{a_g(1-\varepsilon)} \right]^{0.5}$$

Unfortunately most other authors did not present experimental pressure loss data simultaneously with interfacial area data. So in using flow pattern diagrams and in calculating pressure loss with energetic relationships, a comparison of the above equation obtained for 6 mm packings with experimental results of other packings has been proposed by Charpentier |151|. It is concluded that if for pulse flow and spray flow, this equation is a good approximation, whatever the shape and packing size, however, such equation may fail for a trickling flow of liquid. In this case, a better conservative representation *for aqueous liquids such as sulfite or sodium hydroxide solution* is given by,

$$\frac{a}{a_g(1-\varepsilon)} = 0.05 \left[\left(\frac{\Delta P}{Z}\right)_{LG} \frac{\varepsilon}{a_g(1-\varepsilon)}\right]^{1.2}$$

for $\left(\frac{\Delta P}{Z}\right)_{LG} \frac{\varepsilon}{a_g(1-\varepsilon)} < 12\ N/m^2$

This equation is proposed mainly for spheres and pellets packings with the porosity $\varepsilon < 0.50$.

More recently, Morsi et al. |123, 124| compared the interfacial area results obtained in trickle flow with organic and aqueous systems for spherical glass spheres (d_p = 1.16 and 2.4 mm), spherical aluminous catalyst (d_p = 2.4 mm) and glass Raschig rings (d_p = 6.48 mm). Some of the data are presented in Fig. 14 together with the working conditions and the gas-liquid systems used.

It has appeared that for very small packings, the ionic solutions may give interfacial area 5 to 10 times higher than for non foaming organic liquids. So except for small d_p, all the other data are better represented by the correlation (within ± 30 %)

$$a = 7.75\times10^5 \frac{\varepsilon}{a_g(1-\varepsilon)} \delta_{LG}$$

for $3\times10^{-5} < \frac{\varepsilon}{a_g(1-\varepsilon)} \delta_{LG} < 5\times10^{-4}$

When the holdup and pressure drop are correlated with the correlations ψ_L and β versus X, proposed in par. 4.411.

These seem the first published results for organic systems with catalytic packings and confirm the non always applicability of the mass transfer parameters determined with aqueous chemical systems to the case of organic liquids in gas-liquid reactors.

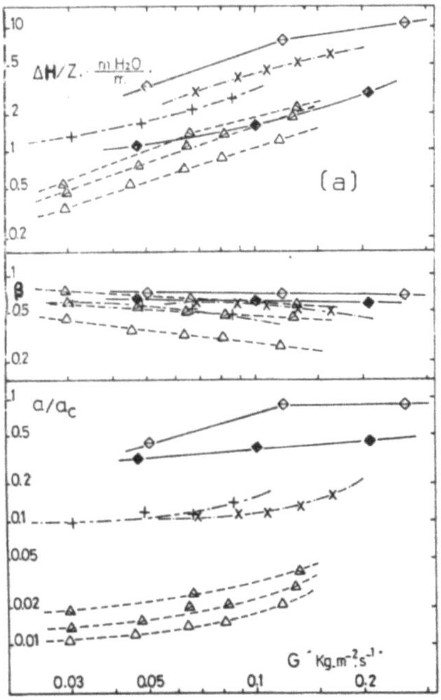

Fig. 14 Head loss, holdup, interfacial area vs G for glass beads $d_p = 1.16$ mm

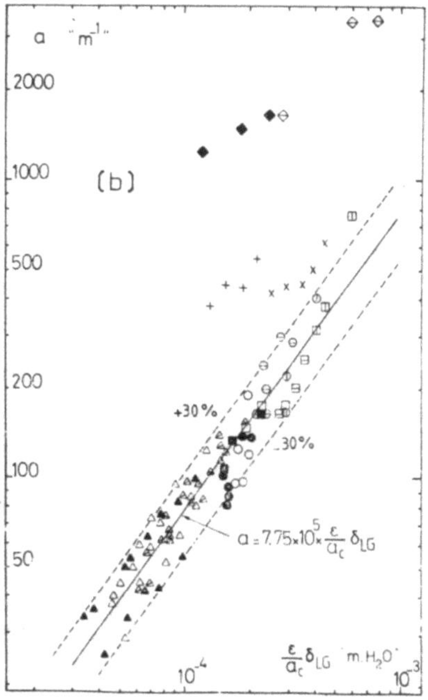

Correlation of interfacial areas

Characteristics of packings and operating conditions

System		CO$_2$–CHA–Toluene + IPA–AIR						CO$_2$–NaOH–AIR		Na$_2$SO$_3$–AIR
C_{BO} (kgmole·m^{-3})		0.2275	0.305	0.350	0.7125	0.2255	0.350	0.20	0.222	0.8 – 0.4
ρ_L (kg·m^{-3})		857	858	859	861	857	858	1003	1005	1079
$\mu_L \times 10^3$ (Pa·s)		0.664	0.656	0.671	0.692	0.644	0.656	1.04	1.08	1.46
$\sigma_L \times 10^3$ (N·m^{-1})		27.6	27.8	28.0	28.2	27.6	27.8	62.8	62.5	66.35
Packing†		G. B.				R. R.		G. B.	R. R.	G. B.
d_p (mm)		1.16				6.48		1.16	6.48	1.16
a_c (m^{-1})		3810				750		3810	750	3810
ϵ		0.263				0.704		0.263	0.704	0.263
L kg·m^{-2}·s^{-1}	0.5		▲	▲	▲					
	1	△	△	△	△	●	●		■	◆
	2	◬	◬	◬	◬	○	○	+	□	
	4					⊖	⊖	×	⊟	◈
	5	◬								
	9						⦶		◫	
G kg·m^{-2}·s^{-1}		0.03	0.03	0.03	0.03	0.103	0.103	0.031	0.10	0.047
		0.132	0.119	0.120	0.128	0.37	0.38	0.163	0.38	0.264
φ kgmole·m^{-2}·s^{-1}		$3.52 \times 10^{-9} \times p \cdot C_{BO}$						$1.07 \times 10^{-9} \times pC_{BO}^{1/2}$		$5.89 \times 10^{-14} \cdot p^{3/2}$

†G. B., glass beads; R. R., Raschig rings

— Mass transfer liquid to solid. Where both gas absorption and particle resistance are significant, it may be necessary to calculate an overall transfer coefficient $k_S a_S$. There are several studies on solid-liquid mass transfer under two-phase downflow conditions (Table XIX). Sato et al. |165| reported values of k_S from the benzoic acid-water system for 5.5 and 12.2 mm particle diameter material in gas and liquid flow rates ranges corresponding to trickle, pulsing and dispersed bubble flow. At low gas flowrates ($G < 0.1\ kg.m^{-2}.s^{-1}$) the results seem to agree reasonably well with an extrapolation of van Krevelen and Krekel's correlation for film like flow |136| :

$$\frac{k_S}{a_g(1-\varepsilon)D_L} = 1.8 \left|\frac{\rho_L \mu_L}{a_g(1-\varepsilon)\mu_L}\right|^{0.5} \left(\frac{\mu_L}{\rho_L D_L}\right)^{0.33}$$

Actually Sato et al. expressed their particle mass transfer coefficients in terms of an enhancement factor that represents the ratio of k_S with two phase flow to k_S at the same liquid flowrate in the absence of gas flow. This enhancement factor was found to be inversely proportional to liquid holdup β, which in turn is a function of the two-phase parameters X or X'. Lemay et al. |166| reported also values of k_S for 6.3 mm diameter benzoic acid spheres in water by use of a fluorescent dye, rhodamine B, dissolved as a 1.0 Wt % solution in molten benzoic acid used to make cast spheres. Their data concern the pulsing flow regime and are in good agreement with the Sato et al's correlation. However they have proposed an alternate representation with a term of power dissipation per unit mass of liquid holdup which necessitates the knowledge of two-phase pressure loss and liquid holdup. Overall, their results suggest that k_S is practically insensitive to radial position.

The results of Sylvester and Pitayagulsarn |167| concerning 3.2 mm pellets of benzoic acid gives higher values for k_S than the values proposed by the two previous studies (a multiplication factor varying between 4 and 12) which may be due to the different shape and size of the particles. However the data concerning $k_S a_S$ seems still in reasonably good agreement with the other studies.

Goto and Smith |162| measured mass transfer rates from particle to liquid for short beds of β-naphtol particles and pellets. Runs were made with smaller particle diameters (0.54 and 2.41 mm) and smaller phase flowrates than the three previous studies. Possibly the entire external surface is not effective for mass transfer in trickle beds containing very small packings and Goto and Smith concluded that the effect of particle size on $k_S a_S$ can change for particles smaller than about 2 mm. Then they proposed two equations that are specific for a particular size and shape and bed packing (Table XIX). It should be noticed that these relationships are independent of the gas flowrate what is understandable so low are the flowrates of their study.

More recent data including the elegant electrochemical technique have been regrouped in Figure 15 |171| where

TABLE XIX

Experimental data available in the literature for liquid-to-solid particle mass transfer for cocurrent gas–liquid downflow in packed beds [151]

Reference	*Packing* d $(10^{-3}$ m)	*Packing* ϵ	*Column* D $(10^{-2}$ m)	*Column* Z (m)	*Solute*	*Liquid*	*Superficial velocity* $(10^{-2}$ m s$^{-1})$ u_L	*Superficial velocity* $(10^{-2}$ m s$^{-1})$ u_G	k_S $(10^{-2}$ m s$^{-1})$	$k_S a$* (s^{-1})	*Relationships* (MKS units)
Sato *et al.* (1972)	5.5 12.2 spheres of benzoic acid	0.41 0.45	6.6 6.6	0.05 0.20	Benzoic acid	H_2O	1–20	1–100	0.003–0.05 0.002–0.03 t = 15–27 °C	0.02–0.16 0.006–0.05	$\frac{k_S}{aD_L} = \frac{1.05}{\beta} Re^{0.5} Sc^{0.33}$
Goto and Smith (1975)	0.54 2.4 β-naphthol and naphthalene pellets and particles	0.52 0.48	2.6 2.6	0.020 0.026	β-naphthol	H_2O	0.05–0.5	0–0.8	0.008–0.04 0.001–0.003 t = 25 °C	0.04–0.2 0.01–0.03	$\frac{k_S}{aD_L} = 69.9 \times 10^3 \times a^{-1.33} Re^{0.67} Sc^{0.33}$ for 0.54 mm particles $\frac{k_S}{aD_L} = 34.1 \times 10^3 \times a^{-1.44} Re^{0.56} Sc^{0.33}$ for 2.4 mm particles
Lemay *et al.* (1975)	6.3 spheres of benzoic acid	0.45	7.6	0.76	Benzoic acid with rhodamine B as dye	H_2O	1.2–4	27–55	0.007–0.012 t = 24 °C	0.04–0.07	$k_S = 0.2\, Sc^{-0.66} \left(\frac{E'_L \mu_L}{\rho_L}\right)^{0.25}$ $E'_L = \left(-\frac{\Delta P}{Z}\right)_{LG} \frac{u_L}{\beta \rho_L}$
Sylvester and Pitayagulsarn (1975)	3.2 benzoic acid pellets	–	15	0.026	Benzoic acid	H_2O	0.3–1.8	18–72	0.016–0.07	0.01–0.07	$k_S = 0.13 \times 10^{-4} \frac{G^{0.38}}{L^{0.78}} \left(1 - \frac{e^y}{1 + e^y}\right)$ $y = L/8$, L and G in lb ft^{-2} h^{-1}

* a calculated by assuming that the particles have a spherical diameter d.
$a = a_c = 6(1 - \epsilon)/d = a_g(1 - \epsilon)$; $Re = \rho_L u_L / a\mu_L$; $Sc = \mu_L / \rho_L D_L$.

$$Sh\Phi = Sh' = \frac{k_S a_S}{a_g(1-\varepsilon)} \frac{d_p}{D_L} \quad \text{and} \quad Re_L = \frac{\rho_L u_L d_p}{\mu_L}$$

D_L is the liquid diffusivity and u_L is the superficial liquid velocity. It may be seen that there still exists a great discrepancy and each correlation should be carefully studied before being used. It has also to be noticed that all the data concern aqueous liquids in absence of chemical reaction.

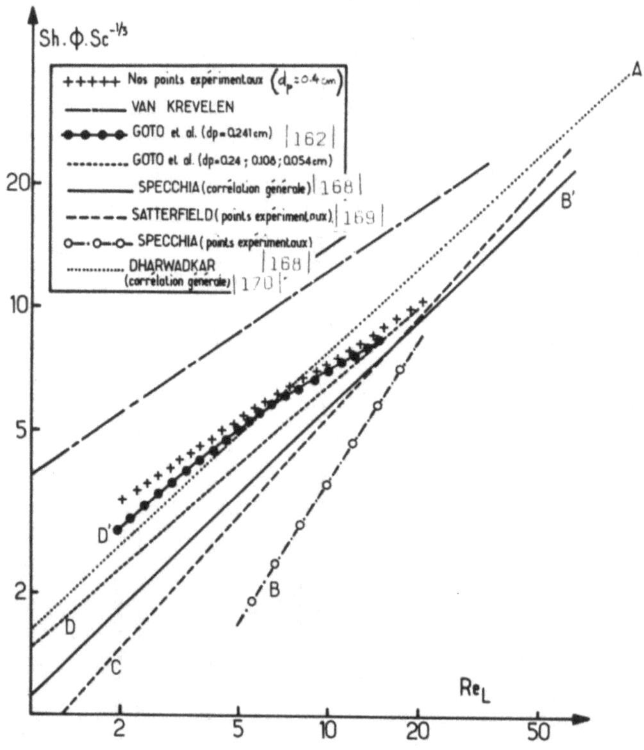

Figure 15 - $k_S a$ values in trickle-bed reactors |171|

4.42 Mechanically agitated bubble reactors.

Mechanically agitated bubble contactors are very effective gas-liquid reactors when viscous liquids or slurries, or very low gas flowrates or when a large volume of liquid are necessitated for carrying out the chemical process. They are also used for the ease with which the intensity of agitation and the residence time of the liquid can be varied and the heat can be removed. Their principal disadvantage is that both the liquid and the gas phase are almost completely backmixed. A variety of agitators are employed, but the most common types are the turbine with straight or inclined blades and the flow turbine. The gas may be introduced through a perforated tube or through a porous or perforatéd plate. The surface aerators will not be considered here.

4.421 Hydrodynamics. The three main parameters are the flooding of the stirrer (the stirrer speed N must be sufficient to assure a good dispersion of the gas inside the whole reactor |172|), the agitation power P_g and the gas holdup $\alpha = 1-\beta$.

If the correlations are well established to determine the agitation power in absence of a gas phase, P_o, for different stirrers |173|, the situation is more complex for the case of aerated tanks. Most authors propose to represent the decrease of the agitation power in presence of a gas phase at the same stirring velocity, by correlations between P_g/P_o and the aerated number $N_A = Q_G/Nd_A^3$. Q_G is here the volumetric gas flowrate and d_A is the agitator diameter. However recently Dickey |174| has reviewed the literature data and experimented air-water systems in a tank of diameter D = 1.52 m with three turbines agitators, d_A = 0.38, 0.51 and 0.61 m for $0.02 < N_A < 0.4$ which corresponds to industrial applications ($N_A > 0.06$ for large diameter tanks). The data, plotted in Fig. 16 are not very encouraging. It may just be concluded that for $N_A > 0.1$, a good approximation is $0.3 < P_g/P_o < 0.6$ with a majority of data $P_g/P_o = 0.4$. For smaller diameter tanks (D < 0.3 m) and for liquids having physico-chemical properties different of water, Hassan and Robinson |175| and Luong and Valesky |176| introduce a Weber number,

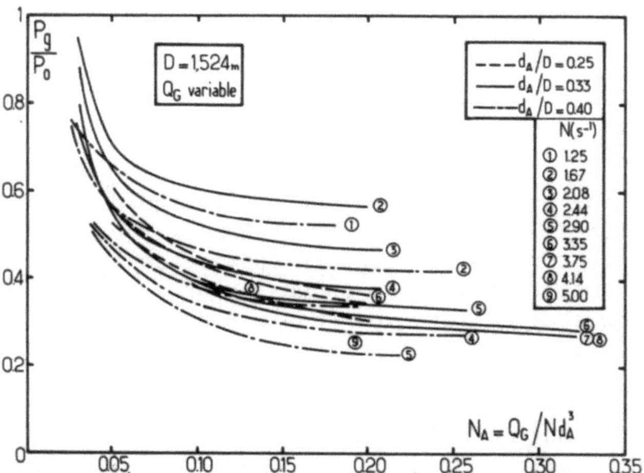

Fig.16 P_g/P_o versus N_A à N et Q_G variables (d'après DICKEY |174|

$$\frac{P_g}{P_o} = CN_A^m We^n, \qquad We = N^2 d_A^3 \rho_L / \sigma_L$$

with C = 0.497 ; m = -0.38 ; n = -0.18 for newtonian liquids and C = 0.514 ; m = -0.38 ; n = -0.194 for non newtonian liquids (with $N_A < 0.06$). A purely empirical but more general correlation is the Michell and Miller |177| plot represented by

$$P_g = \text{Const}\left(\frac{P_o^2 N d_A^3}{Q_G}\right)^m = \text{Const } M^m \qquad \text{for} \quad 1 < M < 10^9$$

the constant and m depend on the tank-stirrer system used |173|. For example Midoux and Charpentier |173| have shown that for coalescing liquids, m = 0.45 and moreover Const = $0.34\sqrt{n_p}$ if the agitator is of turbine type with n_p blades. It has also been shown that this correlation could be used for scaling-up to 800 times in volume, for non newtonian liquids and for light suspensions as long as the agitation regime is turbulent ($R_A > 2000$) |173, 178|.

Besides for ionic liquids inhibiting coalescence or foaming or for a gas distribution by-passing the stirrer (a perforated plate distributor), it is observed that $m \simeq 0.33$.

The variation of the gas holdup with the working parameters is also very complex and depend on the coalescence or foam ability of the gas-liquid dispersion. *For pure coalescing liquids*, Calderbank |179| has proposed,

$$\alpha = \left(\frac{u_s}{u_t}\alpha\right)^{0.5} + 2.16\times10^{-4}\left(\frac{\varepsilon_A \rho_L^{0.2}}{\sigma_L^{0.6}}\right)\left(\frac{u_s}{u_t}\right)^{0.5}$$

where ε_A is the agitation power per unit liquid volume, u_t is the terminal velocity of a single bubble and u_s the gas superficial velocity. This correlation concern tanks diameter up to 100 l. Sridhar and Potter |184| have shown recently that this correlation could be applied for experiments under pressure (up to 10 Atm) with a gas density correction and using the total energy ε_T given to the dispersion by the agitator, the inlet gas kinetic energy and the expansion of the sparged gas. So the second term of the previous equation has to be multiplied by $(\varepsilon_T/\varepsilon_A)(\rho_g/\rho_A)^{0.16}$, ρ_A being the density of air at the operating conditions.

For larger diameter tanks with one or several stirrers and for non foaming liquids with a large range of physico-chemical properties different of air-water, the literature data can be regrouped as

$$\alpha \sim \varepsilon_T^m u_s^n$$

with m and n depending on the agitator geometry |173|.

For ionic and inhibiting coalescence system the results may be regrouped as

$$\alpha \text{ or } \frac{\alpha}{1-\alpha} = 0.0051\ \varepsilon_T^{0.57}\ u_s^{0.24}$$

This correlation gives α or $\alpha(1-\alpha)^{-1}$ depending that the dispersion height or the clear liquid height is equal to the tank diameter |173|. The ionic strenght does not seem to have a great influence and actually only the surface tension gradients in the liquid film located between two bubbles govern the behaviour of the dispersion.

Moreover the gas holdup is very much influenced by additives such as coalescence inhibitors or micronic particles that decrease the sizes of the bubbles and increases α |180, 181|.

4.422 Mass transfer. Most of the data concern the liquid-side mass transfer coefficient k_La and the interfacial area a reported either to the dispersion (or tank) volume or to the liquid volume.

For k_La values, the data may be regrouped as $k_La \sim \varepsilon_A^m u_s^n$ or $k_La \sim \varepsilon_T^p u_s^q$, the values of exponents m, n, p and q depending on the coalescing ability of the dispersion and the geometry of the stirrer. So recently Van Riet |132| has proposed for the literature data concerning non viscous liquids two correlations : one *for pure water* (within 40 % and for tank diameter D up to 2.6 m^3) :

$$k_La = 0.026\ \varepsilon_A^{0.4}\ u_s^{0.5}$$

the second for *ionic solutions inhibiting coalescence* (within 40 % and for tank diameter up to 4.4 m^3)

$$k_La = 0.002\ \varepsilon_A^{0.7}\ u_s^{0.2}$$

The data concern $0.5 < H/D < 1.5$; $0.5 < \varepsilon_A < 10$ kw/m^3. They are independent of the agitator geometry, its position in the tank and concern a well dispersed solution. The agitation power and k_La are reported to the clear liquid volume. H is the tank height.

Taking into consideration the shape of the distributor, the gas-liquid physico-chemical properties, the foaming effects and high gas flowrates, Midoux and Charpentier |173| have proposed comparable correlations in which the specific agitation energy ε_A was replaced by the specific total energy ε_T and a coefficient B was introduced,

$$B = \left(\frac{\rho_L D_A^2}{\mu_L}\right)^{1/3} \cdot \left(\frac{\rho_L D_A^2}{\mu_L}\right)^{-1/3}_{O_2\text{-water}}$$

to precise the influence of the nature of the dispersion on k_L (Fig. 17). ε_D is the energy per unit liquid volume due to the expansion of the sparged gas.

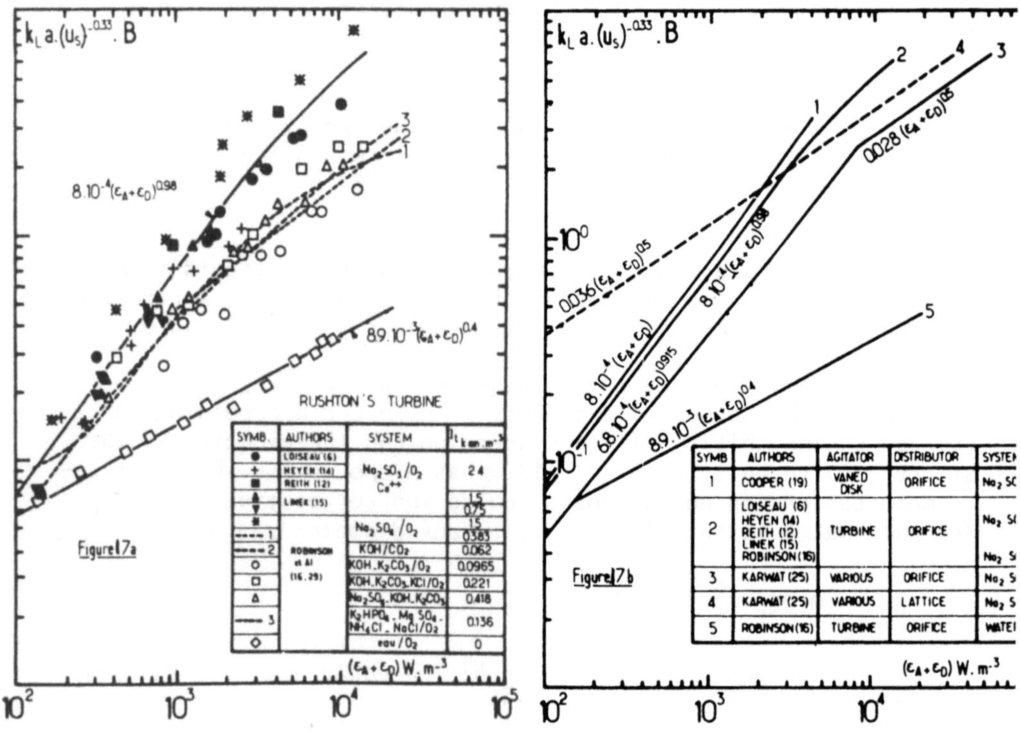

Figure 17 – Liquid-side mass transfer coefficient versus mechanical power |173|

It must be said that it is very difficult to predict with a good accuracy the k_La values for foaming systems which are much higher than the values obtained with water or non foaming liquids. The behaviour is comprised between the two extreme curves of figure 17a. The difficulty is also important when solids are present. For particles diameter 50-200 μ, it has been observed that k_La is markly decreased when the solid volumetric concentration leads to an apparent viscosity which is more than four times the viscosity of the liquid alone |183|. Note that the presence of polymers additifs leads also to a considerable decrease of k_La |180|.

The interfacial areas are correlated by relations involving either the specific agitation energy, or the total energy or the minimum efficient stirring speed. Together with the gas holdup, Calderbank |179| proposed,

$$a(1-\alpha) = 1.44 \left(\frac{\varepsilon_A^{0.4}\rho_L^{0.2}}{\sigma_L^{0.6}}\right)\left(\frac{u_s}{u_t}\right)^{0.5}$$

for large ranges of surface tension σ_L and density ρ_L including water, alcohol, CCl_4, ethylacetate, nitrobenzene and toluene. This correlation was modified by Sridhar and Potter |185| for reactors working under pressure up to 10 Atm. The data concern small volume reactors (less than 100 litres). *More generally, many literature results can be regrouped as* $a \sim \varepsilon_T^m u_s^n$ *with* $m=0.4$ *;* $n < 0.5$ *for coalescing fluids and* $m = 0.8-1$ *and* $n = 0$ *for foaming and inhibiting coalescence fluids.* For example Figure 18 present some data concerning ionic solutions in different size tanks |173|.

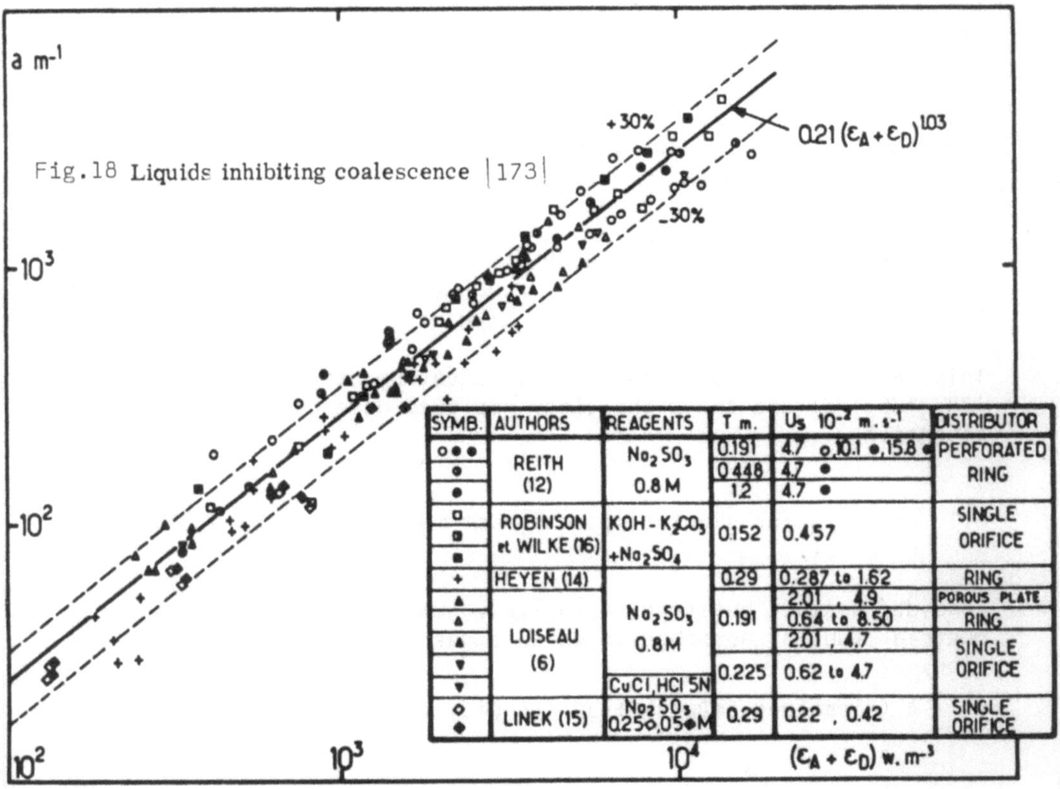

Fig.18 Liquids inhibiting coalescence |173|

Finally for both a and k_La, it would be advisable to carry out experiments with a new gas-liquid system in a small volume reactor (less than 100 ℓ) and the scaling up must be led at the same superficial velocity with a and $k_La \sim \varepsilon_T^{0.8-1}$ for foaming and ionic dispersions and a and $k_La \sim \varepsilon_T^{0.4}$ for pure and non coalescing organic liquids.

For trickle-bed reactors and for well stirred tank reactors, we have seen for both ionic and organic liquids that very often the correlations to use to determine either the hydrodynamic parameters or the mass transfer parameters depend on the foaming or coalescing ability to the dispersion. The "a priori" knowledge of this ability could be obtained by the two bubbles coalescence equipment described in Ref. |152|.

5. PREDICTION OF THE PERFORMANCES OF INDUSTRIAL PILOT-SCALE GAS-LIQUID ABSORBERS BY LABORATORY-SCALE MODELS : SIMULATION

The scaling of the experimental data carried out in a gas-liquid laboratory scale reactor up to the pilot or industrial scale is an important problem. Indeed, as already lengthly explained, the determination of the solubility and the diffusivity of one or several solutes in a reacting solution with sometimes unknown or complicated kinetics is very often a challenging problem. Moreover, it is not always possible to experiment in reactors of continuously increasing sizes because of time and/or equipment cost. These difficulties have justified the idea of making relatively simple laboratory models (or mock-ups) with well-defined interfacial areas and gas and liquid phase contact times (Fig. 19) and carrying out experiments to obtain information in a form that could be directly applicable to design. *The aim is thus to predict the effect of the chemical reaction in an industrial absorber from tests in the laboratory model with the same gas-liquid reactants, or to predict the reactor length for a specified duty using data from the laboratory model, even though the means of agitating both phases in the two types of equipment is quite different.*

This technique has been intensively developed these last years |78, 185-195| and the methodology employed deserves a particular attention.

5.1 Criteria for simulating an industrial absorber by a laboratory model

Consider a small but representative volume element Ωdh of an industrial tubular absorber in which a gas-liquid reaction is occuring,

A (dissolved solute) + zB (liquid reactant) → Products

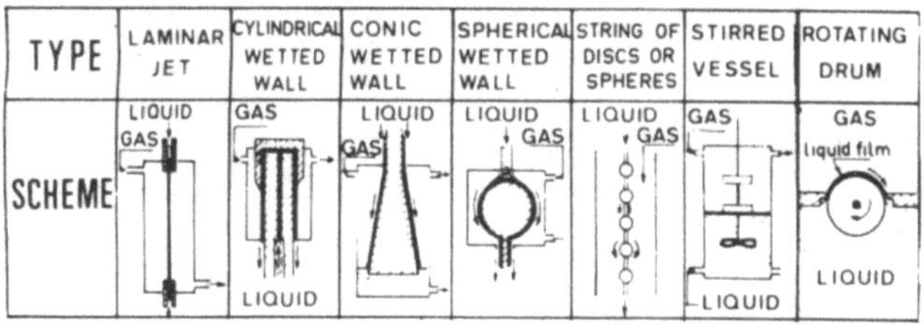

MODEL	ROTATING DRUM	MOVING BAND	LAMINAR JET	WETTED-WALL COLUMN SPHERE	WETTED-WALL COLUMN CYLINDER	WETTED-WALL COLUMN CONE	STRING OF DISCS - SPHERES	STIRRED VESSEL
TIME OF CONTACT (s)	2.10^{-4} 10^{-1}	6.10^{-4} 6.10^{-2}	10^{-3} 10^{-1}	10^{-1} 1	10^{-1} 2	2.10^{-1} 1	10^{-1} 2	6.10^{-2} 10
INTERFACIAL AREA A_m (cm^2)	2 100	0.6 6	0.3 10	10 40	10 100	80	30 360	4 80
k_L ($cm.s^{-1}$)	0.016 0.356	0.021 0.210	0.016 0.160	0.005 0.016	0.0036 0.016	0.005 0.011	0.0036 0.016	0.0016 0.021
$10^5\ k_g$ ($mole.cm^{-2}.s^{-1}.atm^{-1}$)	—	—	10 40	1 — 9			1 25	1 15
A_m/V_m (cm^{-1})	100 1250	80 400	20 80	20 60	25 60	40 70	20 60	0.002 0.540

Figure 19 - Main Laboratory Models |1|

For piston flow, the material balance equations (1a and 3a) for the case when the gas contains only one soluble component and the liquid only one reactant may be presented as

$$\varphi(k_L,k_G,C_A^*,C_{Ao},C_{Bo})a\Omega dh = -\frac{\Omega u_G}{V_M P}dp_h = u_L\Omega dC_{Ao}+r(C_{Ao},C_{Bo})\beta\Omega dh \quad (53)$$

$$-u_L\Omega dC_{Bo} = zr(C_{Ao},C_{Bo})\beta\Omega dh \quad (54)$$

where P is the pressure in the absorber, u_L and u_G are the superficial liquid and mean gas velocity, β is the liquid holdup, p_h is the bulk partial pressure of solute gas at length h, V_M is the molar volume of the gas, $r(C_{Ao},C_{Bo})$ is the reaction rate, C_{Ao} and C_{Bo} are the bulk concentration of reactants, and $\phi(k_L,k_G,C_A^*,C_{Ao},C_{Bo})$ is the rate of absorption per unit interfacial area in the volume element (Eq. 5).

Rearranging these two equations leads to the double numerical integration over the length or the height H of the absorber,

$$\int_{in}^{out} \frac{a\ dh}{u_L} = \int_{in}^{out} \frac{dC_{Ao}}{\phi(k_L,k_G,C_A^*,C_{Ao},C_{Bo})} - \frac{1}{z}\int_{in}^{out} \frac{dC_{Bo}}{\phi(k_L,k_G,C_A^*,C_{Ao},C_{Bo})}$$

$$= -\frac{1}{V_M P}\int_{in}^{out} \frac{u_G dp_h}{u_L\phi(k_L,k_G,C_A^*,C_{Ao},C_{Bo})} \tag{55}$$

$$\int_{in}^{out} \frac{\beta dh}{u_L} = \frac{1}{z}\int_{out}^{in} \frac{dC_{Bo}}{r(C_{Ao},C_{Bo})} \tag{56}$$

Consider now a laboratory model absorber in which the liquid and gas are agitated in such a way as to give rise to mass transfer coefficients, k_G and k_L, of the same magnitude as those occuring in the industrial absorber. Also assume that the ratio u_G/u_L, the concentrations C_A^*, C_{Ao}, C_{Bo}, the temperature and the pressure in the laboratory absorber are the same as those of the volume element located at the height, h, of the industrial absorber. Thus the specific rate of absorption $\phi = \phi(k_L,k_G,C_A^*,C_{Ao},C_{Bo})$ is the same in both apparatuses and the right hands of Eq. (55) are also the same if they are concerning the same concentration limits or partial pressure limits (i.e. entrance and exit bulk concentrations) in both apparatuses.

It comes that the ratio $\frac{ah}{u_L}$ is identical for both absorbers, or,

$$\frac{h}{h_m} = \frac{u_L/a}{(u_L/a)_m} \tag{57}$$

where subscript m is concerning the laboratory model. This defines the scaling ratio. Similar considerations applied to Eq. (56) lead to,

$$\frac{\beta h}{u_L} = \frac{\beta_m h_m}{(u_L)_m} \tag{58}$$

which means that the space time is the same in the volume element of the industrial absorber and in the laboratory model.

The combination of Eqs. (57) and (58) gives,

$$\frac{a}{\beta} = \frac{a_m}{\beta_m} = \frac{A_m}{\beta_m V_m} = \frac{A_m}{V_L} \tag{59}$$

where A_m is the geometrical interfacial area of the model and V_L is the liquid volume of the model of volume V_m. Equation (59) means that by unit volume of absorber or reactor, the ratio of interfacial area to liquid holdup is the same in the laboratory model and in the industrial absorber.

Therefore the three equipment criteria for simulating are identical values of k_L, k_G and a/β in the industrial and the laboratory absorber. If this is combined with identical values of u_L/a and u_G/u_L (or u_G/a), *the word "simulation" means that if the bulk compositions of gas and liquid in the laboratory absorber are the same as those of a volume element of this industrial absorber, the absorption rate per unit interfacial area* ϕ *in this volume will be the same as that in the laboratory model* ϕ_m *whatever the means of agitating the gas and the liquid in both absorbers. Thus* $\phi_m = \phi$ *can be determined experimentally as a function of* C_{Ao}, C_{Bo} *or* p *and Eq. (54) integrated numerically step by step between the limit compositions at the entrance and the exit of the industrial absorber to find the length, H of the absorber.* Note that the third criterion for simulation is necessitated for instance when the reaction between the dissolved gas and a reactant in solution is slow. Indeed there will be reactions proceeding in the bulk of the liquid and the rate of absorption in industrial absorber or in laboratory absorber will depend on the volume of bulk liquid available per unit area of interface.

5.2 Practical selection of a laboratory model to simulate an industrial absorber

5.21 Absorption and reaction in the liquid-film with a pure gas. The first criterion for simulation involves identical values of k_L in both apparatuses. Indeed the average rate of absorption per unit interfacial area of a (pure) gas solute into a liquid with which it reacts is determined partly by physico-chemical factors (solubility, diffusivity and kinetics...) and partly by hydrodynamics factors (gas and liquid flowrates, shape and size of the absorber...). Besides, it is known that various simple theoretical models of the absorption with chemical reaction process lead to conclusions about the effect of the chemical reaction with simple kinetics that are numerically in very close agreement whether the film model or the various surface renewal models are used as the basis of calculation. This means that for a fixed value of k_L, the different theories give approximately the same value for the average rate of absorption and it follows that if a laboratory model of an industrial pure gas absorber is to be designed, the essential feature to be reproduced is the true liquid side mass transfer k_L (i.e. the same mean contact time). Various laboratory models designed for simulating absorbers have been presented in

Figs. 7 and 19. Besides the ranges of contact times and k_L values for different industrial gas-liquid absorber and laboratory apparatus are presented in Table XX.

Laboratory model	Contact time (s)	Industrial absorber	Contact time (s)	Laboratory model to be used for simulation
Cylindrical wetted wall	0.1 - 2	Plate column	0.019 - 0.04	Laminar jet Rotating drum
Spherical wetted wall	0.1 - 1	Packed column	0.03 - 1	Cylindrical wetted wall String of discs or of spheres Stirred vessel
Conical wetted wall	0.2 - 1	Venturi	0.004	Laminar jet
Laminar jet	0.001 - 0.1	Spray column	0.03 - 0.1	Laminar jet Rotating drum
String of discs or of spheres	0.1 - 2	Well stirred tank	0.01	Laminar jet Rotating drum
Rotating drum	0.0002 - 0.1	Bubble column	0.01	Laminar jet Rotating drum
Stirred vessel	0.06 -10			

Table XX - Models for simulating absorbers based on equality of k_L

This table allows for the selection of the laboratory equipment to simulate the industrial absorber in comparing the contact time (or k_L). For instance, it may be seen that any wetted wall or stirred vessel can simulate a packed column and that a laminar jet or a rotating drum can simulate a plate column.

5.22 Absorption and reaction in the liquid-film with a dilute gas solute.

Now when the absorption process is concerned with a dilute solute in the gas phase, there is resistance in the gas phase as well. Then simulation principle involves also identical values of the true gas side mass transfer k_G in both equipments. In the laboratory model, the values of k_G may be adjusted to be the same as that of industrial equipment whether in varying the gas flowrate |7| or in stirring the gas phase independently of its flowrate and of its velocity |7, 186, 195, 196|. Some experimental values of k_G in laboratory models are presented in the Table of figure 19.

5.23 Absorption and complex reaction with a dilute gas solute.
In the two previous cases the reaction between the dissolved gas and the liquid reactant was assumed fast enough to occur only within the liquid film. But when the reaction and mass transfer resistance are of the same magnitude,that is,when reaction occurs both within the film and within the main body of the liquid or when the chemical reaction is low and proceeds only in the bulk of the liquid, the third equipment criterion for simulation is necessary, i.e. equality of the ratios of the interfacial area to the liquid holdup. This criterion may be satisfied by the laboratory double mixed contactor suggested by Levenspiel and Godfrey |100| and schematically presented in the right side of Fig. 7. In essence the proposed device is a steady flow contactor having uniform gas composition and uniform liquid composition throughout by stirring independently both phases, as well as having provisions for independently changing and controlling interfacial area and the surface to volume ratio of the phases by using interface plates having holes so as to give 2 to 30 per cent open areas. A string of spheres with pools full of liquid at the top of each sphere is another such type of laboratory model |190, 192|.

5.3 Practical examples of simulation

5.31 A stirred cell to simulate a "point" in a packed column : use of a point or differential model.

Several examples of the use of a stirred cell to simulate a packed column are found in literature. Let us now describe the practical application of this laboratory model to predict the absorption rates at different levels of a column when an absorption with fast second order irreversible chemical reaction is occuring with mass transfer resistance in both phases.

The variations of k_L and k_G with the respective phase stirring-rates have first to be defined. In the liquid phase, the experimental determination of k_L as a function of liquid side stirrer speed N_L is carried out by physical absorption or by absorption with a slow chemical reaction or by absorption with an instantaneous chemical reaction of a pure gas for a well known geometrical interfacial area (generally the cross sectional area of the stirred cell minus that of the blade type stirrer). In order to determine k_G as a function of gas side stirrer speed N_G, absorption with an instantaneous chemical reaction of a dilute solute is carried out. Note that these values of k_L and k_G are specific to the geometry of the laboratory apparatus and to the absorbed solute gas. The value of k_G for other gases under the same conditions can be estimated by correcting for the diffusivity D_G of solute gas and assuming that the physico-chemical properties of the gas stream are about the same . k_G may be assumed to vary as the square root of the diffusivity. In the same way the values of k_L for other liquids under the same conditions can be estimated by correcting for the diffusivity D_A of the dissolved gas and may be assumed to vary as the square root of this diffusivity.

Similarly when the values of a, k_La and k_Ga are known for the packed column, it is possible then to determine the variations of k_L and k_G as a function of the respective flowrates u_L and u_G. Such predicted values are well established now |2|. One can thus tabulate the stirring speeds of each phase in the laboratory model corresponding to the phase flowrates in the packed column for which k_L and k_G are respectively identical. The walk leading to such tabulation is schematically presented in Fig. 20. So each couple u_L and u_G is "simulated" by each couple N_L and N_G.

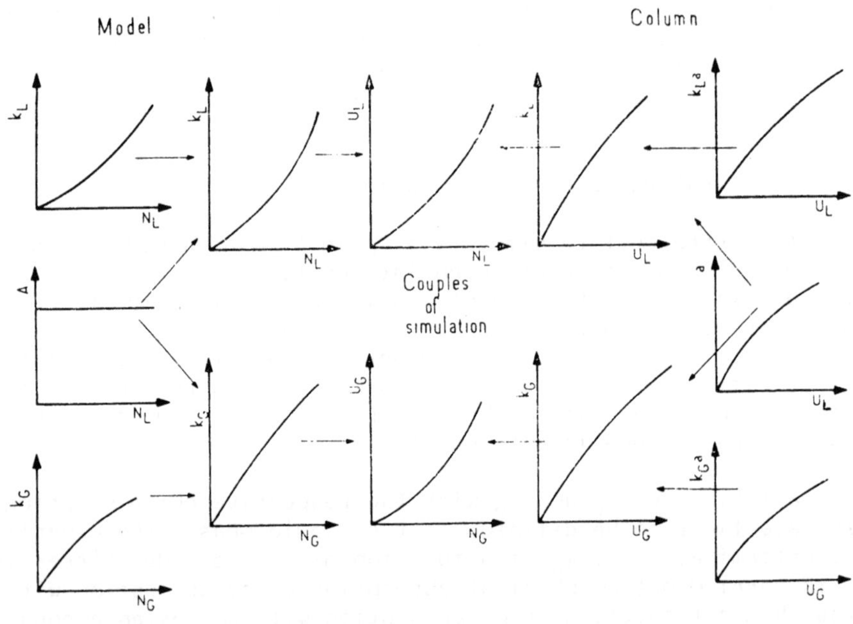

Figure 20 - Schematic procedure to establish model column couples for simulation between (u_L,N_L) and (u_G,N_G).

The following step is the measurement or the theoretical calculation, when possible, of the average rates of absorption per unit of interfacial area ϕ_m of the chemical system in the laboratory model where the values of k_L and k_G are adjusted to be the same at those of the packed column. These measurements are carried out for different liquid compositions and various gas compositions representative of those of different levels of the column and presented as plots ϕ_m vs p for different values of the concentrations of the reactants. The knowledge of these absorption rates is indeed necessary for the previsional calculation of the length of the column H as these relevant values of ϕ_m from the stirred cell are used to integrate Eq. (55).

$$H = \frac{u_G}{aV_M P} \int_{out}^{in} \frac{dp_h}{\phi_m(k_L, k_G, C_A^*, C_{Bo})} = \frac{u_L}{za} \int_{out}^{in} \frac{dC_{Bo}}{\phi_m(k_L, k_G, C_A^*, C_{Bo})} \quad (60)$$

between the inlet and outlet conditions of the packed column.

An illustration of this in using a 10 cm I.D. double stirred cell to simulate a 30 cm I.D. column packed with 20 mm glass Raschig rings on a 1.92 m height has been presented by Laurent |101, 191|. The values of k_L and k_G of the model leading to the simulating couple N_L, u_L and N_G, u_G on Fig. 20 have been determined by absorption of pure CO_2 into different liquids (water, sodium sulfite, sodium hydroxide) and absorption of dilute SO_2 into sodium hydroxide. Then CO_2 from air was absorbed into sodium hydroxide and sodium carbonate solutions in the stirred cell and in the packed column. The objective was to compare the predictions obtained from the model with the results of the packed column experiments for different values of u_L, u_G and gas and reactant concentrations. The experimental values of ϕ_m measured in the stirred cell for different bulk compositions of gas and liquid are presented in Fig. 21 and have been used to integrate Eq. (60) between the given inlet and outlet conditions of the packed column. The resulting values of h = H are presented as "Predicted height" in the Table of this figure. The predicted and actual heights differed by less than 20 % in all cases indicating that the method of using such a laboratory model to design a packed absorber is quite sound. Danckwerts et Alper |78| have even obtained better prediction (within 10 %) in using different packing heights in a 10 cm I.D. column where the fluid flowrates were kept constant.

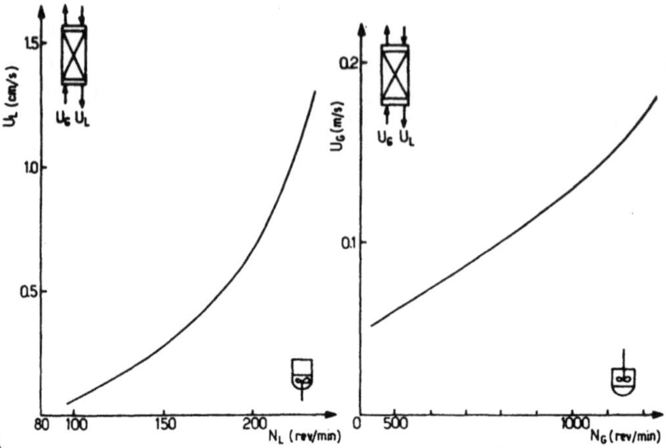

Figure 21 : Simulation of a packed column by a stirred cell

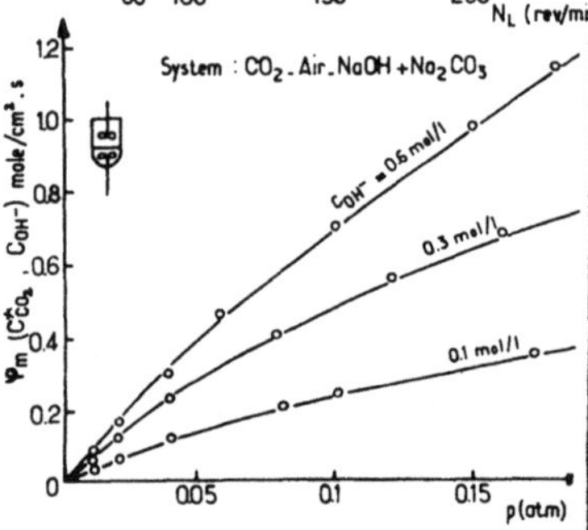

u_G (cm.s^{-1})	$p_{CO_2} \times 10^2$ (atm)		u_L (cm.s^{-1})	$C_{BO} - C_{OH^-}$ (mol.l^{-1})		$\int_{IN}^{OUT} \frac{dC}{\varphi/\sqrt{m}}$ (s.m^{-1})	H (m)	Predicted height (m)	Difference (%)	Error in Material Balance (%)
	Inlet	Outlet		Inlet	Outlet					
7.5	6.2	0.4	0.5	0.60	0.52	654		1.72	10.4	9.9
7.5	3.2	0.01	0.5	0.60	[illegible]	578		1.52	20.8	25
6.9	2.9	0.05	0.23	0.59	0.52	1 120		1.84	4.2	4
6.9	3.5	0.6	0.23	0.59	0.47	1 102		1.81	5.7	3.5
6.9	7.3	0.75	0.23	[illegible]	0.42	1 144		1.88	2.1	2.4
7.6	2.5	0.4	0.5	[illegible]	0.27	673	1.92	1.77	7.8	11.9
7.5	2.8	0.45	0.23	0.30	0.24	1 096		1.80	6.2	6.7
10.9	2.0	0.5	0.23	0.30	0.24	1 145		1.88	2.1	0
7.5	2.6	0.3	0.5	0.28	0.25	692		1.82	5.2	4
7.4	4.2	0.4	0.5	0.28	0.23	722		1.90	1.0	5.9

Results of absorption of CO_2-Air into aqueous solutions of NaOH + Na_2CO_3 experiments in a packed column (D = 30 cm, d = 20 mm, T = 25°C, P = 1 Atm) ; a, k_L and k_G experimentally determined.

5.32 A string of spheres to simulate a complete packed column : use of an integral model.

Differential simulation is not applicable in cases where the absorption rate is influenced by complex reactions occuring partly in liquid-bulk or between several reagents and gases. In this case the height of the model must be related to that of the full sized reactor by a known scaling factor h/h_m and the model simulates all the essential features of the reactor when the inlet and outlet compositions are the same in both equipment without any assumption about the transfer mechanism or reaction kinetics (integral simulation). A classical example is the simulation of a "complete" packed column of known u_L, u_L/u_G, H, k_L, k_G and a/β (as regard both gas and liquid side phenomenae and bulk reactions) by a string of spheres with a cylindrical pool at the top of each sphere to increase the liquid holdup and to vary the values of the simulating criterion a/β in the range $10 < a/\beta < 25$, encountered in the countercurrent gas absorption pilot-scale and industrial scale packed columns and not in the classical laboratory-scale models presented in Fig. 19.

Once the gas and liquid volumetric flowrates $q_L = \Omega_m u_{Lm}$ and $q_G = \Omega_m u_{Gm}$ (in $cm^3 \times s^{-1}$) are determined in the model so that k_L and k_G are the same as in the packed column for the given working conditions u_L and u_G (in $cm^3 \times cm^{-2} \times s^{-1}$) and once the number of the spheres and the dimension of the pools on the top of the spheres are calculated so that a/β is the same (which leads to the scaling ratio, H/H_m), the model now simulates all the essential features of the column. And if the inlet compositions are the same, so are the outlet compositions. The total absorption rate Φ_c of the packed column is thus predicted from the measurement of the total absorption rate in the model Φ_m. An excellent illustration of this technique has been presented by Alper and Danckwerts |190| where a 10 cm diameter column packed with 1.27 cm ceramic Raschig rings over 158 cm height is simulated by a string of hollow 3.72 cm diameter spheres column, 49 cm height. The experiments are been carried out in cases where the bulk concentration of the reactants at a given level in the column could not be calculated using a material balance equation thus leading to theoretical predictions either too complicated, if indeed possible, or involving large errors.

An example in the case of an absorption of a pure gas into a liquid where a fast reaction takes place near the interface and is followed by a reaction in the bulk of the liquid is presented in Table XXI. In these experiments, the absorption rate was dependent on the bulk reaction and it was therefore necessary to make the two criteria k_L and $a/\beta = A_m/V_L$ the same in both absorbers. The last column of Table XXI shows the percentage difference between the predicted $\Phi_p = (u_L\Omega/q_L)\Phi_m$ and measured Φ_c total absorption rates and the maximum difference is less than $\pm$ 7 %.

Alper and Danckwerts |190| also applied this technique to complex chemical cases of simultaneous absorption of CO_2 and NH_3 into water (identical values of k_L and k_G in both equipment), of simultaneous absorption of CO_2 and SO_2 into amine solutions (identical k_L and k_G) and of CO_2 into a solution that contains two amines (identical k_L).

Each time the data in the packed column could be predicted from the experiments carried out in the string of spheres column within a $\pm$ 7 % accuracy. And in all cases, theoretical predictions would have been also hard to make, as the bulk concentration of the reactants at a given level of the packed column could not be calculated using a material balance equation.

Table XXI: Experimental results of absorption of pure carbon dioxide into aqueous monoethanolamine solutions containing sodium arsenite |190|

$$CO_2 + 2C_2H_4OHNH_2 \rightleftharpoons C_2H_4OHNCOO^- + C_2H_4OHNH_3^+ \qquad \text{(fast)}$$

$$C_2H_4OHNCOO^- + H_2O \longrightarrow C_2H_4OHNH_2 + HCO_3^- \qquad \text{(slow)}$$

COLUMN	q_L or u_L	h	PACKING	$10^3\ k_L$	A_m/V_L
	$cm^3 \times s^{-1}$ or $cm \times s^{-1}$	cm	cm	$cm \times s^{-1}$	cm^{-1}
Sphere	$q_L = 1.15$	49	Sphere 3.72	7.5	15.9
Packed	$u_L = 0.6$	133	Raschig ring 1.27		

Run No.	$NaAs_2O_3$	total amine	Carbonation ratio Inlet	Carbonation ratio Outlet Sphere column	Carbonation ratio Outlet Packed column	Total absorption rate Φ_m Sphere column	Total absorption rate Φ_c Packed column	$\frac{\Phi_c - \Phi_p}{\Phi_c}$
	g mole/l			cm³/s				%
1	0.110	1.09	0.00	0.52	0.52	15.10	640	-0.50
2	0.110	1.16	0.00	0.52	0.53	16.70	715	+0.40
3	0.100	1.16	0.40	0.59	0.59	5.85	261	+4.40
4	0.100	1.16	0.57	0.65	0.66	3.13	134	+0.50
5	0.150	1.16	0.69	0.76	0.77	2.39	99	+1.50
6	0.090	1.58	0.30	0.51	0.51	8.50	370	+0.25
7	0.090	1.58	0.48	0.55	0.56	3.27	145	+3.45
8	0.070	1.10	0.37	0.53	0.53	4.90	208	+1.00
9	—	1.28	0.42	0.54	0.54	4.20	190	+6.30
10	—	1.25	0.29	0.46	0.46	6.90	290	+0.30
11	0.070	1.25	0.30	0.51	0.51	7.50	320	+0.90
12	0.070	1.25	0.56	0.61	0.61	2.40	104	+1.90
13	0.155	1.25	0.61	0.68	0.68	2.50	110	+3.80
14	0.145	1.25	0.62	0.67	0.67	2.10	95	+6.30
15	0.065	1.25	0.61	0.65	—	1.80	—	—
16	0.100	1.25	0.61	0.66	—	2.00	—	—
17	0.130	1.25	0.61	0.67	—	2.20	—	—
18	0.155	1.25	0.61	0.68	—	2.50	—	—
19	0.075	1.10	0.46	0.55	0.55	3.40	150	+4.00
20	0.075	1.10	0.54	0.61	0.62	2.50	107	+1.90

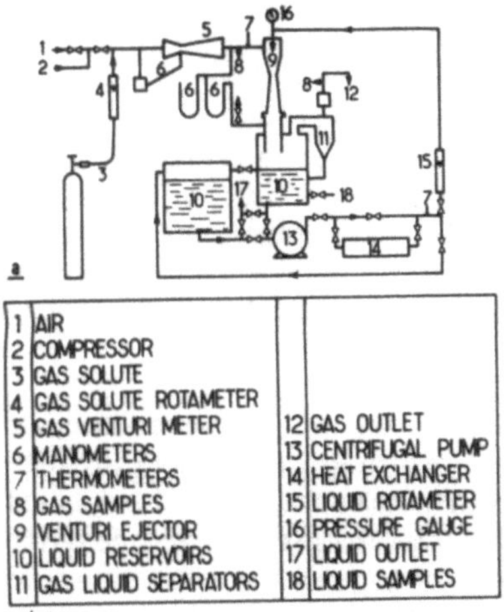

Figure 22a. Schematic diagram of the equipment.

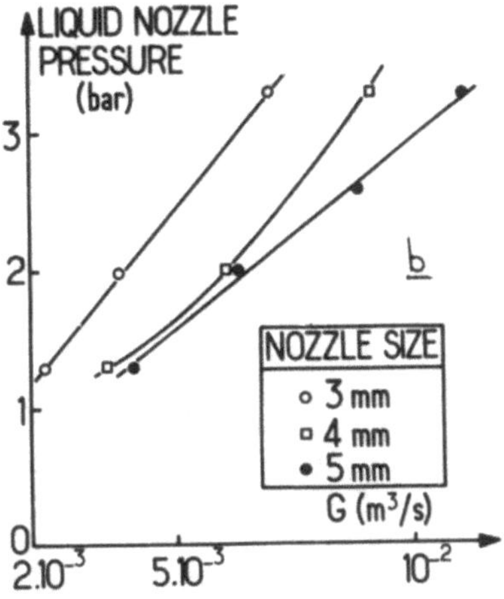

Figure 22b. Characteristics of the ejector nozzles.

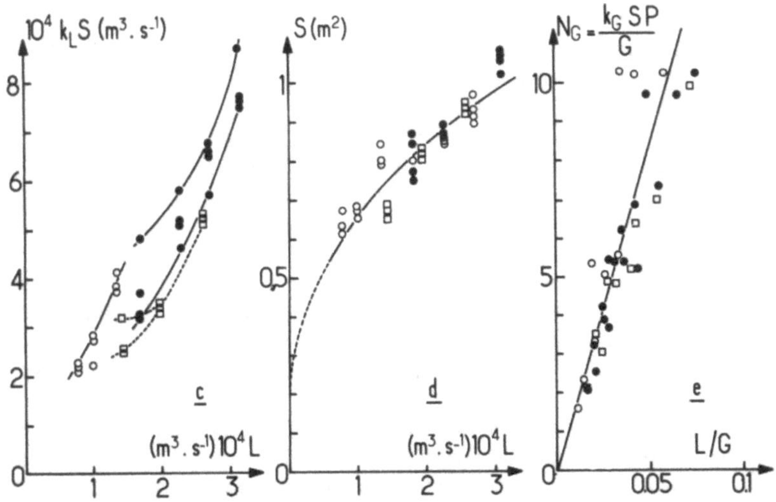

Figures 22c, 22d, 22e. Liquid side mass transfer coefficient (2c), effective interfacial area (2d), and gas side mass transfer coefficient (2e) in the Venturi jet scrubber.

5.33 A Laminar jet to simulate a pilot-scale motivated venturi jet scrubber : use of an integral model.

The simulation technique has most often been applied to the case of packed columns. However recently Laurent et al. |193| applied the integral simulation to the case of a quite different pilot equipment that is a turbulent venturi-jet scrubber.

A schematic diagram of the equipment is presented in Figure 22a. The venturi-jet scrubber is a device that permits the pumping of large quantities of gas by the action of the kinetic energy of a jet of liquid at the throat of a venturi, acting as the motive fluid. This simple mode of action has encouraged its more and more frequent use in the fight against air pollution (dust collection, cooling, and absorption). While the first two functions have received considerable attention, data concerning gas-liquid mass transfer are scarce, particularly for units producing a gas flow rate by their own pumping action alone.

A SAPS venturi jet scrubber of 50 mm nominal diameter was used which comprises a standard venturi diffuser of 33 mm throat diameter and 700 mm length. The hydrodynamic characteristics of the three ejector nozzles of 3, 4 and 5 mm diameter are presented in Figure 22b. The ejector nozzles are designed so that the liquid is injected and pulverized as a conical sheet entirely filling the cross section of the diffuser, giving a good contact between the gas and the liquid and producing a satisfactory aspiration effect by a good conversion of the kinetic energy of the motive liquid. At the exit of the diffuser, the two fluids flowing downwards together are separated in a gravity separator equipped with a jet breaker. The data reported here concern the complete assembly of the scrubber and the separator.

It has been measured that the volume of liquid inside the contactor varies with the liquid flow rate in the range 3 to 5 % of the total volume of the scrubber. The mass transfer efficiency and the interfacial parameters were determined with the use of the chemical method at room temperature, and the kinetics of each reaction employed has been previously studied in a laboratory model of known interfacial area and contact time such as a cylindrical wetted wall, a laminar jet, or a mechanically stirred vessel.

The product k_LS was determined in the slow chemical regime by the absorption of dilute carbon dioxide (2 %) with air into an aqueous solution of potassium carbonate and bicarbonate in presence of sodium hypochlorite. Figure 22c shows the variation of k_LS with q_L=L for each nozzle. This parameter varies mainly with L so long as the device acts as an autoaspirator. But when the gas phase is forced through the scrubber, the influence of q_G = G becomes significant and more and more important as the nozzle size increases.

The global interfacial area S in the scrubber and the separator was determined by the absorption with fast pseudo first-order chemical reaction of dilute carbon dioxide (2 %) with air into a sodium hydroxide aqueous solution. The variations of S as a function of L, G, and the nozzle size are presented in Figure 22d. In the region studied, the interfacial area is independent of gas rate for any given liquid rate. In fact, when the scrubber is operating in the autoaspiration mode, the mean gas velocity is always less than 10 $m.s^{-1}$. In these conditions, the interfacial area is determined solely by the structure of the liquid jet at the exit of the nozzle. The gas-liquid interactions are too weak to break up the films and streams of the liquid flow into drops and to produce an increase in the interfacial area.

The product k_GS was determined in the surface instantaneous chemical regime by the absorption of dilute sulfur dioxide (1 %) with air into a sodium hydroxide aqueous solution. The results are shown in Figure 22e, where k_GSP/G is plotted vs. L/G for the three nozzle sizes. These last results are directly indicative of the practical use of the venturi jet scrubber for the treatment of gaseous effluents to prevent atmospheric pollution. Indeed, the ordinate of Figure 22e is directly proportional to the definition of the number of transfer units N_G of the gas phase which is related to the efficiency $E_G = 1 - (y_s/y_e)$ by the relation $\ln(1-E_G) = -N_G$. Thus it is possible to conclude from the results shown that the efficiency of the scrubber improves as the ratio L/G is increased. Over the operating range studied, the removal efficiencies of the scrubber varied from 78 to 99.99 %. In practice, for a given liquid injection pressure, a satisfactory scrubbing efficiency will be obtained in the part of the capacity curve corresponding to small gas flow rates (G/L < 15-20) |189|.

From the data presented in Figure 22, it is observed for the scrubber that $3 < k_L \times 10^2 < 8$ $cm.s^{-1}$, $8 < K_G \times 10^5 < 26$ $gmole.cm^{-2}.s^{-1}.atm^{-1}$, and $S/V_L = A_m/V_L \simeq 150$ cm^{-1}. Among the various laboratory-scale models, the laminar jet is the only one allowing for the same range of the simulation criteria.

An experimental and theoretical study of this equipment was thus undertaken to determine the limits within which the jet can be regarded as a cylindrical rod in uniform motion of known values of contact times and interfacial area. Several diameter ejector nozzles of the liquid through the gas space are used (d_o = 0.3, 0.5, 0.7, 0.8, 0.9, 1, 1.5 and 2 mm). The jet receivers are either capillary tubes with an internal diameter slightly greater than the diameter of the jet (0.1 mm) or a vaseline oil pool. Several liquids, water, pure isopropanol, pure polyethylene glycol 200, aqueous solution of polyethylene glycol 6 000, aqueous solution of sodium sulfite, and solutions of lauric alcohol in water were tested in presence of stagnant air as the gas phase. The liquid properties are $789 < \rho < 1\,124$ kg/m^3, $0.986\times10^{-6} < \nu < 19\times10^{-6}$ $m^2.s^{-1}$, $21.8\times10^{-3} < \sigma < 72.4\times10^{-3}$ N/m. Provision is also made for varying the length of the jet $0.013 < Z < 0.15$ m and the liquid flow rate $0.2\times10^{-6} < L < 3.2\times10^{-6}$ m^3/s.

The correct working of the laminar jet was observed by means of a cathetometer and photographs and the corresponding values of k_L

and k_G are determined by the absorption of pure CO_2 into water and of dilute SO_2 (1.8 %) with air into a NaOH aqueous solution.

The experimental results are well represented by the equations

$$k_L = 4 \left| \frac{D_{CO_2} L}{\pi^2 d_o^2 Z} \right|^{0.5} \quad \text{and} \quad k_G = \frac{10^{-4}}{d_o} \left(\frac{L}{Z} \right)^{0.5}$$

If now all the simulation criteria k_L, k_G, S/V_L, L/S, and G/S have identical values in both devices, the laboratory model will simulate all the essential features of venturi jet scrubber, and if the inlet compositions are the same in both equipment, so would be the outlet compositions.

The equality of the criteria S/V_L could not be fulfilled with any practical laminar jet. Indeed, for the scrubber this ratio is of the order of 150 cm^{-1}, that would correspond to a laminar jet of 0.25 mm diameter. Stable operation in a laminar jet with a vaseline oil pool as jet receiver would be practically impossible, and moreover the values of the criterium k_L would not be compatible with those in the scrubber. Anyway, jet scrubbers are usually employed for rapid or instantaneous reaction that does need an important liquid holdup but great values of S and k_G and/or k_L.

The values of k_L and k_G in both equipment have been determined by the method of absorption with or without chemical reaction. Then the relation between L and G in the model and in the scrubber giving the same k_L and k_G in both may be established by inspection of the diagrams showing the variation, in each device, of k_L and k_G when changing the flow rates. Thus it has been observed that it is also impossible to achieve simultaneously the same values of k_L and k_G in both equipment because these parameters are principally influenced by the liquid flow rate |189|.

Therefore, it is necessary to restrict the integral simulation to operating ranges for which the influence of one or more of the criteria may be neglected. This leads to perform the simulation with the equality of criteria k_G and G/S or k_L and G/S which corresponds, as previously explained, to the potential ordinary uses of the venturi jet scrubber in gas-liquid absorption with rapid and instantaneous reactions.

Figure 23 gives an example of such a technique of integral simulation in k_G and G/S. Figures 23a and 23b present the variations of k_G vs. the flow rates and the nozzle diameter in each device. Note that k_G is independent of G in the laminar jet in the range 0 to 10^{-3} m^3/s. Figure 23c that is obtained from the two previous cur-

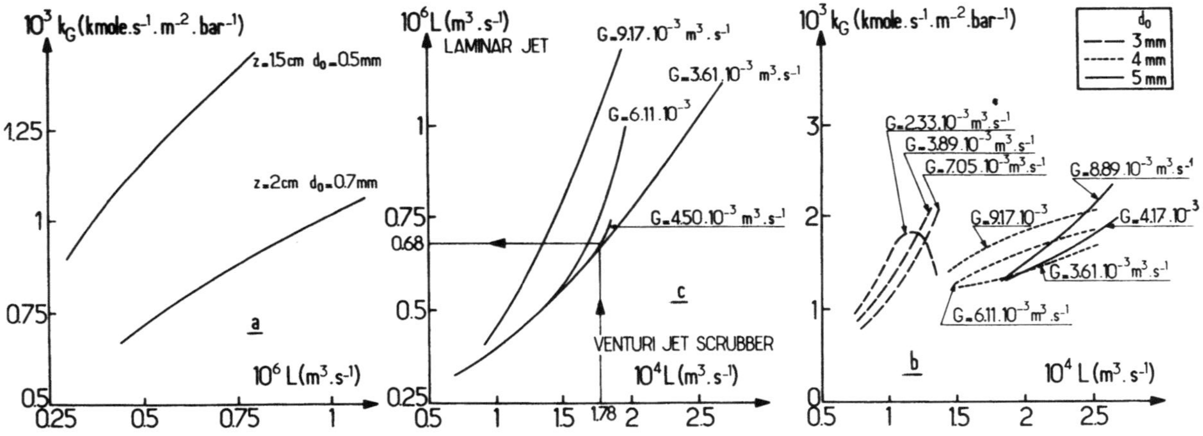

VENTURI JET SCRUBBER					LAMINAR JET					DEVIATION
10^4 L	10^3 G	y_E	y_E/y_S	E_G	E_G	y_E/y_S	y_E	10^7 G	10^7 L	
m^3.s^{-1}	m^3.s^{-1}	%	—	%	%	—	%	m^3.s^{-1}	m^3.s^{-1}	%
2.028	4.833	4.4	51	98.04	98.08	52	4.4	1.361	8.472	2
1.861	6.389	4.3	44	97.72	97.87	47	4.4	1.861	8.472	7
1.777	4.500	4.4	57.9	98.27	98.28	58.3	4.4	1.333	6.777	1
2.000	5.694	0.58	178	99.44	99.44	181	0.65	1.667	8.472	2

Figure 23 **Gas side mass transfer coefficient in the laminar jet (*a*) and in the Venturi jet scrubber (*b*). Figure (*c*) Relation between liquid flow rates in the laminar jet and in the Venturi jet scrubber (couples of simulation criteria k_G).**

ves, is the simulation couple that allows for the same values of k_G in the present venturi jet scrubber fitted with a 4 mm nozzle and in a laminar jet of 0.5 mm diameter and 1.5 cm height. For instance, the scrubber operating with $L = 1.8\times10^{-4}$ m^3/s and $G = 4.5\times10^{-3}$ m^3/s has the same k_G as this laminar jet operating with $L = 0.68\times10^{-6}$ m^3/s.

If now we want to have an idea of the potential, the validity, and the accuracy of the integral simulation, let us arry out experiments with gas and liquids quite different from those used to establish the values of the mass transfer parameters. The absorption of dilute hydrogen sulfide (0.5 to 4 %) with air into a 1.2 N aqueous diethanolamine solute was tested. This is a surface instantaneous reaction necessitating the equality of the criteria k_G and G/S. The gas flow rate in the laminar jet is thus evaluated by

$$G_m = G \times \frac{S_m}{S} = 4.5\times10^{-3} \times \frac{\pi d_o Z}{0.78} = 1.361\times10^{-7}\ \mathrm{m^3.s^{-1}}$$

As the model simulates all the essential features of the scrubber, the integral simulation predicts that if the inlet composition y_E is the same in both equipment, so will be the outlet composition y_S. The results of the experiments carried out in the model and in the scrubber are presented in the Table of Figure 23. The deviations in the prediction is within 10 % which is very promising if it is observed that an error of about 5 % must be accounted for the accuracy of the analyzer. Note also the ratios of approximately 250 and 34 000, respectively, for the values of liquid and gas flow rates encountered in both equipment. Complementary data with the same gas-liquid system tested in the scrubber with different nozzles or with other gas-liquid systems needing the equality of the criteria k_L and G/S in both devices are presented by Fonteix |197| and Laurent et al. |191|. In spite of the limitations of the present technique, the results reported here are representative of the general use of the venturi jet scrubber and have extended the use of the integral simulation to a gas-liquid reactor different from a packed column.

6. CONCLUSIONS : SOME SCALE-UP PROBLEMS

We have presented here some considerations on characteristics of multiphase gas-liquid reactors, design equations, thermodynamic and kinetic data... Moreover the results of the mass transfer parameters (most often measured by the chemical method, i.e. the integral values of a, k_La and k_Ga), and their dependence on the fluid and equipment hydrodynamics have been outlined. In practise, three possibilities to use them are offered to the designer, depending on the fixed aim :

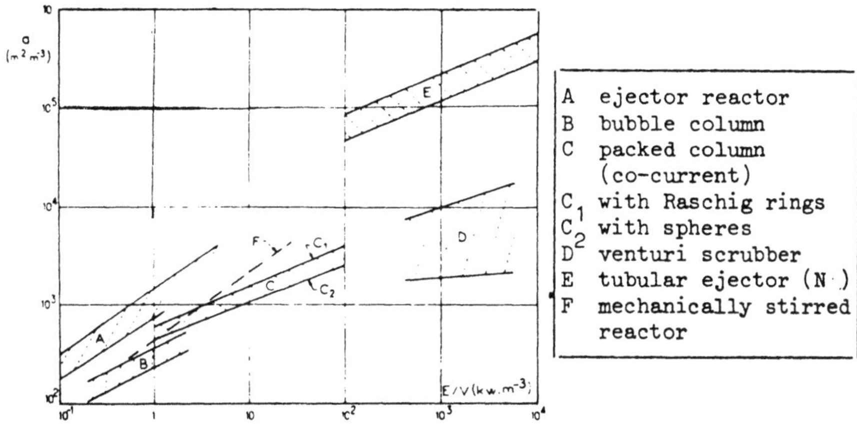

FIGURE 24 : Interfacial area in several types of reactors |198|

. *either* the gas-liquid equipment has already been operated but the problem is to change the gas-liquid couple inside the existing reactor for economical reasons. In such a case, although this procedure is time-consuming, it seems now reasonably possible to predict the performance obtained with any new gas-liquid system in the reactor from experiments carried out in a small-scale laboratory apparatus with the same k_L, k_G and a/β as the existing industrial equipment (laminar jet, wetted wall, stirred cell...) that is, in simulating the classical industrial reactor by laboratory scale apparatuses having a quite different shape. In most practical cases, these laboratory experiments will simultaneously provide the required knowledge of the reaction kinetics.

. *either* the type of the gas-liquid reactor has been choosen (packed column, plate column, mechanically agitated tank...) and the problem is thus the sizing of the equipment in using the published mass transfer data. In such case, the diagrams and correlations published in literature |125, 139, 199, 200| for contactors with diameters most often smaller than 0.4 m should be used with a fair degree of confidence in order to scale tubular packed, spray and plate columns up to 2 or 3 m in diameter. Moreover for mechanically agitated or bubble reactors, it is recommended to carry out small scale experiments with the given gas-liquid couple in laboratory apparatus (D = 5-20 cm) geometrically similar (in shape, agitation and contact time) to the choosen type of reactor and then to do the scale-up to diameters larger than 1 m on the basis of the total energy input and the superficial gas velocity as seen in par. 4.

. *or* the suitable type of reactor has to be choosen prior to the sizing. Then, this is often a technico-economical problem with a competition between the value of the interfacial to create and the energetical expense to create it. In such case, Table I and the diagram presented by Nagel et al. |198| will be of great help to make a comparison between the order of magnitude of interfacial area provided by the main gas-liquid equipment and the energetic cost (Fig.24).

REFERENCES

1. J.C. Charpentier, "Gas-Liquid Reactors" in Chem. React. Engng. Reviews, Edt by Dan Luss and V.W. Weekman, ACS Symposium Series 72, 223, 1978.
2. J.C. Charpentier, "Mass Transfer Rates in Gas-Liquid Absorbers and Reactors", Advances in Chemical Engineering, Edt by Drew and Vermeulen, Academic Press, Vol. 11, 1980.
3. N. Barona, Hydrocarbon Processing, 1979, July, 179.
4. H. Hikita, S. Asai, Int. Chem. Engng., 1964, 4, 332.

5. K. Onda, E. Sada, T. Kobayashi, M. Fujine, Chem. Eng. Sci., 1970, 25, 753 ; 761 ; 1023.
6. G. Astarita, "Mass Transfer with Chemical Reaction", Elsevier, Amsterdam, 1967.
7. P.V. Danckwerts, "Gas-Liquid Reactions", Mc Graw Hill, 1970.
8. Y.T. Shah, "Gas-Liquid-Solid Reactor Design", Mc Graw Hill, 1979.
9. W.F. Linke, A. Seidell, "Solubilities : Inorganic and Metal Organic Compounds", 4th Edition, Vol. I (1958) ; Vol. II (1965), D. Van Nostrand Company Inc., New-York.
10. A. Seidell, W.E. Linke, "Solubilities of Inorganic and Organic Compounds", Supplement to 3rd Edition, D. Van Nostrand Company Inc., New-York, 1952.
11. J.H. Hildebrand, J.M. Prausnitz, R.L. Scott, "Regular and Related Solutions", Van Nostrand-Reinhold, New-York, 1970.
12. E. Wilhem, R. Battino, Chem. Rev., 1973, 73, 1.
13. G.T. Preston, E.W. Funk, J.M. Prausnitz, Phys. Chem. Liquids 1971, 2, 193.
14. P.M. Cukor, J.M. Prausnitz, J. Phys. Chem., 1972, 76, 598.
15. C.C. Chappelow, J.M. Prausnitz, AIChE J., 1974, 20, 1097.
16. R. Cysewski, J.M. Prausnitz, I.E.C. Fund., 1976, 15, 304.
17. T.J. Edwards, G. Maurer, J. Newman, J.M. Prausnitz, AIChE J. 1978, 24, 966.
18. O.R. Rivas, J.M. Prausnitz, I.E.C. Fund., 1979, 18, 289.
19. E. Wilhem, R. Battino, J. Chem. Thermodyn., 1973, 5, 117.
20. L.R. Field, E. Wilhem, R. Battino, J. Chem. Thermodyn., 1974, 6, 237.
21. J.E. Byrne, R. Battino, W.F. Danforth, J. Chem. Thermodyn., 1974, 6, 245.
22. H.L. Clever, R. Battino, in "Solutions and Solubilities", Edited by M. Dack, p. 379, Wiley-Interscience, New-York 1975 and "The Solubilities of gas in liquids", Edited by M. Dack, Wiley-Interscience, 1977.
23. J.H. Hildebrand, R.H. Lamoreaux, I.E.C. Fund., 1974, 13, 110.
24. D. Fleury, W. Hayduck, Canad. J. Chem. Engng., 1975, 53, 195.
25. R. Battino, H. Clever, Chem. Rev., 1966, 60, 395.
26. J.M. Perry, "Chemical Engineers' Handbook", 5th Ed., Mc Graw Hill, 1973.
27. V.M. Ramm, "Absorption of gases", Israel Program for Scientific Translation, Jerusalem (1968).
28. R.C. Reid, J.M. Prausnitz, T.K. Sherwood, "The Properties of Gases and Liquids", 3rd Ed., Mac Graw Hill, 1977.
29. G. Maffiolo, J. Vidal, H. Renon, I.E.C. Fund., 1972, 11, 100.
30. J. Meisner, I.E.C. Fund., 1972, 11, 83.
31. D.P. Maloney, J.M. Prausnitz, AIChE J., 1976, 22, 74.
32. L. Stiel, D.F. Harnish, AIChE J., 1976, 22, 117.
33. M. Orentlicher, J.M. Prausnitz, Chem. Eng. Sci., 1964, 19, 775.
34. I. Kritchevsky, A. Ilinskaya, Acta Physicochim. URSS, 1945, 20, 327.

35. J.P. O'Connell, J.M. Prausnitz, I.E.C. Fund., 1963, 3, 347.
36. W. Hayduck, H. Laudie, AIChE J., 1973, 19, 1233.
37. D. Beutier, H. Renon, AIChE J., 1978, 24, 1122.
38. C.L. de Ligny, N.G. Van der Veen, J.C. Van Houwelingen, I.E.C. Fund., 1976, 15, 336.
39. A. Tamir, J. Wisniak, Chem. Eng. Sci., 1975, 30, 335.
40. J. Tokunaga, J. Chem. Engng. Japan, 1975, 8, 7 and 326.
41. D.W. Van Krevelen, P.J. Hoftijzer, Chimie Industrie, 1948, Septembre, 168.
42. P.V.L. Barrett, PhD, Cambridge, England, 1966.
43. K. Onda, E. Sada, T. Kobayashi, S. Kito, K. Ita, J. Chem. Engng. Japan, 1970, 3, 18.
44. E.W. Tiepel, K.E. Gubbins, Ind. Eng. Chem. Fund., 1973, 12, 18 and 1975, 14, 143 and 137.
45. A. Schumpe, I. Adler, W.D. Deckwer, Biotechnology and Bioengng., 1978, 20, 145.
46. M.R. Kamal, L.N. Canjar, Chem. Eng. Prog., 1966, 62, 82.
47. A.W. Nienow, Brit. Chem. Engng., 1965, 10, 827.
48. S. Bretsnajder, "Prediction of Transport and other Properties of Fluids", Pergamon Press, New-York, 1971.
49. H. Ertl, R.K. Ghai, F.A. Dullien, AIChE J., 1974, 20, 1.
50. R.K. Ghai, H. Ertl, F.A. Dullien, AIChE J., 1973, 19, 881.
51. J. Simons, A.B. Ponter, Canad. Journ. Chem. Engng., 1975, 53, 541.
52. A.H.P. Skelland, "Diffusional Mass Transfer", Wiley Interscience, New-York, 1974.
53. C.R. Wilke, P. Chang, AIChE J., 1955, 1, 264.
54. E.G. Scheibel, Ind. Eng. Chem., 1954, 46, 2007.
55. R. Sitaraman, S.H. Ibrahim, N.R. Kuloor, J. Chem. Engng. Data, 1963, 8, 198.
56. K.A. Reddy, L.K. Doraiswamy, I.E.C. Fund., 1967, 6, 77.
57. M.A. Lusis, G.A. Ratcliff, Canad. Journ. Chem. Engng., 1968, 46, 385.
58. D.F. Othmer, M.S. Thakar, Ind. Eng. Chem., 1953, 45, 589.
59. A. Akgerman, J.L. Gainer, I.E.C. Fund., 1972, 11, 373.
60. H. Sovova, Coll. Czech. Chem. Comm., 1976, 41, 3715.
61. J. Leffler, H.T. Cullinan, I.E.C. Fund., 1970, 9, 84 and 88.
62. P.A. Jonnson, A.L. Babb, Chem. Rev., 1956, 56, 387.
63. G.A. Davies, A.B. Ponter, K. Graine, Canad. Jour. Chem. Engng., 1967, 45, 372.
64. A. Dim, G.R. Gardner, A.B. Ponter, T. Wood, J. Chem. Engng. Jap., 1971, 4, 92.
65. M.T. Tyn, Chem. Engng. Journ., 1971, 6, 27.
66. D. De Kee, H. Laudie, Hydroc. Process., 1974, Sept., 224.
67. A.R. Gordon, J. Chem. Phys., 1937, 5, 522.
68. J.R. Vinograd, J.W. Mac Bain, J. Amer. Chem. Soc., 1941, 63, 2008.
69. P.L.T. Brian, R.F. Baddour, D.C. Matiatos, AIChE J., 1964, 10, 727.
70. T.K. Sherwood, J.L. Wei, AIChE J., 1955, 1, 522.

71. R.A.T.O. Nijsing, R.H. Hendriksz, H. Kramers, Chem. Eng. Sci., 1958, 8, 81 and 1959, 10, 38.
72. R.A. Robinson, R.H. Stokes, "Electrolyte Solutions", 2nd Ed. Academic Press, New-York, 1959.
73. D.L. Wise, G. Houghton, Chem. Eng. Sci., 1966, 21, 999.
74. A.L. Shrier, Chem. Eng. Sci., 1967, 22, 1391.
75. W. Hayduck, H. Laudie, AIChE J., 1974, 20, 611.
76. A.F. Mazarei, D.C. Sandall, AIChE J., 1980, 26, 154.
77. G.A. Ratcliff, J.G. Holdcroft, Trans. Inst. Chem. Engrs., 1963, 41, 315.
78. P.V. Danckwerts, E. Alper, Trans. Inst. Chem. Engrs., 1975, 53, 34.
79. A. Yasunishi, F. Hoshida, Int. Chem. Engng., 1979, 19, 498.
80. T. Reith, S. Renken, B.A. Israel, Chem. Eng. Sci., 1968, 23, 619.
81. J.M. Burgess, P.H. Calderbank, Chem. Eng. Sci., 1975, 30, 743 and 1107.
82. V. Linek, J. Mayrhoferova, Chem. Eng. Sci., 1969, 24, 481.
83. T. Vermeulen, G.M. Williams, G.E. Langlois, Chem. Eng. Prog. 1955, 51, 85.
84. J.F. Cameron, "Fluid Density Measurements in Enclosed Systems", Internat. Confer. on Radio Isotopes in Scientific Research, Vol. 1, 426 (1957).
85. P.H. Calderbank, Trans. Inst. Chem. Engrs., 1958, 37, 443.
86. W. Kawecki, T. Reith, J.W. Van Heuven, W.J. Beek, Chem. Eng. Sci., 1967, 22, 1519.
87. E.K. Todtenhauft, Chem. Ing. Techn., 1971, 43, 336.
88. J. Landau, H.G. Gomaa, A.M. Al Taweel, Trans. Inst. Chem. Engrs., 1977, 55, 212.
89. P.H. Calderbank, F. Evans, J. Rennie, Proceed. Int. Symp. on Distil. (Int. Chem. Engrs.), p. 51, 1960.
90. H. Kürten, P. Zehner, German. Chem. Engng., 1978, 1, 347.
91. T. Sridhar, O.E. Potter, Chem. Eng. Sci., 1978, 33, 1347.
92. M.M. Sharma, P.V. Danckwerts, Brit. Chem. Engng., 1970, 15, 522.
93. V. Linek, Chem. Eng. Sci., 1972, 27, 627.
94. C.W. Robinson, C.R. Wilke, AIChE J., 1974, 20, 285.
95. A.A. Beenackers, W.P.M. Vna Swaaij, Proceed. Eur. Chem. Engng. Symp., Heidelberg, 1976.
96. B.D. Prasher, AIChE J., 1975, 21, 407.
97. E.E. Isaacs, F.D. Otto, A.E. Mather, J. Chem. Eng. Data, 1977 22, 71 and 317.
98. E.E. Isaacs, F.D. Otto, J.I. Lee, A.E. Mather, Canad. J. Chem. Engng., 1976, 54, 214 and 1977, 55, 210.
99. E. Sada, H. Kumezawa, M.A. Butt, J. Chem. Eng. Data, 1977, 22, 277.
100. O. Levenspiel, J.H. Godfrey, Chem. Eng. Sci., 1974, 29, 1723.
101. A. Laurent, Thesis, Institut National Polytechnique de Lorraine, Nancy, France, 1975.

102. A. Laurent, J.C. Charpentier, C. Prost, Jour. Chimie Physique, 1974, 71, 613.
103. T. Reith, W.J. Beek, Chem. Eng. Sci., 1973, 28, 1331.
104. A. Yasunishi, Chem. Engng. Japan, 1977, 3, 215.
105. U. Onken, W. Schalk, Ger. Chem. Engng., 1978, 1, 191.
106. S. Yagi, H. Inoue, Chem. Eng. Sci., 1962, 17, 411.
107. G. Astarita, G. Maruci, L. Coletti, La Chimica e l'Industria, 1964, 9, 1021.
108. K.J.A. De Waal, J.C. Okeson, 1966, 21, 559.
109. J.A. Wesselingh, A.C. Vant'Hoog, Trans. Inst. Chem. Engrs., 1970, T49, 48.
110. K.R. Westertep, L.L. Van Dierendonck, J.A. De Kraa, Chem. Eng. Sci., 1963, 18, 157.
111. V. Linek, J. Mayrhoferova, Chem. Eng. Sci., 1966, 21, 777, 1969, 24, 481 ; 1970, 25, 787.
112. K. Onda, H. Takeuchi, Y. Maeda, Chem. Eng. Sci., 1972, 27, 449.
113. V. Linek, J. Turdik, Biotech. Bioengng., 1971, 13, 353.
114. T. Reith, Brit. Chem. Engng., 1970, 15, 1559.
115. V. Nyvlt, F. Kastanek, Coll. Czech. Chem. Commun., 1975, 40, 1853.
116. G. Paradis, G.A. L'Homme, Mémoire de fin d'Etudes, Université de Liège, Belgique, 1972.
117. E. Alper, Trans. Inst. Chem. Engrs., 1973, 51, 159.
118. K. Sridharan, M.M. Sharma, Chem. Eng. Sci., 1976, 31, 767.
119. K. Ganguli, H. Van den Berg, Chem. Eng. Sci., 1978, 33, 27.
120. C. Alvarez-Fuster, N. Midoux, A. Laurent, J.C. Charpentier, Chem. Eng. Sci., 1980, 35.
121. K. Sridharan, Ph. D. Thesis, University of Bombay, India, 1975.
122. C. Alvarez-Fuster, Thesis, Institut National Polytechnique de Lorraine, Nancy, France, 1980.
123. B.I. Morsi, A. Laurent, N. Midoux, J.C. Charpentier, Chem. Eng. Sci., 1980, 35, 1467.
124. B.I. Morsi, N. Midoux, A. Laurent, J.C. Charpentier, Entropie, 1980, n° 91, 40.
125. A. Laurent, J.C. Charpentier, The Chem. Engng. Jour., 1974, 8, 85.
126. G.E.H. Joosten, P.V. Danckwerts, Chem. Eng. Sci., 1973, 28, 453.
127. E.R. Matheron, O.C. Sandall, AIChE J., 1979, 25, 332.
128. G. Baldi, S. Sicardi, Chem. Eng. Sci., 1975, 30, 617 and 769.
129. J.M. Veteau, "Mesure des aires interfaciales dans les écoulements diphasiques", Rapport C.E.A.-R.5005, CENG, France, 1979.
130. V.S. Patwardhan, Cand. J. Chem. Engng., 1978, 56, 56 and 558.
131. N. Midoux, A. Laurent, J.C. Charpentier, AIChE J., 1980, 26, 157.

132. B. Gal-Or, W. Resnick, I.E.C. Proc. Des. Devel., 1966, 5, 15.
133. I.M. Hassan, C.W. Robinson, Chem. Eng. Sci., 1980, 35, 1277.
134. R. Botton, D. Cosserat, J.C. Charpentier, Chem. Eng. Journ., 1978, 16, 107 ; 1980, 20.
135. K.M. Kan, P.F. Greenfield, IEC Proc. Des. Devel., 1978, 17, 482 ; 1979, 18, 740.
136. J. Van Krevelen, T.C. Knekels, Recueil des Travaux Chimiques des Pays Bas, 1948, 67, 512.
137. S.J. Parulekar, Y.T. Shah, Chem. Engng. J., 1980, 20, 21.
138. A. Vignes, I.E.C. Fundam., 1966, 5, 189.
139. J.C. Charpentier, "A review of the data on mass transfer in most of gas-liquid reactors", in "Two-Phase Flow and Heat Transfer", Edt by Kakac and Veziroglu, Hemisphere Publishing Corp., 1977, pp. 869-910.
140. M. Zlokarnik, Adv. in Biochem. Engng., 1978, 8, 133 and 1979, 11, 157.
141. K. Schügerl, J. Lücke, U. Oels, Adv. in Biochem. Engng., 1977, 7, 1.
142. H. Hofmann, Int. Chem. Eng., 1977, 17, 19.
143. R.V. Chaudhari, P.A. Ramachandran, AIChE J., 1980, 26, 177.
144. H. Blenke, Adv. Biochem. Engng., 1979, 13, 121.
145. J. Stichlmaier, A. Mersmann, Int. Chem. Eng., 1978, 18, 223.
146. H. Brauer, "Problems and advances in gas absorption", 3rd Int. Summer School in "Modelling of heat and mass transfer processes and chemical reactors", Varna, May 21-30, 1979 (Bulgarie).
147. J.C. Charpentier, M. Favier, AIChE J., 1975, 21, 1213.
148. A. Gianetto, G. Baldi, V. Specchia, S. Sicardi, AIChE J., 1978, 24, 1087.
149. J.L. Turpin, R.L. Huntington, AIChE J., 1967, 13, 1196.
150. R.P. Larkins, R.R. White, D.W. Jeffrey, AIChE J., 1961, 7, 231.
151. J.C. Charpentier, Chem. Eng. Journ., 1976, 11, 161.
152. J.C. Charpentier, Journ. Powder Bulk Solids Technology, 1978, 2, 53.
153. J.C. Charpentier, M. Favier, AIChE J., 1975, 21, 1213.
154. N. Midoux, M. Favier, J.C. Charpentier, J. Chem. Eng. Jap. 1976, 9, 350.
155. Y. Sato, T. Hirose, F. Takahashi, M. Toda, J. Chem. Eng. Jap., 1973, 6, 147 and 315.
156. B.I. Morsi, N. Midoux, J.C. Charpentier, AIChE J., 1978, 24, 357.
157. V. Specchia, G. Baldi, Chem. Eng. Sci., 1977, 32, 515.
158. L.D. Clements, P.C. Schmidt, AIChE J., 1980, 26, 314 and 317.
159. L.P. Reiss, I.E.C. Proc. Des. Devel., 1967, 6, 486.
160. C.N. Satterfield, AIChE J., 1975, 21, 209.
161. S. Goto, J. Levec, J.M. Smith, I.E.C. Proc. Des. Devel., 1975, 14, 473.

162. S. Goto, J.M. Smith, AIChE J., 1975, 21, 706.
163. A. Gianetto, V. Specchia, G. Baldi, AIChE J., 1973, 19, 916.
164. B.W. Shende, M.M. Sharma, Chem. Eng. Sci., 1974, 29, 1763.
165. Y. Sato, T. Hirose, T. Ida, Kagaku, Kogaku, 1974, 38, 534.
166. Y. Lemay, G. Pineault, J.A. Ruether, I.E.C. Proc. Des. Devel., 1975, 14, 280.
167. N.D. Sylvester, P. Pitayagulsarn, I.E.C. Proc. Des. Devel. 1975, 14, 473.
168. V. Specchia, G. Baldi, A. Gianetto, I.E.C. Proc. Des. Devel. 1978, 17, 362.
169. C.N. Satterfield, M.W. Van Eek, G.S. Bliss, AIChE J., 1978, 24, 709.
170. A. Dharwadkar, N.D. Sylvester, AIChE J., 1977, 23, 376.
171. G. Delaunay, A. Storck, A. Laurent, J.C. Charpentier, Entropie, 1980, n° 92.
172. K.R. Westertep, L.L. Van Dierendonck, J.A. De Kraa, Chem. Eng. Sci., 1963, 18, 1157.
173. N. Midoux, J.C. Charpentier, Entropie, 1979, n° 88, 5.
174. D.S. Dickey, AIChE Meeting, Paper 116d, 72d Annual Meeting, San Francisco, Nov. 1979.
175. I.T. Hassan, C.W. Robinson, AIChE J., 1977, 23, 48.
176. H.T. Luong, B. Volesky, AIChE J., 1979, 25, 893.
177. B.J. Michel, S.A. Miller, AIChE J., 1962, 8, 262.
178. K. Kurten, P. Zehner, Ger. Chem. Engng., 1979, 2, 220.
179. P.H. Calderbank, Trans. Inst. Chem. Engrs., 1958, 36, 443.
180. V.R. Ranade, J.J. Ulbrecht, AIChE J., 1978, 24, 796.
181. K.L. Ganguli, Ph. D. Thesis, Delft, Holland, 1975.
182. K. Van't Riet, I.E.C. Proc. Des. Devel., 1979, 18, 357.
183. G.E.H. Joosten, J.G.M. Schilder, J.J. Janssen, Chem. Eng. Sci., 1977, 32, 563.
184. T. Sridhar, O.E. Potter, Chem. Eng. Sci., 1980, 35, 683.
185. P.V. Danckwerts, A.J. Gillham, Trans. Inst. Chem. Engrs., 1966, 44, 42.
186. P.V. Danckwerts, S.F. Rizvi, Trans. Inst. Chem. Engrs., 1971, 49, 24.
187. A. Laurent, C. Prost, J.C. Charpentier, Chemische Technik, 1974, 26, 471, ref. 52/74.
188. J.C. Charpentier, A. Laurent, AIChE J., 1974, 20, 1029.
189. C. Fonteix, A. Laurent, G. Besson, J.C. Charpentier, "Hydrodynamics and mass transfer in a pilot-scale turbulent venturi jet scrubber : prediction and simulation of its performances by a laboratory-scale laminar jet", 3rd Int. Summer School in "Modelling of heat and mass transfer processes and chemical reactors", Varna, May 21-30, 1979 (Bulgarie).
190. E. Alper, P.V. Danckwerts, Chem. Eng. Sci., 1976, 31, 599.
191. A. Laurent, J.C. Charpentier, J. Chimie Physique, 1977, 10, 1001.

192. A. Laurent, J.C. Charpentier, Proceed. 3rd Conf. on Applied Chemistry, Unit Operations and Process, Veszprem, 29-31 Aug. 1977, p. 419, Hungary.
193. A. Laurent, C. Fonteix, J.C. Charpentier, AIChE J., 1980, 26, 282.
194. A. Laurent, C. Fonteix, G. Besson, J.C. Charpentier, CHISA'78 Sect. Distillation and absorption, 21-25 Aug. 1978, Praha, Tchecoslavaquie.
195. I. Bjerle, S. Bengtsson, K. Färnkvist, Chem. Eng. Sci., 1972, 27, 1853.
196. H. Hikita, S. Asai, H. Ishikawa, Y. Saito, Chem. Eng. Sci., 1975, 30, 607.
197. C. Fonteix, Thesis, Institut National Polytechnique de Lorraine, Nancy, France, 1978.
198. O. Nagel, H. Kurten, R. Sinn, Chem. Ing. Techn., 1972, 44, 899.
199. R. Botton, D. Cosserat, J.C. Charpentier, Chem. Eng. Sci., 1980, 35, 82.
200. V.V. Mahajani, M.M. Sharma, Chem. Eng. Sci., 1980, 35, 941.

HYDRODYNAMICS OF MULTIPHASE REACTORS

Giancarlo Baldi

Istituto Chimica Industriale Politecnico of Torino, Italy

GENERAL

It is well recognized that hydrodynamics is decisive in reactor design. This is particularly true for multiphase reactors, where the simultaneous presence of several phases gives rise to very complex hydrodynamic situations which can affect the reactor performance very strongly.
The conversion rate in these reactors is a function of the reaction kinetic rate and mass and heat transfer between the phases, that depend on the hydrodynamic regimes. These affect also the dissipated energy, a factor of primary importance for economical operations. To make an accurate account of the interaction of kinetics and transfer phenomena, we should know the reactants concentration and the temperature in each phase as a function of the spatial coordinates; for this, we should know the volumetric distribution of each phase in the reactor, the flow patterns, the local values of mass and heat transfer area and coefficients.
Flow regimes, energy dissipation, holdup of the phases, flow patterns are therefore the important hydrodynamic factors which can make it possible to build up a model for design and scale–up.
Among these factors, the flow patterns are of utmost importance for modelling the reactor.
Multiphase reactors have a peculiar hydrodynamic heterogeneity that can hardly be described in terms of average conditions. The local hydrodynamic situation at the phases contact area (the rate of surface renewal, if one speaks in terms of the well know Danckwerts' theory) can change remarkably from point to point. This makes difficult to extrapolate data obtained in certain conditions (i.e. mass transfer without chemical reaction) to other conditions (i.e. mass transfer coupled with a chemical reaction), because of the different "average" hydrodynamic situation "effective" to the process.
In this lecture, we will take a loock at the hydrodynamics of some types of three–phase reactors, namely: trickle bed, bubble packed and stirred slur-

ry reactors.

2. TRICKLE–BED REACTORS

Trickle–bed reactors are catalytic packed bed reactors with a cocurrent downward two–phase flow of liquid and gas (fig. 1). The reactants are partially in the gas phase and partially in the liquid phase.

Cocurrent downward two–phase flow through a packed bed is also used for absorption of gas into a reacting liquid, when the ratio of the liquid to the gas flow rate is either very low or very high. In these cases, in fact, the usual countercurrent operations may suffer of throughput limitations by flooding.

Another use of trickle–bed reactors is in the field of biological reactions with immobilized enzymes or microorganisms on an inert support.

The packings employed in catalytic reactors are pellets, extrudates, spheres with a size less than 5-6mm. In absorption operations, larger sizes of packing with a high void fraction are employed (Raschig rings, Berl saddles). For biological reactions, a variety of packings are used, from small beads to wood chips. Furthermore, a very large range of liquid and gas flow rates per unit empty area is used. In petroleum processing, the representative limiting flow conditions for industrial units are (1):

$$0.8 < L < 25\ \mathrm{kg/m^2 \cdot s}$$
$$0.010 < G < 2\ \mathrm{kg/m^2 \cdot s}$$

If one also considers pilot-plant and laboratory reactors, the liquid mass velocity may be lower than 0.1 kg/m^2s.

Owing to this range of fluids flow rate and the different packing sizes and shapes employed, several hydrodynamic conditions can be met in practice.

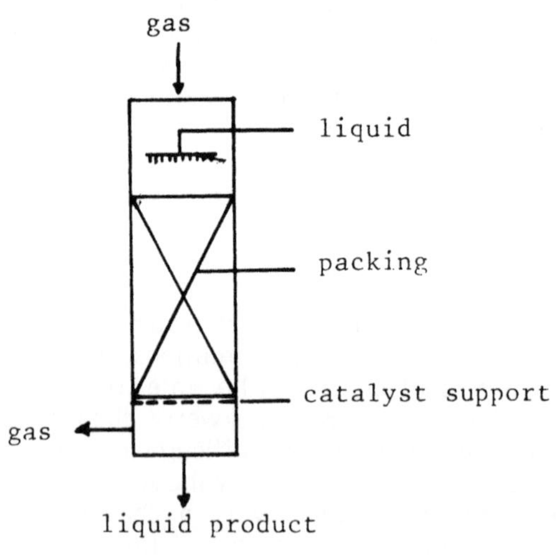

Fig. 1 - Trickle-bed reactor.

2.1 Flow regimes

On increasing the gas flow rate at constant liquid flow rate, several flow regimes set up (fig. 2). Typically, at intermediate liquid flow rate ($5 < L < 30$ kg/m^2s) and low gas flow rate, a gas continuous or trickling flow exists, where the liquid trickles over the packing particles and the gas flows through the void space.
At very high liquid flow rates ($L >$ 35–40 $kg/m^2.s$) and low gas flow rates the dispersed bubble flow regime exists; the gas in form of bubbles is carried out by the continuous liquid phase.
On increasing the gas flow rate, first the pulsing flow is encountered, characterized by pistons of dense gas-liquid mixtures.
It is due to flooding of the packing channels, followed by their emptying when the local pressure drop reaches a suitable value. The flooding is probably due to large superficial waves on the liquid film, produced by the interaction of the gas flow with the liquid film (4) (fig. 3).
When the pulsing flow appears, remarkable oscillations of the pressure drop are observed
A further increase of the gas flow rate leads to an increase of the frequency of the pistons, and lastly to the spray flow regime, where a uniform mist flows along the column: the shear stress of the gas on the liquid surface breaks the liquid in small drops.
The transition between pulsing and spray flow occurs gradually; the oscillations of the pressure drop increase in frequency and decrease in amplitude as the gas flow rate is increased, and lastly tend to disappear.
At very low liquid flow rates ($L < 5$ $kg/m^2.s$), the pulsing flow does not appear; an increase of G leads directly to the spray flow.
This phenomenology is valid for non foaming systems only; with foaming sistems, a so called foaming flow regime arises before pulsing flow. It is characterized by the formation of foam in all the packed bed; the set up of this regime gives rise to a sudden increase of the pressure drop.
The flow regime mainly involved in laboratory and pilot-plant reactors is the gas continuous flow; the industrial units can work in the pulsing flow regi-

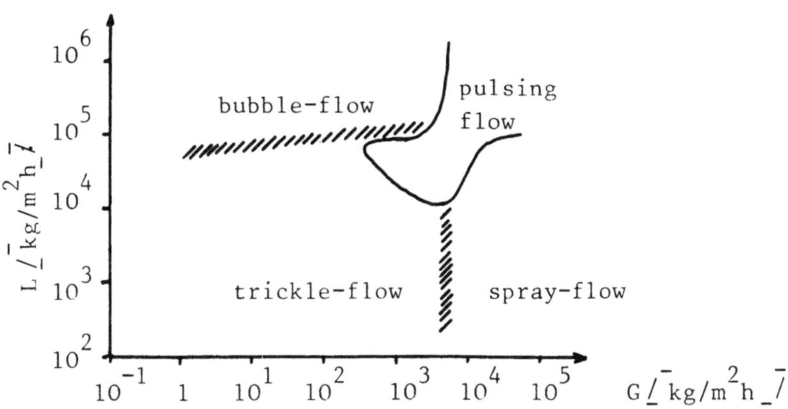

Fig. 2 - Flow regime map.

Fig. 3 - Occlusion of a channel due to the formation of a large wave.

me also. Absorption operations may also be carried out in the spray regime. Dispersed bubble flow has a little importance in practice.
As concernes the influence of the flow regime on hydrodynamic (pressure drop and liquid holdup) and transfer parameters (mass and heat transfer coefficients), it was observed that only two hydrodynamic regimes can be considered (2, 3): a low gas-liquid interaction regime (gas continuous or tricking flow regime) and a high interaction regime (pulsing flow, foaming flow, spray flow regime).
The transition between these two regimes depends on several factors: packing shape and size, bed porosity, physical properties of the fluids, wetting characteristics of the packing.
For non foaming flow and $d_p \leq 6$ mm the transition is well enough defined (2,5) and can be determined from fig. 4 (4).
The parameters λ and ψ are so defined (6):

$$\lambda = \left(\frac{\rho_G}{\rho_{air}} \cdot \frac{\rho_L}{\rho_{wt}} \right) \tag{1}$$

$$\psi = \frac{\sigma_{wt}}{\sigma_L} \left[\frac{\mu_L}{\mu_{wt}} \cdot \left(\frac{\rho_{wt}}{\rho_L} \right)^2 \right]^{1/3} \tag{2}$$

and they should take into account the physical properties of the fluids, in comparison with those of water and air at room temperature.
The transition from the low to the high interaction regime for foaming systems is not yet well defined. Foaming flow occurs at somewhat lower liquid and gas flow rate than pulsing flow for non foaming systems, but a correlation of the data is quite difficult since it is not easy to define "a priori" when a system is foaming or not: in other words, the usual physical properties are not able to take account of the foaming characteristics.
Froth is formed by gas bubbles entrapped in liquid lamellae, that hinder the coalescence of bubbles. The coalescence rate is affected by the viscosity and the surface tension of the liquid, but also strongly by gradients of concentration in the liquid which arise in proximity of the gas-liquid interface. This latter effet is difficult to evaluate a priori, so that an experimental study is needed to see whether a gas-liquid system is or is not foaming.

2.2 Liquid distribution

The liquid distribution is an important design parameter, since it affects the performance of the reactor. A bad distribution gives rise to insufficient wetting of the particles, and hence decreases the catalyst utilization.
The liquid distribution is measured by a collector device at the bottom of the packed bed, where the liquid flow rate through several small areas is separately collected. These areas must be sufficiently large (at least 2 $d_p x 2d_p$) to overcome the intrinsic discontinuity of the liquid flow through a packed bed. When the liquid flow rate collected in each section of the collector, referred to the collecting area, is constant in the column section, the liquid distribution is defined as uniform.
In the upper zone of the packed bed, a developing flow exists, where the liquid distribution changes with the axial coordinate.
The depth of this zone depends on the liquid distributor geometry, the packing shape and size and the fluids flow rate. With a good distributor, a half a meter calming section is generally sufficient to achieve a steady distribution, that does not change any more with the axial coordinate.
The evolution of the initial to the steady distribution is due to the radial spread of the liquid at the packing contact points. By assuming the packed bed as a homogeneous medium, the evolution of the velocity profile is given by:

$$\frac{\partial v^*}{\partial z} = D_R \frac{1}{r} \left(r \frac{\partial v^*}{\partial r} \right) \tag{3}$$

where D_R is the radial spread factor. At zero gas flow rate, it increases with the packing size, and is almost independent from the liquid flow rate. In the presence of a cocurrent gas flow rate, the radial spread factor decreases as v_G increases, mainly in the high interaction regime (8, 9) (fig. 5). This is due to the fact that the gas flow breaks off the liquid bridges between the packing particles, through which the radial movement of liquid occurs.
To examine the steady distribution, one must consider separately a wall zone (generally half a particle diameter thick) and a packed zone.
In the packed zone, the liquid distribution is generally uniform in all hydrodyna-

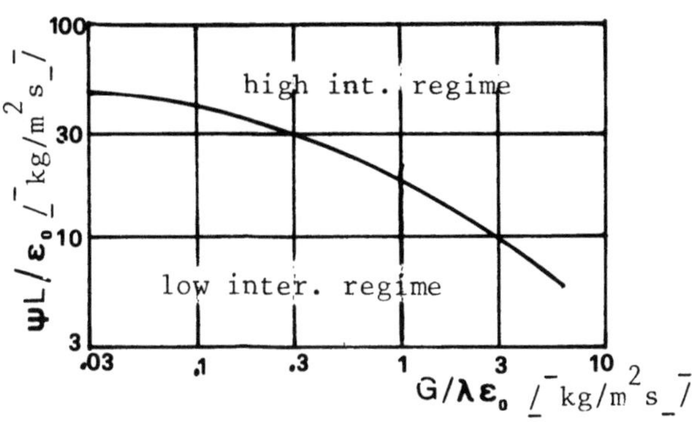

Fig. 4 - Transition between low and high interaction regime.

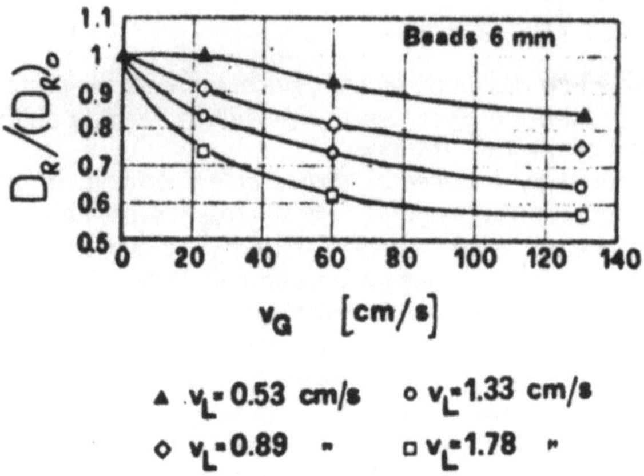

Fig. 5 - Effect of gas and liquid flow rate on the radial spread factor (after Baldi and Specchia (9)).

mic regimes for $d_p < 6$ mm and $D_c/d_p > 12$ (8–11). At higher d_p values, preferential paths can be observed.

In the wall zone, generally a higher liquid flow rate per unit section area exists, due to the higher bed porosity.

The liquid flowing in this zone is much less effective to the reaction; the fraction of wall flow rate with respect the total flow rate is given by:

$$\frac{v_{L,w} \cdot \pi \cdot D_C \cdot d_p/2}{v_L \cdot \pi \cdot D^2{}_C/4} = 2\,\frac{v_{L,w}}{v_L}\,\frac{d_p}{D_C} \tag{4}$$

The ratio $v_{L,w}/v_L$ depends mainly on d_p and d_p/D_C: it decreases by decreasing both these parameters. The increase of the cocurrent gas flow rate also reduces the liquid velocity at the wall, especially in the high interaction regime.

In the gas continuous regime, it was found that for $d_p < 6$ mm and $D_c/d_p > 18$, $v_{L,w}/v_L$ equals 1 (10, 11). Then, from Eq. (4) it can be seen that for $D_c/d_p > 20$, the wall flow rate is less than 10% of the total flow rate: for $D_c/d_p > 10$, this percentage would be much higher than 20%.

From this it can be seen that the wall flow is an important factor in laboratory reactors, where it can affect remarkably the conversion yield. It must also be considered, however, that there is a continuous exchange between the wall and the packing zone through the wall-packing contact points, where the wall flow is diversed toward the packed zone and viceversa. In this way, the height of the packed bed by-passed by the wall flow depends on the number of contact points per unit column height, so that it is smaller and smaller as the pellets size decreases.

2.3 Pressure drop and liquid holdup

These hydrodynamic parameters are strictly connected: the pressure drop is affected by the void space through which the gas flows, and by the tortuosity of the packing channels. The liquid holdup reduces the area available for the gas flow and changes the tortuosity of the bed. In turn, the pressure drop acts as a driving force

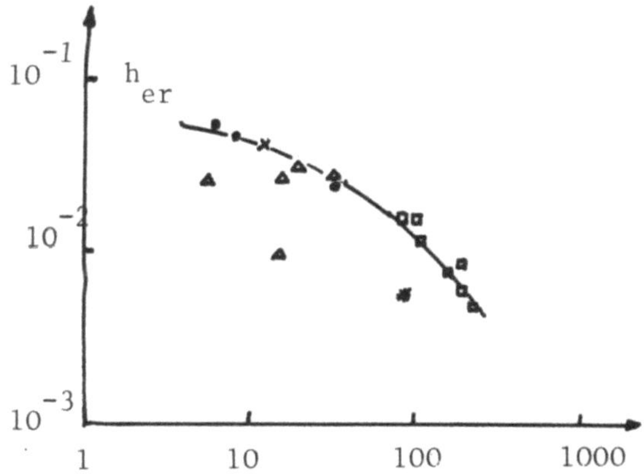

$$Eo = \frac{g\,d_p^2\,\rho_L}{\sigma_L}$$

Fig. 6 - Residual holdup correlation (after Charpentier et al. (12)).

for draining the liquid in the bed.
First, some definitions about the liquid holdup must be given.
The total holdup h_t is the volume of liquid held in the column for unit column volume. If the packing is porous, there is an intraparticle holdup h_i, given by the liquid filling the pores, and an interparticles holdup h_e, constituted by the liquid in the bed void space.
This can be divided into two parts: the draining holdup h_{ed}, that is the liquid volume which one can collect at the bottom of the bed after a sudden shutoff of the liquid feed, and the residual holdup h_{er}, that is the volume which remains in the column after draining due to capillary forces.
The intraparticle holdup will be considered in a later section. Here the interparticle holdup will be examined.
The distinction between residual and draining holdup arises from the fact that they are affected by different parameters.
The residual liquid volume held in the column is a result of the balance between the gravity forces ($\rho_L \cdot g$) and the capillary forces (depending on the surface tension σ_L and the diameter of the capillaries, this being a function of d_p). A combination of these parameters in a dimensionless number gives the Eötvos number Eö:

$$Eö = \frac{\rho_L g\, d_p^2}{\sigma_L} \tag{5}$$

This number was used to correlate the residual holdup in several packings by Charpentier et al. (12): the correlating line in shown in fig. 6. For packings like those used in trickle-bed catalytic reactors, Eö is less than 5; according to the figure, h_{er} is about 6% of the reactor.
The draining holdup is a result of a balance between a driving force and a resistance. The resistance is given by the friction forces caused by the liquid flow over the packing. Either viscous or inertia forces are present, but the latter ones generally predominate. These are due to turbulence arising at sudden changes of the flow path at the packing contact points.

The driving forces are the gravity force and, in a concurrent two-phase, the pressure drop.

For free liquid trickling (that is, when G = O), Buchanan (13) pointed out that the draining holdup is affected by the Reynolds number ($Re_L = \frac{v_L \rho_L d_p}{\mu_L}$), the Froude number ($Fr = v_L^2/gd_p$) and the shape of the packing. The Froude number represents the ratio between the turbulent drag forces, proportional to $\rho_L v_L^2$, and the gravity, responsible for the draining. The Froude and Reynols numbers can be combined to give the Galileo number Ga:

$$G_a = \frac{Re_L^2}{Fr} = \frac{d_p^3 \rho_L^2 g}{\mu_L^2} \tag{6}$$

that does not depend on the liquid velocity.

Many correlations have been proposed in the literature for the draining liquid holdup whithout a gas flow rate. A critical review is shown in the paper by Gianetto et al. (5). A correlation that seems to have a wide validity is the following (3):

$$\frac{h_{de}}{\varepsilon_o} = 3.86\,(Re_L)^{0.55}\,(Ga)^{-0.42}\,\left(\frac{a_v d_p}{\varepsilon_o}\right)^{0.65} \tag{7}$$

As a cocurrent gas flow rate is fed, the draining holdup decreases, due to the shear stress on the liquid surface.

The effect of the gas flow is little in the gas continuous regime; at the onset of the pulsing flow, the decrease of h_{de} is steeper because of the high gas-liquid interaction; when the spray flow takes place, the variation of h_{de} with v_G becomes very little again. For foaming systems a sharp decrease of h_{de} is observed at the inception of foaming flow.

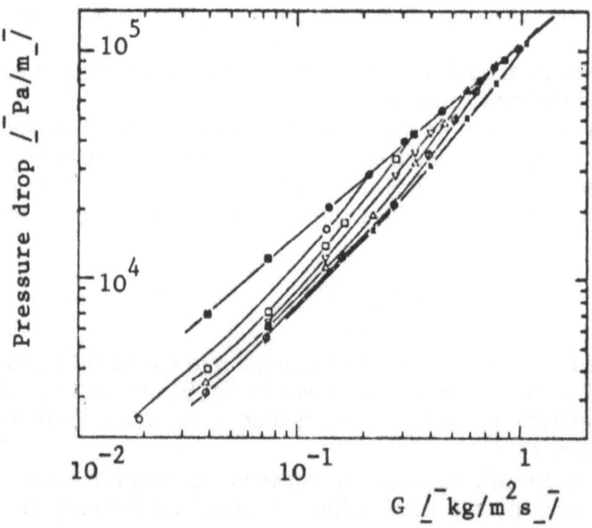

Fig. 7 - Multiplicity of hydrodynamic states (after Kan and Grenfield (14)).

Pressure drop of a gas flow rate in dry packings is usually interpreted by an Ergun type equation:

$$\delta_G = C_1 \frac{(1-\varepsilon_o)^2}{\varepsilon_o^3} \mu_G v_G + C_2 \frac{1-\varepsilon_o}{\varepsilon_o^3} \rho_G v_G^2 \qquad (8)$$

C_1 and C_2 are experimental coefficients that depend on the tortuosity and the shape of the packing channels, i.e. on the shape and size of the packing.
When liquid flows cocurrently, the pressure drop increases. In the gas continuous flow, the effect of the liquid is mainly to reduce the void space available for the gas flow. The presence of liquid changes also the shape of the packing channels and their tortuosity. In the high interaction regime, the effect of the liquid is more complicated: due to the periodical flooding of some channels, sudden accelerations of both gas and liquid take place, which contribute considerably to the dissipation of energy. In spite of this, no sudden changes of pressure drop are observed on passing from the low to the high interaction regime, probably because the increase of pressure drop due to acceleration phenomena is compensated by a remarkable decrease of the liquid holdup and an increase of the void space. At the onset of the pulsing flow, pulsation of the pressure drop can only be observed.
For foaming systems, instead, the pressure drop increases remarkably at the foaming flow onset, depending on the rate of coalescence of the gas bubbles: the higher the rate of coalescence, the lower the pressure drop (7).
For very small packings ($d_p < 1{,}5$ mm), a multiplicity of hydrodynamic states exist in the low interaction regime, dependent on the past flow history. If, at a given liquid flow rate, the gas flow rate is increased from zero to a maximum value and then reduced, lower pressure drop values are obtained (fig. 7). The higher the maximum gas flow rate previously attained, the lower will be the pressure drop (14). As it has been shown, the increase of the gas flow rate breaks off the liquid bridges between the packing particles and the tortuosity of the channel decreases. When the gas velocity is subsequently reduced the liquid bridges do not form again.
From the foregoing discussion, it is evident that the simultaneous flow of a liquid and a gas through a packed bed is rather complex, and hence it is difficult to find general correlations of pressure drop and liquid holdup.
Many authors have proposed correlations valid in a certain range of hydrodynamic situations; a review of these correlations can be found in the work of Gianetto et al. (5).
Following an approach for cocurrent two-phase flow in empty tubes, some correlations are based on the parameter defined as follows (15):

$$\chi = \sqrt{\delta_L / \delta_G} \qquad (9)$$

where δ_L and δ_G are the pressure drops for a single phase flow (liquid and gas respectively) through the packed bed, at the same liquid and gas flow rate of the two phase flow.
A modification of the χ parameter was suggested on the basis of an energy balance (16):

$$\chi' = (\xi_L / \xi_G)^{0.5} \qquad (10)$$

where:

$$\xi_L = L/\varepsilon_o \tag{11}$$

$$\xi_G = \left(\frac{1}{\rho_G g}\,\delta_G + 1\right)(G/\varepsilon_o) \tag{12}$$

In other empirical correlations the liquid and gas Reynolds numbers raised to suitable exponents are used.
All these types of correlations are valid in the high interaction regime.
Among the most recent correlations of pressure drop in the high interaction regime the following are suggested (2):
a) non-foaming systems:

$$\left(\frac{\delta_{LG}}{\delta_L}\right)^{0.5} = 1 + \frac{1}{\chi} + \frac{1.14}{\chi^{0.54}} \tag{13}$$

b) foaming systems:

$$\left(\frac{\xi_{LG}}{\xi_L}\right)^{0.5} = 1 + \frac{1}{\chi'} + \frac{6.55}{(\chi')^{0.42}} \tag{14}$$

where:

$$\xi_{LG} = \frac{1}{\varepsilon_o g}\left[\frac{L}{\rho_L} + \frac{G}{\rho_G}\right]\delta_{LG} + \frac{L+G}{\varepsilon_o} \tag{15}$$

For liquid holdup in the high interaction regime the following empirical correlations may be suggested (3):
a) non-foaming systems:

$$\varepsilon_o^{-1} h_{ed} = 0.125\,(Z/\psi^{1.1})^{-0.312}\left(\frac{a_v d_p}{\varepsilon_o}\right)^{0.65} \tag{16}$$

where:

$$Z = (Re_G)^{1.167} / (Re_L)^{0.767} \tag{17}$$

b) foaming systems:

$$\varepsilon_o^{-1} h_{ed} = 0.0616\,(Z/\psi^{1.1})^{-0.172}\left(\frac{a_v d_p}{\varepsilon_o}\right)^{0.65} \tag{18}$$

In the low interaction regime, a flow model has been proposed (3), that allows a theoretical approach.
It is assumed that a relationship between pressure drop δ_{LG} and v_G similar to that in a dry packing is valid. One must consider that the liquid in the column reduces the area available for the gas flow, and furthermore changes the tortuosity of the packing channels. It can be written:

$$\delta_{LG} = C_1^* \frac{[1-(\varepsilon_o - h_e)]^2}{(\varepsilon_o - h_e)^3}\,\mu_G v_G + C_2^* \frac{1-(\varepsilon_o - h_e)}{(\varepsilon_o - h_e)^3}\,\rho_G v_G^2 \tag{19}$$

The coefficients C_1^* and C_2^* differ somewhat from the dry packing coefficients; it was shown that they are independent from the liquid flow rate (3); therefore, they can be determined from pressure drop data for a gas flow through a "wetted packing" at zero liquid flow rate, i.e. when only the residual liquid holdup is present in the bed.
The liquid holdup, in its turn, is affected by the pressure drop, since this makes draining easier; but its dependence on operating parameters and physical properties in essentially the same as at zero gas flow rate (Eq. 7). Keeping in mind the meaning of Froude number as the ratio between the drag turbulent forces and the draining forces, for two phase flow it can be defined:

$$Fr^* = \frac{\rho_L v_L^2}{(\rho_L g + \delta_{LG})\, d_p} \tag{20}$$

and a modified Galileo number Ga* can be defined:

$$Ga^* = \frac{d_p^3 \, \rho_L (\rho_L g + \delta_{LG})}{\mu_L^2} \tag{21}$$

From Eq. (7), the liquid holdup can then be calculated as follows:

$$\varepsilon_o^{-1} h_{ed} = 3.86\, (Re_L)^{0.545}\, (Ga^*)^{-0.42} \left(\frac{a_v\, d_p}{\varepsilon_o}\right)^{0.65} \tag{22}$$

If the C_1^* and C_2^* coefficients for pressure drop are known, Eq. (19) and (22) can be used simultaneously to calculate by trial-and-error both δ_{LG} and h_{ed} in the low interaction regime, with a certain accuracy (± 30%).
It was demonstrated that an equation of the following form:

$$h_e = a(Re_L)^b\, (Ga^*)^c$$

Is valid, for a given packing, even when a multiplicity of hydrodynamic states exist (very small packings) (14).

2.4 Liquid texture of trickling flow

It was observed (12) that the whole liquid holdup in a packed column can be divided into an anisotropic fraction (narrow rivulets with vertical continuity), an isotropic fraction (a laminar film partially covering the packing particles) and an isolated fraction with a very low liquid renewal (semistagnant liquid pockets).
Accordingly, there exist a variety of local hydrodynamic situations of the liquid flow, ranging from a nearly turbulent (rivulets) to a very slow motion (semistagnant pockets).
It is interesting to examine the response curve from an inlet tracer disturbance in the liquid. The response curve shown in fig. 8 was obtained with a step decrease of the inlet tracer concentration.
The whole response curve can be divided in three zones: a time delay zone, due to a fraction of liquid having the same residence time in the column; a rapid fall zone and a tail zone, due to a fraction of liquid having a spread of residence times in the reactor. One can assume that the time delay zone corresponds to the most active

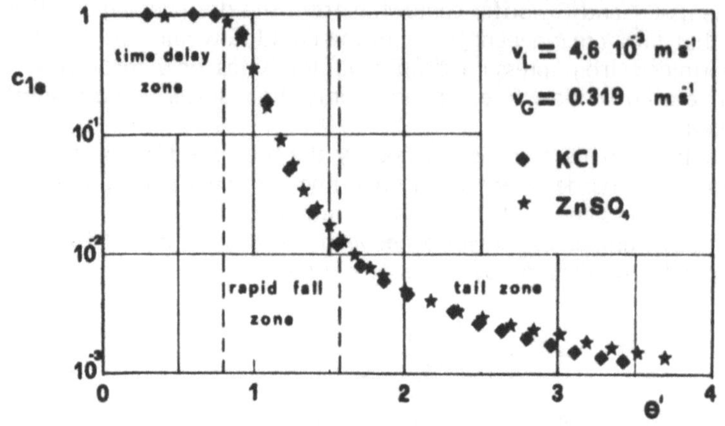

Fig. 8 - Response curve in a trickling bed for a step decrease of the inlet tracer concentration; θ'= dim.less time (after Sicardi et al. (17)).

fraction of holdup, due to the rivulets, the rapid fall zone to the laminar film and the tail zone to the stagnant pockets, which exchange matter with the flowing liquid.

Many models have been proposed to interpret the hydrodynamics of a trickling flow. It is hard to choose the "best" model, since it is rather difficult define what is the "best".

It is well known that an increase of the number of the model parameters leads to a better fit of the esperimental response curve. On the other hand, several parameters make more difficult pratical use of the model and their identification: furthermore, as the parameters number increases, their physical meaning becomes less clear.

The dispersion model does not allow a good interpretation of the response curve, and moreover does not introduce the important physical concept of the semistagnant pockets.

Two models can be suggested: the piston-exchange model (PE) and the piston-dispersion-exchange model (PDE). The first one considers the liquid flowing in plug flow, and assumes a mass exchange between flowing liquid and stagnant volume. The second one is similar to PE model but axial dispersion is superimposed to the plug flow.

The PDE model is able to describe well the three zones of the response curve shown in fig. 8. With this model, the operating or dynamic holdup h_{edy}, through which the liquid flows, is almost equal to the total interparticle holdup and the stagnant holdup h_{es} is very small.

The PE model considers as dynamic holdup that corresponding to the time delay zone; h_{edy} in this case is very close to the draining holdup and the stagnant holdup is a little smaller than the residual one.

Fig. 9 shows the trend of h_{es} calculated with the two models as a function of v_L for various v_G (17). As can be seen, static holdup values are strongly dependent on the hydrodynamic model adopted.

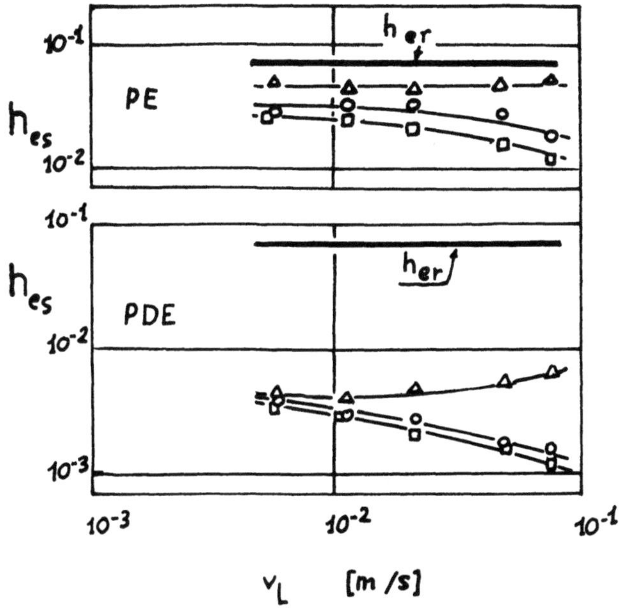

Fig. 9 - Static holdup determined from PE and PDE model; △: v_G=0; ○: v_G=0.32; □: v_G=0.61; 3 mm cylindrical packing (after Sicardi et al. (17)).

3. PACKED BUBBLE REACTORS

Packed bubble reactors are fixed bed reactors with upward gas and liquid flow. Their use is not as wide as trickle-bed reactors, since the pressure drop is higher. Nevertheless, they are advantageous with respect to trickle-bed reactors in the cases where the liquid reactants are treated with a small amount of gas, or when a high liquid residence time is required. The liquid holdup in these reactors, in fact, can be much higher. For the same reason, they can be succesfully employed in highly exothermic reactions. The large liquid volume offers a big heat capacity that makes easier temperature control. Furthermore good total wetting of the pellets ensures a good heat release from the inside of the catalyst particles and avoids hot spots.
Bubble packed columns are also employed for chemical absorption of a gas into a reacting liquid. Upflow is preferred to downflow when the reaction kinetic rate is slow, so that a large liquid volume is needed. For absorption, usual packings are used like Berl saddles or Rasching rings; but also unusual packings with a high void space (90%) are employed (wire packings, for instance).

3.1 Hydrodynamic regimes

Three distinct flow regimes can be observed, termed bubble flow, slug or pulsing flow, and spray flow (fig. 10) (18). At low gas flow rate, bubble flow exists, characterized by gas bubbles moving in the continuous liquid phase at a slightly higher velocity than the liquid.

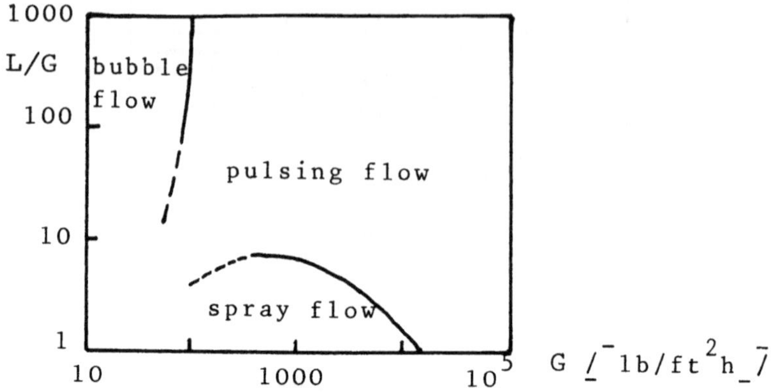

Fig. 10 - Flow map for upward two-phase flow through packed beds.

On increasing the gas flow rate at a given liquid flow rate, a nonhomogeneous flow regime, called slug or pulsing flow, is first encountered, characterized by alternate pistons of more and less dense mixtures of the two phases.
A further increase of the gas flow rate leads to the disappearance of the slugs, and a gas continuous flow with the liquid suspended as a mist in the gas stream is produced. This is called spray flow. The flow map shown in fig. 10 regards air and water flow through small packings. The lines in the figure must be considered as zones of transition, since there is not a sudden change between the regimes. This is particularly true for the transition to the spray flow, that occurs gradually. Information about systems different from water/air are missing. The bubble flow regime is that usually encountered in pactice. In the other flow regime, the upwflow columns behave like downflow columns in the corresponding regimes. It was observed (19) that higher gas-liquid areas are obtained in upward than in downward flow at the same liquid and gas flow rate. This is achieved, however, with a higher pressure drop, so that the increase of the energy costs reduces the benefits of a more effective mass transfer.

3.2 Pressure drop

The total pressure drop is the sum of a frictional pressure drop and a static pressure drop, this being due to the liquid column:

$$(\delta_{LG})_t = (\delta_{LG})_f + g\,\rho_L\left(1 + \frac{\varepsilon_G}{\varepsilon_o}\right) \qquad (23)$$

The frictional pressure drop in upflow is a little lower than in downflow, especially at low liquid flow rates in the bubble flow regime (19). This is due to the buoyancy force that acts as a driving force for the gas flow. By increasing the fluids flow rate, the frictional pressure drop in upward flow becomes very similar to that in downward flow (fig. 11).
In upward conditions the flow of both phases is, in all regimes, affected remarkably by the presence of the other phase, so that is difficult to propose a flow model. Few correlations have been proposed for evaluating the frictional pressure drop. The most commonly used correlation is the empirical one proposed by Turpin and

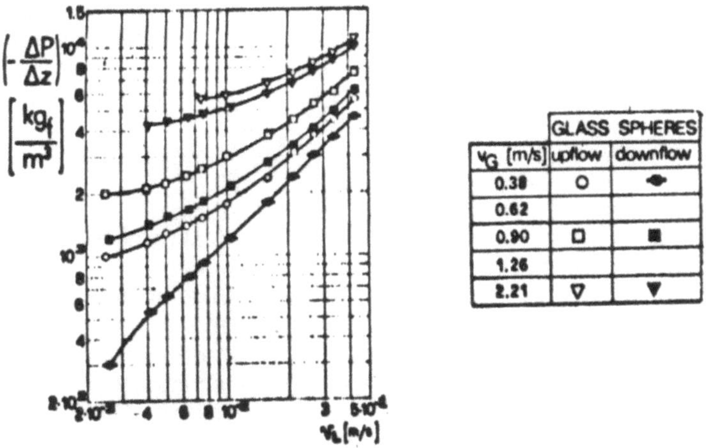

Fig.11 - Pressure drop in upflow and downflow.

Huntington (18), based on the two phase-friction factor concept. This is defined:

$$f_{LG} = \frac{(\delta_{LG})_f \, d_{pe}}{2 \, \rho_G \, v_G^2} \tag{24}$$

where d_{pe} is an equivalent particle diameter defined as follows:

$$d_{pe} = \frac{2}{3} \, d_p \, \frac{\varepsilon_o}{1 - \varepsilon_o} \tag{25}$$

f_{LG} was found to depend only on the gas and liquid Reynolds numbers, calculated with the superficial velocities of the fluids and the nominal diameter of the packing particles.
The Turpin and Huntigton' correlation is the following:

$$f_{LG} = 8.0 - 1.12 \, (\ln Z) + 0.0152 \, (\ln Z)^2 \tag{26}$$

where Z is defined:

$$Z = \frac{(Re_G)^{1.167}}{(Re_L)^{0.767}} \tag{26 bis}$$

A similar correlation is also valid for downflow, in the high interaction regime (3).

3.3 Gas holdup

In the bubble flow regime, the gas holdup ε_G is usually considered. In general the gas holdup in packed bubble columns at low gas flow rates is much higher than in empty bubble columns. This is due to the lower rising velocity of the bubbles in a

packed column than in an empty tube, and also to a "residual" gas holdup ε_{Gr} in the column at zero gas flow rate. ε_G, in fact, decreases by decreasing the gas superficial velocity, but tends to a residual value when v_G tends to zero.
Accordingly, the total gas holdup can be expressed as a sum of a residual and an operating part:

$$\varepsilon_G = \varepsilon_{Gr} + \varepsilon_{Gd} \tag{27}$$

Fig. 12 shows the behaviour of the gas holdup in several packed bubble columns, compared with that in an empty column, at zero liquid flow-rate.
The residual holdup depends on shape and size of the packing, on the column diameter and on the surface tension of the liquid: ε_{Gr} increases by decreasing the packing size, the column diameter and the surface tension, and increasing the bed porosity ε_o (20). ε_{Gr} reduces remarkably as the liquid velocity is increased, approaching to zero at the minimum fluidization velocity of the bed (21).
The residual gas holdup seems to be an important factor, since it was found to be ineffective in gas-liquid mass transfer (21): i.e. it has the behaviour of a true "static" holdup, with a very low renewal of the volume of the gas. The operating gas holdup can thus be considered also a "dynamic" holdup, in the sense that it represents the gas volume through which the gas mainly flows. It increases as the gas flow rate increases and the liquid surface tension decreases. It seems to be almost independent of the packing size and shape. The cocurrent liquid velocity should raise the bubbles velocity, and hence should decrease the operating holdup. In the same time, however, the increase of turbulence reduces the size of the bubbles, and the result is a constant operating holdup till $v_L = 2\mathrm{x}10^{-2}$m/s. At higher liquid velocity, the acceleration effect predominates, and ε_G decreases.
The operating gas holdup in packed catalyst reactors is proportional to the gas superficial velocity raised to 0.66, as quoted by Hofmann (20).
With packings used for absorption processes, the residual holdup was not observed. The total gas holdup is affected by the gas velocity as follows (22, 23):

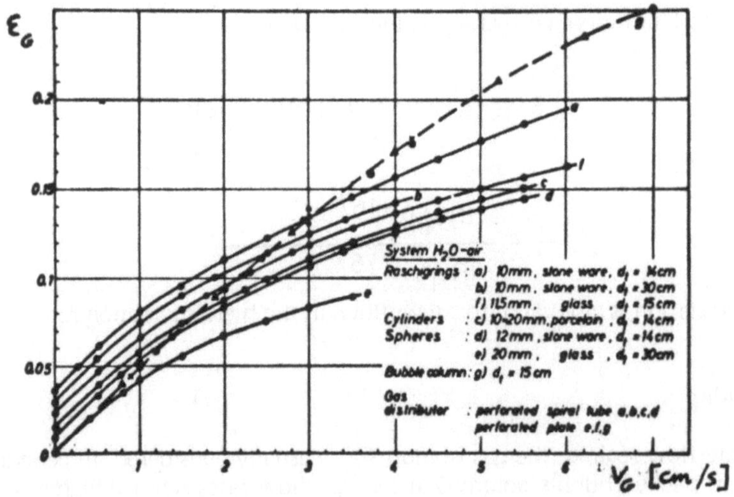

Fig. 12 - Gas holdup vs. v_G in bubble packed columns.

$$\varepsilon_G \sim v_G^{0.5} \tag{27}$$

The proportionality constant depends on the packing and physical properties of the liquid.

3.3 Flow patterns

The gas and liquid flow is generally far away from a plug flow, showing a remarkable backmixing or spread of residence time.
This, for the gas phase, is due to the fact that the bubbles have a certain size distribution. Since the rising velocity is connected with the size of bubbles, the largest bubbles remain in the reactor for shorter a time than the smallest one. These latter rapidly become almost ineffective to the process, when the gaseous reactant is mixed with an inert component. Due to their long residence time, in fact, the gaseous reactant can absorb almost totally into the liquid, leaving the inert component alone in the bubbles. Hence these small bubbles and the bubbles entrapped in the packing (residual holdup) constitute an ineffective volume of gas.
The circulation of the liquid due to the rising bubbles is the major responsible of the liquid backmixing. It seems, however, that a part of the backmixing is probably due to semistagnant liquid pockets as for trickle-bed reactors. The research work on this topic is scarce; an indirect confirmation of the existence of this static volume can be drawn from some results of liquid-solid area, obtained with dissolution of solid particles into a reacting liquid. The measured values of solid-liquid area are always considerably less than that of the packing area (fig. 13), whereas it is intuitive that physical wetting of particles is total. If the measured values are considered as "effective" areas for mass transfer, it can be concluded that part of the liquid volume inside the reactor is constituted by semistagnant portions, with a very poor contribution to the liquid-solid mass exchange.
This may be of relevant importance in processes where the conversion rate is affected remarkably by mass transfer phenomena, since the reactants supply to the catalyst and the catalyst utilisation may be considerably limited.

4. SLURRY STIRRED REACTORS

Slurry stirred reactors are three-phase reactors in which gas and solid catalyst are dispersed in a continuous liquid phase. They are very widely used for several chemical processes (hydrogenation, oxidation, alogenation etc.) and also for biochemical reactions. The commercial sizes range from several tens of cubic meters to less than 1 m^3.
The amount of solid for unit reactor volume is usually very low: for catalytic reactions, typically 10-20 g/l of catalyst are used, or even less. The catalyst particles are very small, about 100—200 μm and can have a rather high density; in biochemical reactions large and light solid agglomerates are encountered.
Some times, the reaction rate is largely affected by gas-liquid mass transfer, some other times, heat transfer is critical for reaction control.
A variety of requirements can thus be encountered, which fulfilment needs a variety of geometries, that affect considerably the hydrodynamic conditions.
The shape of the stirrer is one of the most important factors affecting hydrodyna-

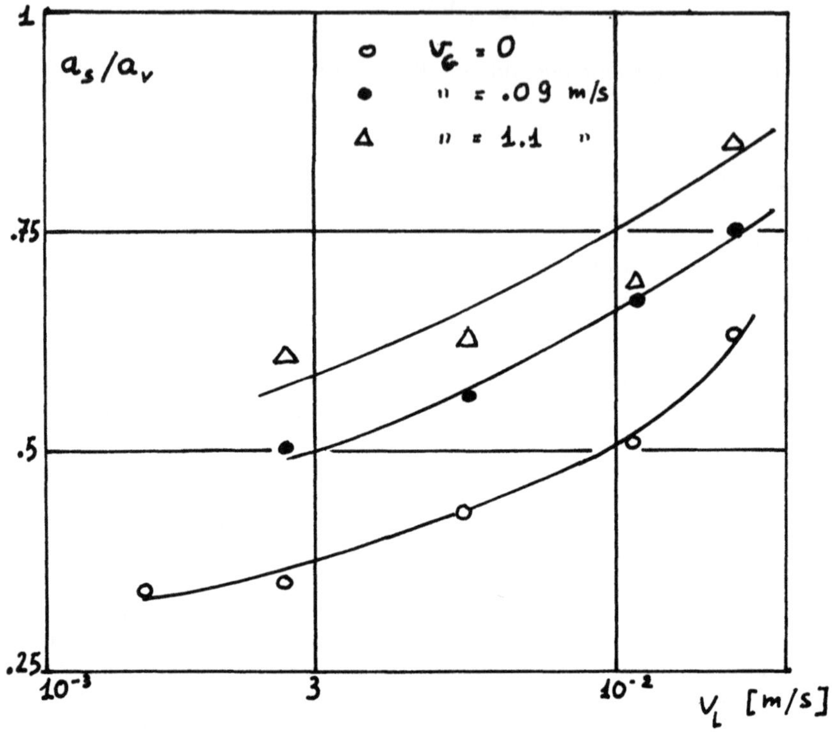

Fig. 13 - "Effective" solid-liquid area in upward two-phase flow.

mics. Stirrer must circulate the liquid in the reactor, breack off the gas in small bubbles, keep in suspension the solid. Often much care must be taken to avoid damnage of the suspended solids (i.e. in fermentation), or abrasion of the solid catalyst.

Some times, the best solution from one point of view is not suitable from the other so that a compromise solution must be adopted.

According to the pumping direction, the stirrers can be grouped into two classes: radial and axial stirrers. The first class is represented by flat and curved blade impellers and disc turbines (Rushton turbines). They are chatacterized by a high power dissipation, high shear stress in the impeller zone, uneven distribution of turbulence, relatively low pumping rate. Marine propellers and inclined blade turbines belong to the second class, characterized by a low power dissipation, more uniform turbulence distribution in the vessel, high pumping rate.

The radial stirrers give rise to two recirculation loops of liquid, one below and the other above the stirrer; with the axial stirrers instead, only one big recirculation loop takes place (fig. 14).

The gas is conventionally sparged below the stirrer; however, the conversion of the gas phase in a large number of gas-liquid reactions is low, due to the limited residence time and/or the low rate of chemical reaction. In these cases the unreacted gas may be recycled externally. Alternatively, surface aeration can be succesfully adopted, wherein the surface turbulence of the liquid entraines a certain amount of gas into the liquid.

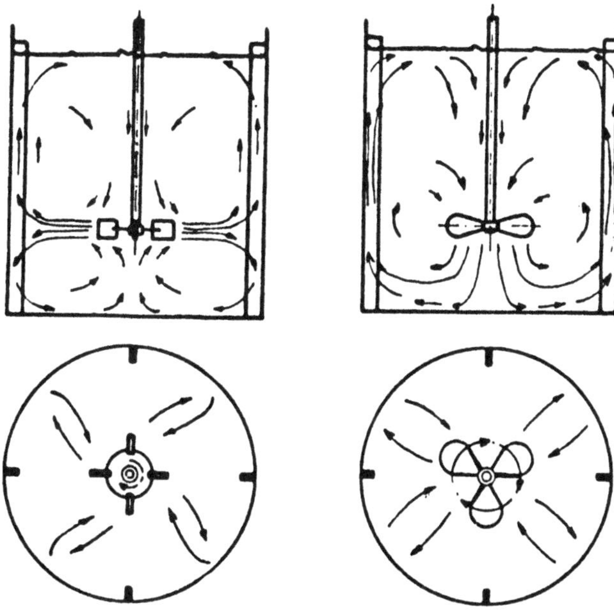

(a) (b)

Fig. 14 - Flow patterns: (a):radial stirrer; (b): axial stirrer.

Surface aeration can be improved by the use of gas inducing-type agitators, that essentially consist in a hollow shaft with a hollow impeller where the gas flows. When the impeller rotates, the sudden acceleration of the fluid elements causes reduction of the static pressure behind the blades. If the reduced pressure area is connected to the gas space, the gas is induced into the liquid in finely dispersed bubbles (fig. 15).

The important hydrodynamic parameters for design of stirred reactors are:

— power input;

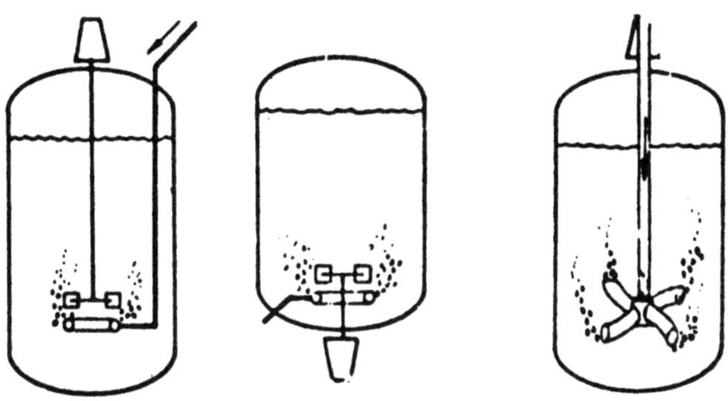

a) b) c)

Fig. 15 - Gas dispersion methods: a):top stirrer; b): bottom stirrer; c): gas inducing stirrer.

- pumping flow rate of the stirrer, for evaluating the recirculation of liquid in the reactor; this is connected to homogenization of the phases in the reactor;
- gas holdup, that allows the determination of the mean residence time of the gas in the reactor;
- mean bubble diameter, which is connected to the gas-liquid area.

The parameters affecting hydrodynamics are:

- physico-chemical parameters: liquid density, liquid viscosity, interfacial tension;
- operating parameters: gas flow rate, liquid flow rate, liquid volume, stirrer speed;
- geometrical parameters: stirrer shape and size, number of stirrers, vessel diameter, height of liquid, bottom shape, clearance of stirrer, number and width of baffles.

The suspended solid does not affect remarkably hydrodynamics, since solid load is usually very low. Neverthless, a good reactor performance is some times connected with a good suspension of solid, so that the minimum stirrer velocity for complete suspension must be considered also.

These parameters are usually grouped into dimensionless numbers; the most important ones for hydrodynamics are:

$$(N_P)_L = \frac{P_L}{\rho_L N^3 D^5}$$ Power number for liquid stirring

$$(N_P)_G = \frac{P_G}{\rho_L N^3 D^5}$$ Power number for gassed liquid stirring

$$Re_N = \frac{\rho_L N D^2}{\mu_L}$$ Reynolds number

$$Fr_N = \frac{N^2 D}{g'}$$ Froude number

$$N_A = \frac{Q_G}{N D^3}$$ Aeration number

$$We_N = \frac{\rho_L N^2 D^3}{\sigma_L}$$ Weber number

Many geometrical ratios are also used to characterize the hydrodynamic conditions: for example D/T, H/T, C/T, etc... (fig. 16).

A standard mixing apparatus is defined, for which H/T = 1, D/T = 1/3, C/T = 1/3, b/T = 1/10.

All these dimensionless numbers and geometrical ratios, however, are hardly sufficient to allow a good general extrapolation of results from one size to another. This gives an idea of the complexity of hydrodynamics in stirred reactors.

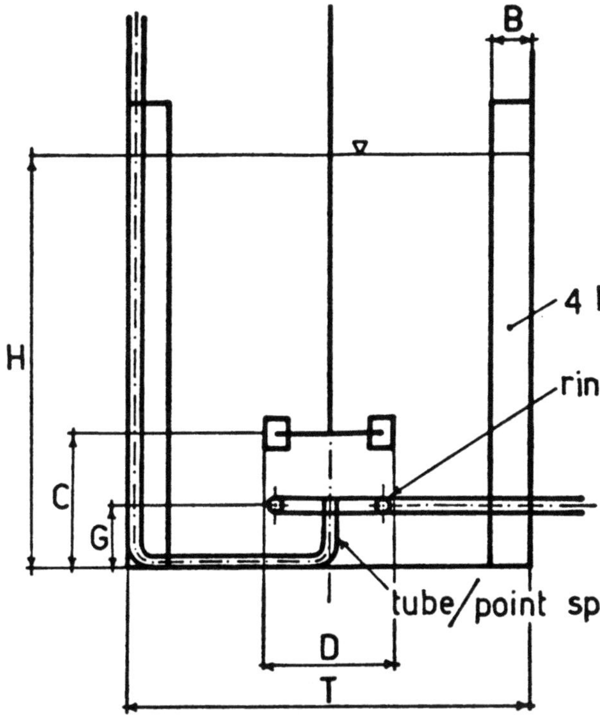

Fig. 16 - Standard stirred slurry reactor.

4.1 Mechanical dispersion of sparged gas

The gas phase is subdivided into small bubbles by the turbulence existing in the stirrer zone, then the bubbles are dispersed into the liquid by the recirculating flow.
According to the classical approach of Hinze (25) and Calderbank (26), a bubble can exist in a turbulent field when its diameter is so small that the conservative forces, due to the superficial tension, balance the turbulent forces, tending to break it off. On the basis of the isotropic turbulence assumption, the turbulent force acting on a unit area of a bubble, τ , is given by:

$$\tau \sim \rho_L \bar{u}^2 = \rho_L \left(\frac{d_b}{\rho_L}\right)^{2/3} \left(\frac{P}{V}\right)^{2/3} \qquad (29)$$

where $\bar{u}$ is the mean square root fluctuating velocity, and P/V is the dissipated power per unit volume. A bubble can therefore exist if the ratio $\tau\, d_b/\sigma_L$ is equal or lower than a critical value. From Eq. (29) the maximum size of a bubble is given by:

$$d_{b,\,max} \sim \frac{\sigma_L^{\,0.6}}{\rho_L^{\,0.2}\,(P/V)^{0.4}} \qquad (30)$$

From this equation, it can be seen that the bubbles size decreases by increasing the dissipated power per unit volume. That means that, at the same stirrer velocity, the

radial stirrers, giving a higher dissipated power, form smaller bubbles. Smaller bubbles have a lower rising velocity (and hence a longer residence time in the vessel) and a larger interfacial area per unit gas volume.
For this reason, the radial stirrers are usually employed when gas-liquid mass transfer affects largely the reaction rate; among these stirrers, the Rushton type turbine is largely employed.
Several works have clearly defined the dispersion mechanism for this type of stirrer (27, 28, 29). In the presence of the liquid alone, behind each blade there are two vortices (fig. 17a). When a very low gas flow rate is fed below the stirrer, little cavities form, in the eye of the vortices, called vortex cavities (fig. 17b). An increase of the gas flow rate, leads to an increase of the volume of cavities, which adhere to the blades (clinging cavities, fig. 17c). This regime is relatively unstable, especially at low stirrer speeds; further increase of the gas flow rate leads to the formation of stable large cavities (fig. 17d), which seem to be the most effective for gas dispersion. These large cavities do not form simultaneously behind all blades, but occupy successively the different blades. A further increase of the gas flow rate leads to an increase of the volume of the large cavities, so that at low stirrer velocity, the stirrer may be almost completely envelopped by the gas. This situation is called flooding by some authors, and is very ineffective for dispersion.
If we observe the evolution of the dispersion state as the stirrer speed increases at constant gas flow rate, we can see successively the situation shown in fig. 18. At a

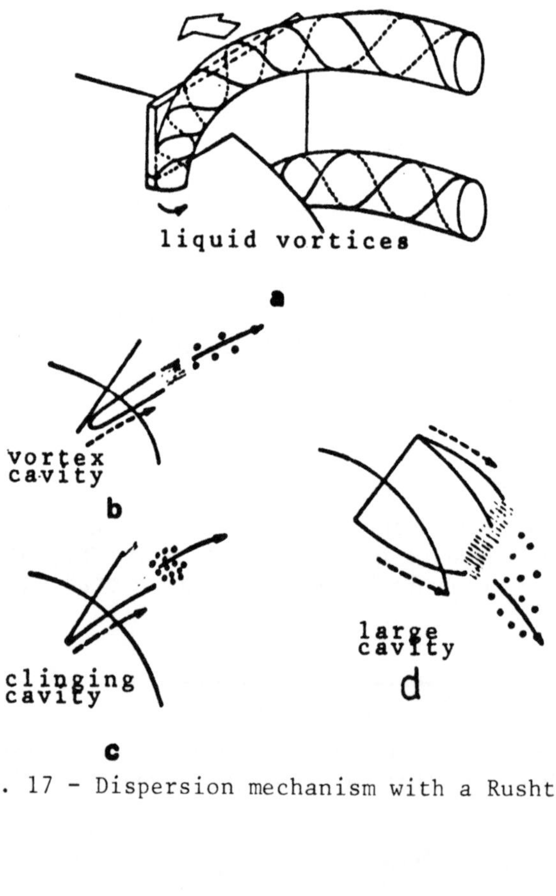

Fig. 17 - Dispersion mechanism with a Rushton turbine.

low stirrer speed, the gas is not dispersed and large bubbles rise around the shaft (fig. 18a). As N increases, the large cavities appear and the bubbles are more and more dispersed in the reactor zone above the stirrer (fig. 18b). A state is then reached where the bubbles are dispersed radially (fig. 18c): the large cavities tend to change into the clinging cavities. A further increase of N leads first to a dispersion of bubbles also in the zone below the stirrer (fig. 18d); then, at very high N values, a secondary recirculation loop can be observed (fig. 18e). The hydrodynamic situations shown by fig.s 18d and e correspond to the clinging cavities regime.

Let us observe the effect of the hydrodynamic regime on the gas holdup ε_G (volume of gas per unit dispersion volume). Fig. 19 shows qualitatively the behaviour of ε_G as a function of N. At very low values of N, the stirrer is "flooded" and ε_G is very low and almost constant. At a certain stirrer velocity (N_F), the gas begins to be dispersed and ε_G increases. This corresponds to the transition between the situation shown in figs. 18a and b; N_F is the "flooding stirrer speed". As N increases, the increase of hold-up is first considerable, then becomes lower. Another characteristic velocity N_L is defined, which corresponds roughly to the transition between the situation shown in figs. 18c and d. A further increase of N does not affect remarkably the holdup. N_L is called the loading point (33).

Of course, these hydrodynamic situations affect also the power consumption. In a

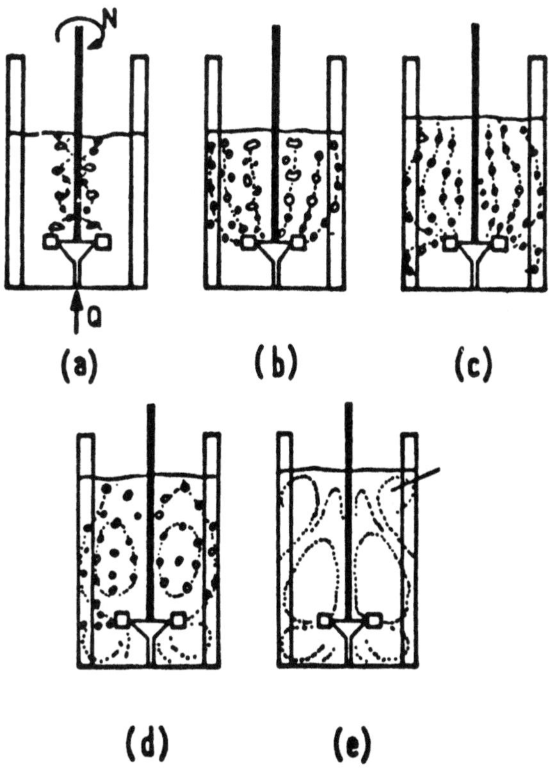

Fig. 18 - Dispersion evolution as N increases at Q_G = cons.

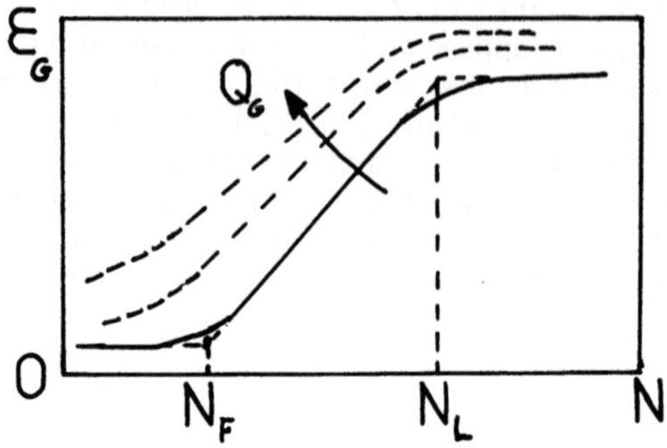

Fig. 19- Gas holdup as a function of the stirrer speed.

stirred liquid, the power dissipated P_L is given by the well know relationship:

$$P_L = (N_P)_L \cdot \rho_L N^3 D^5 \tag{31}$$

where $(N_P)_L$ is the power number, that depends on the system geometry and the Reynolds number. In turbulent regime, i.e. at $Re_N > 10.000$, $(N_P)_L$ is constant.
In an aerated system, the power dissipated by the stirrer decreases, since a two-phase fluid with a lesser density is pumped (or alternatively, the liquid flow rate pumped by the stirrer decreases, since the delivery area is partially occupied by the gas cavities). A gassed power number can be defined:

$$(N_P)_G = \frac{P_G}{\rho_L N^3 D^5}$$

that is usually plotted as a function of the aeration number N_A:

$$N_A = \frac{Q_G}{N D^3}$$

N_A is the ratio between the gas flow rate that must be handled by the stirrer and the maximum liquid flow rate that can be pumped by the stirrer (proportional to $N\ D^3$).
Fig. 20 shows the behaviour of $(N_P)_G$ at constant Q_G as N increases (i.e. N_A decreases) for a 6 blade Rushton turbine (32). It can be seen that, as N increases, $(N_P)_G$ decreases, due to formation of large cavities behind the blades. These cavities reduce the form drag of the blades, by hindering the formation of turbulent liquid vortices. At a certain stirrer speed, a minimum of $(N_P)_G$ is attained, which corresponds to the situation shown by fig. 18c. Then $(N_P)_G$ increases again, owing to the formation of clinging cavities.
The minimum of $(N_P)_G$ occurs approximately at the loading point.
From the foregoings, it can be seen that the "best" hydrodynamic regime for gas-liquid dispersion with a Rushton turbine is that shown by fig. 18c, corresponding to a minimum of the gassed power number and to a high value of the gas holdup (loading conditions). Pollard (32) showed that around these conditions, a maximum of the volumetric mass transfer coefficient referred to the dissipated power is also obtained, at a given gas flow rate.
It was stressed (33) that in a diagram of N versus Q_G, the loading and flooding conditions are shown by two curves, so that there are two well defined relationships

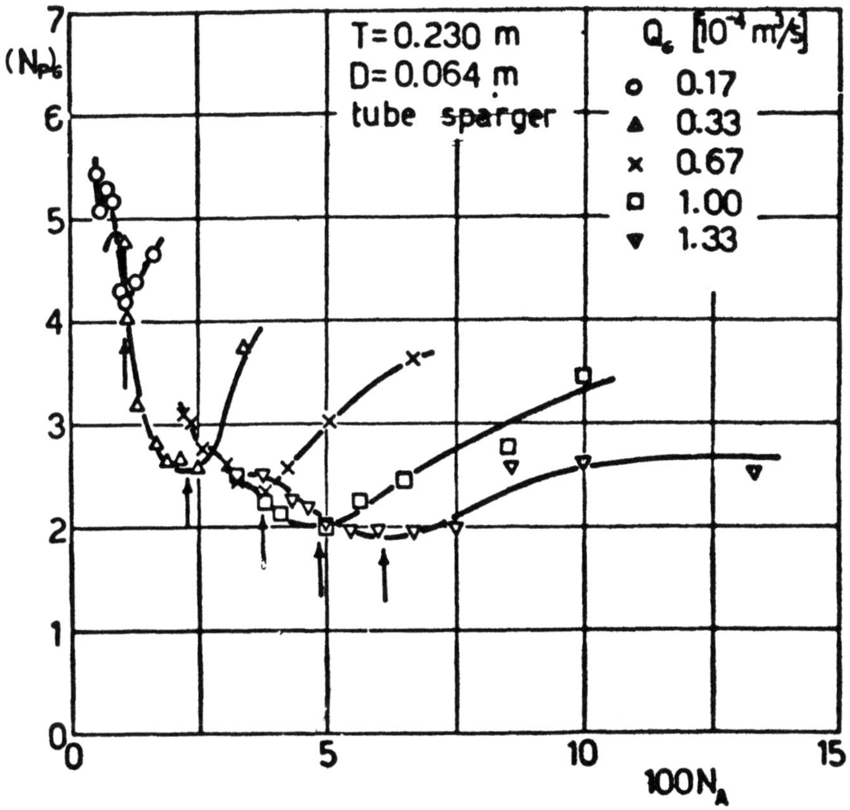

Fig. 20- Influence of N_A on $(N_P)_G$ at several Q_G values.

between these parameters for both the hydrodynamic regimes, for a given system geometry.

On the basis of $(N_P)_G$ minimum, Nienow et al. (30) proposed the following correlation equation for loading:

$$N_L D^2 = 4Q_G\ T^{0.2} \qquad \text{(SI units)} \tag{32}$$

Bruxelmane's correlation (35) is the following:

$$\frac{Q_G}{N_L D^3} = 2.45\ (Fr_N)^{0.78}\ (\frac{D}{T})^2\ (\frac{l}{D})^{0.41}\ (\frac{q}{D})^{1.20}\ n_p^{\ 1.35} \tag{33}$$

For N_F, the following two correlations may be suggested: after Bruxelmane (33):

$$\frac{Q_G}{N_F D^3} = 3.93\ (Fr_N)^{0.72}\ (\frac{D}{T})^{1.90}\ (\frac{l}{D})^{0.47}\ (\frac{q}{D})^{1.30}\ n_p^{\ 1.50} \tag{34}$$

after Roustan (35) (valid for 6-blade turbines):

$$N_F = 0.94\ \frac{Q_G^{\ 0.4}\ T^{0.5}}{D^{2.5}} \qquad \text{(SI units)} \tag{35}$$

For low gas flow rate ($v_G < 0.03$ m/s), van Dierendonck et al. (36) proposed correlations, valid for several pure liquids, according to which N_F is independent from the gas flow rate:

$$\frac{N_F D^2}{(gT^3)^{1/2}} = 0.07 \qquad (T < 1 \text{ m}) \qquad (36\text{ a})$$

$$\frac{\mu_L N_F D^2}{T \sigma_L} \left(\frac{\rho_L \sigma_L^3}{\mu_L^4} \right)^{1/4} = 2 \left[\frac{H - C}{T} \right]^{1/2} \qquad (T > 1) \qquad (36\text{ b})$$

As can be seen, the agreement between different workers is not good: N_L and N_F are considered proportional to Q_G and T^2 about by Bruxelmane, and to $Q_G^{0.4 \div 0.5}$ and $T^{0.2 \div 0.5}$ by others. Van Dierendonck et al. do not consider the effect of Q_G on flooding onset. The number of parameters affecting hydrodynamics can be one of the reason of this disagreement: little variations of geometrical ratios can have a by no means negligible effect. Another reason may be the different experimental procedure adopted: the minimum of $(N_P)_G$ and the changes of slope of ε_G vs. N, and furthermore the evaluation of holdup at constant liquid volume or dispersion volume. Moreover, a great influence on the determination of flooding and loading onset can have the superficial aeration (entrainment of gas from the liquid surface), that occurs expecially in large vessels. It is interesting to know that the onset of the superficial aeration is given, according to van Dierendonck et al., by the same equations (36a) and (36b) that are assumed valid for flooding.

All these considerations are valid for coalescing systems, like air in water. With other systems, things may be different. For example, foaming can occur before saturation of holdup, so taht ε_G increases remarkably. The secondary recirculation loop or the formation of frooth give rise to a second noticeable decrease of the gassed power number as N increases. Unfortunately, very few data exist on liquid systems different from water, so that no correlations of hydrodynamic regimes are available.

4.2 Dissipated power

As said before, power dissipated by a stirrer in a non gassed liquid can be calculated from the power number, that depends on the geometry of both stirrer and vessel.

In the presence of a sparged gas flow rate, as previously observed, a reduction of the dissipated power takes place, due to the fact that a two-phase flow with a lower density than that of the liquid is pumped. Therefore, the decrease of the dissipated power is a function of the gas flow rate handled by the stirrer, referred to the pumping capacity of the stirrer (proportional to ND^3 for geometrically similar systems). It must be considered, however, that there is a recirculation of gas bubbles, so that the actual gas flow rate $(Q_G)_e$ passing through the stirrer is greater than Q_G (fig. 21):

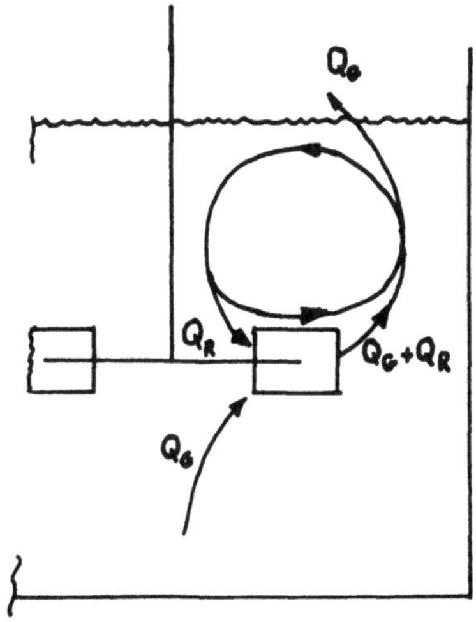

Fig. 21 - Recirculation of gas.

$$(Q_G)_e = Q_G \quad\quad Q_R \tag{37}$$

where Q_R is the recirculated gas flow rate.
In spite of this, several correlations of P_G/P_L based on $N_A = \dfrac{Q_G}{ND^3}$ only have been proposed for Rushton turbines. For example Calderbank (39) proposed for 6-blades turbines:

$$\begin{aligned} P_G/P_L &= 1 - 12{,}6\ N_A \qquad (N_A \leq 0{,}035) \\ P_G/P_L &= 0{,}62 - 1{,}85\ N_A \qquad (N_A \geq 0{,}035) \end{aligned} \tag{38}$$

This correlation is valid for pure, non foaming liquids, in a quite wide range of physical properties of liquid, at N values well above N_F.
The most widely used correlation, however, is an empirical correlation proposed by Michel and Miller (40), valid for geometrically similar systems:

$$P_G \sim \left[\frac{P_L^2\ N\ D^3}{Q_G^{\ 0.56}}\right]^n \tag{38}$$

where the proportionality constant and the exponent n depend on the system geometry and on the characteristics of the liquid.
The paper by Midoux and Charpentier (35) gives an exaustive review of the values of these parameters from several works.
For non foaming liquid the most frequent value for n is 0.45; with this value, Eq. (38) can be rielaborated and the following relationship can be obtained for a given gas-liquid system:

$$P_G/P_L \sim (N_A)^{-0.25} \cdot \left(\frac{Re_N}{Fr_N}\right)^{0,1} \tag{39}$$

This points out clearly that the major parameter of the gassed power reduction is the aeration number, but other dimensionless numbers must be introduced to have

a good correlation.
The Re_N and Fr_N numbers probably take account of the recirculated gas flow rate. In the presence of liquids that hinder bubbles coalescence or show foaming characteristics, the exponent n may be markedly lower ($n \leq 1/3$).

4.3 Gas holdup

The behaviour of the gas holdup as a function of N for several gas flow rates has been seen in fig. 19.
The shown trend is usual for coalescing systems; when coalescence is hindered or foaming takes place, the holdup increases remarkably at the same operating conditions; typically, the characteristic stirrer velocity N_L is not clearly marked and the holdup continues to increase considerably.
A physical property of the liquid that play an important role is the surface tension: a decrease of surface tension gives rise generally to an increase of gas holdup in pure liquids. In the presence of liquid solutions the effect of surface tension is not straightforward, because of interfacial phenomena. The liquid components may contain in their molecule atomic groups that are more or less hydrophilic, so that in vicinity of the gas surface a separation of components and a preferential orientation of the molecules of some components may occur. This leads to changes of local surface tension and even of the local viscosity, so that coalescence phenomena may differ considerably from those in a pure liquid with the same physical properties as the liquid solution. As a result, the average size of bubbles may be quite different in the two cases.
The effect of viscosity is not clear. Some authors (34) found ε_G to increase with μ_L until a maximum, and then decrease. Some others (39), instead, did not find any effects of the viscosity in the range $5 \cdot 10^{-4} - 2{,}8 \cdot 10^{-2}$ Pa · s.
A theoretical approach to the holdup correlation can start from the intuitive consideration that holdup increases as the bubble size decreases. In fact smaller bubbles have a lower rising velocity and furthermore can be entrained more easily by the liquid flow streams, so that they can recirculate several times before leaving the vessel. The size of the bubbles, as already seen [eq. (30)] is mainly a function of the dissipated power per unit volume, and the surface tension.
Several holdup correlations are therefore based on the power dissipated. For pure liquids, they have the general form:

$$\varepsilon_G \sim (P/V)^{\alpha_1} \cdot \sigma_L^{\alpha_2} \cdot v_G^{\alpha_3} \qquad (40)$$

with $\alpha_1 = 0{,}4 \div 0{,}3$; $\alpha_2 = -0{,}6 \div 0{,}3$; $\alpha_3 = 0{,}5 \div 0{,}7$.
As regards the dissipated power, some authors use the shaft power P_G; some others (41) take into account also the power supplied to the system by the gas phase. This is partly due to the kinetic energy of the gas flow at the outlet of the sparger, partly to isothermic expansion of the gas phase along the reactor. The former is usually negligible; the latter, (P_E), can be calculated as follows:

$$P_E/V = \rho_G Q_G \frac{RT}{M} \ln\left[\frac{p_o}{p_s}\right] \qquad (41)$$

Several correlations have been proposed; the most famous one has been proposed by Calderbank (39):

$$\varepsilon_G = 1.94\,(V_G\ \varepsilon_G)^{1/2} + 4.19\times10^{-3}\left[\frac{(P_G/V)^{0.4}\ \rho_L^{\ 0.2}}{\sigma_L^{\ 0.6}}\right] v_G^{\ 0.5} \qquad (42)$$

(SI units)

The first term on the right side takes account of the holdup at $N = 0$; it becomes rapidly negligible as N increases.
All these correlations are valid till $N = N_L$ or a little further.
For ionic solutions, different correlations have been proposed, because of the interfacial phenomena that hinder the coalescence. It was found (42) that ε_G increases as the following parameter increases:

$$\beta = C\left(\frac{d\sigma_L}{dC}\right)^2 \cdot \left[1 + \frac{d(\log\ f)}{d(\log\ C)}\right] \qquad (43)$$

where C is the concentration of the salt and f is its activity in solution.
Significant effects on ε_G have little concentrations of surfactants and also the presence of very fine suspended solids. This is particularly interesting for three-phase reactors: Ganguli (43) found an increase of ε_G as the suspended solid concentration increased.
The phenomenon is not yet clear, but it seems that the fine suspended particles affect considerably the coalescence of the bubbles. The solid tested by Ganguli was constituted for 55% by very fine particles, with $d_p \leq 0.4\ \mu m$.

4.4 Solid suspension

A good suspension of solid catalyst is required in order to have a good solid-liquid mass transfer and to avoid deposit on the bottom. This latter, in fact, owing to the local high concentration of catalyst, may lead to side reactions and very high local heat generation rates.
Two main solid suspension states can be defined, namely: complete suspension, in which all particles are just suspended, and homogeneous suspension, in which the particles concentration is uniform throughout the reactor. The complete suspension state is attained at a certain stirrer velocity N_{js}, called the minimum stirrer velocity for complete suspension. This state is considered as optimum as far as solid-liquid mass transfer is concerned. For $N < N_{js}$, there are heavy deposits of solids on the vessel bottom and the solid-liquid interfacial area effective for mass transfer decreases steeply, causing a remarkable decrease of mass transfer rate. On the other hand, for $N > N_{js}$ the increase of dissipated power is not balanced by the increase of the mass transfer coefficient.
In the complete suspension state, particles previously suspended drop and stop in certain zones of the bottom of the vessel, then (after 1-2 seconds) are suspended again.
This is due to turbulent phenomena. According to a proposed model (45, 46), the turbulent forces acting on the particle surface, assuming isotropic turbulence, are expressed as follows:

$$d_p^2\ \rho_L\ \bar{u}^2 \sim d_p^2\ \rho_L \left(\frac{d_p}{\rho_L}\right)^{2/3} \cdot \left(\frac{P}{V_b}\right)^{2/3} \qquad (44)$$

where $(P/V)_b$ is the dissipated power in the bottom zone.
These turbulent forces are balanced by the gravity forces, so that:

$$\rho_L\, d_p^2 \left(\frac{d_p}{\rho_L}\right)^{2/3} \cdot \left(\frac{P}{V}\right)_b^{2/3} \sim d_p^3\, \Delta\rho \cdot g \qquad (45)$$

The local dissipated power by turbulent phenomena changes remarkably in a stirred vessel: it is higher in the stirrer zone, and decreases with the distance from the stirrer. If, as a first approximation, $(P/V)_b$ is assumed proportional to the average dissipated power per unit volume of liquid, it can be written:

$$M = \sqrt{\frac{\Delta\rho \cdot g}{\rho_L}} \cdot \frac{T\, d_p^{1/6}}{N_p^{1/3}\, D^{5/3}\, N_{js}} = \text{const} \qquad (46)$$

M, a dimensionless number, represents the ratio $\left[(P/V)_b / (P/V)\right]^{1/3}$. In practice, the dissipated power distribution changes with stirrer velocity and the geometrical dimensions, so that M is a function of the Reynolds number $Re_L^{*} = \dfrac{\rho_L D^3 N_{js}}{\mu_L T}$ and of the ratio C/D. It is also strongly affected by the shape of the stirrer: axial stirrers with downward pumping direction give M numbers much higher, and hence N_{js} much lower, than the radial ones. M is also slightly affected by the concentration of solid.
The most extensive examination of the conditions for complete suspension in non gassed liquids was given by Zwietering (44) who proposed the following empirical equation:

$$N_{js} = S\, \frac{\mu_L^{0.1}\, (g\, \Delta\rho)^{0.45}\, d_p^{0.2}}{\rho_L^{0.55}\, D^{0.85}}\, B^{0.13} \qquad (47)$$

where S is a constant depending on the sistem geometry and B is the mass fraction of solid.
In the presence of a sparged gas phase, the dissipated power at constant stirrer speed decreases: therefore one can expect that N_{js} increases. Some results (31) show in fact that not only N_{js} is higher, but its increase is so remarkable that also the shaft power required for suspension is higher than in liquid-solid systems. The increase of the power for complete suspension is more considerable in full aeration than in loading conditions. This can be explained, according to the model previously discussed, as a strong reduction of the dissipated power in the bottom zone, when the bubbles are dispersed also below the stirrer. In the loading conditions, the bubbles are mainly distributed in the zone above the stirrer, so that the local power distribution below the stirrer does not differ much from that in liquid mixing.
In the complete suspension state there exist generally axial gradients of solid concentration, that can give rise to some problems in continuous stirred reactors. In this case, the catalyst is continuously fed and withdrawn; if the suspension is homogeneous, the concentration of catalyst inside the reactor is the same as in the outlet stream, and the amount of catalyst in the reactor can be evaluated accurately. If gradients of concentration exist in the vessel, the solid concentration in the outlet stream may be very different from the average one in the reactor. This makes difficult the control of the catalyst amount in the reactor and hence the

conversion yield.
The concentration profile depends on the settling velocity of particles and the stirrer speed: a low settling velocity and a high stirrer speed reduce the dishomogeneity degree. Very few data exist on this subject: fig. 22 shows the solid concentration profile in a vessel stirred with a Rushton turbine at $N = N_{js}$ as a function of d_p (47); it can be seen that a pseudo-homogeneous zone exists above the stirrer, i.e. a zone where the solid concentration is close to the average one. An increase of the stirrer speed decreases the degree of dishomogeneity of the suspension, but the rate of increase of dissipated power is higher than that of homogeneity degree (46).
In practice, it is better to place the withdrawal tube in corrispondence of the pseudohomogeneous zone. No data exist till now about the effect of a sparged gas on the solid concentration profiles. It was observed, hovewer, that in the loading conditions the suspended particles stay preferentially in the zone below the stirrer, whereas in the full aeration conditions the suspension seems more homogeneous.

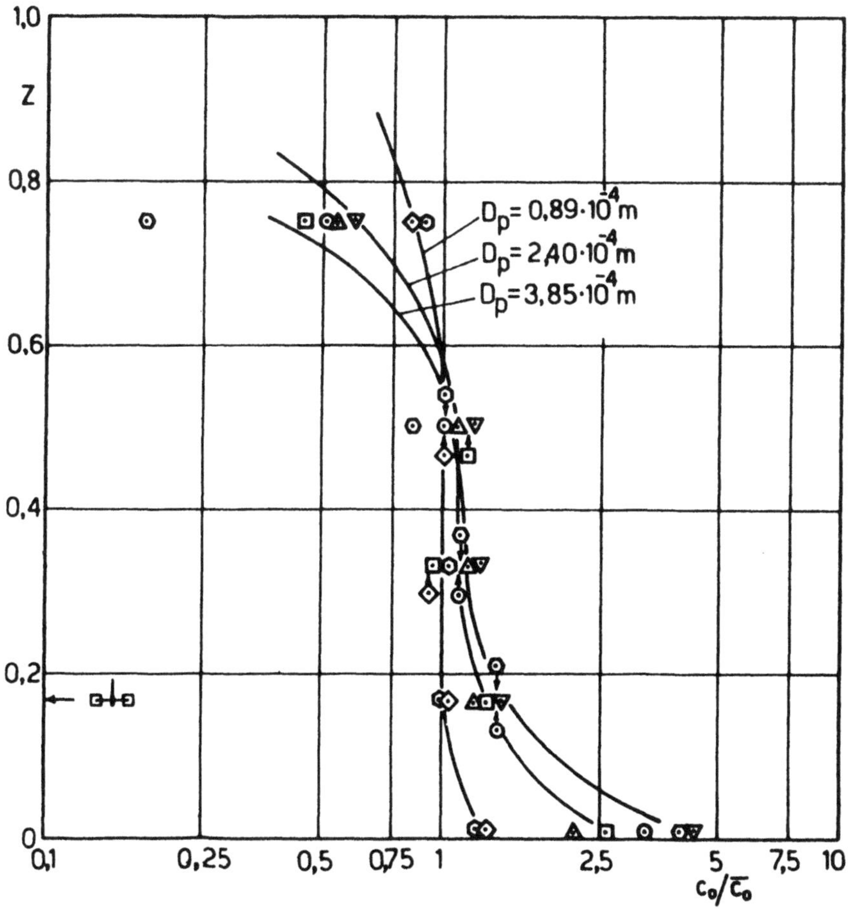

Fig. 22 - Axial concentration profile of suspended particles.

NOTATIONS

a_s = Solid-liquid interfacial area (m^{-1})

a_v = dry packing surface area per unit volume of empty column (m^{-1})

b = baffels width (m)

c = stirrer clearance (m)

d_b = bubbles diameter (m)

d_p = particles size (m)

D = stirrer diameter (m)

D_c = column diameter (m)

D_r = radial spread factor of the liquid (m)

$(D_r)_o$ radial spread factor of the liquid at zero gas flow rate (m)

Eö = $\rho_L g d_p^2 / \sigma_L$, Eötvos number (dim.less)

Fr = v^2/gd_p, Froude number (dim.less)

Fr_N = N^2D/g, Froude number in stirred vessels (dim.less)

G = gas mass velocity (kg/m^2s)

Ga = $d_p^3 \rho_L^2 G/\mu_L^2$, Galileo number (dim.less)

h_i = intraparticle liquid holdup (dim.less)

h_e = interparticle liquid holdup (dim.less)

h_t = total liquid holdup (dim.less)

h_{ed}, h_{er} = draiming and residual liquid holdup (dim.less)

h_{es}, h_{edy} = static and dynamic liquid holdup (dim.less)

H = height of liquid (m)

l = length of a blade (m)

L = liquid mass velocity (kg/m^2h)

n_p = blades number

N = stirrer speed (s^{-1})

N_A = Q_G/ND^3, aeration number (dim.less)

N_F = flooding stirrer speed (s^{-1})

N_L = loading stirrer speed (s^{-1})

(N_P) = $P/\rho_L N^3 D^5$, power number (dim.less)

N_{js} = stirrer speed for complete suspension (s^{-1})

P = shaft dissipated power; local dissipated power by turbulent phenomena (w)

q = heigh of a blade (m)

Q_G = gas flow rate (m^3/s)

Q_R = recirculated gas flow rate (m^3/s)

r = radial coordinate (m)

Re_L, Re_G = liquid and gas Reynolds numbers, based on the superficial velocity (dim.less)

Re_N = $\rho_L ND^2/\mu_L$, Reynolds number in stirred vessels (dim.less)

T = vessel diameter (m)

$\overline{u}$ = mean square root fluctuating velocity (m/s)

v^* = local superficial velocity (m/s)

V = superficial velocity (m/s)

$v_{L,W}$ = volumetric flow rate of the liquid per unit empty area at the wall (m/s)

V = volume (m^3)

Greek Symbols

δ = pressure drop (Pa/m)

ε_o = bed porosity m

ε_G = gas holdup

ε_{Gr}, ε_{Gd} = residual and operating gas holdup

λ = defined in eq. (1) (dim.less)

μ = viscosity kg/ms

ρ = density kg./m^3

σ_L = surface tension kg/s^2

ψ = defined in eq. (2) (dim.less)

χ = defined in eq. (8) (dim.less)

f_L, f_G = defined in eqs. (11) and (12) (fg./m^2 · s)

$\Delta\rho = \rho_L - \rho_s$ (kg/m^3)

Subscrits

wt = water

G = gas

L = liquid

LG = two-phase

S = solid

REFERENCES

1) – Satterfield C.N. – A.I. Ch. E. J. 21, 209 (1975)
2) – Midoux N., Favier M., Charpentier J. C. – J. Chem. Eng. Japan, 9, 350 (1976)
3) – Specchia V. Baldi G. – Chem. Eng. Sci., 32, 515 (1977)
4) – Sicardi S., Gerhard H., Hofmann H. – Chem. Eng. J., 18, 173 (1979)
5) – Gianetto A., Baldi G., Specchia V., Sicardi S. – A.I.Ch.E.J., 24, 1087 (1978)
6) – Charpentier J.C., Favier M. – A.I.Ch.E.J., 21, 1213 (1975)
7) – Charpentier J.C. – J. Powder and Bulk Solid Techn. 2, 53 (1978)
8) – Specchia V., Rossini A., Baldi G. – Ing. Chim. Ital., 10, 171 (1974)
9) – Baldi G., Specchia V. – Ing. Chim. Ital., 12, 107 (1976)
10) – Herskowitz M., Smith J.M. – A.I.Ch.E.J., 24, 439 (1978)
11) – Herskowitz M., Smith J.M. – A.I.Ch.E.J., 24, 450 (1978)
12) – Charpentier J.C., Proust C., van Swaaij W., Le Goff P. – Chim. Ind. Gén. chim., 99, 803 (1968).
13) – Buchanan J.E. – Ind. Eng. Chem. Fundamental, 6, 400 (1967)
14) – Kan K., Grenfield P.F. – Ind. Eng. Chem. Process Des. Devel, 18, 740 (1979).
15) – Larkins R.P., White R.R., Jeffrey D.W. – A.I.CH.E.J., 7, 231 (1961)
16) – Charpentier J.C., Proust C., Le Goff P. – Chem. Eng. Sci. 24 1777 (1969)

17) – Sicardi S., Baldi G., Specchia V., – In press on Chem. Eng. Sci.
18) – Turpin J.L., Huntington R.L. – A.I.Ch.E.J., 13, 1196 (1967)
19) – Specchia V., Sicardi S., Gianetto A. – A.I.Ch.E.J., 20, 646 (1974)
20) – Hofmann H. – Cat. Rev. 17, (1), 71 (1978)
21) – Ohshima S., Takematsu T., Kuriki Y., Shimada K, Suzuki M., Kato J. – J. Chem. Eng. Japan, 9, 29 (1976)
22) – Sawant S.B., Pangarkar V.G., Joshi J.B. – Chem. Eng. J. 18, 143 (1979)
23) – Pexidr V., Charpentier J.C. – Collect. Czech. Chem. Comm., 40, 3130 (1975)
24) – Stiegel G.I., Shah Y.T. – Ind. Eng. Chem. Process Des. Dev., 16, 37 (1977)
25) – Hinze K.O. – A.I.Ch.E.I., 1, 289 (1955)
26) – Calderbank P.N. – Brit. Chem. Eng., 1, 206 (1956)
27) – Van't Riet K., Smith J.M. – Chem. Eng. Sci., 28, 1031 (1973)
28) – Van't Riet K., Smith J.M. – Chem. Eng. Sci., 30, 1093 (1975)
29) – Bruijn W., Van't Riet K., Smith J.M. – Trans. Instn. Chem. Engrs. 52, 88 (1974)
30) – Nienow A.W., Wisdom D.I., Middleton J.C. – II Eur. Conf. on Mixing, Cambridge March – April 1977
31) – Sicardi S., Conti R., Baldi G., Franzino E. – Mycrosymposium on Mixing, March 25, 1980, Koln (West Germany)
32) – Pollard G.I. – Proc. Int. Symposium on Mixing, 25-28 February 1978 – Mons (Belgium)
33) – Bruxelmane M. – Proced. Int. Symposium on Mixing, 25-28 February 1978, Mons (Belgium)
34) – Rushton J.H., Bimbinet J.J., Can.I. Chem. Eng., 46, 16 (1968)
35) – Midoux N., Charpentier I.C. – Entropie, n. 88, pag. 5 (1979)
36) – Van Dierendonck L.L., Fortuin J.M.F., Vanderbos D. – Proceed. IV Europ. Symp. Chem. React. Eng., 9-11 September 1968, Bruxelles
37) – Loiseau B., Midoux N., Charpentier J.C. – A.I.Ch.E.J., 23, 931 (1977)
38) – Van't Riet K., Boom J.M., Smith J.M. – Trans. Instn. Chem. Engrs, 54, 124, 1976
39) – Calderbank P.H. – Trans. Instn. Chem. Engrs, 36, 443, (1958)
40) – Michel B.I., Miller S.A. – A.I.Ch.E.J. 8, 262 (1962)
41) – Miller D.N. – A.I.Ch.E.J., 20, 445 (1974)
42) – Lee J.C., Meyrick D.L. – Trans. Inst. Chem. Engrs, 48, 37 (1970)
43) – Ganguli K.L. – Chem. Eng. Thesis, Technische Hogeschool, Delft, Holland (1975)
44) – Zwietering T.N. – Chem. Eng. Sci., 8, 244 (1958)
45) – Baldi G., Conti R., Alaria E. – Chem. Eng. Sci., 33, 21 (1978)
46) – Conti R., Baldi G. – Proceed. Intern. Symp. on Mixing. 25-28 February 1978, Mons (Belgium)
47) – Baldi G., Conti R., Gianetto A. – To be published on A.I.Ch.E.I.

HEAT TRANSFER IN GAS-LIQUID-SOLID REACTORS

Giancarlo Baldi

Istituto di Chimica Industriale, Politecnico, Torino, Italy

1. GENERAL

In three-phase gas-liquid-solid reactors, the reaction heat is generated inside the catalyst pellets, then is transferred to the surrounding fluid phase, and lastly is removed from the reactor by either cooling surfaces or outlet fluids flow rates (adiabatic reactors). Three transport processes must therefore be considered.

a) intraparticle heat transfer, within individual catalyst pellets;

b) interphase heat transfer, between the catalyst and the phases in contact;

c) heat transfer through the reactor.

2. INTRAPARTICLE HEAT TRANSFER

A good review of intraparticle transport processes has been given by Satterfield |1|. Under steady-state conditions, the diffusion flux across a boundary surface surrounding a portion of a porous catalyst pellet equals the reaction rate. The heat released by the reaction must all be transferred across the same boundary surface. It can be written therefore for a sphere:

$$- D_i \frac{dC_i}{dr} (-\Delta H) = \lambda_p \frac{dT}{dr} \qquad (1)$$

An integration of this equation leads to:

$$T - T_s = \frac{(-\Delta H) D_i}{\lambda_p} (C_s - C_i) \qquad (2)$$

This equation, coupled with the differential equation for mass diffusion:

$$\frac{D_i}{r^2} \frac{d}{dr} \left(r^2 \frac{dC_i}{dr}\right) = k_\infty \exp\left(- \frac{E}{RT}\right) . C_i^n \qquad (3)$$

allows determination of concentration and temperature as a function of the spacial coordinates.

It is known that heat effects can affect remarkably the reaction rate inside a pellet.

In practice, thermal effects are taken into account by an effectiveness factor, that depends on the Thiele modulus, and on the following parameters β and γ:

$$\beta = \frac{C_s(-\Delta H) D_i}{\lambda_p T_s} \qquad (4)$$

$$\gamma = \frac{E}{RT_s} \qquad (5)$$

β represents $\Delta T_{max}/T_s$, the maximum percentage temperature difference between the centre and the outer surface of a pellet when the reaction is so fast that C_i at the centre of the pellet is zero.

γ represents the sensitivity of the reaction rate to the temperature change.

The graphs of the non isothermal effectiveness factor can be easily found in the literature (see for example reference |1|. For exothermic reactions ($\beta>0$) the effectiveness factor is higher than that corresponding to the isothermal case; it increases by increasing both β and γ, and can reach values much higher than 1. The opposite occurs for endothermal reactions.

Whether or not thermal effects are important is primarily governed by β. This parameter depends on the ratio D_i/λ_p between intraparticle diffusivity and thermal conductivity of the porous solid. Heat conduction in a porous solid occurs partially through the solid medium, partially through the fluid filling the pores. This latter is so significant that no correlations of λ_p with the thermal conductivity of solid is possible. It was found |2,3,4|, in fact, that the values of λ_p for pellets in air decrease by decreasing the pressure; at atmospheric pressure, λ_p ranges from 10^{-4} to 10^{-3} cal/(s. cm. ºC); under vacuum, λ_p is much lesser (even an order of magnitude), because of the lesser thermal conductivity of the gas in the pores. In slurry and trickle bed reactors, the pores of catalyst are supposed to be filled by liquid; since thermal conductivity of liquids is 10 to 100 times greater than that of gases, the values of λ_p would be higher than those for pellets in a gaseous medium.

Let us consider an example to evaluate the order of magnitude of thermal gradients inside a pellet.

Kenney and Sedriks |5| examined the hydrogenation of liquid crotonaldehyde over a palladium on allumina catalyst at atmospheric pressure.

At the tested conditions (70°C), the saturation hydrogen concentration in the liquid C_s=2,5 10^{-6} gmole/cm^3, D_i=5,8.10^{-5} cm^2/s, (-ΔH)=22.000 cal/gmole.

Assuming a λ_p value of 5.3.10^{-4} cal/cm.s.°C, as given by Sher |2| for Pt - allumina catalyst in air (conservative value), and C_i=0 at the centre of the pellet, the maximum temperature difference between the centre and the outer surface of a pellet, given by eq. (2) would be 0.006 °C.

Suppose instead the same reaction in the gas phase; the maximum concentration of hydrogen is $\frac{273}{343}\frac{1}{22400}$ = 3,55 x 10^{-5} gmoles/cm^3; the intraparticle diffusivity would increase considerably and a value of 5.10^{-2} cm^2/s may be conservative. In these conditions:

$$\Delta T_{max} = 75°C \ ; \quad \beta = 0.22$$

If E is not too low, this value of β would give rise to a significant increase of the reaction rate in the pellet. In general, the assumption of isothermal conditions is largely valid when the pores of the catalyst are totally filled up of liquid. Significant intraparticle temperature gradients may exist, instead, when the pores are full of gas and a catalysed gas phase reaction takes place.

A criterion for isothermal operations inside a pellet is as follows |6|

$$|\gamma\beta| < 0.05.n$$

where n is the exponent in power law rate espression. γ values of 10 to 30 are usual, so that β must be less than 5.10^{-3} n in order to avoid intraparticle heat effects.

3. INTERPHASE HEAT TRANSFER

Because of the heat generation rate inside a pellet, a temperature difference between the outer surface of the pellet and the bulk of surrounding fluid can exist; the heat transfer balance for a pellet is written as follows:

$$R\, V_p (-\Delta H) = h_s A_p (T_s - \bar{T}) \qquad (7)$$

where R is the observed reaction rate per unit catalyst volume and h_s is the heat transfer coefficient.

The same equation can be written in another way:

$$-\lambda_p \left.\frac{dT}{dr}\right|_{r_p} = h_s (T_s - \bar{T}) \qquad (8)$$

Whether intraparticle or interphase resistance prevails is governed by the particle Biot number $Bi_p = h_s\, d_p/\lambda_p$. Interphase heat transfer is controlling when |6|:

$$\frac{h_s d_p}{\lambda_p} < 10 \qquad (9)$$

condition that can be easily met not only in laboratory reactors, because of the low flow rate, but also in industrial reactors. In the case of highly exothermal reactions, this would lead to a particle temperature significantly greater than that of the bulk stream.

In three phase reactors, the catalyst in usually surrounded by liquid; solid-gas contact takes place when there are unwetted parts of the catalyst outside area, that may occur in trickle-bed reactors only, in particular conditions.

Very few data are available for heat transfer between particles and liquid. Some experimental results can be found for particles suspended in an agitated liquid |7|. A satisfying analogy between heat and mass transfer was noted, so that similar relationships for heat and mass transfer coefficients have been proposed |7|:

$$k_s (Sch_s)^{2/3} = \frac{h_s}{c_{pL}\rho_L} \cdot (Pr_L)^{2/3} = 0.13 \left|\frac{(P/V)\mu_L}{\rho_L^{\,2}}\right|^{1/4}$$

(S.I. units) (10)

Let us consider how important is the difference of temperature between a pellet and the surrounding liquid in s stirred reactor. The heat balance can be written:

(i) $k_s A_p (C_L - C_s)\ (-\Delta H) = h_s A_p (T_s - \bar{T})$

From eq. (10), this equation becomes:

(ii) $$\frac{h_s(-\Delta H)}{C_{pL}\rho_L}\left(\frac{Pr_L}{Sch_L}\right)^{2/3}\cdot(C_L - C_s) = h_s(T_s - \bar{T})$$

and hence:

(iii) $$T_s - \bar{T} = \frac{(-\Delta H)}{C_{pL}\rho_L}\left(\frac{Pr_L}{Sch_L}\right)^{2/3}(C_L - C_s)$$

Assume $(-\Delta H) = 3.10^4$ kcal/kmole; $c_{pL}\rho_L = 10^3$ kcal/m^3.C; Pr_L/Sch_L = = 1/100; with these values:

$$T_s - \bar{T} = 6.6(C_L - C_s) \quad \text{(S.I. units)}$$

Bearing in mind that the concentration of the gaseous reactant in the liquid phase is usually very low, it can be seen that the temperature difference between the catalyst and the surrounding liquid in a stirred slurry reactor is always very low.

Things would be very different if the pellet will be in contact with the gas. Assuming the analogy between mass and heat transfer coefficients, the same equation (iii) can be written. In this case, however, $Pr_G/Sch_G = 1$ and $\rho_G\, C_{pG} = 1$ kcal/m^3 ºC can be assumed; this leads to:

$$T_s - \bar{T} = 3.10^4\,(C_G - C_s)$$

Even in the case of $C_G \cong C_s$ (no external mass transfer limitations) the temperature difference between the pellet and the surrounding gas may be significant.

From the example, it may be concluded that the difference of temperature between the particle and the external fluid is not remarkable when the catalyst pellet is completely surrounded by liquid, as in stirred slurry and upflow packed bed reactors. On the contrary, the removal of heat from the pellets can be critical when the gas is in contact with dry zones of the outer catalyst surface. This can occur in trickle-bed reactors, expecially at the laboratory scale.

Due to the particular structure of the liquid flow, the local solid-liquid heat transfer coefficients have very different values around a pellet. Stagnant liquid pockets exist, with a very low liquid renewal; moreover, at the low liquid flow rate used in laboratory reactors, some part of the outer pellet area may be covered by a very thin liquid film almost motionless. All these zones are very ineffective for heat transfer. If the reaction is highly exotermal and fast, the heat generated in the catalyst near these zones can hardly be removed; therefore local increases of temperatu re take place, that may eventually, in the presence of a volatile liquid, give rise to evaporation of the thin outside liquid film.

Dry zones of catalyst are generated in this way, which are very ineffective for heat transfer and very effective for mass transfer. The local temperature is further increased, and evaporation of liquid inside the pores may even occur. The gaseous reactant diffuses very fast in the dry pores and reaches the wetted zones of the catalyst, increasing further and further the average reaction rate and the heat generation rate. Some times a faster gas phase reaction can also take place. It is quite easy, in this case, to have marked hot spots of temperature, dangerous for the catalyst activity , and instability of the reactor.

Germain et al. |8| found, in the hydrogenation of α-methilstyrene in a laboratory trickle-bed reactor, unstable hot spots travelling along the column.

This thermal instability would generally disappear by suitably increasing the liquid flow rate, because of a better wetting of the particles.

4. INTERPARTICLE HEAT TRANSFER IN TRICKLE-BED REACTORS

Heat released by the catalyst is carried throughout the reactor by the fluids motion.

Usually, trickle-bed reactors are considered as adiabatic; moreover, gas temperature in each section is considered to equal that of the liquid, because of the much lower heat capacity of gas.

Assuming the plug flow model, the axial temperature profile is given by:

$$\frac{dT}{dz}\,(L\,c_{pL} + Gc^{*}_{pG}) = (-\Delta H)\,R\,(z,\,T) \qquad (11)$$

In the presence of volatile liquids, it is better to evaluate the gas heat capacity with a fictitious specific heat c^{*}_{pG}, defined as the rate of variation of the enthalpy of the gas saturated with the liquid vapour:

$$c^{*}_{pG} = \left(\frac{dH_G}{dT}\right) \qquad (12)$$

The liquid vapour increases remarkably the enthalpy of the gas phase, because of the latent heat of evaporation.

The axial increase of temperature for exothermal reactions

augments downstream the probability of instability phenomena discussed in the previous section. In order to control the raise of temperature in commercial units, several methods are adopted: injection of cold gas along the reactor (Fig. 1a), recycling of cooled liquid products (Fig. 1b), cooling coils between sections of catalytic bed (Fig. 1c).

The injection of cold gas is commonly adopted in hydrotreating reactions of mineral oils; the only disadvantage is given by the marked increase of gas velocity in the bed, so that the different catalyst sections may operate in different hydrodynamic regimes.

Recycle of part of the outlet liquid is very efficient for temperature control, because of the high capacity of liquid; however, the method has many disadvantages, as dilution of the reactants, with negative effects on kinetics, decrease of the hourly space velocity of the liquid, and increase of pressure drop. The third method seems an attractive one; unfortunately gas to coil heat

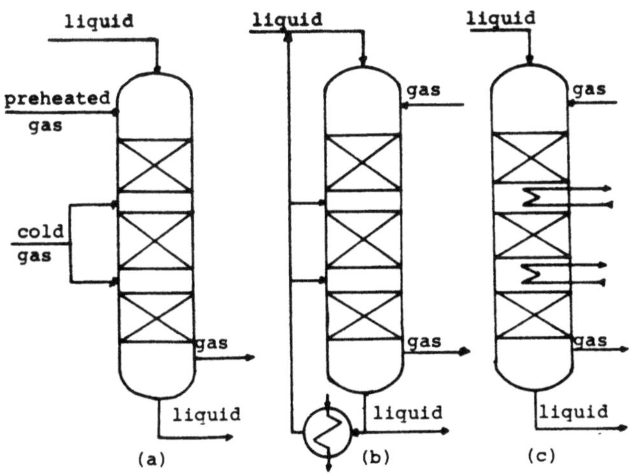

Fig.1 : Temperature control in trickle-bed reactors; a): cold gas injection; b): liquid recirculation; c): cooling coils.

transfer coefficient is very low, so that a large transfer area is required; moreover internal coils make more complicated reactor building and substitution of catalyst.

In laboratory trickle-bed reactors, control of temperature is achieved by external cooling; the gas flow rate may eventually be used as a controlling variable, but some care must be taken when volatile liquids are treated.

An increase of the gas flow rate favours vaporization of liquid. When very low liquid flow rates are adopted, this can lead to a dramatic change of thermal conditions. Fig. 2 shows the axial temperature profile in a laboratory trickle-bed reactor for cyclohexene hydrogenation |10|. An increase of G decreases first the temperature gradients; at a certain value of G however, some zones of the catalyst become dry, and a much faster gas phase reaction takes place. This in turn increases the heat generation rate, and the reaction passes in all the reactor from a liquid regime to a gaseous regime, with a step increase of temperature.

For jacketed laboratory reactors, radial heat transfer must be considered; radial temperature gradients exist also in industrial reactors because of heat losses from the wall.

Radial heat transfer can be interpreted by a two-parameter model in which an equivalent heat conductivity in the bed k_e and a heat transfer coefficient h_w at the wall are considered |9|. According

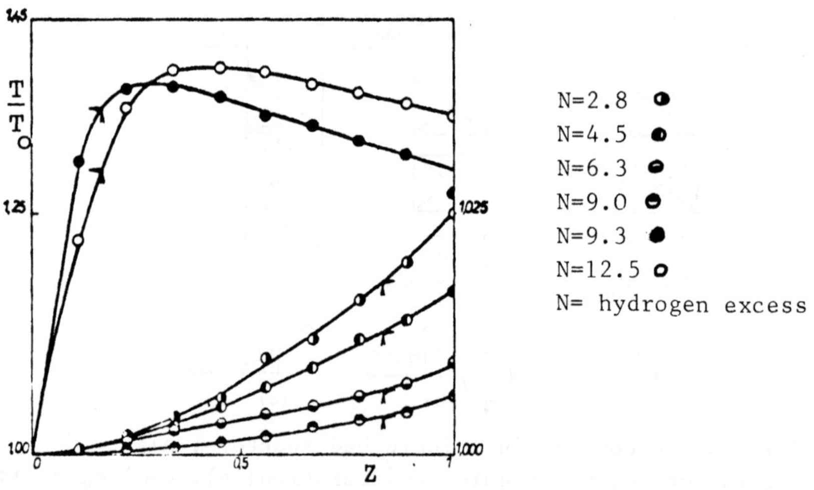

Fig. 2: Effect of the gas flow rate on the axial temperature profile for hydrogenation of cyclohexene (after Hanika et al. /10/)

to this model, the temperature profile in the bed is given by;

$$(Lc_{pL} + Gc_{pG}) \frac{\partial T}{\partial z} = k_e \left(\frac{1}{r}\frac{\partial T}{\partial r} + \frac{\partial^2 T}{\partial r^2}\right) + (-\Delta H) \cdot R \qquad (13)$$

Equal temperature of the three phases are assumed at each point of the reactor. The boundary condition at the wall is:

$$h_w (T_w - T_R) = k_e \left.\frac{\partial T}{\partial r}\right|_{r = r_w} \qquad (14)$$

Fig. 3 shows the radial profile of temperature in a trickle-bed when heat is supplied from the wall. It can be observed the discontinuity of the temperature profile near the wall, interpreted as due to a resistance in the wall zone.

The usual experimental method for evaluating the two parameters consist in supplying heat to the two-phase flow from the outside at constant wall temperature, and measuring the radial temperature profile as a function of the axial coordinate. The identification of h_w and k_e can be carried out by several methods; a good review can be found in the work by Tsang et al. |12|. A very convenient method is to solve Eq. (13) (without the reaction rate term) with appropriate boundary conditions, and to search the optimal h_w and k_e values for the best fit of the experimental profiles of temperature.

Few data of h_w and k_e have been published |9|. From these, k_e appears to be a complex function of the fluids flow rates and the packing size and materials.

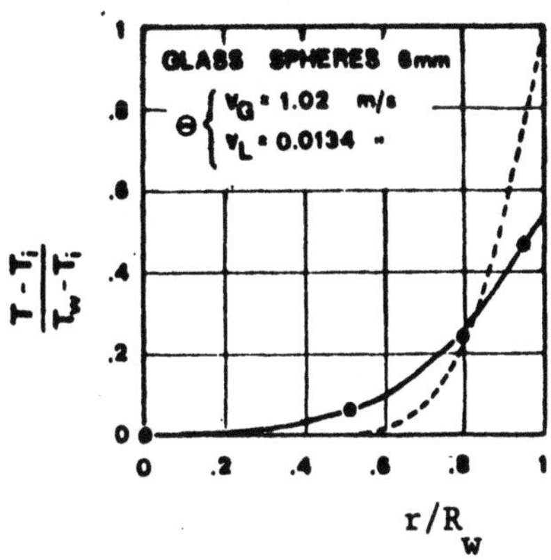

Fig. 3: Radial temperature profile; full line: two-parameter model (k_e, h_w); dotted line: one-parameter model (k_e).

It has been interpreted in terms of a stagnant contribution $(k_e)_o$ and radial mixing contributions of gas $(k_e)_G$ and liquid $(k_e)_L$:

$$k_e = (k_e)_o + (k_e)_G + (k_e)_L \quad (15)$$

The stagnant contribution is due to heat conduction through the solid bed and the stagnant fluids around the pellets. The mixing contributions are due to the radial component of the velocity of the fluids.

$(k_o)_e$ and $(k_e)_G$ are usually lower than $(k_e)_L$), that predominates largely at high liquid flow rates as those adopted in industrial reactors.

In laboratory reactors where very low liquid flow rates are adopted, probably $(k_e)_o$ and $(k_e)_G$ are also important.

$(k_e)_o$ and $(k_e)_G$ can be evaluated by the same correlation equations proposed for gas flow through packed beds, at least in the low interation regime. The most recent ones are |11|:

$$\frac{\rho_G v_G c_{PG} d_p}{(k_e)_G} = 8.65 \left[1 + 19.4 \left(\frac{d_p}{D_c}\right)^2\right] \quad (16)$$

$$\frac{(k_e)_o}{\lambda_G} = \varepsilon_o + \frac{(1 - \varepsilon_o)}{0.220 \, . \, \varepsilon_o^2 + 0.666(\lambda_G/\lambda_s)} \quad (17)$$

The evaluation of $(k_e)_G$ and $(k_e)_o$ from eqs. (15) and (16) allows the determination of $(k_e)_L$ from k_e data. It was found that $(k_e)_L$ so determined is strictly connected with the mass radial dipersion, as shown in Fig. 4. This constitues a physical support of the model.

Correlation of $(k_e)_L$ have been proposed for air-water systems |10| in connection with the hydrodynamic regimes:

- low interaction regime:

$$\frac{v_L \rho_L c_{PL} d_p}{(k_e)_L} = 0.041 \frac{(v_L \rho_L d_p)^{0.87}}{h_e \mu_L} \quad (18)$$

- high interaction regime:

$$\frac{v_L \rho_L c_{PL} d_p}{(k_e)_L} = 338 \frac{(v_L \rho_L d_p)^{0.675}}{\mu_L} \, . (h_e)^{0.29} \left(\frac{a_v d_p}{\varepsilon_o}\right)^{-2.7} \quad (19)$$

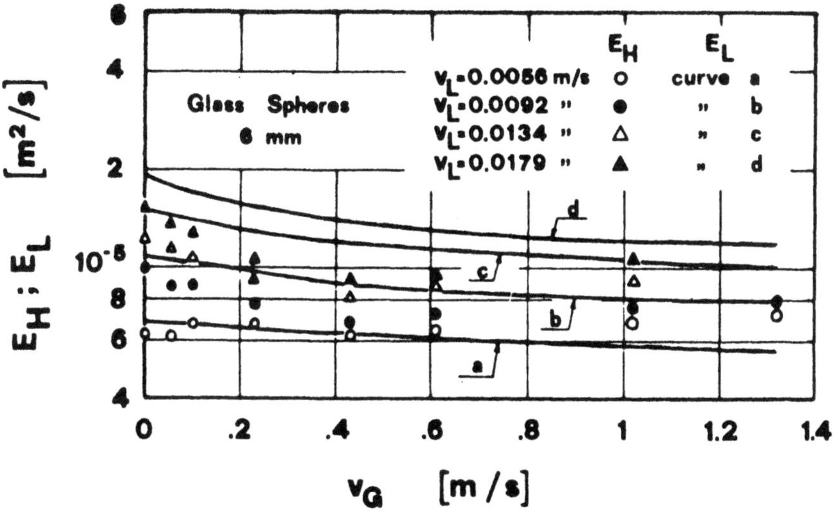

Fig. 4: Comparison between radial mass and heat dispersion coefficients (after Specchia and Baldi /9/).

The effect of the gas flow rate is taken into account by the liquid holdup h_e that is a function of v_G also. An evaluation of the order of magnitude of h_w in the low interaction regime can be obtained by the following equations:

$$\frac{h_w d_p}{\lambda_L} = 0.057 \left(\frac{v_L \rho_L d_p}{h_e \mu_L}\right)^{0.89} (Pr_L)^{1/3} \qquad (20)$$

In the high interaction regime, it seems than h_w is almost independent from the fluids flow rate, and may have a value as high as 2000 kcal/h.m^2.ºC.

From these correlations, one can estimate how important are the radial temperature profiles, As a rough approximation, a uniform heat generation in the reactor and no axial temperature profile can be considered. In this case, the radial temperature profile is parabolic, and it is easy to demonstrate that:

$$\frac{T_i - T_m}{T_m - T_w} = \left(1 + \frac{8k_e}{D_c h_w}\right)^{-1} \qquad (21)$$

where T_i=temperature at the centre of the reactor and T_m is the average temperature in the bed.

The temperature profile depends therefore on the Biot number $\frac{h_w D_c}{k_e}$

From eqs. (15) and (18), neglecting the effect of v_G on liquid holdup, assuming $h_e \sim v_L^{0.3}\, d_p^{-0.9}$, one can writes:

(i) $k_e \cong (k_e)_L \sim v_L^{0.4}\, d_p^{-0.8}$

(ii) $h_w \sim v_L^{0.6}\, d_p^{0.8}$

(iii) $\dfrac{h_w D_c}{k_e} = Bi_c \sim D_c\, v_L^{-0.2}$

As can be seen, the Biot number is greatly affected by D_c and little by d_p and v_L. An increase of D_c leads to an increase of Biot number, and an increase of the radial temperature gradients.

In a 14 cm column packed with 6 mm spheres, a Biot number of about 4 was found |9|. For a 1 m diameter reactor, the Biot number would be of the order of 30, and $(T_i-T_m)/(T_m-T_w)$, according to eq. (21), would be 0.78, that means a strong temperature difference between the centre and the wall zone.

For a 3 cm diameter reactor (laboratory size) and 1 mm pellets, according to eq. (iii), the Biot number would be 1.32 and $(T_i-T_m)/(T_m-T_w)=0.14$.

5. HEAT TRANSFER IN SLURRY STIRRED REACTORS

Slurry stirred reactors are usually isothermal;owing to the vigorous liquid mixing,the temperature is about constant throughout the reactor. Temperature control is obtained by heat transfer surfaces, like coils and jacket.

Due to the variety of geometry of mixing apparatuses, several correlations for heat transfer coefficient have been proposed.

For non gassed systems, these correlations have the general form:

$$Nu\,(Pr)^{-0.33} \sim (Re_N)^{\alpha}\left(\frac{\mu}{\mu_w}\right)^{\beta} \cdot f\ \text{(geometrical parameters)} \qquad (22)$$

The exponent α ranges between 0.5 and 0.7, with an average value of about 0.65.

When a gas flow rate is fed to the reactor, at low stirrer speeds, the heat transfer coefficient increases as a function of v_G. This increase is lesser and lesser as N increase. With radial stirrers, a critical velocity exists, beyond which an increase on the gas flow rate decreases the heat transfer coefficient |13|.

The bubbles of gas cause a destruction of the laminar layer of liquid around the heat transfer surface, Beside this positive effect, the gas bubbles also influence the heat transfer coefficient in a negative way. The presence of gas in the vicinity of the transfer surface tends to reduce heat transfer, because of the much higher resistance of gas than that of the liquid. At low impeller speeds, the positive effect predominates; the opposite occurs at higher N values.

Some correlations have been proposed, that unfortunately are not comparable because of the different experimental conditions. It seems interesting that by Pollard and Topiwala |14|.

$$Nu = C \left(\frac{P}{V}\right)^{\alpha} (\varepsilon_G)^{\beta} \qquad (23)$$

where C=f (Pr_L, $\mu/\mu w$, Re_N, dimensionless geometrical ratios).

The gas holdup ε_G depends itself mainly on P/V, so that this is the major parameter affecting Nu.

Constant dissipated powder per unit volume is often adopted as a criterion for scale-up. This criterion, together with a constant catalyst load, leads to an almost constant reaction rate per unit volume R. By adopting this scale-up criterion, from eq. (23), Nu=const. results, i.e. $h_N \sim 1/T_V$, as the size of the reactor is changed and a geometrical similarity in maintained.

The transfer area in a geometrically similar system would increase as T_V^2, so that the transferred heat is at the maximum proportional to T_V.

On the other hand, R is constant in scale-up, and the heat generation would be R $(-\Delta H)T_V^3$, i.e. proportional to T_V^3. It can be seen that for reactions highly exothermal, heat transfer through cooling surfaces is quite critical in scale-up; other technical solutions must be considered, like external heat transfer, evaporation, etc.

Laboratory reactors do not suffer usually of heat transfer limitations, because of their small size.

NOTATION

a_V - packing area (m^2/m^3)

A_p - pellet outer area (m^2)

C_i, C_s - concentration inside a pellet and at the solid interface, respectively (k-moles/m^3)

C_L, C_G - concentration in the bulk of liquid and gas, respectively (k-moles/m^3)

c_{pL}, c_{pG} - specific heat of liquid and gas (kcal/kg°C)

d_p - pellet diameter (m)

D_i - intraparticle diffusivity (m^2/s)

E - activation energy (kcal/kmole)

h_e - total interparticle liquid holdup (-)

h_N, h_s, h_w - heat transfer coefficients, respectively: in stirred reactors, particle-to-fluid, at the wall in trickle--bed reactors (kcal/s m^2°C)

$(-\Delta H)$ - heat of reaction (kcal/kmoles)

k_e - equivalent heat conductivity in the bed (kcal/s.m.°C)

$(k_e)_o, (k_e)_G, (k_e)_L$ - stagnant and radial mixing contribution of the gas and liquid to k_e (kcal/s.m.°C)

k_s - solid-liquid mass transfer coefficient (m/s)

$Nu = \frac{h_N T_V}{L}$ - Nusselt number for stirrer vessel (-)

P - dissipated power (w)

$Pr_L = \frac{c_{pL}\mu_L}{\lambda_L}$ - Prandtl number of the liquid phase (-)

r - radial coordinate (m)

r_w - reactor radius (m)

R - reaction rate (kmoles/s.m^3)

Re_N - rotational Reynolds number (-)

Sc_L - Schmidt number of the liquid phase (-)

T - temperature (K)

T_s - temperature at the outer surface of pellets (K)

$\bar{T}$ - bulk temperature (K)

T_m, T_i - average and axial temperature in the bed (K)

T_v - diameter of a stirred reactor (m)

v_L, v_G - superficial liquid and gas velocity (m/s)

V - volume (m^3)

GREEK LETTERS

β - defined by eq. (4) (-)

γ - defined by eq. (5) (-)

ε_G - gas holdup (-)

$\lambda_p, \lambda_L, \lambda_G, \lambda_s$ - heat conductivity of pellet, liquid, gas, solid respectively (kcal/s.m°C)

μ_L, μ_W - bulk and wall viscosity of liquid (kg/m.s)

ρ_G, ρ_L - gas and liquid density (kg/m^3)

REFERENCES

1) Satterfield C.N. - "Mass transfer in heterogeneous catalysis". M.I.T. Press. Cambridge, Massachussets (1970)
2) Sehr R.A. - Chem. Eng. Sci.,9,145 (1958)
3) Mischke R.A., Smith J.M. - Ind. Eng. Chem., Fundamentals,1, 288 (1962)
4) Masamune S., Smith J.M. - J. Chem. Eng. Data,8,54 (1963)
5) Kenney C.N. Sedriks W. - Chem. Eng. Sci.,27,2029 (1972)
6) Mears D.E. - Ind. Eng. Chem. Process Des. Develp.,10,541 (1971)
7) Calderbank P.H. "Mass transfer" in "Mixing: theory and practice", edited by Uhl V.W. and Gray J.B., Academic Press. New York, 1967
8) Germain A.H., Lefebre A.G., L'Homme G.A. - Chemical Reaction Engineering II, ACS Monograph Series, No. 133, pag. 164 (1974)
9) Specchia V., Baldi G., - Chem. Eng. Commun.,3,483 (1979)
10) Hanika J., Sporka K., Ruzicka V., Hrstka J. Chem. Eng. J., 12,193 (1976)
11) Specchia V., Baldi G., Sicardi S. - Chem. Eng. Comm.,4,361 (1980)
12) Tsang T.H., Edgar T.F., Houghen J.O. - Chem. Eng. J.,11,57 (1976)
13) De Mearteheire E. - Proc. Int. Symposium on Mixing, 21-24 February 1978, Mons, Belgium
14) Pollard R., Topiwala, H.H. - Biotech. Bioeng.,18,517 (1976)

DESIGN AND SCALE-UP OF TRICKLE BED REACTORS. SOLID-LIQUID CONTACTING EFFETIVENESS

Giancarlo Baldi

Istituto di Chimica Industriale, Politecnico, Torino
Italy

1. HETEROGENEOUS REACTOR MODEL

The classical methodology of chemical reaction engineering for design and scale-up of catalytic reactors consists in a separate analysis of microkinetics and macrokinetics.

Microkinetics concerns the chemical and physical kinetics at the scale of the pellets and their coupling in order to get the apparent reaction rate. Macrokinetics refers to the transport of momentum, mass and heat at the scale of the reactor, and its aim is to establish the model of the reactor. Combination of micro and macrokinetics leads to the design equations of the reactor.

Microkinetics in a trickle-bed reactor consists of several phenomena. Typically, a gaseous reactant A reacts with a liquid reactant B on the active sites of the catalyst to give a product C. The processes are (Fig. 1): transfer of A from the gas to the liquid, transfer of A and B through the liquid phase to the solid external surface, diffusion and reaction of both reactants into the pores. The product C undergoes similar mass transfer steps from the active sites of the catalyst to the liquid phase. When heat effects are present, heat transfer steps associated to the mass transfer steps must be considered.

Macrokinetics concerns hydrodynamics (flow regimes, pressure drop, radial distribution of fluids, interfacial areas, liquid holdup), macromixing of fluids, interparticle heat transfer.

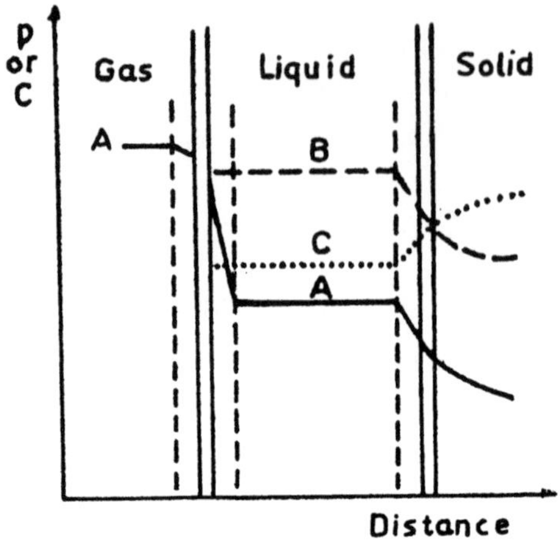

Fig 1: Microkinetic processes in trickle-bed reactors.

The model of the reactor, therefore, should take into account mass and heat transfer inside the catalyst and between the phases, axial and radial heat and mass transport, effect of hydrodynamics.

We can extend to the trickle-bed reactor the classical model adopted for gas-solid reactors, by basically assuming the liquid phase as a continuous one and the solid phase in contact with the liquid only.

For a bimolecular reaction between a gaseous reactant A and a liquid non-volatile reactant B, the reactor can be described by the following set of equations (continuum-heterogeneous model):

Mass balance in the gas phase:

$$-D_G \frac{d^2C_{AG}}{dz^2} + \frac{dC_{AG}}{dz} + (k_L a_L)_A \, (HC_{AG} - C_{AL}) = 0 \tag{1}$$

Mass balance in the liquid phase:

$$-D_L \frac{d^2C_{AL}}{dz^2} + v_L \frac{dC_{AL}}{dz} + (k_s a_s)_A (C_{AL} - C_{AS}) - (k_L a_L)_A (HC_{AG} - C_{AL}) = 0 \tag{2}$$

$$-D_L \frac{d^2C_{BL}}{dz^2} + v_L \frac{dC_{BL}}{dz} + (k_s a_s)_B (C_{BL} - C_{BS}) = 0 \tag{3}$$

Mass balance in the solid phase:

$$(k_s a_s)_A (C_{AL} - C_{AS}) = (k_s a_s)_B (C_{BL} - C_{BS}) = (1-\varepsilon_o) k_r \eta_t C_{AS}^{\alpha} C_{BS}^{\beta} \quad (4)$$

Heat balance in the reactor:

$$-k_e \frac{d^2T}{dz^2} + (v_L \rho_L C_{PL} + v_G \rho_G C_{PG}) \frac{dT}{dz} -$$

$$- (1-\varepsilon_o) k_r \eta_t C_{AS}^{\alpha} C_{BS}^{\beta} (-\Delta H) = 0 \quad (5)$$

The model assumes that the phases at each point of the reactor are at the same temperature, and ignores radial gradients of concentration and temperature. This latter is a relevant assumption: radial profiles of temperature are likely to be found in commercial units, due to heat losses and their large diameter (up to 3 m), so that some times their effects may be important. Even with these simplifications, the model is quite complicated and must be reasonably simplified for pratical purposes.

Before proposing simpler models, an analysis of the continuum - heterogenous model must be made.

Axial mass and heat dispersion have a relevant effect on the reactor yield only when the catalytic bed is short and the axial gradients of temperature and concentration are strong, i.e. the conversion rate is high and the reaction heat in considerable.

A criterion for negligible (<5%) axial dispersion influence was given by Mears |1| in the case of a first order reaction:

$$\frac{Z}{d_p} > \frac{20}{Bo_p} \ln \frac{C_{in}}{C_{out}} \quad (6)$$

where Bo_p is the pellet Bodenstein number:

$$Bo_p = \frac{v_L d_p}{D_L} \quad (7)$$

Fig. 2 shows the range of Bo_p values from literature data; the spread is quite wide, and depends on several reasons; difficulty of the experimental determination of D_L, the use of different identification methods, the interpretation of liquid RTDF with different hydrodynamic models (PD or PDE model).

At any rate, the lowest Bo_p value is about 0.1. With this value of Bo_p, for a 90% conversion, the length of the reactor must be higher than $460 d_p$ in order to neglect the axial dispersion effects.

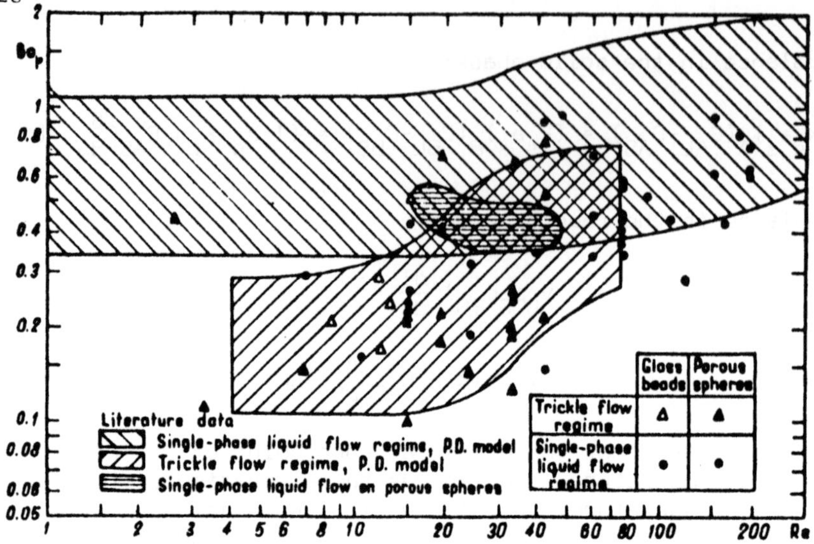

Fig. 2: Bodenstein number for axial dispersion (after Charpentier et al. /28/).

The Mears' criterion is quite restrictive: actually, the effects of the axial dispersion are not negligible for much shorter beds than those determined from eq. (6) |2|. In practice, axial dispersion can be neglected in industrial reactors, whereas it may affect the reaction yield in laboratory reactors. The effects of axial dispersion can be made negligible by employing beds of catalyst pellets suitably diluted with inert particles of the same size.

The same could be said about axial thermal conductivity. It is ignored in commercial units, where a too strong temperature axial gradient is generally avoided to prevent instability phenomena as those discussed in the section on heat transfer.

In hydrotreating reactions, the temperature raise in the reactor is about 30-50°C only. In laboratory reactors, when heat effects are relevant, usually the temperature raise in controlled by external cooling.

Let us now examine the effect of external mass transfer (gas-to--liquid and liquid-to-solid). As it is well known, external mass transfer phenomena can effect the conversion rate remarkably when the reaction is reasonably fast.

The role of external mass transfer in trickle-bed reactors has been widely discussed by Goto et al. |3|. Specchia et al. |4| showed that, at the conditions usually met in commercial reactors (v_L>4 mm/s, v_G>0.5 m/s) the external mass transfer can affect appreciably the conversion rate if the Thiele modulus is larger than 10. In laboratory reactors, in which much lower superficial velocities of the phase are adopted, external mass transfer may play a relevant role for much lower Thiele modulus.

Both gas-to-liquid and liquid-to-solid mass transfer volumic coefficients can be evaluated from the correlations proposed in the literature, based on physical mass transfer data. A question arises however: how are valid these values for reacting systems? In other words, are mass transfer effective interfacial areas and coefficients equal in reacting and non reacting systems? Physical mass transfer volumic coefficients have been obtained from the interpretation of gas absorption or solid dissolution data in packed columns by the plug flow model, i.e. by assuming an average hydrodynamic condition. Actually liquid hydrodynamics is quite complex, and there are local hydrodynamic situations very different from the average one: semistagnant liquid pockets exist, that contribute very littue to the physical phenomenon, but they may be effective when a chemical reaction is present, as shown in the literature |5,6|

Furthermore, the porosity of catalyst may play a significant role on interfacial phenomena (liquid film distribution) and even on liquid-solid mass transfer itself: it was proposed, on the basis of experimental results |7|, a reduction of the effective liquid-solid area for mass transfer to take account of the fact that only a part of the outer solid surface, corresponding to the pores mouth, is available for supplying the liquid reactant to the catalyst.

Moreover, it must be considered that the literature data of mass transfer were obtained by using aqueous solutions as liquid phase, whereas in trickle-bed reactors organic liquids are mostly present.

For all these reasons, the mass transfer coefficient values determined from literature data may be considered only as a very rough approximation of the actual ones.

Lastly, there is the important problem of liquid distribution, In the model, a total wetting of particles is assumed (solid phase in contact with the liquid only) and at the same time, an effectiveness factor η_t is introduced, that in principle should take account of the intraparticle resistance to mass and heat transfer, but in practice depends also on several other factors, such as wetting of particles, uneven concentration of reactants at the solid-liquid interface, local solid-liquid heat and mass transfer. η_t is generally assumed to be a function of the solid-liquid contacting effectiveness, a concept introduced to take account of a complex phenomenology, peculiar of trickle-bed reactors, due to the discontinuous nature of the liquid phase and the uneven distribution of liquid.

2. SOLID-LIQUID EFFECTIVENESS FACTOR

A good distribution of liquid in the packed bed is essential for stability of a trickle-bed reactor. The liquid distribution must be considered on both a macroscopic and microscopic scale.

Macroscopic liquid distribution concerns the local liquid superficial velocity in a section of the reactor. It is affected by liquid superficial velocity, packing size, and distributor design. A good design of this device is necessary in industrial reactors to avoid a bad irrigation of the catalyst; moreover, sufficiently smalll pellets must be employed, to avoid channelling and wall flow, and sufficiently high liquid flow rates per unit empty area must be adopted, to irrigate conpletely the pellets. On the other hand, smaller pellets and higher liquid velocities mean a higher pressure drop, so that a compromise must be found. Industrial reactors are packed with 6 to 2 mm pellets, and operate with a liquid velocity between 4 and 10 mm/s about. These operating conditions would ensure a good distribution of the liquid after a certain height of the bed, according to the literature data obtained in laboratory size apparatuses. In industrial reactors a bad distribution may neverthless be met, due to several reason. Owing to their large diameter, the initial bed height necessary to reach the steady good distribution may be remarkable, unless a very good liquid distributor is employed. On the other hand, it is difficult to design a liquid distributor for an even distribution of the flow rate on such a large section area as that of a commercial unit. To avoid uneven distribution in the top section of the catalytic bed, a considerable layer of inert particles may be needed. Besides, side reactions may give rise to formation of solids, that occlude some channels of the bed, especially when very small pellets are used; this results in a very poor irrigation of zones of the bed downstream the occlusion.

Laboratory reactors, instead, operate at very low liquid velocities (as low as 0.1 mm/s). To prevent channelling and a relevant wall flow, very small particles are generally used; moreover a tall calming section of inert packing and sofisticated liquid distributors are often adopted.

Microscopic liquid distribution concerns the scale of the particles.

Even with a good macroscopic distribution, the liquid around a pellet is unevenly distributed, in the sense that the local liquid velocity and film thickness may be very different along the external surface of the solid particle. This affects the local value of the rate of supply of reactants to the catalyst surface. As a result, different values of reactants concentration can exist at the solid-liquid interface.

This means that the concentration profiles inside a pellet are not symmetrical, as usually assumed, and the catalyst effectiveness factor cannot be calculated in the usual way.

Moreover, at very low liquid superficial velocities, as those adopted in laboratory reactors, a complete wetting of the external catalyst area may be not ensured, especially when not negligible heat effects are present; in this case, in fact, evaporation of the liquid film may arise, and dry zones on the catalyst outer surface are formed.

As a result, an "effective" solid-liquid interfacial area lower than the geometrical area of packing must be considered for the liquid reactant supply to the catalyst.

When the total external wetting of particles is not ensured, questions may arise about the degree of internal wetting, i.e. the amount of pores volume occupied by the liquid phase.

The active internal area in contact with the gas phase is not effective for the reaction, when the liquid reactant is not volatile; on the other hand, it may be much more effective than the wetted portion, when a gas phase reaction can take place.

All these facts should be taken into account by the liquid-solid contacting effectiveness, that, up to now, has not been yet defined clearly in quantitative terms.

Poor contacting effectiveness has been invoked by several workers to explain increases of the apparent kinetic constant with the liquid flow rate |8-11| and, on the other hand, reaction rate per unit catalyst volume in trickle-bed reactors higher than those in slurry reactors (probably due to a gas phase reaction on dry zones of the catalyst) |12-14|.

In general terms, contacting effectiveness can be defined as the ratio between the actual catalyst effectiveness factor η_t and that (η) in an ideal trickle-bed reactor, in which the pellets are evenly and totally in contact with the liquid:

$$\eta_c = \eta_t/\eta \tag{8}$$

When negligible heat effects and no gas phase reactions are present, η_c is generally lower than 1, and represents the degree of catalyst utilization in a trickle-be reactor.

For highly exothermal reactions and when gas phase reactions can take place, η_c may be higher than 1, owing to a faster gas reaction on the dry zones of the catalyst.

Another definition of η_c can be found in the literature.

For a first order reaction with respect to the liquid reactant, assuming: no external mass transfer limitation, plug flow of the liquid, no heat effects, the relationship between the inlet and the outlet concentration is given by:

$$\ln \frac{C_{in}}{C_{out}} = \frac{Z\, k_v (1-\varepsilon_o)\eta}{v_L} \qquad (9)$$

When a poor solid-liquid contacting effectiveness is present, eq. (9) must be rewritten:

$$\ln \frac{C_{in}}{C_{out}} = \frac{Z\, k_{app} (1-\varepsilon_o)\eta}{v_L} \qquad (10)$$

with $k_{app} \lesssim k_v$.

Satterfield |15| defined solid-liquid contacting effectiveness the ratio:

$$\eta_C = \frac{K_{app}}{k_v} \qquad (11)$$

The diagram of Fig. 3 has been determined from experimental values of k_{app} from laboratory and pilot plants, for hydrotreating reactions of mineral oils.

Dudukovic |16| proposed the following model to evaluate η_t:

$$\eta_t = \eta_i \cdot \eta^* \qquad (12)$$

Here η_i is the fractional pore volume filled by the liquid, and η^* the effectiveness factor of a pellet totally wetted (inside and outside) with the diameter proportional to V_{eff}/S_{eff} (effective volume divided to effective external surface).

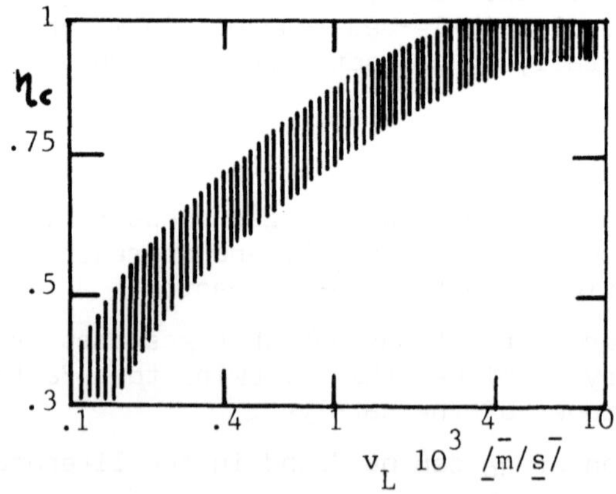

Fig. 3: Contacting effectiveness from hydrotreating reaction processes (after Satterfield /15/).

The Thiele modulus of such a pellet for a first order reaction is given by:

$$\phi_t = \frac{V_{eff}}{S_{eff}} \sqrt{\frac{k_v}{D_i}} = \frac{\eta_i V_p}{\alpha S_p} \sqrt{\frac{k_v}{D_i}} \qquad (13)$$

where α is the fraction of the external area effectively wetted. From eq. (13):

$$\phi_t = \frac{\eta_i}{\alpha} \phi \qquad (14)$$

where ϕ is the Thiele modulus of the actual pellet, supposed totally wetted. Hence:

$$\eta^* = \frac{\tanh\left(\frac{\eta_i}{\alpha}\phi\right)}{\frac{\eta_i}{\alpha}\phi} \qquad (15)$$

In the limitting cases of ϕ very little and ϕ very high from eqs. (12) and (15) one obtains:

$$\eta_t \simeq \eta_i \qquad (\phi \ll 1) \qquad (16)$$

$$\eta_t \simeq \alpha \cdot \eta \qquad (\phi \text{ very high}) \qquad (17)$$

From comparison with the definition of η_c given by eq. (8), it follows in these two cases respectively:

$$\eta_c \simeq \eta_i \qquad (18)$$

$$\eta_c \simeq \alpha \qquad (19)$$

Another approach to the definition of η_c was given by Colombo et al. |17| in the case of η_i=1. It is based on the consideration that in a pellet completely and effectively wetted, the maximum path of the molecules in the pores is of the order of the pellet radius; when the effective external surface is only partially wetted, the maximum internal path increases, and diffusion rate decreases.

An apparent intraparticle diffusivity $(D_i)_{app}$ lower than the actual one D_i results, which is connected with the extent of effective wetting. Fig. 4 shows the behaviour of the ratio $(D_i)_{app}/D_i$ as a function of v_L in a trickle-bed reactor packed with porous carbon particles.

The ratio is much lower than 1 at low liquid velocities as those used in laboratory reactors, showing that large part of the particle outer area is not effectively wetted. At $v_L \simeq 4$ mm/s, $(D_i)_{app}/D_i$ becomes 1, and the particles may be considered totally effectively wetted.

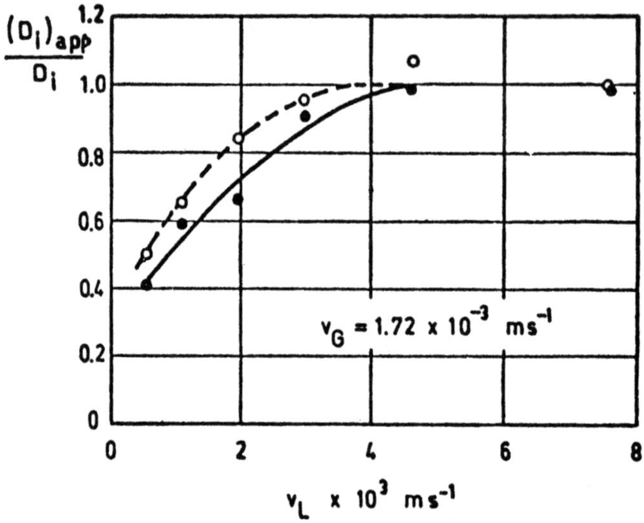

Fig. 4: Apparent intraparticle diffusivity in trickle conditions; ○: 1 mm packing; ●: 3mm packing (after Colombo et al./17/).

becomes:

$$\phi_t = \frac{V_p}{S_p}\sqrt{\frac{k_v}{(D_i)_{app}}} = \sqrt{\frac{D_i}{(D_i)_{app}}} \cdot \phi \qquad (20)$$

From comparison with eq. (14), assuming $\eta_i=1$, the effective external area fraction is given by:

$$\alpha = \sqrt{\frac{(D_i)_{app}}{D_i}} \qquad (21)$$

Notice that both approaches assume spherical symetry inside a pellet, that seems unrealistic because of the gradient of concentration of reactants along the catalyst external surface, due to the existence of different local hydrodynamic conditions of the liquid. This asymetry is taken into account by the concepts of effective area or $(D_i)_{app}$, that for this reason cannot be defined on the basis of hydrodynamics only but depend on the considered process itself. It is intuitive in fact that the gradient of reactants concentration at the solid-liquid interphase depends strongly on the demand of reactants supply, and hence on the reaction rate.

3. CATALYST WETTING

As it has been seen, the contacting effectiveness is a function of the wetting of the catalyst. Two kinds of wetting have to be considered, namely:

- Internal wetting, or pore filling η_i
- External effective wetting, that is, the amount of external catalyst surface effectively wetted by the flowing liquid.

Internal wetting can easily be determined, when heat effects are not present, from dynamic tests with an inert tracer dissolved in the liquid. If the tracer is not volatile and does not adsorb on the porous solid, the first moment of the response curve from a pulse inlet disturbance is proportional to the total intraparticle plus interparticle liquid volume held in the reactor:

$$\mu_1 = \frac{\int_0^\infty t\, C(t)dt}{\int_0^\infty C(t)dt} = \frac{Z}{v_L}(h_e + h_i) \tag{22}$$

where h_e and h_i are the interparticle and intraparticle liquid holdup respectively.

From dynamic tests with non porous particles of the same size of the catalyst particles, h_e can be evaluated; the difference between the total and interparticle holdup gives h_i.

From particle porosity ε_i and bed porosity ε_o, pore filling can be calculated as follows:

$$\eta_i = \frac{(1-\varepsilon_o)}{h_i}\varepsilon_i \tag{23}$$

Although the intraparticle holdup in beds of porous and non porous particles is not exactly the same, owing to the effect of a porous surface on the falling film of liquid, the method is sufficiently accurate.

As concerns the experimental technique, a pulse disturbance may give rise some times to errors, due to the long tail of the response curve, because of the low intraparticle diffusivity of liquid tracers.

This means that very low concentrations in the tail zone, probably affected by relevant analytical errors, are weighted by t values very high, so that they can have a not negligible influence on the first moment. A step decrease disturbance is easier from an experimental point of view; furthermore the liquid volume in the reactor is given by the zero-th moment of the response curve, and this means that the lowest C(t) values are not multiplied by an increasing weighing value, so that they affect very little the experimental results.

Values of η_i very close to 1 were found in model studies |17,18|, even at very low liquid velocities where the complete physical wetting of particles is questionable, probably due to capillary phenomena. By constrast, very low values of η_i (from 0.3 to 0.6) were determined in a laboratory trickle-bed reactor with hydrotreating catalyst |19|. This may be due to the large molecular size of the adopted tracer and the small pores of the catalyst, so that a relevant fraction of pores cannot be reached by the tracer; but, perhaps, this result may be due to the adopted pulse disturbance technique.

External effective wetting of particles could be determined by the same chemical method used for gas-liquid interfacial area, i.e., dissolution of slightly soluble particles in a reacting liquid.

This method can only be adopted in model studies and not with the actual liquid and solid phases used in industrial processes. Furthermore the obtained results are strongly affected by the semistagnant liquid pockets, that in such fast reaction processes are almost ineffective.

Another mothod is based on the relationship (eq. 21) between α and the ratio $(D_i)_{app}/D_i$. Although this relationship is based on a model that needs yet experimental confirmation, it appears up to now the only method to be adopted in process development, because it may be used in laboratory and pilot plant reactors, with the same (or similar) catalyst particles and liquid phase employed in the industrial process.

Fig. 5 shows a comparison between values determined with the two methods |20|. The difference is quite high, and is probably due to the effect of semistagnant liquid pockets. Assuming that their volume equals the static volume of the liquid, as determined in hydrodynamic modelling, and supposing this volume to collect at the particles contact points, it is possible to evaluate the static area fraction α_s in contact with the semistagnant pockets |21|. It can be seen that at low v_L values, the values determined with the chemical method (α_c) plus the α_s values equal the external wetted area from $(D_i)_{app}/D_i$ data. As v_L increases, the semistagnant liquid pockets become more and more effective for mass transfer, so that $\alpha_c+\alpha_s$ becomes higher than 1.

The good agreement between $\alpha_c+\alpha_s$ and $\sqrt{(D_i)_{app}/D_i}$ (at low v_L values) is an indirect confirmation of the validity of the determination of the effective external wetting from intraparticle diffusivity data.

Independently of the adopted experimental method, the "effective"

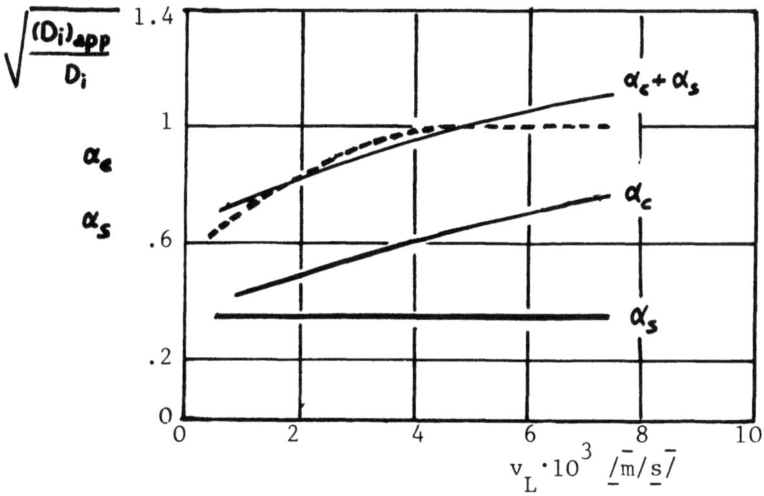

Fig. 5: Solid-liquid interfacial area fraction: α_s : static area; α_c : chemically determined area; ----: $\sqrt{(D_i)_{app}/D_i}$ (after Sicardi et al. /21/).

wetted area decreases as v_L decreases, till very low values at the liquid velocities used in laboratory reactors. Recent results show a neglibible influence of the gas flow rate when the reactor operates in the continuous gas flow regime |26|.

As a conclusion, some considerations must be made on external wetting.

The "static" area α_s seems to be as high as about 40% of the total catalyst area. "Effective" area is the area in contact with the flowing liquid plus an unknown part of static area. The extent of this area effective for the process depends on the process itself: for very slow processes, all the static area may be effective; an increase of the reaction rate leads probably to a decrease of the effective area. "Physically" wetted area, i.e. the catalyst outer area in contact with the liquid phase seems to be total, when evaporation is negligible, even at liquid velocities as low as 0,1 mm/s. At lower v_L, rupture of the thin liquid film due to interfacial phenomena is quite easy, especially with porous irregular shaped particles.

4.. SCALE-UP

From the previous discussion, it is apparent that the continuum-heterogeneous model contains such a number of unknown, or very badly known, parameters that seems unrealistic its today use for scale-up and design purposes, so that simplified models have to be used in practice. THe objective of any simplification can be characterized through the following points:

- the model should not be more detailed as absolutely required for the particular purpose;
- the model should contain as few parameters as possible to make their identification from experimental data as easy possible.

When heat effects are not important, or the reactor is maintained almost isothermal, an usual simplification is to lump all the transfer parameters into one parameter only, to be evaluated from experimental data.

In the so called "pseudo-equilibrium" model, proposed by Silvester et al. |22|, mass transfer resistance is replaced by a suitable overall term, and axial dispersion is neglected. Assuming the gaseous reactant as a key reactant, it can be written, ignoring the variation of its concentration in the liquid phase:

$$R = k_L a_L (HC_{AG} - C_{AL}) = k_s a_s (C_{AL} - C_{AS}) =$$

$$= \eta_t k_v (1 - \varepsilon_o) C_{AS} \qquad (24)$$

An overall resistance can therefore be defined:

$$\frac{1}{K_T a_v} = \frac{1}{K_L a_L} + \frac{1}{k_s a_s} + \frac{1}{\eta_t K_v (1-\varepsilon_o)} \qquad (25)$$

so that:

$$R = K_T a_v \; HC_{AG} \qquad (26)$$

The overall coefficient must be determined from experimental runs in a pilot reactor. It must be observed, to this regard, that usually in pilot runs the same conversion yield as in the industrial unit is required.

For this, the same liquid hourly space velocity (LHSV) criterion is adopted:

$$LHSV = \frac{v_L}{Z} \cdot 3600 \qquad (27)$$

Due to the much lower catalytic bed height in pilot than in industrial reactors, a much lower v_L value is also adopted. On the contrary, to evaluate $K_T a_v$ values useful for scale-up, it is necessary that the tests in the pilot plant will be carried out at the same hydrodynamic conditions as those supposed for the industrial unit, since mass transfer parameters and contacting effectiveness are affected strongly by hydrodynamics.

It is almost impossible to maintain both hydrodynamic conditions and LHSV in pilot and industrial reactors: this would require an equal height of the catalytic bed in both reactors. Therefore, scale-up of trickle-bed reactors must face two conflicting requirements.

What requirement is more important depends on the process itself; for complex reaction networks (which are very often met in practice), the same conversion criterion is obviously much more important, because it allows an evaluation of the whole reaction process.

The same hydrodynamic conditions criterion means necessarily a much lower conversion in the pilot reactor; in the extreme case, this might be considered as a differential reactor. With such a situation, the reaction rate is generally not well reproduced at various conversion stages, because of the difficulty in matching the actual fluids compositions; furthermore heat effects are not taken well into account. For these reason, the Silvester et al. model can be used for scale-up only in the case of a simple reaction process, where intrinsic kinetics is known, to evaluate the influence of hydrodynamics on the conversion rate. Eq. (25) can in particular be employed to evaluate the influence of gas-liquid mass transfer. In a bed diluted with inert particles of the same size of the catalyst particles, it can be rewritten:

$$\frac{1}{K_T a_V} = \frac{1}{K_L a_L} + \frac{1}{m k_s a_s} + \frac{1}{\eta_t k_V (1-\varepsilon_o) m} \qquad (28)$$

where m is the catalyst wheight per unit solid weight.

By plotting $(1/K_t a_V$ vs. $1/m)$, the overall resistance should be a straight line, with an intercept at $(1/m)=0$ equal to $1/K_L a_L$.

For hydrotreating processes of mineral oils, interparticle mass transfer is usually considered as negligible, so that a simpler "homogeneous" model is adopted.

Assuming a first order reaction of the liquid reactants (sulfur, nitrogen, metals), the relationship between the inlet and the outlet concentration is given by eq. (10), that can be rewritten as follows:

$$\ln \frac{C_{in}}{C_{out}} = \frac{3600\ k_{app}(1-\varepsilon_o)\eta}{LHSV} \qquad (29)$$

To take into account the contacting effectiveness, which is included in $k_{app} (k_{app}=\eta_c\ k_V)$, it is assumed that the wetting of the particles, in laboratory and pilot plant reactors, is proportional to the intraparticle holdup, that is:

$$k_{app} \sim k_v h_e \tag{30}$$

Assuming $h_e \sim v_L^{\omega}$, from eqs. (29) and (30) one obtains, for given liquid physical properties and catalyst size:

$$\ln \frac{C_{in}}{C_{out}} \sim \frac{Z^{\omega}}{(LHSV)^{1-\omega}} \tag{31}$$

From available holdup correlations, ω should have a value close to 1/3. Actually, ω must be evaluated experimentally, since it represents the interaction between hydrodynamics and chemical reaction. From literature results |10-11|, the conversion in laboratory trickle-bed reactors for hydroprocessing reactions was found well interpreted by an equation like eq. (31), but the ω values were quite different on changing the considered reactions (desulfurization, denitrogenation, demetallation reactions).

To reduce the effect of contacting effectiveness in laboratory tests, it has been suggested to dilute the catalytic bed with very small inert particles (100-500 μm particles) |19|.

These small particles increase the liquid holdup, ensuring a better wetting of the catalyst pellets. Moreover, the dilution of the bed decreases the conversion rate per unit reactor volume, and reduces in this way the influence of gas-to-liquid mass transfer.

According another model, the hydrodynamic factors affectting the reactor performance are lumped into axial dispersion |10|. For small deviations from plug flow and for a first order reaction, it can be written:

$$\ln \frac{C_{in}}{C_{out}} = \frac{3600k_v(1-\varepsilon_o)\eta}{LHSV} - \frac{3600^2 k_v^2 (1-\varepsilon_o)^2 \eta^2}{(LHSV)^2} \frac{d_p}{Z\, Bo_p} \tag{32}$$

By assuming:

$$Bo_p \sim (LHSV)^{\alpha} Z^{\alpha} \tag{33}$$

we obtain:

$$\ln \frac{C_{in}}{C_{out}} = \frac{3600k_v(1-\varepsilon_o)\eta}{LHSV} - \frac{|3600k_v(1-\varepsilon_o)\eta|^2}{(LHSV)^{2+\alpha}} \frac{d_p}{Z^{1+\alpha}} \tag{34}$$

According to Hochmann and Effron |27|, the α values range between 1 and 0,5; as the exponent ω in the holdup criterion, however, it is actually a factor to be determined experimentally.

All these models are (more or less) valid when heat effects are negligible, the intrinsic reaction rate is moderately fast, gas-

-phase reaction is absent. In the other cases, the reactor model would be much more complicated. Some models tending to a more accurate description of the phenomenology in a trickle-bed reactor have been recently proposed by academic workers |21,24|.

These models introduce several parameters, namely:

- dry fraction of the catalyst outer area;
- fraction of the catalyst outer area wetted by semistagnant and flowing liquid portions;
- liquid-solid mass transfer coefficients in the various zones;
- gas-solid mass transfer;
- transport between stagnant and flowing parts of liquid;
- intrinsic reaction rate for gas-phase and liquid-phase reaction.

As can be seen the parameters are quite numerous, so that they seem almost impossible to identify with a reasonable number of tests.

The chemical engineer is always compelled by two conflicting needs: accurate analysis of a phenomenon, description of the phenomenon with a simple model containing a little number of parameters to determine experimentally. This compulsion is particulary severe in the case of a trickle-bed reactor, where a further complication exist: an accurate and complete analysis of the reactor is still lacking.

As a conclusion, we can repeat the words of van de Vusse and Wesselingh |25| that a considerable safety factor followed by a prayer is quite understandable in design an scaling-up of a trickle-bed reactor.

NOTATION

a_v, a_L, a_s - geometrical, gas-liquid and liquid-solid area per unit reactor volume (m^{-1})

$Bo_p = v_L d_p / D_L$ -particle Bodenstein number (-)

C_G, C_L, C_s - concentration in the gas and liquid phase, and at the solid interface (kmoles/m^3)

C_p - specific heat (kcal/kg.K)

d_p - particle diameter (m)

$D_i, (D_i)_{app}$ -actual and apparent intraparticle diffusivity ($m^2 s^{-1}$)

D_L, D_G - axial dispersion coefficient ($m^2 s^{-1}$)

h_e, h_i - inter and intraparticle liquid holdup (-)

H - gas-liquid solubility constant (-)

$(-\Delta H)$ - reaction heat (kcal/kmoles)

k_e - equivalent heat conductivity of the bed (kcal/s.m.K)

k_v, k_{app} - actual and apparent first order kinetic constant (s^{-1})

k_L, k_s - gas-liquid and liquid-solid mass transfer coefficients ($m.s^{-1}$)

K_T - overall mass transfer coefficient ($m.s^{-1}$)

$LHSV = 3600.v_L/Z$ - liquid hourly space velocity (h^{-1})

R - reaction rate per unit reactor volume ($kmoles/m^3.s$)

S_p - external area of a pellet (m^2)

t - time (s)

T - temperature (K)

v_G, v_L - gas and liquid superficial velocity (m/s)

V_p - pellet volume (m^3)

z - longitudinal coordinate (m)

Z - bed height (m)

GREEK LETTERS

α - effectively wetted fraction of the catalyst outer area

α_c - solid-liquid area fraction determined with a chemical method

α_s - catalyst outer area fraction in contact with stagnant liquid

$\varepsilon_o, \varepsilon_i$ - bed and particle porosity

η, η_t - effectiveness factor for a catalyst pellet totally and partially wetted respectively

η_c - liquid contacting effectiveness

η_i - pore filling

ρ_G, ρ_L - gas and liquid density (kg/m^3)

ϕ, ϕ_t - Thiele modulus for a totally and partially effectively

SUBSCRIPTS

A - gaseous reactant

B - liquid reactant

REFERENCES

1) Mears D.E. - Chem.Eng.Sci.,26,1361(1971)
2) Goto S., Smith J.M. - A.I.Ch.E.J.,21,714(1975)
3) Goto S., Watabe S., Matsubara M. - Can.J.Chem.Eng.,54,551 (1976)
4) Specchia V., Baldi G., Gianetto A. - Proceedings of IV ISCRE, Heidelberg, p.656 (April 6-8, 1976)
5) Joosten G.E.H., Danckwerts P.V. - Chem.Eng.Sci.,28(1973)
6) Baldi G., Sicardi S. - Chem.Eng.Sci.,30,617(1975)
7) Herskowits M., Carbonell R.G., Smith J.M. - in press on A.I.Ch.E.J.
8) Henry G.H., Gilbert J.B. - IEC Process Des.Devel.,12328(1973)
9) Koros R.M. - Proceedings IV ISCRE, Heidelberg (April 6-8, 1976)
10) Montagna A.A., Shah Y.T. - IEC Process Des.Devel.,14,479(1975)
11) Paraskos J.A., Frayer J.A., Shah Y.T. - IEC Process Dev. Devel.,14,315(1975)
12) Sedriks W., Kenney C.N. - CHem.Eng.Sci.,28,559(1973)
13) Satterfield C.N., Ozer F. - I.A.Ch.E.J.,19,1259(1973)
14) Germain A.H., Lefebre A.G., L'Homme G.A. - Chem.React.Eng.II, ACS Monograph Series, n.133,p.174(1974)
15) Satterfield C.N. - A.I.Ch.E.J.,21,209(1975)
16) Dudukovic M.P. - A.I.Ch.E.J.,23,940(1977)
17) Colombo A.J., Baldi G., Sicardi S. - Chem.Eng.Sci.,31,1101 (1976)
18) Schwartz J.G., Dudukovic M., Weger E. - Proceedings of IV ISCRE, Heidelberg, p.382 (April 6-8, 1976)
19) Van Klinken J. - Proceedings of Int.Symp.on Chem.Eng.Gas--Liquid-Solid Catalyst Reactions,p.172, Liège (March 1-3, 1978)
20) Baldi G., Gianetto A. - Proceedings of Int.Symp.on Chem.Eng. Gas-Liquid-Solid Catalyst Reactions,p.109, Liège (March 1-3, 1978)
21) Sicardi, S., Baldi G., Specchia V., Gianetto A., Chem.Eng. Sci.,35,67(1980)
22) Sylvester N.D., Kulkarni A.A., Carberry J.J. - Chem.Eng.,53, 313(1975)
23) Henry G.H., Gilbert J.B. - Ind.Eng.Chem.Process Dev.Devel., 12,328(1973)
24) Crine M., Marchot P. - Proceedings of Int.Symp. on Chem.Eng. Gas-Liquid-Solid Catalyst Reactions,pag.134, Liège (March 1-3, 1978)
25) Van de Vusse J.G., Wesselingh J.A. - Proceedings of IV ISCRE, Survey Papers, pag.561, Heidelberg (April 6-8, 1976)
26) Sicardi S., Baldi G., Specchia V., Mazzarino I., Gianetto A. - In press on Chem.Eng.Sci.
27) Hochmann J.M., Effron E. - Ind.Eng.Chem.Fundamentals,8,63(1969)
28) Charpentier J.C., Bakos M., Le Goff P. - Proc.2nd Conference on Appl Physical Chem.,vol.2,pag.31,Vesprem,Hungary, 1971

THREE-PHASE FLUIDIZATION

C. G. J. Baker

Separation Processes Service,
AERE Harwell, Oxfordshire, UK

ABSTRACT. This paper describes various features of three-phase fluidized beds in which the solids are suspended by co-current flows of a liquid and a gas. The former constitutes the continuous phase and the latter a dispersed bubble phase.

Following a brief description of the fundamentals of fluidization, some industrial applications of three-phase fluidized beds are described. The current scientific knowledge of the following aspects are then reviewed: phase holdups, wakes, bubble dynamics, mixing, heat transfer, and mass transfer. Finally, an attempt is made to assess possible areas in which three-phase fluidized beds can perform a useful role in multiphase contacting operations.

1. THE FLUIDIZED STATE

1.1 Introduction

Consider a single fluid flowing upwards through a bed of solid particles. As the velocity of the fluid is increased, several different flow regimes can be observed. At low flowrates, the fluid merely percolates through the interstices between stationary particles in the bed. In this, the fixed bed regime, the drag on the solids in the bed increases with velocity. This manifests itself as an increase in the pressure drop across the bed. As illustrated in Figure 1, there is generally a power law relationship between the pressure drop ΔP_b and the superficial velocity U.

With increasing flowrate, the particles move apart somewhat

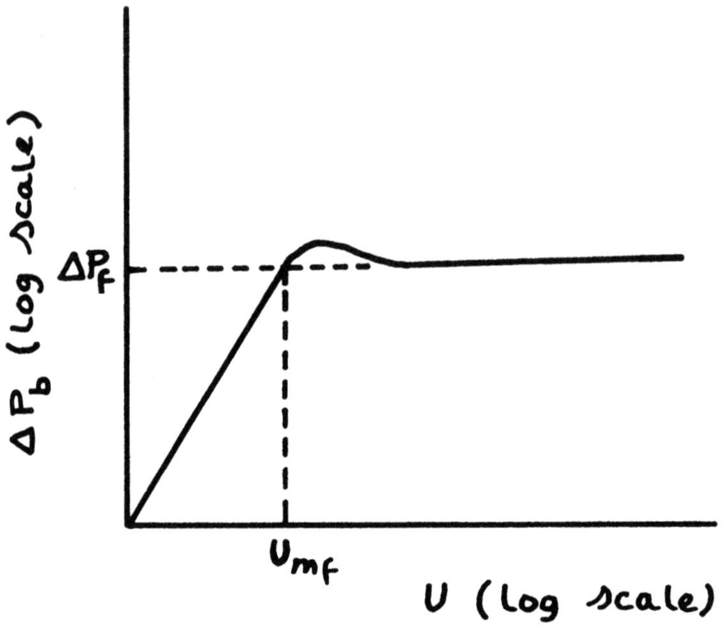

Fig. 1: Plot of bed pressure drop against fluid velocity.

and a few vibrate and circulate in very restricted regions. At a still higher fluid velocity, a point is reached where all the particles are just suspended in the upward flowing gas or liquid. At this point, the drag exactly counterbalances the effective weight of the particles. This condition is referred to as the point of incipient fluidization and the velocity at which it occurs is termed the minimum fluidizing velocity U_{mf}. At this point, the bed pressure drop becomes equal to ΔP_f, the effective weight of the bed divided by the cross-sectional area of the column:

$$\Delta P_f = H_o(\rho_s - \rho_f)g. \tag{1}$$

Here H_o is the height of the fixed bed, ρ_s the bulk density of the solids, ρ_f the density of the fluid, and g the acceleration due to gravity.

An increase in the fluid velocity above U_{mf} normally results in an expansion of the bed as the particles move apart in order to maintain the drag constant. Consequently, in the fluidized state, the bed pressure drop is essentially constant as shown in Figure 1. Expansion of the bed increases with increasing velocity until the latter attains the terminal velocity of the smallest particles, which are then entrained by the fluid and form a dilute phase above the main bed. As the velocity is increased still further, the surface of the bed becomes diffuse as it enters the pneumatic or hydraulic transport regimes.

The above description is equally applicable when the fluid phase is a gas or a liquid. However, as discussed below, the appearance of the bed is generally quite different in these two cases.

1.2 Aggregative and particulate fluidization

The characteristics of both the solid and the fluid determine whether the fluidization is aggregative or particulate. Aggregative fluidization, which normally occurs when the bed is fluidized by a gas, is characterized by the formation of large bubbles in the

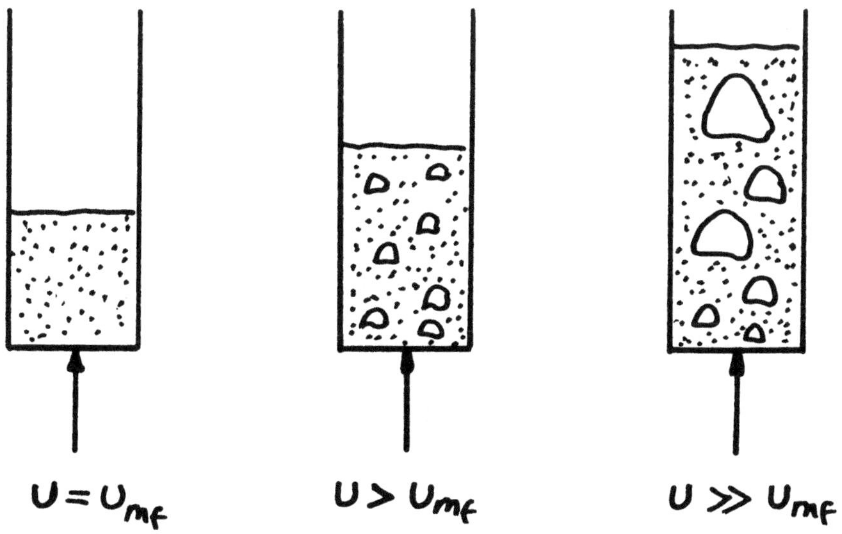

Fig. 2: Aggregative fluidization.

bed (Figure 2). The particles appear to act not as individual units but rather as aggregates having no discernible free path. At higher gas flowrates, the bubbles may grow in size until their frontal diameters become essentially equal to the diameter of the containing vessel. This state, which is termed slugging, is especially pronounced in beds having large height-to-diameter ratios.

When particles are fluidized by a liquid or a gas under high pressure, the result is commonly particulate fluidization (Figure 3). Here the bed height increases smoothly with velocity but large-scale bubbling is not observed. Under these conditions, the particles are discretely separated from eath other and exhibit a mean free path which appears to increase with velocity.

On the basis of experimental evidence, it has been suggested that the value of the Froude number U_{mf}^2/d_pg determines whether aggregative or particulate fluidization is observed. For Froude numbers less than unity, particulate fluidization is normally observed. Froude numbers exceeding unity are characteristic of

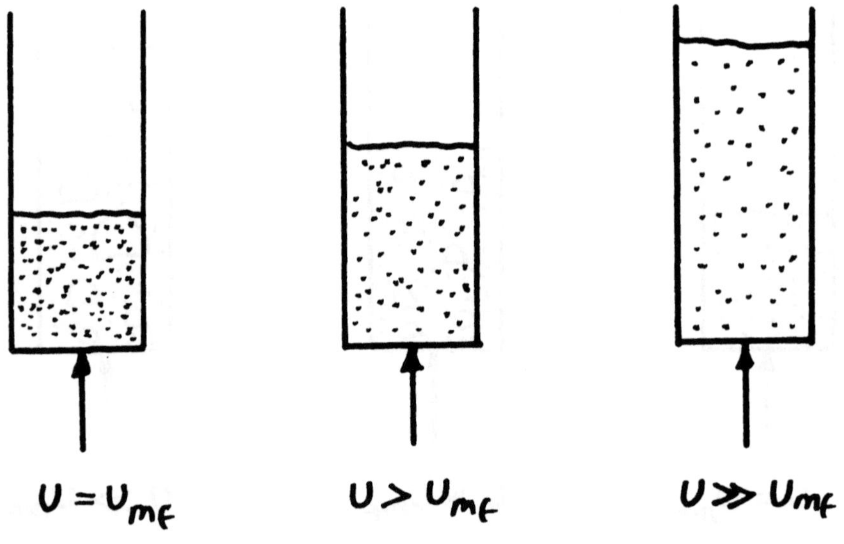

Fig. 3: Particulate fluidization.

aggregative fluidization. However, there does not appear to be a sharp demarcation between these two types of fluidized bed behaviour.

1.3 Three-phase fluidization

We now turn our attention to three-phase fluidization in which a bed of solids is fluidized by the co-current flow of a liquid and a gas. As shown in Figure 4, the liquid forms a continuous phase containing the particles and the gas a dispersed bubble phase. This operation may be used for the continuous processing of liquid as well as gaseous reactants and relatively large particles can be employed.

Three-phase fluidized beds promote intimate contacting between a solid, a liquid and a gas. Consequently, their primary use is as chemical reactors. In the right circumstances, three-phase fluidized beds possess several advantages over other types of multi-phase reactor. Vigorous mixing throughout the bed leads to

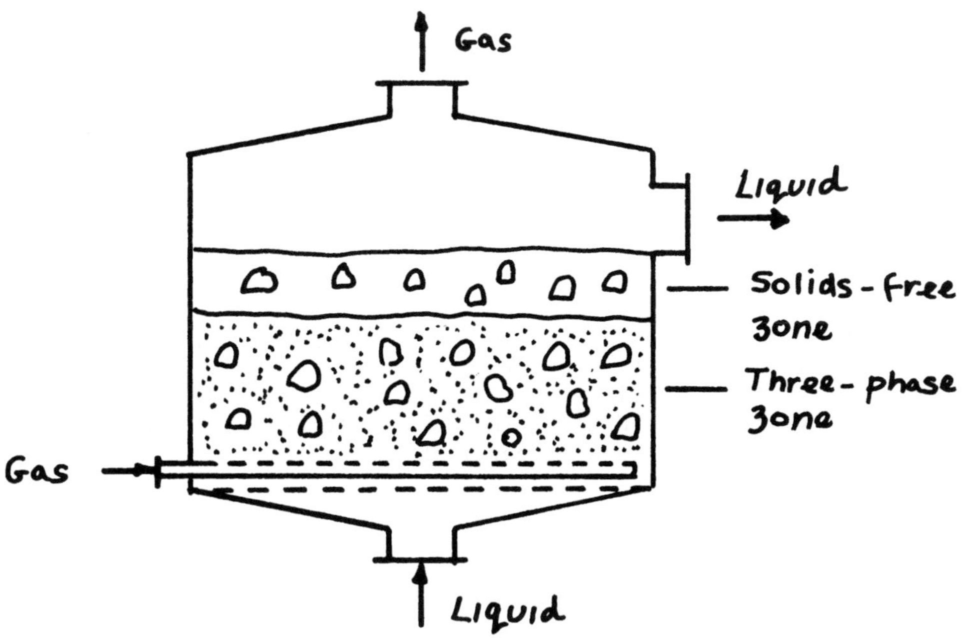

Fig. 4: Three-phase fluidized bed.

essentially isothermal conditions. Moreover, the bed temperature is easily controlled as a result of the high thermal capacity of the liquid. Solids can be added or withdrawn continuously from the bed, thereby eliminating the need for costly shutdowns for catalyst replacement. Attrition of the particles is generally low as they are cushioned by the liquid. The rate of heat transfer between the bed and the column wall or immersed exchanger tubes is extremely high; moreover, scaling is minimized as a result of the scouring action of the solids. Finally, the three-phase fluidized bed, like its two-phase counterpart, is ideally suited to large-scale continuous operation.

There are, however, a number of disadvantages of three-phase fluidization which may make its use undesirable. If catalyst is added and withdrawn on a continuous basis, the excellent mixing of solids in the bed will result in a wide distribution of catalyst residence times and hence activity. Backmixing of reactants and products will also occur while they are adsorbed on the moving catalyst. This results in loss of yield and a lowering of conversion. Three-phase fluidized beds are also characterized by a relatively high pressure drop in comparison with other forms of equipment. This may make their use uneconomical. Finally, erosion of the vessel walls and internals may occur as a result of the solids motion. This may make maintenance costs higher than in fixed bed units.

2. INDUSTRIAL APPLICATIONS OF THREE-PHASE FLUIDIZED BEDS

Three-phase fluidized beds have been employed for the treatment of residual feedstocks and other heavy fractions using the H-Oil and Hy-C processes {1, 2}. In these, the residuum is catalytically hydrogenated and desulphurized to yield distillates. Basically, the two processes are very similar. The catalyst bed within the three-phase reactor is expanded somewhat in excess of its settled volume by the co-current flow of oil and hydrogen. Typical reaction conditions are a temperature of 350-600 ^{o}C and a pressure of around 10^7 N/m^2. The main difference between the two processes lies in the type of catalyst used.

A three-phase fluidized bed reactor is also employed in a similar manner in the H-COAL process for catalytic coal liquefaction. Here the feedstock is a coal-in-oil slurry which is able to pass through the reactor without plugging it.

Chem Systems has developed a process for the methanation of gas streams containing hydrogen and carbon oxides which is centred around a three-phase fluidized bed {3}. The methanation reaction is carried out in the presence of an inert circulating oil which both fluidizes the catalyst and removes the large exothermic heat of reaction as both sensible and latent heat. The reaction is carried out at about 350°C and 7 x 10^6 N/m^2. This process was considered by the company to be superior in the areas of control and safety to those processes involving gas phase methanation.

The three-phase fluidized bed also has great potential in biological treatment processes. This is evidenced by the Dorr-Oliver Hy-Flo fluidized bed system {4}. At the heart of this process is a vertical column filled with inert solids such as sand. During start-up, the surface of these particles is coated with a population of bacteria and other micro-organisms. Pre-oxygenated wastewater is introduced into the base of the column via a distributor plate. The flowrate of the water is sufficient to assure fluidization and thereby eliminate the possibility of clogging the reactor. The fluidized bed is claimed to provide over 33.3 cm^2 of surface for biological growth per cm^3 of reactor volume. In comparison, rotating discs and trickling filters have surface areas of 1.6 and 0.8 cm^2/cm^3 respectively. Pilot plant studies on treatment systems handling up to 300 m^3/day over a 20-month period indicated an 85-90 % BOD removal in 15 minutes, as compared to a 6-8 hour treatment time required by conventional units.

The use of a three-phase fluidized bed in the production of calcium bisulphite for the pulp and paper industry has been proposed by Volpicelli and Massimilla {5} on the basis of pilot plant tests. They investigated the behaviour of limestone (0.6-10 mm) fluidized by the co-current flow of sulphur dioxide and water. They observed limestone dissolution rates which were about two orders of magnitude greater than those obtained in conventional fixed bed absorbers. In addition, pH control was better and there was a lower risk of the unit becoming plugged. Other advantages were that limestone fines, unsuitable for use in packed towers, could be properly utilized and that start-up and shut-down were easier.

Østergaard {6} and Shah {7} have published reviews on three-phase fluidization, including sections on its application. Undoubtedly, the industrial use of this device will continue to grow as it becomes more familiar and as further research provides suitable data for its design and scale-up.

3. LIQUID AND GAS HOLDUPS

3.1 Experimental techniques

A knowledge of the holdups (or fractions of the bed volume) occupied by the solid, liquid and gas is of fundamental importance in obtaining an understanding of the bed hydrodynamics. These quantities cannot be measured directly by, say, analysing photographs of the bed as the flow is normally too turbulent to enable the individual phases to be distinguished accurately. Consequently, an indirect method is invariably employed.

The holdups may be easily determined experimentally from a knowledge of the mass of solids in the bed and the pressure profile up the column {8, 9}. The latter can be obtained by means of manometers attached to a line of pressure taps up the column wall. One practical difficulty is frequently encountered. At high gas rates, the measured pressures tend to fluctuate somewhat about a mean value as a result of small variations in the height of the liquid in the bed. Therefore, in order to obtain a true profile, it is advisable to photograph the manometer bank in order to 'freeze' the liquid levels {10}.

Typical pressure profiles for a three-phase fluidized bed and a bubble column, in which there are no solids, are illustrated in Figure 5. In the latter case, the profile is normally found to be linear over most of the height of the test section. This

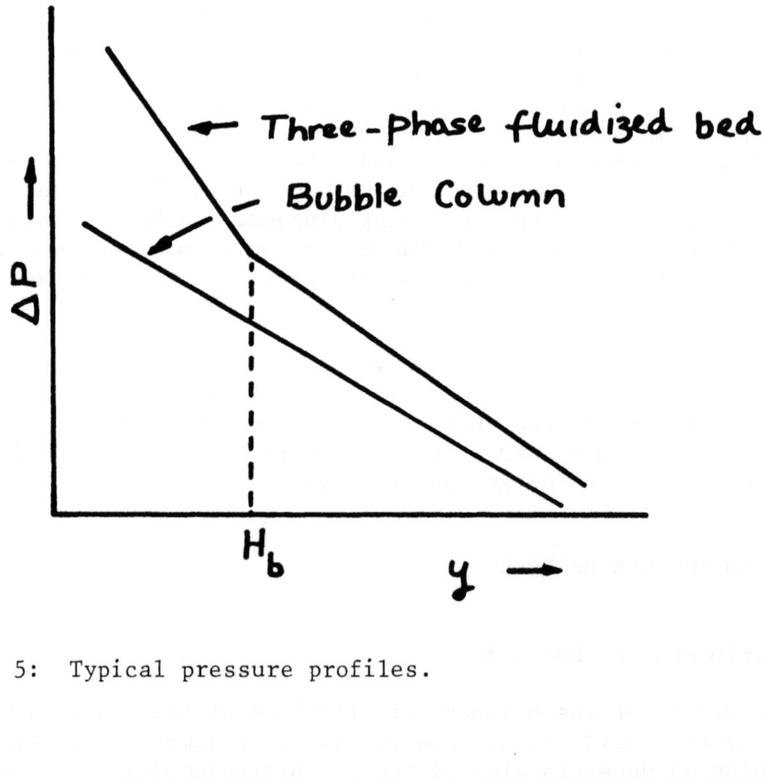

Fig. 5: Typical pressure profiles.

indicates that the gas is uniformly dispersed throughout the liquid phase. In the case of three-phase fluidized beds the profile consists of two straight lines which intersect at height H_b above the distributor. The lower portion of the profile corresponds to the three-phase bed while the upper portion depicts the solids-free zone which surmounts it. Thus, the effective height of the three-phase bed can be taken as H_b.

The individual phase holdups can be determined by the following analysis. In the three-phase bed, the sum of the liquid, gas and solids holdups, ε_L, ε_g and ε_s respectively, is unity:

$$\varepsilon_L + \varepsilon_g + \varepsilon_s = 1. \tag{3.1}$$

The static pressure drop P_b across height H_b of the three-phase bed is:

$$P_b = H_b g(\varepsilon_L \rho_L + \varepsilon_g \rho_g + \varepsilon_s \rho_s) \tag{3.2}$$

where ρ_L, ρ_g and ρ_s are the densities of the liquid, gas and solid respectively and g is the acceleration due to gravity. The solids holdup may be calculated directly from a knowledge of the mass of particles in the bed M_s:

$$\varepsilon_s = \frac{M_s}{\rho_s H_b A} \tag{3.3}$$

where A is the cross-sectional area of the column. Therefore, from eqns. (3.1)-(3.3):

$$\varepsilon_L = ((P_b/H_b g) - (1 - M_s/\rho_s H_b A)\rho_g - (M_s/H_b A))/(\rho_L - \rho_g) \tag{3.4}$$

and

$$\varepsilon_g = (-(P_b/H_b g) + (1 - M_s/\rho_s H_b A)\rho_L + (M_s/H_b A))/(\rho_L - \rho_g). \tag{3.5}$$

The validity of eqns. (3.4) and (3.5) is clearly dependent upon the fact that the dynamic component of the measured pressure and the friction losses are minimal. In order to ensure that this is the case, Kim et al {9} filled their column with liquid alone. Pressure profiles were then recorded at a variety of liquid flowrates ranging from zero upwards. No significant differences between the profiles were observed, even with liquids having a viscosity as high as 80 cp. Consequently, it may be concluded that this approach for measuring holdups in three-phase beds is indeed valid.

An alternative approach may be employed if the three-phase bed occupies the whole of the test section. It involves simultaneously shutting off the liquid and gas flows to the column by means of fast-acting valves and subsequently draining the remaining liquid from the column (Figure 6). The volume of gas contained in the bed is equal to the reduction in the height of the liquid level ΔH after shutting off the flows multiplied by the column cross-sectional area A. The gas holdup is therefore given by:

$$\varepsilon_g = \Delta H/H_b. \tag{3.6}$$

Similarly, if V_L is the volume of liquid drained from the column,

$$\varepsilon_L = V_L/H_bA. \tag{3.7}$$

The solids holdup is determined, as before, from a knowledge of

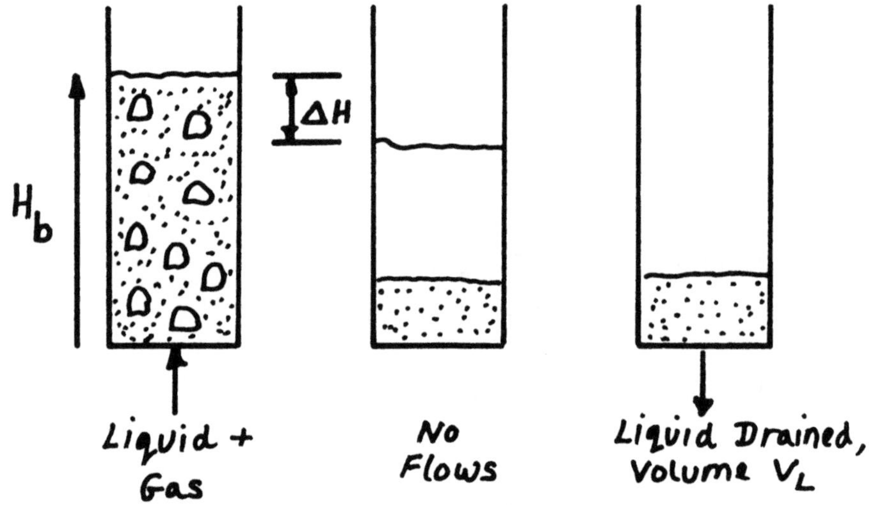

Fig. 6: Measurement of phase holdups by shutting off liquid and gas flows.

the mass of solids in the bed.

This technique suffers from a number of disadvantages. First, it can only be employed when there is no solids-free zone above the three-phase bed. Second, the valves will never close simultaneously and this will undoubtedly introduce some error. The problem is most pronounced in beds of large diameter in which the flowrates are relatively large.

3.2 Bed porosity

The bed porosity, or the fraction of the bed volume occupied by voids, is $(\varepsilon_L + \varepsilon_g)$ and is directly proportional to the expanded bed height H_b. In this section, the published experimental studies on H_b and $(\varepsilon_L + \varepsilon_g)$ are reviewed.

Consider a bed of particles fluidized by, say, water alone. The expanded bed height is $H_b{}^o$. Gas is now introduced into the bed while the liquid velocity is maintained constant. Typical plots of H_b against gas velocity U_g are illustrated in Figure 7. With large particles the bed height increases monotonically with U_g as the gas occupies an increasing volume within the bed. However, with small particles, the bed height initially decreases but subsequently increases at higher gas rates.

Stewart and Davidson {11} studied the expansion characteristics of beds of glass ballotini and of iron and lead shot fluidized by water and air. They attributed the initial contraction observed with small particles to the presence of liquid wakes travelling behind the gas bubbles. This phenomenon is discussed at some length in section 4. Briefly, the liquid in the wake travels at a much greater velocity than that moving through the interstices of the bed. This results in a lowering of the average superficial liquid velocity and a consequent reduction in the liquid holdup which, at low gas rates, more than offsets the added holdup of the gas.

Dakshinamurty et al {12, 13} measured the porosity of three-phase fluidized beds of rockwool shot, glass beads, iron shot and sand. The size and density of these particles ranged from 1 to 6.8 mm and from 2400 to 7700 kg/m^3, respectively. Two liquids were employed, namely, water and kerosene. Air was used as the gas phase. The liquid and gas flowrates ranged from 1-20 cm/s and from 0.1-7.5 cm/s, respectively. The reduction in porosity on injecting gas into the bed was observed with rockwool shot and small glass beads fluidized by both liquids. In beds of 1.0 and 2.1 mm sand particles, the contraction was observed only with water while with the other particles, no such contraction occurred.

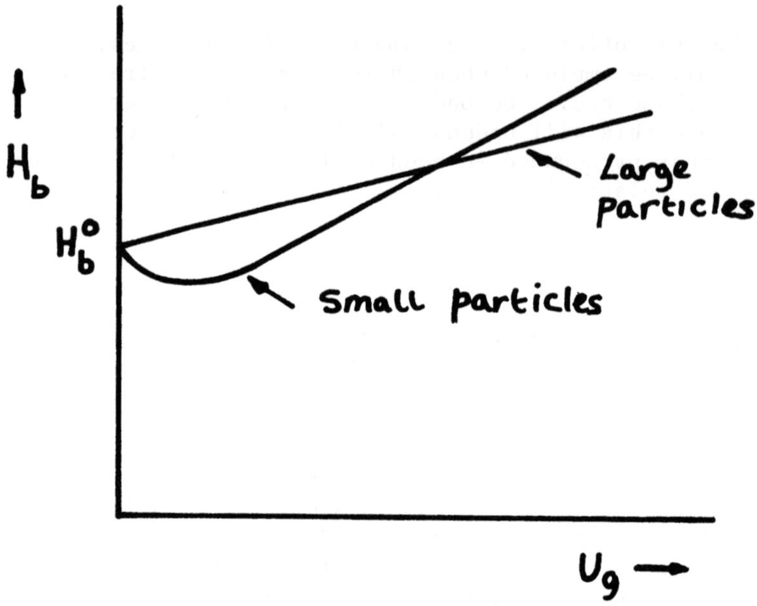

Fig. 7: Expanded bed height as a function of gas velocity at constant liquid velocity.

The authors developed two correlations to predict the porosity of three-phase fluidized beds:

for $Re_t > 500$,

$$(\varepsilon_L + \varepsilon_g) = 2.65\ (U_L/U_t)^{0.6}\ (\mu_L U_g/\sigma)^{0.08} \tag{3.8}$$

for $Re_t < 500$,

$$(\varepsilon_L + \varepsilon_g) = 2.12\ (U_L/U_t)^{0.41}\ (\mu_L U_g/\sigma)^{0.08} \tag{3.9}$$

where U_t, U_L, μ_L and σ are the particle terminal velocity, liquid velocity, viscosity and surface tension, respectively. Also, Re_t is the particle Reynolds number based on the terminal velocity U_t. Equations (3.8) and (3.9) showed average deviations of 4.6 and 4.4 % respectively from the experimental data.

An extensive investigation of bed porosity was undertaken by

Kim et al {9}. Experiments were carried out in a large two-dimensional bed. Three solids ranging in size from 1-6 mm were employed and air served as the gas phase. The effect of viscosity (1-70 cp) was determined using solutions of sugar and carboxymethyl cellulose of different concentration. Various acetone-water mixtures were employed to examine the effect of surface tension in the range 40-73 dyne/cm. Liquid superficial velocities were varied between 2.7 and 10.2 cm/s where possible; the air velocity ranged from 0.7 to 16.1 cm/s.

Kim et al observed that beds containing the larger beads (> 2.6 mm) normally expanded upon the introduction of gas. Exceptions were, however, noted with the higher viscosity solutions. In these cases, the height of the liquid fluidized bed initially diminished. In the case of 1 mm beads, an initial bed contraction was observed in all systems studied.

From an analysis of their own and other published data, the authors concluded that an initial contraction upon introducing gas was a characteristic of beds containing particles having a minimum fluidizing velocity less than 1.28 cm/s. In contrast, beds of particles having a minimum fluidizing velocity exceeding 1.28 cm/s invariably expand. One exception to this general rule must, however, be noted. Beds of small, non-wettable particles have been observed to expand on introducing gas whereas, on the basis of their minimum fluidizing velocity, they would have been expected to contract. The reason for this is discussed in section 4.

Kim et al correlated their bed porosity data by means of empirical relations involving the Froude and Weber numbers:

For beds which initially expand on introducing gas,

$$(\varepsilon_L + \varepsilon_g) = 1.40\ (Fr_L)^{0.170}\ (We)^{0.078} \tag{3.10}$$

Standard error of estimate = 0.040.

For beds which initially contract on introducing gas,

$$(\varepsilon_L + \varepsilon_g) = 1.301\ (Fr_L)^{0.128}\ (We)^{0.073} \exp\left[0.031(U_L/U_g)\varepsilon_{L_2}\right] \tag{3.11}$$

Standard error of estimate = 0.035.

In these equations, $Fr_L = U_L^2/d_p g$ and $We = U_g \nu/\sigma$. Here ν is the generalized viscosity constant $g_c K\ 8^{m-1}$, where K is the fluid consistency and m the fluid behaviour index. For Newtonian fluids,

ν can, of course, be replaced by the viscosity μ_L. In eqn. (3.11), ε_{L_2} is the liquid holdup in the corresponding liquid-solid fluidized bed which was given by

$$\varepsilon_{L_2} = 1.353\ (Fr_L)^{0.206}\ (Re_L)^{-0.100} \qquad (3.12)$$

Standard error of estimate = 0.038.

Alternatively, ε_{L_2} can be estimated from the Richardson-Zaki equation {14}.

Equation (3.11) predicts a bed contraction at lower gas rates and an expansion at higher velocities. In general, the correlations of Dakshinamurty yield porosities somewhat higher (∿ 17%) than those observed by Kim. However, the former worker employed a cylindrical bed and the latter a two-dimensional one. Baker et al {10} also measured porosities in a cylindrical column which were about 10% higher than Kim's. These observations would suggest therefore that the porosity is only weakly dependent on the bed geometry.

3.3 Liquid and gas holdups

A number of workers have measured the individual phase holdups in three-phase fluidized beds. These include Michelsen and Østergaard {15, 16}, Efremov and Vakrushev {17}, Vail et al {18} and Razumov et al {19}. The studies of Michelsen and Østergaard are worthy of particular note since radioactive tracer techniques were employed to measure the liquid and gas holdups independently. However, the most extensive study of individual phase holdups in three-phase fluidized beds was again published by Kim et al {9}. The ranges of variables studied were the same as those employed in the study of bed porosity described above.

Kim et al observed that ε_L increased with liquid velocity and decreased with gas velocity. At a given liquid rate, ε_L was greater in the liquid-solid bed than in the corresponding three-phase beds. In all cases, the liquid holdup increased with viscosity, the effect being more marked the smaller the particle size. The effect of surface tension on ε_L was not large.

The liquid holdup data were correlated in terms of Froude, Reynolds and Weber numbers:

$$\varepsilon_L = 1.504\ (Fr_L)^{0.234}\ (Fr_g)^{-0.086}\ (Re_\ell)^{-0.082}\ (We)^{0.092} \qquad (3.13)$$

Standard error of estimate = 0.039.

In this equation, (Fr_g) denotes the gas phase Froude number, $U_g/d_p g$.

The gas holdup, which is generally considerably less than the liquid holdup, may be obtained by difference from eqns. (3.11) and (3.13). However, at low gas rates in particular, the values obtained are relatively inaccurate.

Kim's values of ε_L and ε_g agreed well with those of Michelsen and Østergaard at low gas rates. However, at values of the latter exceeding the maximum employed by these authors, there was considerable discrepancy between the results of the two studies. In general, for small beads (say 1 mm) Michelsen and Østergaard's correlation predicted generally higher liquid holdups than found by Kim et al. With large (eg, 6 mm) beads, their correlation predicted higher values of ε_L for the lower viscosity solutions. However, the reverse trend was observed at higher viscosities. As Michelsen and Østergaard's correlations were derived from experiments in which only water was used as the liquid phase, they cannot be expected to describe accurately the variation of ε_L and ε_g with liquid viscosity and surface tension. The same is true for other published correlations (eg, that of Razumov et al {19}).

4. WAKES

As noted above, the initial contraction observed on introducing gas into liquid fluidized beds of small particles was attributed by Stewart and Davidson {11} to the existence of wakes moving behind the fast-rising bubbles. Stewart {20} and Efremov and Vakrushev {17} independently proposed a model of the three-phase fluidized bed to account for this phenomenon. They considered the bed to consist of two separate regions, namely a bubble and wake region and a particulate region (Figure 8). The former is composed of the gas bubbles (holdup ε_g) and their associated wakes (holdup ε_w). The latter contains the solids (holdup ε_s) and the liquid which travels through the interstices of the bed (holdup ε_{Lp}). Note that ε_{Lp} excludes any liquid associated with the wake phase. The total liquid holdup is denoted, as before, by ε_L. In the present treatment, ε_g, ε_L, ε_s and ε_w are based on the total bed volume; in contrast, ε_{Lp} is based on the volume of the particulate region alone.

In the analysis which follows, it will be assumed that

i) the wakes consist only of liquid,

ii) the wakes move at the bubble rising velocity,

iii) the absolute velocity of the rising bubbles is given by

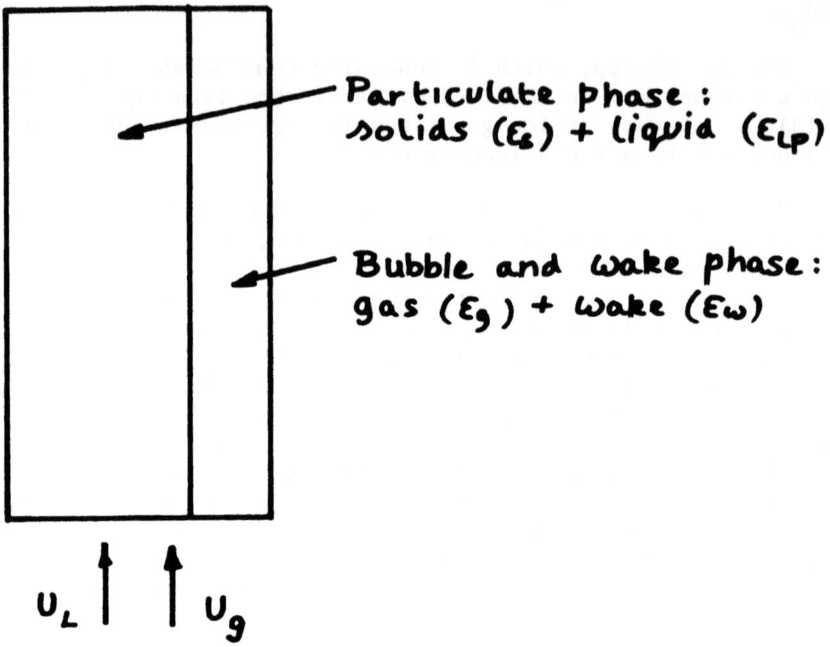

Fig. 8: Phases in a three-phase fluidized bed.

the expression

$$U_b = U_g/\varepsilon_g. \tag{4.1}$$

The validity of these assumptions is discussed below.

Now, the volumetric flowrate of the liquid entering the bed $= U_L A$. The volumetric flowrate of liquid leaving the bed in the wake phase

$$= \varepsilon_w \, A \, U_b$$

$$= \varepsilon_g \, kA \, U_g/\varepsilon_g = kA \, U_g$$

where k is the ratio of the wake to the gas volume, $\varepsilon_w/\varepsilon_g$.

∴ Flowrate of liquid leaving in particulate phase

$$= U_L A - kA \, U_g = A(U_L - k \, U_g).$$

Now, the area occupied by the particulate phase

$= A(1 - \varepsilon_g - \varepsilon_w)$

$= A(1 - \varepsilon_g - k\varepsilon_g)$.

$\therefore$ The liquid velocity in particulate phase is

$$U_{Lp} = \frac{U_L - k\,U_g}{1 - \varepsilon_g - k\varepsilon_g}. \qquad (4.2)$$

The liquid holdup in the particulate phase is calculated as follows.

Volume of liquid in the bed = HA ε_L.

Volume of liquid in the wake phase = HA ε_w = HA $k\varepsilon_g$.

$\therefore$ Volume of liquid in particulate phase

= HA ε_L - HA $k\varepsilon_g$ = HA$(\varepsilon_L - k\varepsilon_g)$.

The total volume of the particulate phase

= HA$(1 - \varepsilon_g - \varepsilon_w)$ = HA$(1 - \varepsilon_g - k\varepsilon_g)$

$$\therefore \varepsilon_{Lp} = \frac{\varepsilon_L - k\varepsilon_g}{1 - \varepsilon_g - k\varepsilon_g}. \qquad (4.3)$$

Equation (4.3) may be manipulated to yield

$$\varepsilon_L = \varepsilon_{Lp}(1 - \varepsilon_g - k\varepsilon_g) + k\varepsilon_g. \qquad (4.4)$$

The relationship between ε_{Lp} and U_{Lp} is given by, for example, the Richardson-Zaki equation {14}:

$$\varepsilon_{Lp} = (U_{Lp}/U_t)^{1/n}. \qquad (4.5)$$

$\therefore$ From eqns. (4.2) and (4.5):

$$\varepsilon_{Lp} = \left[\frac{U_L - kU_g}{U_t(1 - \varepsilon_g - k\varepsilon_g)}\right]^{1/n} \qquad (4.6)$$

and from eqns. (4.4) and (4.6):

$$\varepsilon_L = \left[\frac{U_L - kU_g}{U_t(1 - \varepsilon_g - k\varepsilon_g)}\right]^{1/n} (1 - \varepsilon_g - k\varepsilon_g) + k\varepsilon_g$$

or

$$\varepsilon_L = \left[(U_L/U_t) - k(U_g/U_t)\right]^{1/n}$$

$$\cdot\ (1 - \varepsilon_g - k\varepsilon_g)^{1-1/n} + k\varepsilon_g. \qquad (4.7)$$

Equation (4.7) provides a relationship between ε_L and ε_g for a given three-phase fluidized bed with k, the ratio of the wake to gas holdup, as a parameter. As may be seen from this equation, the presence of wakes ($k > 0$) reduces ε_L for a given ε_g. Consequently, provided k is sufficiently large, ($\varepsilon_L + \varepsilon_g$) in the three-phase bed may be less than ε_{L_2} for the liquid fluidized bed and a contraction will be observed.

Equation (4.7), or a variant of it, has been employed by a number of workers to estimate k. Efremov and Vakrushev {17} carried out their experiments in beds containing approximately monosized glass beads having diameters ranging from 0.32 to 2.15 mm. They correlated their data by means of the equation

$$k = 5.1\ \varepsilon_{L_2}^{\ 4.85}\left[1 - \tanh\ (40(U_g/U_L)\varepsilon_{L_2}^{10} - 3.32\ \varepsilon_{L_2}^{\ 5.45})\right]. \qquad (4.8)$$

Bhatia and Epstein {21} proposed the following correlation:

$$k = k'(1 - \varepsilon_s)^3 \qquad (4.9)$$

where k' is an estimate of the wake to gas holdup ratio in the absence of solids:

$$k' = 0.61 + 0.037/(\varepsilon_g + 0.013). \qquad (4.10)$$

Darton and Harrison {22} suggested that the following equation be used to calculate k:

$$k = 1.4(U_L/U_g)^{0.33} - 1. \qquad (4.11)$$

The above correlations were derived from data obtained with relatively small particles using water as the fluidizing medium. Baker et al {23} extended the range of particle sizes studied and performed experiments using a variety of liquids. They also employed their own experimental values of the bubble rising velocity U_b as opposed to values calculated from eqn. (4.1). The use of this equation has been shown {24} to yield values of U_b as much as one-third larger than those measured experimentally. Baker et al {23} proposed the use of the following correlation to calculate k:

$$k = 1.617(U_L/U_g)^{0.610}\sigma^{-0.654}. \tag{4.12}$$

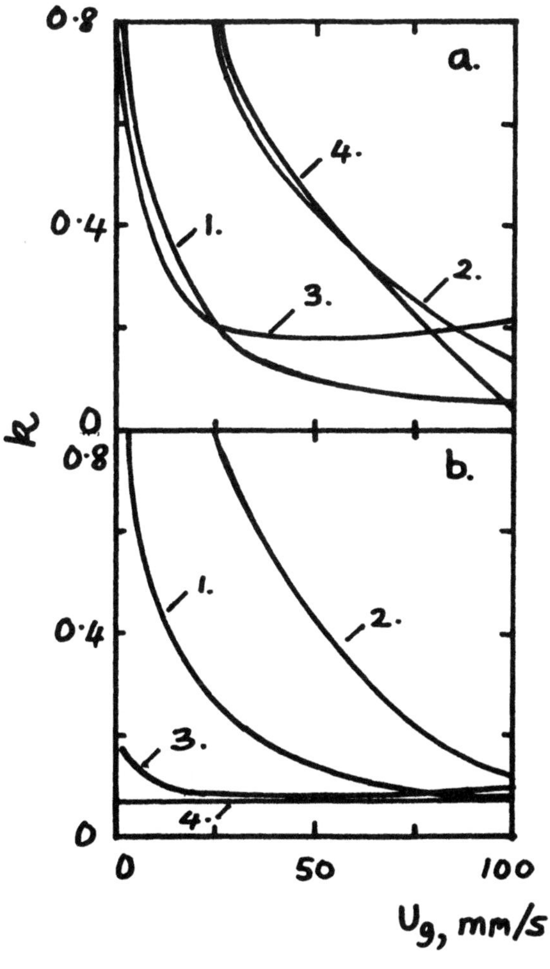

Fig. 9: Comparison of the different correlations for k (U_L=50 mm/s, (a) d_p=1 mm, (b) d_p=6 mm):
1. Baker et al {23}; 2. Darton and Harrison {22};
3. Bhatia and Epstein {21}; 4. Efremov and Vakruschev {17}.

A comparison of the predictions of eqns. (4.8)-(4.12) is shown in Figure 9 for beds of 1 mm and 6 mm glass beads fluidized by water. As may be seen, there is a wide discrepancy between the different correlations. That of Darton and Harrison is identical in form to the correlation of Baker et al, but it predicts consistently higher values of k. This may result from the use of much more extensive data in the latter study. Efremov and Vakrushev's correlation agrees closely with that of Darton and Harrison for the beds of 1 mm beads but deviates from it considerably for the beds of 6 mm beads.

Any attempt to describe the characteristics of wakes in three-phase fluidized beds by a single parameter, k, is obviously a gross simplification. In practice, the wakes appear to consist of two distinct regions. Immediately behind the bubble and separated from it by a horizontal interface there is often a clear liquid region. This is followed by a more diffuse portion of the wake in which the solids concentration may range from zero to the value pertaining in the particulate region of the bed. Finally, as observed by Rigby and Capes {25}, vortices may be shed from the stable wake thereby contributing to k.

Epstein and coworkers {26-29} have published a number of theoretical papers dealing with the characteristics of wakes in three-phase fluidized beds. Of particular interest are those works {28, 29} in which the influence of the solids concentration in this phase is examined. The objective of these studies was to attempt to explain the experimental observation {21, 30} that reducing the liquid surface tension or rendering the solids non-wettable either reduced or reversed the tendency for liquid fluidized beds to contract on injecting gas.

El-Temtamy and Epstein {28} assumed that bubbles in cylindrical three-phase fluidized beds are of the spherical cap type, the sphere-completing volume of which constitutes the effective wake volume (see Figure 10). The analogues in a two-dimensional bed are circular and elliptical-cap bubbles. The included angle θ is, in general, a function of the bubble Reynolds number. However, over the range of conditions commonly encountered in the water-air fluidization of small (< 2 mm) particles, Henriksen and Østergaard {31} found θ to be essentially a function only of the kinematic viscosity of medium surrounding the bubble. Thus, for a given liquid fluidized bed, θ and hence the ratio of the wake to gas volume for a single bubble, denoted by k_o, can be easily determined. The relationship between k_o and the wake to gas volume ratio for a multi-bubble system k was assumed to be given by the following equation, which was derived from results obtained with liquid-liquid systems:

$$k = k_o\, e^{-5.08\, \varepsilon_g}. \qquad (4.13)$$

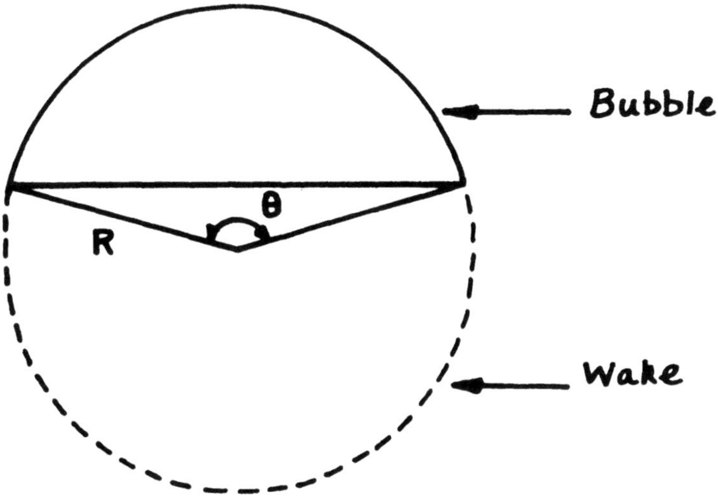

Fig. 10: Sphere-completing wake model.

The technique of El-Temtamy and Epstein enables the wake volume to be determined independently of the bed expansion characteristics. These authors also derived a set of equations analogous to (4.7) in which the solids content of the wakes was allowed to vary from zero to the value existing in the particulate phase. They solved these equations by trial and error for x, the ratio of the solids holdup in the wake phase to that in the particulate phase.

From an analysis of data published in the literature, El-Temtamy and Epstein concluded that, with water as the fluidizing medium, the assumption of solids-free wakes ($x = 0$) was valid for particles larger than 1 mm. However, in the case of particles smaller than this, the values of k obtained were considerably in excess of those based on the assumption that $x = 0$. In general, k was found to increase with increasing particle size (for a given bed voidage), with increasing liquid velocity, and with decreasing gas velocity.

The solids content of the wakes was found to increase with

decreasing particle size, increasing gas velocity and increasing liquid viscosity. The following correlation for calculating x was recommended:

$$x = 1 - 0.877\, U_L'/U_{gL}. \tag{4.14}$$

Standard error of estimate = 0.101.

In this equation, $U_L' = U_L/\varepsilon_{L2}$ and U_{gL} is the velocity of the gas relative to that of the liquid:

$$U_{gL} = (U_g/\varepsilon_g) - (U_L/\varepsilon_L). \tag{4.15}$$

Note that for $U_L'/U_{gL} > 1.14$, eqn. (4.14) predicts negative values of x. The equation is not valid under these circumstances. In any event, the bubble wakes play a diminishing role under these conditions.

In a later paper, El-Temtamy and Epstein {29} reexamined the question as to whether a three-phase fluidized bed will initially expand or contract. However, they relaxed the assumption common to most earlier treatments of this subject that the wakes were solids-free. They derived the following expression for the critical solids content of the wakes:

$$x_c = 1 - \frac{(1 + \frac{1}{k})U_L'\left[n - (n - 1)\varepsilon_L\right]}{U_L' + U_{gL}}. \tag{4.16}$$

When $x < x_c$, the bed will initially contract; however, when $x > x_c$, it will expand. The authors noted that a bed contraction continued to occur at values of x up to 0.73. They also pointed out that with low surface tension liquids and non-wettable solids, the particles are able to fall through the rising bubbles into the wake, thereby increasing x (see Figure 11). This reduces the tendency for the bed to contract.

Vortex shedding was first suggested by Rigby and Capes {25} as a mechanism responsible for the contraction observed in beds of small particles in which the volume of the stable wake was observed to be negligible. On shedding, the vortices were observed to consist of pure liquid and their velocity was somewhat less than that of the bubble. During the life of a vortex, its velocity decayed to that of the liquid in the particulate phase while its solid content increased to that of its surroundings. Thus, vortices act in a similar manner to wakes in that they abstract liquid from the particulate phase, thereby reducing its velocity. Rigby and Capes suggested that vortex shedding should be of importance when the

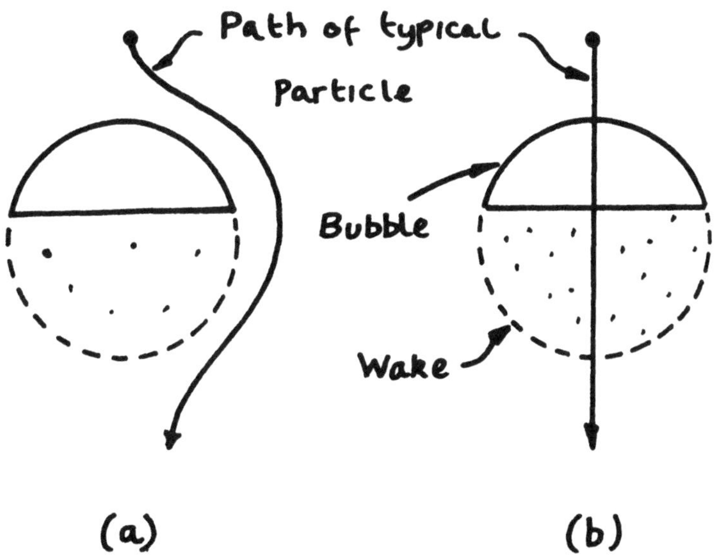

Fig. 11: Wake structure in beds of (a) wettable and (b) non-wettable particles.

bubble Reynolds number lies between about 200 and 700.

Using a simplified analysis, Rigby and Capes showed that the ratio of the total equivalent volume of stable wake due to all the vortices V_e' to the bubble volume V_b was given by

$$(V_e'/V_b) = H_b R_v V_v \, \varepsilon_v \, \bar{U}_v / U_b^2 \, V_b . \qquad (4.17)$$

In this equation, R_v is the rate of vortex shedding, V_v, ε_v and $\bar{U}_v$ are the volume, average voidage and average rising velocity of the vortex, respectively. The other symbols have been defined above. In order to test their theory, the authors injected bubbles of known volume into a bed of 775 µm particles fluidized by water at 4 U_{mf}. From photographs of the bubbles, wakes and vortices, they were able to estimate the values of the parameters in eqn. (4.17) and consequently to calculate the contribution of the vortices. In qualitative terms, their findings were consistent with the predictions of the above equation. Consequently, it is probable that vortex shedding also plays a significant role in determining the hydrodynamic behaviour of three-phase fluidized beds.

5. BUBBLE DYNAMICS

Apart from providing a clearer understanding of the hydrodynamics of three-phase fluidized beds, a knowledge of the characteristics of bubbles in these devices is of practical interest for several reasons. Firstly, it enables the liquid-gas interfacial area to be estimated. This may be of importance in situations where the bed is employed as an absorber or in cases in which a chemical reaction is mass-transfer controlled. Secondly, the degree of axial mixing in the liquid phase, and hence the performance of the bed, appears to be related to the characteristics of the bubbles. These aspects are discussed in later sections of this paper. Finally, it is probable that the degree of elutriation from the bed is also dependent on the size and rising velocity of the bubbles.

Published studies of bubbles in three-phase fluidized beds can be divided into three categories. The first includes those studies aimed at establishing the relationship between the size and rising velocity of single bubbles. In the second category, the objective of the investigations is to determine the mechanism of bubble breakup in three-phase fluidized beds. Finally, in the third category, the characteristics of bubbles in freely-bubbling beds are described.

Results of experiments in category one were first reported by Massimilla et al {32} who measured the rising velocity of single bubbles of known volume injected into liquid fluidized beds. Water and air were employed as the liquid and gas phases and the solids were either 0.22 mm sand, 0.79 or 1.08 mm glass beads, or 0.2 mm iron sand. The authors concluded that the relationship between the bubble rising velocity and its size was essentially dependent upon bed expansion and was relatively insensitive to particle diameter.

Henriksen and Østergaard {31} also measured the rising velocity of large single bubbles in a two-dimensional bed. They employed water, methanol and glycerol solutions to vary the surface tension and viscosity of the fluidizing liquid between 22.7 and 72 dyne/cm and 1.0 and 170 cP respectively. Particle diameters ranged from 0.2 to 3 mm. Where circular cap bubbles were observed, a correction was applied to the measured value of the rising velocity U_b to account for the drag due to the walls in the two-dimensional bed. These authors concluded that the relationship between U_b and the radius of the circular cap R was of the form described by Davies and Taylor {33} for large bubbles rising through inviscid liquids:

$$U_b = \alpha(gR)^{\beta}. \tag{5.1}$$

In this equation, α is a constant which was shown by Davies and Taylor to equal 2/3 for an inviscid liquid. Under the same conditions, the exponent β was shown to be 1/2. The values of α observed by Henriksen and Østergaard in three-phase beds ranged from 0.42 to 0.60. They concluded that surface tension had no effect on the rising velocity, but that the latter was somewhat reduced in beds of higher viscosity.

The second category of studies is concerned with the mechanism of bubble breakup in three-phase fluidized beds. Lee, Sherrard and Buckley {34} postulated that bubbles are split when they are penetrated by a solid particle. In the absence of viscous forces, this was shown to occur if the Weber number $We = \rho_s U_b^2 dp/\sigma$ exceeded three. Photographic evidence was presented which, the authors claimed, supported their hypothesis.

Henriksen and Østergaard {35} studied the breakup of 20 mm bubbles in beds of water and methanol. Three solids were employed, namely 5 mm steel spheres and 3 and 6 mm glass spheres. A bubble was held stationary by a downward flow of liquid and a particle was allowed to fall through it. In no case was the bubble observed to disintegrate. They therefore concluded that the bubbles are broken up as a result of Taylor instability {36} of their roofs. In this process, a perturbation at the interface between two fluids grows without bound and results in the collapse of that interface. Taylor showed that this can occur at a horizontal interface if the upper fluid is more dense than the lower one. This is the case with rising bubbles in a three-phase fluidized bed. However, since the minimum sized particle capable of splitting a bubble in air-water beds was shown to be about 8.5 mm, it was concluded that the instability was generated by fingers of liquid projecting down through the roof of the bubble.

The work of Kim et al {24}, described in detail below, tended to support Henriksen and Østergaard's proposed mechanism of bubble breakup. They observed that the size of the bubbles in three-phase fluidized beds is not particularly sensitive to the properties of the solids. This would tend to eliminate the mechanism proposed by Lee, Sherrard and Buckley {34}. Also, the average bubble size was shown to increase with the liquid viscosity. As demonstrated by Bellman and Pennington {37}, the introduction of a viscosity term into Taylor's original equations eliminates the tendency for disturbances at the interface to grow without bound, thereby reducing bubble breakup.

Studies in the third category enable the effect of interactions between bubbles to be assessed. Massimilla et al {32} also measured the size distribution of bubbles emerging from the top of a freely-bubbling bed. Their results showed that the mean bubble diameter increased with the gas flowrate, the distance from the injection nozzle, the initial bubble size, and the fractional expansion of the bed. The results of Østergaard {38} qualitatively support these findings. He concluded that the rate of coalescence was relatively high near the point of incipient fluidization and decreased progressively as the liquid rate and hence the bed height increased.

Page and Harrison {39} also measured the size distribution of bubbles leaving a bed of 0.5 mm sand particles. Water and air were again employed as the fluid phases. The bubble size was observed to decrease with increasing liquid velocity and, in contrast to the results of Massimilla et al, with an increase in the gas rate at low liquid velocities. At high liquid rates it was observed to be independent of the gas flow.

Rigby et al {40} measured the size, frequency, rising velocity and size distribution of gas bubbles within three-phase fluidized beds by means of an electroresistivity probe. They employed water, air, and four sands ranging in size from 0.120 to 0.775 mm. The results obtained were similar to those of Massimilla et al. The bubble size displayed a general tendency to increase with increasing gas rate and increasing height above the grid. The authors were unsuccessful in correlating their bubble rise velocity data by means of a relationship of the form of eqn. (5.1) with $\beta = 1/2$. They proposed instead a more complex empirical relationship involving bed voidage as well as the fluid superficial velocities.

Darton and Harrison {22} also employed an impedance probe to study bubble characteristics in water-fluidized beds of 0.5 mm sand. They observed that the bubble size distributions were log normal and that the mean size again increased with increasing gas rate and with decreasing water rate.

Kim et al {24} performed an extensive investigation of the bubble characteristics in freely-bubbling beds. Their data were obtained by analysing movie films of the bubbles in their two-dimensional column. As in their earlier studies, the viscosity and surface tension of the liquid phase as well as the fluid velocities were varied over relatively wide ranges.

In common with Darton and Harrison {22}, Kim et al observed that the bubble size distributions were log normal. Geometric standard deviations, which ranged from about 1.20 to 1.30, were somewhat lower than corresponding values obtained in liquid-gas beds. The authors noted that the mean bubble size was primarily a function of the gas rate; it was not unduly affected by either the velocity or physical properties of the liquid. The following correlations were presented:

$$d_b = 13.4\, U_L^{0.052}\, U_g^{0.248}\, \nu^{0.008}\, \sigma^{0.034}. \qquad (5.2)$$

Standard error of estimate = 7.2 mm.

$$U_{br} = 83.1\, U_L^{0.065}\, U_g^{0.339}\, \nu^{0.025}\, \sigma^{0.179}. \qquad (5.3)$$

Standard error of estimate = 164 mm/s.

$$U_{br} = 18.0\, d_b^{0.989}. \qquad (5.4)$$

Standard error of estimate = 178 mm/s.

In these equations, U_L, U_g and U_{br} are in mm/s, d_b is in mm, σ in dyne/cm and ν in mNs^m/m^2. The rising velocity of the bubble relative to that of the liquid, U_{br}, was calculated from the relation:

$$U_{br} = U_b - U_L/\varepsilon_L \qquad (5.5)$$

in which U_b is the rising velocity of the bubbles relative to the column walls.

Equations (5.2) and (5.3) do not include any terms describing the dimensions of the equipment. In particular, the diameter of the gas injection nozzles will undoubtedly influence the bubble size. This aspect was not examined by Kim et al, who employed four 6.3 mm diameter nozzles throughout. It is, however, worthy of further study.

Another notable omission from the above equations is a particle size term. Earlier studies (see, eg, reference {34}) had led to the general understanding that the bubbles were larger in beds of small particles than in beds of large particles. On this basis, the existence of "bubble coalescing" and "bubble disintegrating" regimes of three-phase fluidization was proposed {8}. However, Kim et al concluded that these different regimes existed only at very low gas rates which are of little importance industrially.

Equation (5.4) illustrates that, as with Davies-Taylor bubbles, the rising velocity of bubbles in three-phase fluidized beds is essentially a function only of their size. However, the exponent β was close to unity as opposed to the theoretical value of 1/2. Bubbles in three-phase beds were seen to rise more slowly than their counterparts in solids-free systems.

Kim et al observed bubbles in the size range 15 to 90 mm with rising velocities ranging from about 250 to 1700 mm/s. As noted earlier, the equation

$$U_b = U_g/\varepsilon_g \tag{5.6}$$

derived by Nicklin {41} for ideal bubbly flow in liquid-gas beds was found to hold only approximately for three-phase fluidized beds. Values of the ratio $\varepsilon_g U_b/U_g$, which should ideally be unity, ranged from about 0.63 with 6 mm beads to 1.04 with 1 mm beads. The authors therefore concluded that Nicklin's assumptions that the bubbles are all of equal volume, uniformly spaced, and rise at the same velocity were not valid and lead to significant errors in eqn. (5.6).

As will become apparent from the later discussion of mass transfer in three-phase fluidized beds, the physics of bubble formation is of major importance. This potentially fruitful area of study appears to have been neglected in the literature.

6. MIXING

A number of studies of liquid and gas phase mixing in three-phase fluidized beds have been published and the topic has recently been reviewed by Shah {7}. The purpose of these studies was to obtain information which could lead to the successful modelling of these devices as chemical reactors. In almost every case, the data were interpreted using the axially-dispersed piston flow model. Although this provides a convenient method for quantifying the data, it is questionable as to whether the model is strictly applicable to three-phase fluidized beds in which turbulent mixing of the fluids is readily visible.

6.1 Axial dispersion in the gas phase

Studies of gas phase mixing in three-phase fluidized beds are very limited. Schügerl {42} reported the results of experiments in a 13.5 cm column in which helium was employed as a tracer. Five solids with diameters ranging from 0.04 to 0.5 mm were studied. Schügerl concluded that in both three-phase fluidized beds and bubble columns, axial mixing in the gas phase is caused by non-uniformity of the bubble velocities. This process is diminished by the bubble coalescence and redistribution which occur perpendicular to the direction of flow. The presence of solid particles in the column had little effect on the intensity of mixing in the gas phase.

In a similar study, Michelsen and Østergaard {15, 16} concluded that gas phase axial dispersion was rather small except at high gas flowrates. Mixing was also observed to be more pronounced in beds of small particles.

In the above studies, the gas rates employed were relatively low. In industrial systems operating at higher velocities, gas mixing could be more intense.

6.2 Axial dispersion in the liquid phase

Michelsen and Østergaard {15, 16} also studied liquid mixing in three-phase fluidized beds. They presented their data in terms of the height of a mixing unit, HMU, defined as

$$HMU = 2H_b/Pe_L = 2E_L/U_L. \tag{6.1}$$

In this equation, Pe_L is the liquid phase Peclet number U_LH_b/E_L and E_L is the liquid phase axial dispersion coefficient. Thus, HMU is equal to zero for plug flow and is infinite under perfect mixing conditions.

The above authors concluded that particle size played an important role in governing axial dispersion in three-phase systems. For example, their beds of 6 mm particles were characterized by a very low degree of liquid mixing (HMU < 10 cm) at gas velocities less than 3.6 cm/s. Here, HMU was independent of the gas rate and increased slightly with the liquid rate. In contrast, at U_g = 18 cm/s, HMU decreased from about 35 cm to 20 cm as the liquid velocity was raised from 10 to 15 cm/s. In beds of 3 mm, mixing was also relatively low (HMU < 50 cm). In this case, HMU increased with gas rate but decreased sharply with increasing liquid velocity. In beds of 1 mm particles, mixing was much more intense with values of HMU in the range 40-120 cm being recorded. Again, this parameter was observed to increase with gas rate and to decrease with liquid rate.

Kim et al {8} studied axial liquid mixing in three-phase systems fluidized by water and air. With the beds of 2.6 and 6 mm particles, two tracer techniques were employed, namely pulse and step injection. Good agreement was obtained between these two methods. In the case of the 1 mm particles, only step injection was employed. The height of the mixing unit was found to increase with both liquid and gas velocity. The data can be correlated by the equation {43}:

$$HMU/H_o = 4.55 \times 10^{-3}\, U_L^{0.815}\, U_g^{0.309}. \tag{6.2}$$

Standard error of estimate = 0.102.

In eqn. (6.2), U_L and U_g are both in mm/s. In the original statistical analysis, a particle size term was included. However, it was subsequently omitted as its exponent was small (0.083) compared with the exponents of the U_L and U_g terms.

Both Kim et al and Michelsen and Østergaard concluded that axial mixing increased with the gas rate. This is probably due to the presence of large bubbles in the bed under these conditions. However, there are unexplained differences in the observed trends of HMU with particle size and liquid velocity. For beds of small particles (eg, 1 mm), Michelsen and Østergaard's results generally predict somewhat higher values of HMU than found by Kim. However, with larger particles (eg, 6 mm), the reverse is true.

Kato et al {44} obtained mixing data in three-phase beds containing particles ranging in size from 63 to 177 μm and with particle concentrations up to 0.2 g/cm^3. Experiments were performed in 66, 122 and 214 mm diameter columns. The data were correlated by means of the following equation:

$$Pe_L' = 13\sqrt{Fr_g'}/(1 + 8(\sqrt{Fr_g'})^{0.85}) \qquad (6.3)$$

Here, $Pe_L' = U_g d_c/E_L$, $Fr_g' = U_g^2/gd_c$ and d_c is the column diameter.

7. HEAT TRANSFER

Published data on heat transfer in three-phase fluidized beds are extremely limited. Østergaard {45} noted that the injection of air into a bed of glass spheres fluidized by water resulted in a marked increase in the heat transfer coefficient h. He observed that this increase was largest at the lowest liquid rates and was particularly marked near the minimum fluidizing velocity.

Viswanathan et al {46} measured wall-to-bed heat transfer rates in beds of 649 and 928 μm quartz particles fluidized by water and air. Their experiments were carried out in a column 51 mm in diameter at a constant expanded bed height. This was achieved by simultaneously changing the water and air rates over relatively narrow ranges. The initial marked increase in h resulting from the injection of air into the bed was again noted. The heat transfer coefficient was observed to increase with particle size. Typical values were of the order of 5.1 kW/m^{2}K.

A more extensive investigation of heat transfer between three-phase fluidized beds and an axially-mounted electric heater was carried out by Baker et al {10}. Experiments were performed in a 240 mm diameter column. Beds of 0.5, 1, 3 and 5 mm beads were fluidized by cocurrent flows of air and water. In addition, experiments on the corresponding two-phase (liquid-gas, liquid-solid) beds were also undertaken.

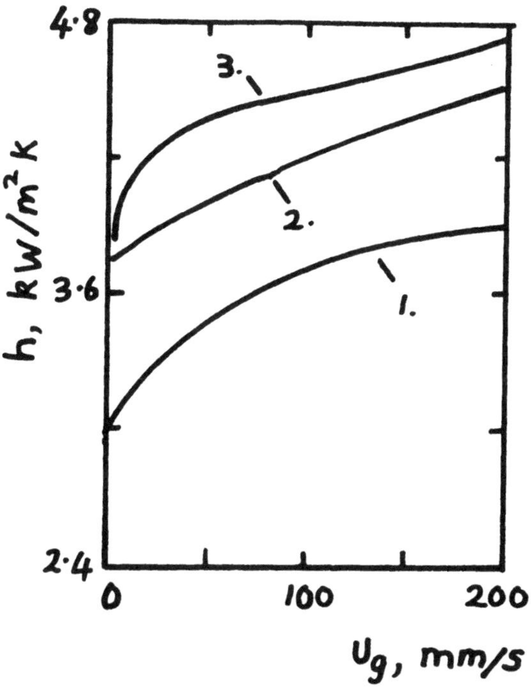

Fig. 12: Plots of heat transfer coefficient against gas velocity in a three-phase bed of 5 mm beads (after Baker et al {10}): 1. U_L=41 mm/s; 2. U_L= 92 mm/s; 3. U_L=126 mm/s.

The heat transfer coefficient was observed to increase with gas velocity for all liquid rates and particle sizes studied. The increase was initially quite rapid (Figure 12). However, at higher gas rates it became less marked and h appeared to approach a maximum. As may be deduced from Figure 12, plots of the heat transfer coefficient against liquid rate exhibited maxima. The same trend was observed with liquid fluidized beds.

Plots of h against particle size d_p are shown in Figure 13. The result of introducing solids into a liquid-gas bed (d_p = 0) was, in general, to increase the value of the heat transfer coefficient. A notable exception to this was observed at the higher gas rates where the coefficient went through a local minimum. Thus, under these conditions, h was smaller in the three-phase bed than in the corresponding liquid-gas bed. As may be seen, this

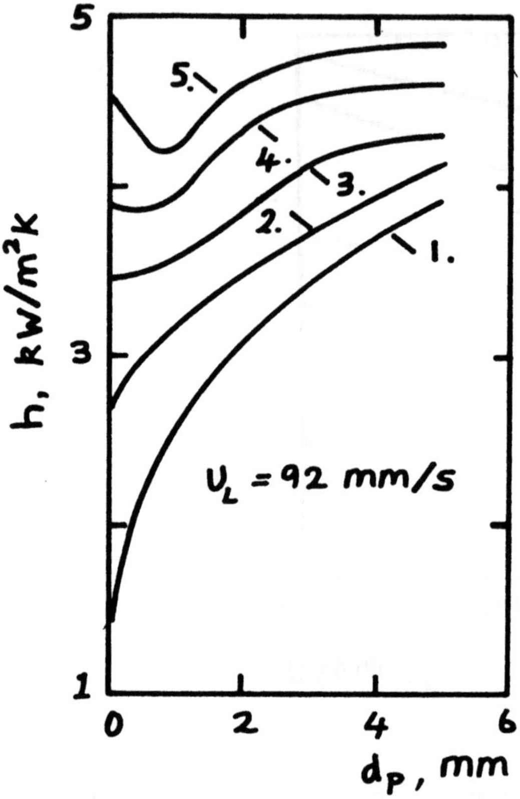

Fig. 13: Plots of heat transfer coefficient against particle size (after Baker et al {10}): 1. U_g = 0 mm/s; 2. U_g = 16 mm/s; 3. U_g = 59 mm/s; 4. U_g = 119 mm/s; 5. U_g = 237 mm/s.

is only true for particle sizes up to about 1.5 mm. At lower liquid rates than shown in the figure, this local minimum was observed to occur at lower gas rates and also to be extended to a larger particle size. It was also generally observed that h was relatively insensitive to d_p at values of the latter exceeding 3 mm, particularly at high gas rates.

As may be seen in Figure 13, three-phase fluidized beds always exhibited larger values of h than liquid-fluidized beds. Although increasing the gas rate at larger particle sizes resulted in a larger value of h in the three-phase fluidized beds, the difference between this value and the corresponding liquid-solid coefficient was essentially independent of particle size. However, for small particles, this difference decreased with increasing particle size up to about 1.5 mm.

Baker et al correlated their data by means of the equation

$$h = 1.98\, U_L^{0.070}\, U_g^{0.059}\, d_p^{0.106} \tag{7.1}$$

in which h is expressed in kW/m^2K, U_L and U_g in mm/s and d_p in mm. The multiple correlation coefficient was 0.9974.

8. MASS TRANSFER

A number of workers have measured the rates of liquid-gas mass transfer in three-phase fluidized beds. Massimilla et al {47} are credited with undertaking the first such study. These workers carried out experiments in columns having diameters of 30 and 90 mm and heights ranging from 100 to 1200 mm. The solids employed were 0.22 mm silica sand and ballotini of average diameter 0.5 and 0.8 mm. A mixture of carbon dioxide and nitrogen constituted the gas phase and water the liquid phase.

The rate of absorption of carbon dioxide increased with increasing particle size at all liquid velocities. The absorption rate was lower in the three-phase beds than in the solids-free bubble columns. This was attributed to a higher rate of bubble coalescence and, consequently, to a lower gas-liquid interfacial area in the three-phase fluidized beds.

Østergaard and Suchozebrski {48} employed the plug flow model to calculate values of the volumetric mass transfer coefficient k_La in beds of particles fluidized by water and a mixture of carbon dioxide and nitrogen. The solid phase consisted of 1 and 6 mm ballotini and experiments were carried out in a column 100 mm in diameter and 290 mm high. Liquid samples were taken at six points in the column and analysed volumetrically.

In the plug flow model, it is assumed that there is no axial or radial mixing in either fluid phase. The differential equation and boundary condition describing the mass transfer process are

$$U_L\, (dC/dy) = k_La\, (C^* - C) \tag{8.1}$$

$$C = C_o \text{ at } y = 0. \tag{8.2}$$

In these equations, C denotes the concentration of the diffusing species, C* the saturation concentration, and y the distance above the distributor. The solution of eqns. (8.1) and (8.2) is

$$C = C^* - (C^* - C_o)e^{-(k_La/U_L)y} \tag{8.3}$$

Beds of 6 mm particles were characterized by high rates of mass transfer, beds of 1 mm particles by low rates and bubble columns by intermediate rates. The values of k_La increased with gas rate but were relatively insensitive to the liquid velocity. They also observed that the mass transfer coefficient decreased with increasing distance from the distributor in bubble columns and in beds of 1 mm beads. In contrast, in beds of 6 mm beads, k_La passed through maxima at different heights above the distributor.

Østergaard and Fosbøl {49} reported measurements on the rate of mass transfer of oxygen in three-phase fluidized beds. They employed the same equipment as Østergaard and Suchozebrski and reported generally similar trends in the data. Typical values of k_La were: 100-200 h^{-1} in beds of 6 mm beads, 30-100 h^{-1} in bubble columns, and 10-25 h^{-1} in beds of 1 mm beads. The liquid velocity had no effect on the absorption rate in the beds of 6 mm particles and in bubble columns. However, in beds of 1 mm beads an increase in liquid velocity caused a marked increase in k_La. Shallow beds of large particles exhibited particularly high mass transfer rates.

Again, k_La was observed to vary with height above the distributor. A twofold decrease from the bottom to the top of the column was not uncommon. Four possible reasons for this variation were given: (i) the decrease in the static pressure with height (this affects C*), (ii) the progressive decrease in the mass flow-rate of gas with height due to the absorption of oxygen, (iii) axial mixing in the liquid phase, (iv) bubble coalescence and disintegration. In all likelihood, (iv) is probably the correct explanation. In any event, the mere fact that k_La varies with height above the distributor makes the data somewhat unreliable for scale-up purposes.

Dakshinamurty et al {50} also measured rates of absorption of oxygen in water in three-phase fluidized beds containing particles of different size (2.1-6.8 mm) and density (2400-11200 kg/m^3). Their experiments were carried out in a 56 mm diameter column, 680 mm high. The mass transfer coefficient was observed to increase with increasing particle size and density as a result, it was claimed, of bubble disintegration. Also, k_La increased with increasing gas velocity but was independent of the liquid rate. In general, their results were qualitatively in accord with those of Østergaard and co-workers {48, 49}.

Lee and Worthington {51} studied the absorption of carbon dioxide in three-phase beds of 6 mm particles. Their experiments were performed in a square column of 40 cm^2 cross-section equipped with six sampling points installed at different heights. Again, volumetric mass transfer coefficients calculated on the basis of plug flow were found to decrease with height. The authors attributed this to a reduction in the superficial gas velocity as a result of absorption. The values of k_La obtained were about 20% higher than those found by Østergaard and Suchozebrski {48}. Lee and Worthington attributed this discrepancy to the use of condensate water rather than tap water. In the distributor region, k_La was approximately proportional to U_g and essentially independent of U_L.

Alvarez-Cuenca {52} studied the mass transfer of oxygen between air and deoxygenated water in three-phase fluidized beds containing 1, 3 and 5 mm glass beads. Experiments were carried out in a two-dimensional column equipped with approximately fifty sampling taps from which liquid samples could be withdrawn and analysed. This technique enabled the oxygen concentration throughout the bed to be mapped. The resulting contour diagrams vividly illustrated the complexity of the flow patterns which exist in multiphase systems.

The shape of typical axial concentration profiles is illustrated in Figure 14. At low gas rates (eg, 4 cm/s), the curves were generally S-shaped. The concentration was essentially constant in the distributor region but increased somewhat in the bulk of the three-phase bed. The steep rise in the curve therefore occurs

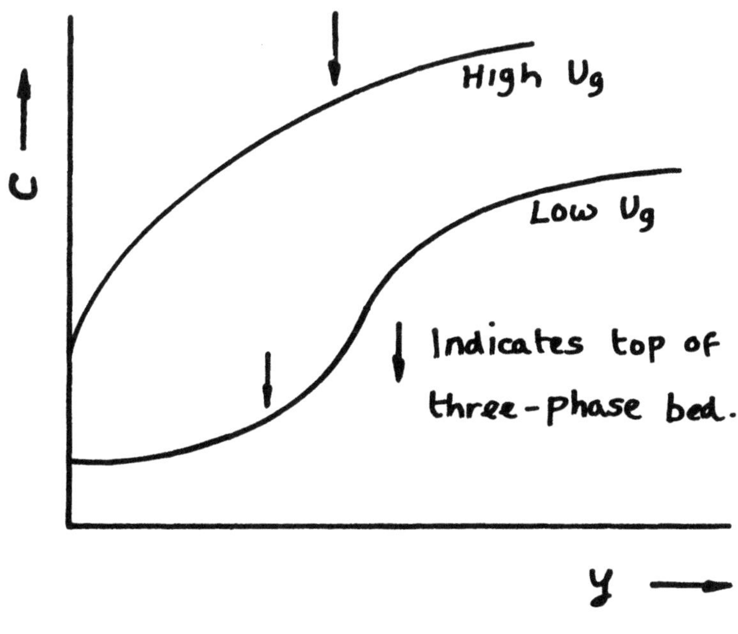

Fig. 14: Shape of typical oxygen concentration profiles in a three-phase fluidized bed (after Alvarez-Cuenca {52}).

in the solids-free zone above the bed. At high gas rates (eg, 28 cm/s), the S-shape became less pronounced and the oxygen concentration increased rapidly with height throughout the three-phase bed. This behaviour was similar to that observed in bubble columns.

Alvarez-Cuenca examined the ability of four flow models to describe his oxygen concentration data. These were (i) the continuous stirred tank reactor model, (ii) the plug flow model with constant k_La, (iii) the axial dispersion model with constant k_La and E_L, (iv) his own two-zone model. Model (i) was eliminated in the first instance because in all cases the oxygen concentration was observed to vary considerably throughout the column. Model (ii) was judged unsatisfactory as the calculated k_La was observed to vary with position in the bed. However, for the sake of comparison with other data, typical point values of k_La at y = 14 cm are shown in Figure 15. These are significantly larger than the values quoted by Østergaard and co-workers {48, 49}. In general, the ability of the plug flow model to describe the data decreased as the particle size increased. The two-parameter axial dispersion model (iii) offered no improvement over the plug flow model.

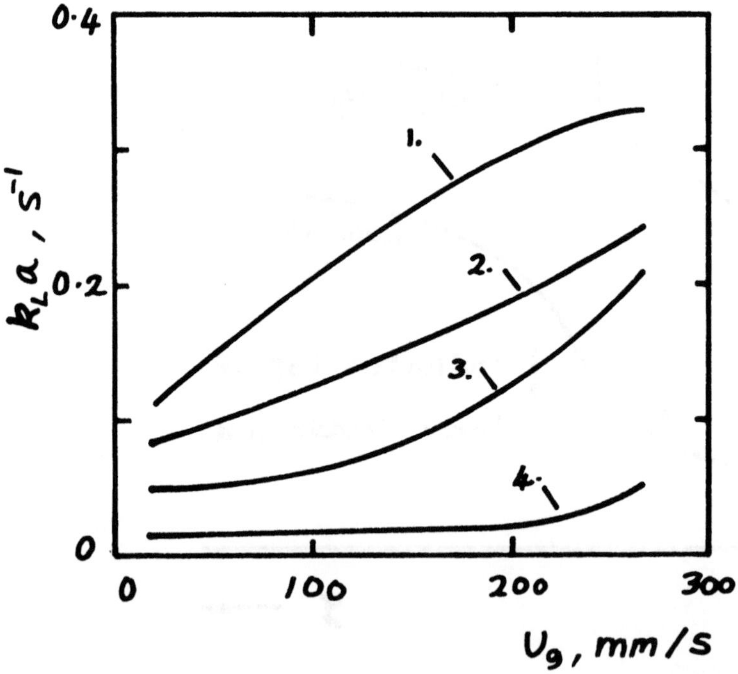

Fig. 15: Mass transfer coefficients as a function of gas velocity (after Alvarez-Cuenca {52}). U_L = 75 mm/s, y = 14.7 cm: 1. liquid-gas bed; 2. d_p = 1 mm; 3. d_p = 3 mm; 4. d_p = 5 mm.

Consequently, the author devised model (iv) which, he claimed, was more realistic of the processes actually occurring in the bed.

In the two-zone model, Alvarez-Cuenca postulated that the bed could be divided into two independent plug flow sections, each of which was characterized by its own mass transfer coefficient. The demarcation between these sections, designated the grid and bulk zones, was located at height H_d above the distributor. In practice, H_d varied between 20 and 40 cm; however, a constant value of 33 cm was assumed. In general, the two-zone model was found to give a better statistical fit to the data than either the plug flow or axial dispersion models. However, none of the models was successful in predicting the S-shaped concentration profiles observed at low fluid velocities.

9. A TECHNOLOGICAL ASSESSMENT OF THREE-PHASE FLUIDIZATION

Various aspects of our current scientific knowledge of three-phase fluidization have been described above in sections 3 to 8. Attempts are now made to determine the adequacy of this knowledge for design purposes and to assess possible areas in which three-phase fluidized beds can make a useful technological contribution.

The scientific studies of three-phase fluidization reviewed above can basically be classified under three headings, namely hydrodynamics, heat transfer and mass transfer. A great deal of the effort has been devoted to the first of these topics. Consequently, the expanded bed height and the individual phase holdups may be calculated with a reasonable degree of confidence. For aqueous systems, at least, eqns. (3.10)-(3.13) are probably the most reliable for this purpose as they are based on a relatively wide range of experimental conditions. In the case of the other hydrodynamic parameters (wake holdup, bubble characteristics and mixing parameters), the picture is not so complete. Consequently, although the trend of these parameters with the main operating variables is fairly well understood, their absolute magnitude is open to question. With this qualification, eqn. (4.12) may be used to estimate the wake holdup, eqns. (5.2)-(5.4) the bubble size and rising velocity and eqn. (6.3) the liquid-phase axial mixing coefficient.

Heat transfer coefficients can be estimated in aqueous systems by means of eqn. (7.1). However, data for other liquids are completely lacking. Further study in this important area is clearly warranted.

In the case of mass transfer, the situation is not so clear. The available experimental evidence suggests that neither the plug flow nor the axial dispersion models with constant coefficients satisfactorily describe the transport processes occurring in three-phase fluidized beds. The two-zone model proposed by Alvarez-Cuenca may prove to be a better representation. However, more work is necessary to establish this. At best, only an order of magnitude estimate of volumetric mass transfer coefficients can be obtained for a very limited number of systems.

In common with other chemical reactors, design procedures for three-phase fluidized beds will undoubtedly require primarily a knowledge of the process kinetics and the flow characteristics. In many instances, mass transfer will be the rate-determining step. Consequently, future research on this topic and on establishing more reliable flow models will undoubtedly prove most valuable for design and scale-up purposes. Research should also be extended to include hydrocarbon systems as many potential applications for three-phase fluidization lie in this area.

In order to be a viable alternative to more conventional multiphase reactors in any particular application, the three-phase fluidized bed must provide either a technical or an economic advantage over its competitors. Published data suggest that reaction (and absorption) rates are significantly higher in three-phase fluidized beds than in fixed bed devices. This enables the fluidized bed reactor to be built smaller with a consequent saving on capital costs. However, this saving is offset by added pumping costs due to the higher bed pressure drop. With the present high costs of energy, this may make the fluidized system uneconomical. However, there is some experimental evidence to suggest that with large superficial gas velocities, the rate of absorption is high at the distributor and progressively diminishes with height. Under these conditions, it might be possible to reduce significantly the reactor height, and hence the pressure drop and operating costs, with little reduction in performance.

There may be compelling technical reasons favouring the choice of a three-phase fluidized bed reactor. For example, a fluidized system would be preferred in cases where a fixed bed reactor might become clogged with fines. Alternatively, it could prove very beneficial to carry out highly exothermic catalytic gas-phase reactions in the presence of an inert moderating liquid which acts as a heat sink.

Three-phase fluidized beds may also find potential applications as heat exchangers, particularly with dirty liquids. The presence of both the gas and solids enhances the heat transfer coefficient considerably. More importantly, the scouring action of the solids will minimize or even eliminate any fouling. Provided the bed height is kept to a minimum, the pressure drop will not be excessive. On the other hand, the addition of inert particles to a bubble column absorber has little or no benefit from the viewpoint of enhancing the mass transfer rate. In certain cases, it appears to be detrimental. Moreover, it may add significantly to the pressure drop.

In conclusion, there are a number of technological areas in which the three-phase fluidized bed can play a useful role. In these situations, technical or capital cost advantages outweigh any economic penalty resulting from the relatively high pressure drop. In these areas, this form of multiphase contactor will undoubtedly play a more important role in the future.

NOMENCLATURE

a Liquid-gas contact area per unit volume of bed

A Column cross-sectional area

C Concentration of diffusing species

C_o Concentration of diffusing species at distributor

C^* Saturation concentration of diffusing species

d_b Bubble diameter

d_c Column diameter

d_p Particle diameter

E_L Liquid-phase axial diffusion coefficient

Fr_g Gas-phase Froude number, U_g^2/d_pg

Fr'_g Gas-phase Froude number, U_g^2/d_cg

Fr_L Liquid-phase Froude number, U_L^2/d_pg

g Acceleration due to gravity

g_c Dimensional constant

h Heat transfer coefficient

H_b Expanded bed height (three-phase bed)

H_b^o Expanded bed height (liquid bed)

H_d Height of grid zone

H_o Height of static bed

HMU Height of the mixing unit, $2H_b/Pe_L$

k Ratio of wake to gas holdup in three-phase bed

k_L Mass transfer coefficient

k_o Ratio of wake to gas holdup for single bubble

k' Ratio of wake to gas holdup in liquid-gas bed

K Liquid consistency

m Liquid behaviour index

M_s Mass of solids in bed

n Exponent in Richardson-Zaki equation

P_b Pressure drop across three-phase fluidized bed

Pe_L Liquid-phase Peclet number, $U_L H_b/E_L$

Pe'_L Liquid-phase Peclet number, $U_g d_c/E_L$

R Radius of spherical cap bubble

R_v Vortex shedding rate

Re_L Liquid-phase Reynolds number, $U_L d_p \rho_L/\mu_L$

Re_t Liquid-phase Reynolds number, $U_t d_p \rho_L/\mu_L$

U Fluid velocity

U_b Bubble rising velocity relative to column walls

U_{br} Bubble rising velocity relative to liquid

U_g Gas superficial velocity

U_{gL} Velocity of gas relative to liquid

U_L Liquid superficial velocity

U_L' Absolute velocity of liquid

U_{Lp} Velocity of liquid in particulate phase

U_{mf} Minimum fluidization velocity

U_t Particle terminal velocity

$\bar{U}_v$ Average rising velocity of a vortex

V_b Bubble volume

V_e' Equivalent volume of stable wake due to all the vortices

V_L Volume of liquid in bed

V_v Volume of vortex

We Weber number, $U_g \nu/\sigma$

x Solids content of wakes

x_c Critical solids content of wakes

y Height above distributor

α Coefficient in Davies-Taylor equation

β Exponent in Davies-Taylor equation

ΔH Reduction in liquid level in column after shutting off flows

ΔP_b Pressure drop

ΔP_f Effective weight of bed per unit area

ε_g Gas holdup

ε_L Liquid holdup

ε_{Lp} Liquid holdup in particulate phase

ε_{L2} Liquid holdup in liquid fluidized bed

ε_s Solids holdup

ε_v Liquid holdup in wake phase

ε_w Wake holdup

μ_L Liquid viscosity

ν Generalized viscosity constant, $g_c K8^{m-1}$

ρ_f Fluid density

ρ_g Gas density

ρ_L Liquid density

ρ_s Solids density

σ Surface tension

Θ Included angle of spherical or circular cap bubble

REFERENCES

1. Chervenak, M.C., Johanson, C.A., Johnson, C.A., Schuman, S.C. and Sze, M., Oil Gas J., 1960, 58, 80.

2. Chervenak, M.C., Wolk, R., Byrd, C.R., Hellwig, L.R. and Van Driesen, R.P., Chem. Engng. Progr., 1963, 59, 53.

3. Blum, D.B. and Toman, J.J., 68th A.I.Ch.E. Meeting, Los Angeles, Nov. 1975.

4. Anon., New Products and Services, Chem. Engng., 1976 (Oct. 25) 87.

5. Volpicelli, G. and Massimilla, L., Chem. Engng. Sci., 1970, 25, 1361.

6. Østergaard, K., Advances in Chemical Engineering, 1968, 7, 71, Academic Press, New York.

7. Shah, Y.T., "Gas-Liquid-Solid Reactor Design", McGraw-Hill, 1979, Chapter 9.

8. Kim, S.D., Baker, C.G.J. and Bergougnou, M.A., Can. J. Chem. Engng., 1972, 50, 695.

9. Kim, S.D., Baker, C.G.J. and Bergougnou, M.A., Can. J. Chem. Engng., 1975, 53, 134.

10. Baker, C.G.J., Armstrong, E.R. and Bergougnou, M.A., Powder Technol., 1978, 21, 195.

11. Stewart, P.S.B. and Davidson, J.F., Chem. Engng. Sci., 1964, 19, 319.

12. Dakshinamurty, P., Subrahmanyam, V. and Rao, J.N., Ind. Engng. Chem. Process Des. Develop., 1971, 10, 322.

13. Dakshinamurty, P., Rao, K.V., Subbaraju, R.V. and Subrahmanyam, V., Ind. Engng. Chem. Process Des. Develop., 1972, 11, 318.

14. Richardson, J.F. and Zaki, W.N., Trans. Instn. Chem. Engrs., 1954, 32, 35.

15. Michelsen, M.L. and Østergaard, K., Symp. Fund. Appl. Fluidization, Tampa, 1968, preprint 31d.

16. Michelsen, M.L. and Østergaard, K., Chem. Engng. J., 1, 37.

17. Efremov, G.I. and Vakhrushev, I.V., Int. Chem. Engng., 1970, 10, 37.

18. Vail, Yu.K., Manakov, N.Kh. and Manshilin, V.V., Int. Chem. Engng., 1970, 10, 244.

19. Razumov, I.M., Manshilin, V.V. and Nemets, L.L., Int. Chem. Engng., 1973, 13, 57.

20. Stewart, P.S.B., Ph.D. Thesis, University of Cambridge, 1965.

21. Bhatia, V.K. and Epstein, N., Proc. Int. Symp. on Fluidization and its Applications, Cepadues-Editions, Toulouse, 1974.

22. Darton, R.C. and Harrison, D., Chem. Engng. Sci., 1975, 30, 581.

23. Baker, C.G.J., Kim, S.D. and Bergougnou, M.A., Powder Technol., 1977, 18, 201.

24. Kim, S.D., Baker, C.G.J. and Bergougnou, M.A., Chem. Engng. Sci., 1977, 32, 1299.

25. Rigby, G.R. and Capes, C.E., Can. J. Chem. Engng., 1970, 48, 343.

26. Epstein, N. and Nicks, D., in Fluidization Technology, D.L. Keairns, ed., Hemisphere Publ. Co., Washington D.C., 1976, Vol. 1, 389.

27. Epstein, N., Can. J. Chem. Engng., 1976, 54, 259.

28. El-Temtamy, S.A. and Epstein, N., Int. J. Multiphase Flow, 1978, 4, 19.

29. El-Temtamy, S.A. and Epstein, N., Can. J. Chem. Engng., 1979, 57, 520.

30. Armstrong, E.R., Baker, C.G.J. and Bergougnou, M.A., in Fluidization Technology, D.L. Keairns, ed., Hemisphere Publ. Co., Washington, D.C., 1976, Vol. 2, 405.

31. Henriksen, H.K. and Østergaard, K., Chem. Engng. J., 1974, 7, 141.

32. Massimilla, L., Solimando, A. and Squillace, E., Brit. Chem. Engng., 1961, 6, 232.

33. Davies, R.M. and Taylor, G., Proc. Roy. Soc. (London), 1950, A200, 375.

34. Lee, J.C., Sherrard, A.J. and Buckley, P.S., Proc. Int. Symp. on Fluidization and its Applications, Cepadues-Editions, Toulouse, 1974, p.407.

35. Henriksen, H.K. and Østergaard, K., Chem. Engng. Sci., 1974, 29, 626.

36. Taylor, G., Proc. Roy. Soc. (London), 1950, A201, 192.

37. Bellman, R. and Pennington, R.H., Quart. Appl. Math., 1954, 12, 151.

38. Østergaard, K., Chem. Engng. Sci., 1966, 21, 470.

39. Page, R.E. and Harrison, D., Powder Technol., 1972, 6, 245.

40. Rigby, G.R., Van Blockland, G.P., Park, W.H. and Capes, C.E., Chem. Engng. Sci., 1970, 25, 1729.

41. Nicklin, D.J., Chem. Engng. Sci., 1962, 17, 693.

42. Schügerl, K., Proc. Int. Symp. Fluidization, Eindhoven, Netherlands University Press, 1967, 782.

43. Kim, S.D., Ph.D. Thesis, The University of Western Ontario, London, Canada, 1974.

44. Kato, Y., Kagaku Kogaku, 1963, 1, 3.

45. Østergaard, K., in Fluidization, J.F. Davidson and D. Harrison, eds., Academic Press, New York, 1971.

46. Viswanathan, S., Kakar, A.S. and Murti, P.S., Chem. Engng. Sci., 1965, 20, 903.

47. Østergaard, K. and Suchozebrski, W., in Proc. 4th Europ. Symp. on Chem. Reaction Engng., Pergamon Press, Oxford, 1971, p.21.

48. Østergaard, K. and Fosbøl, P., Chem. Engng. J., 1972, 3, 105.

49. Dakshinamurty, P., Chiranjevi, C., Subrahmanyam, V. and Rao, P.K., Proc. Int. Symp. on Fluidization and its Applications, Cepadues-Editions, Toulouse, 1974.

50. Lee, J. and Worthington, H., Instn. Chem. Engrs. Symp. Ser., 1974, B2, 1.

51. Alvarez-Cuenca, M., Ph.D. Thesis, The University of Western Ontario, London, Canada, 1979.

DESIGN AND SCALE-UP OF GAS FLUIDIZED BED REACTORS

R. Krishna

Koninklijke/Shell-Laboratorium, Amsterdam
(Shell Research B.V.), Badhuisweg 3, 1031 CM Amsterdam

ABSTRACT. This paper considers the problem of scaling-up gas fluidized bed reactors. Using the interphase mass transfer model of Davidson combined with a bubble growth mechanism, up to a maximum stable bubble size, postulated by Darton, expressions are derived to describe the variation of bubble hydrodynamic parameters (hold-up, rise velocity) and mass transfer parameters ($k_L a'$, mass transfer coefficient per unit bubble volume and height of transfer unit, HTU) as a function of fluidized bed height H and reactor diameter D_T. The predictions of the combined Davidson-Darton model are tested with experimental data published by De Groot (1967), Van Swaaij and Zuiderweg (1972) and De Vries (1972).

The experimental data of these workers show that the Davidson-Darton combined model correctly predicts the variation of the interphase mass transfer coefficient with bed diameter, bed height and superficial gas velocity. It is also demonstrated that the interphase mass transfer coefficient per unit bubble volume is independent of column diameter, as required by theoretical considerations for a freely bubbling bed.

The combined Davidson-Darton theory is used to suggest a strategy for scaling-up fluidized beds. The strategy consists essentially of "cold" experimental studies on hydrodynamics and mass transfer in columns of about 0.5 - 1.0 diameter and 3-4 m in height with simulated gas and solid. The results obtained in this unit can then be extrapolated to commercial scale operation by use of a Davidson-Darton type approach.

Other aspects of fluidized bed reactor scale-up considered in this paper include mathematical modelling, influence of internals and high pressure, high temperature operation.

INTRODUCTION

The purpose of this paper is to suggest a possible strategy for scaling-up gas fluidized bed reactors. The paper has been written with an arbitrary catalytic chemical process in mind, but the greater part of the treatment should also be applicable to gas-solid non-catalytic reactions. The paper considers only bubbling gas fluidized beds; fast fluidization and transfer line operations are outside the scope of the present comminication. Further, it will be implicitly assumed that the reaction rate is slow enough to require deep beds (taller than, say, 4 m) in order to achieve high (> 90 %) conversion levels. In this case the hydro(aero)-dynamics, mass transfer and backmixing all play important roles in determining the achievable conversion in commercial operation.

We assume in the following treatment that the kinetics of the reaction are known. A single reaction with first order kinetics is arbitrarily assumed. If the kinetics of the reaction has been measured in a fixed bed bench scale unit, then the reaction rate may be written as

$$r_A = -\rho_b \, k_m \, c_A \tag{1}$$

where r_A is rate of production of A, $kmol.m^{-3}.s^{-1}$; ρ_b is the bulk density of the catalyst, kg/m^3; k_m is a first order rate constant, $(m^3 \text{ reactant}).(\text{kg catalyst})^{-1}.s^{-1}$ and c_A is the molar concentration of reactant A, $kmol/m^3$.

If ρ_{fb} represents the density of the fluidized bed, kg catalyst/(m^3 fluidized bed), then the rate constant per unit volume of fluidized bed is given as

$$K_r = k_m \rho_{fb} \quad \frac{(m^3 \text{ reactant})}{(m^3 \text{ fluid bed})\ (s)} \tag{2}$$

We will discuss fluidized bed performance in terms of the number of reaction units, NRU, defined by

$$NRU = \frac{K_r H}{U} \tag{3}$$

where H is the total fluid bed height, m; and U is the superficial gas velocity, m/s. If the reactor is equipped with internals, U refers to the velocity based on net free area available for flow. The operating gas velocity will be taken to be at least $5 \times U_{mf}$, where U_{mf} is the minimum fluidization velocity.

Briefly stated, the present paper deals with the following problem:

Given the information on reaction kinetics (i.e. given k_m), what is the height of the reactor required to achieve a specified conversion in a vessel of specified diameter?

The reactor diameter D_T, determined by the chosen operating gas velocity, could be as large as 10 m for near-atmospheric pressure operation.

To answer the question posed above we need to know how the hydrodynamic, mass transfer and mixing parameters vary with height and with diameter. We begin our analysis of the problem by considering the bubble mechanics in a fluid bed and develop a strategy for scale-up using an approach suggested by combination of the interphase transfer model of Davidson (see Davidson <u>et al</u>., 1977) and a bubble growth mechanistic model of Darton (1977, 1979).

Parallel to the discussion on bubble hydrodynamics below we will also indicate how some of the key hydrodynamic parameters may be measured in a "cold" experimental unit, employing a simulated gas and real catalyst or simulated solid.

The model we choose for a fluidized bed is the simple two-phase model (bubble phase and dense phase); also taking account of the variations of the model parameters along the reactor height.

Bubble Size Prediction

We consider the physical picture of a fluidized bed as shown in Fig. 1. Bubbles formed at the distributor rise and coalesce to form larger bubbles. There is sufficient evidence in the literature to suggest that there exists a maximum stable bubble size, $d_{b\ max}$, for a given catalyst size and distribution. To accommodate the concept of maximum stable bubble size in our treatment we follow Werther (1977) and introduce an equilibration height h^* above the distributor (see Fig. 1) where the bubbles reach a maximum size.

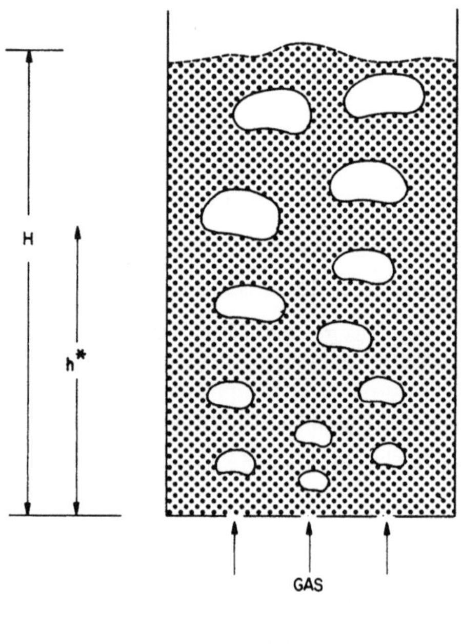

FIG. 1: TWO-PHASE MODEL FOR FLUIDIZED BED

h^* IS HEIGHT ABOVE DISTRIBUTOR WHERE BUBBLES REACH MAXIMUM STABLE SIZE ; H IS TOTAL BED HEIGHT

There are many correlations available in the literature for prediction of bubble size as a function of fluid bed height (e.g. Darton, 1977; Rowe, 1976; Werther, 1977); we choose here the approach of Darton *et al.* (1977). Darton's coalescence model is pictured in Fig. 2. Here it is assumed that coalescence occurs between bubbles of neighbouring streams and that the distance travelled by bubbles before coalescence is proportional to their horizontal separation. The diameter d_b of a sphere having the same volume as the bubble is obtained as

$$d_b = 0.54 \left(U-U_{mf}\right)^{\frac{2}{5}} \left(h + 4\sqrt{A_o}\right)^{\frac{4}{5}} \Big/ g^{\frac{1}{5}} \tag{4}$$

where h is the height above the distributor, m; g is the acceleration due to gravity, 9.81 m/s^2; A_o is the "catchment area" of the bubble stream at the distributor plate, which characterises different distributors, and is usually the area of plate per orifice. The constant 0.54 is obtained from data fitting.

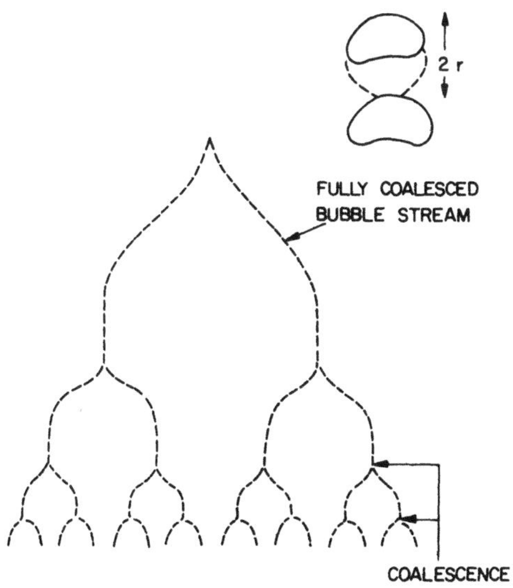

FIG. 2 : BUBBLE PATHS DURING COALESCENCE
(THE PROCESS SHOULD BE VISUALIZED IN THREE DIMENSIONS)

r IS THE RADIUS OF THE ROUGHLY SPHERICAL REGION ENCLOSING THE BUBBLE AND ITS WAKE

Another bubble size correlation is that of Werther (1977) for porous plate distributors:

$$d_b = 0.00853\,[1 + 27.2\,(U-U_{mf})]^{\frac{1}{3}}\,[1 + 6.84\,h]^{1.21} \qquad (5)$$

Both correlations (4) and (5) predict similar d_b values for porous plate distributors (Darton (1977) chooses $A_o = 56\ mm^2$ for use of his correlation for this case).

Fig. 3 illustrates some calculations for d_b using equation (4). This equation holds when the total fluidized bed height $H < h^*$. If $H \geqslant h^*$, we calculate $d_{b\,max}$ from

$$d_{b\,max} = 0.54\left(U-U_{mf}\right)^{2/5}\left(h^* + 4\sqrt{A_o}\right)^{4/5}\Big/ g^{1/5} \qquad H \geqslant h^* \qquad (6)$$

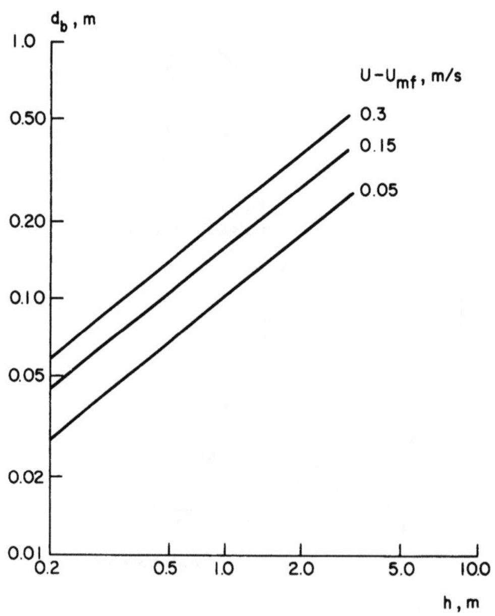

FIG. 3: BUBBLE DIAMETER d_b AS A FUNCTION OF HEIGHT ABOVE DISTRIBUTOR, h, ACCORDING TO THE DARTON (1977) CORRELATION:

$$d_b = 0.54\,(U-U_{mf})^{2/5}\,(h + 4\sqrt{A_o})^{4/5}/g^{1/5}$$

The "equilibration" height h^* ranges from 0.5 to 2.0 m (see the calculations of Werther, 1977), depending on the particle size and distribution. h^* is an important parameter for scale-up purposes, as will be seen later. "Cold" experimental units used for studying hydrodynamics and mass transfer must have diameters at least twice the maximum stable bubble size predicted from equation (6). Also the height of the experimental unit must be greater than h^*. Usually, a choice of column diameter $D_T > 0.6$ m and height $H > 3$ m will satisfy this requirement and yield usable scale-up information.

Bubble Rise Velocity and Bubble Hold-up

On the basis of extensive measurements carried out in beds of different diameters, Werther (1977) suggests the following correlation for the rise velocity V_b of a bubble swarm as a function of the bubble diameter d_b:

$$V_b = \phi \sqrt{g\, d_b} \tag{7}$$

where

$$\left.\begin{aligned} \phi &= 0.64 && \text{for } D_T \leqslant 0.1 \text{ m} \\ \phi &= 1.6\, D_T^{0.4} && \text{for } 0.1 < D_T < 1.0 \text{ m} \\ \phi &= 1.6 && \text{for } D_T \geqslant 1.0 \text{ m} \end{aligned}\right\} \tag{8}$$

The relation (8) is pictured in Fig. 4. The increase in the effective bubble rise velocity in large diameter beds is due to bubble interactions and large scale mixing patterns.

The bubble rise velocity is an important parameter which must be determined in cold engineering experimental studies. While direct measurements with probes are possible, an alternative procedure is the use of the dynamic gas disengagement technique, pictured in Fig. 5. The technique consists of instantaneously shutting off the gas supply and measuring (by video, ciné or electronic techniques) the fluid bed height H as a function of time. The bubble hold-up, ε_b (fraction of bubble phase on total fluid bed volume) can be determined from the disengagement graph.

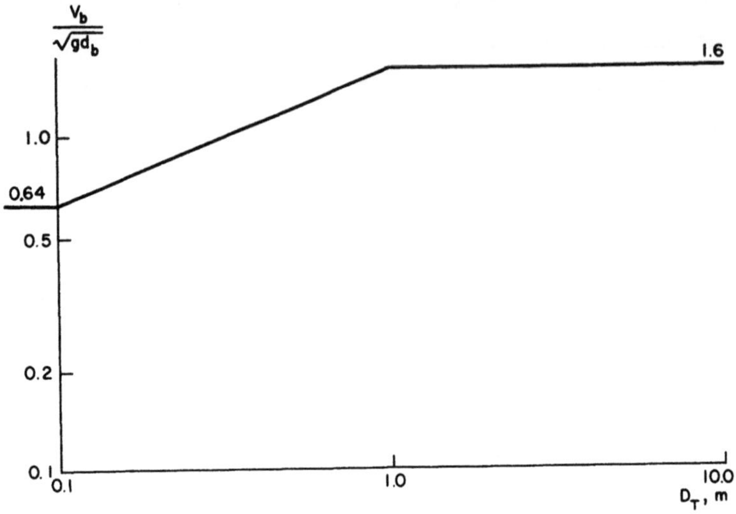

FIG. 4 : WERTHER'S CORRELATION FOR BUBBLE RISE VELOCITY :
$V_b = \phi \sqrt{gd_b}$

The technique works particularly well in the velocity range $0.05 < U < 0.5$ m/s, the usual range of operation of industrial units.

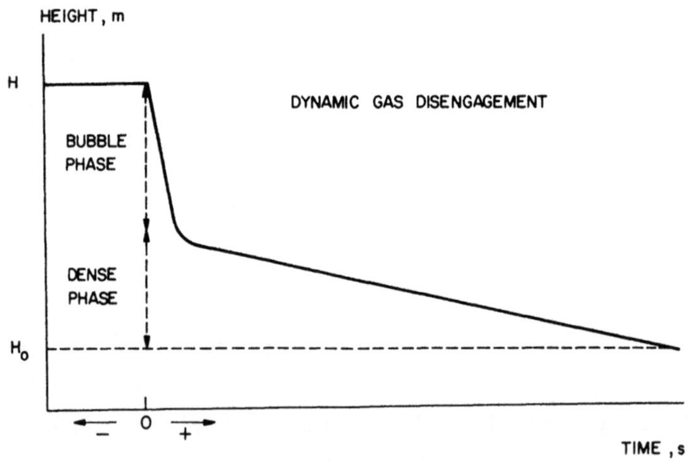

FIG. 5 : SCHEMATIC OF DYNAMIC GAS DISENGAGEMENT PROCESS

The rise velocity of the bubble swarm is then

$$\bar{V}_b = \frac{U-U_{mf}}{\varepsilon_b} \tag{9}$$

where it is assumed that the "visible bubble flow rate" is $U-U_{mf}$. Values of $\bar{V}_b$ determined in the above fashion are integral values over the length of the fluid bed.

In the event we wish to predict ε_b, we may combine equations (4), (7) and (9) to give

$$\varepsilon_b = \frac{1}{H}\int_0^H \frac{(U-U_{mf})}{\phi\sqrt{g\ d_b}}\ dh \tag{10}$$

while integrating we must impose the condition that when $H > h^*$, $d_b = d_{b\ max}$ (equation (6)). Fig. 6 shows typical predictions of equation (10) for $h^* = 0.5$ m and $h^* = 1.0$ m.

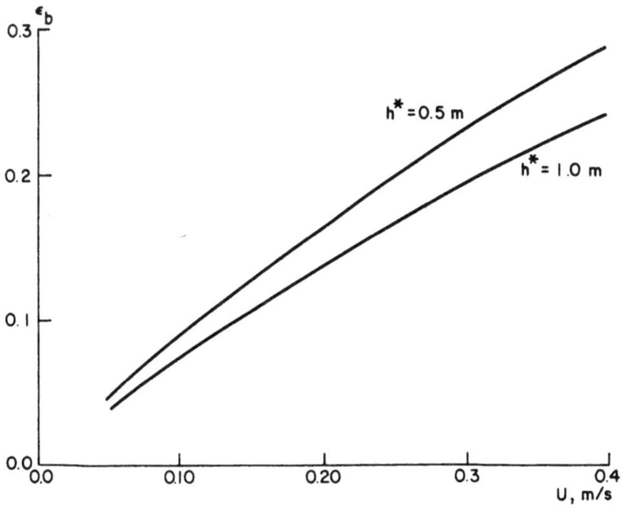

FIG. 6: BUBBLE HOLD-UP ϵ_b AS A FUNCTION OF THE SUPERFICIAL GAS VELOCITY U AND EQUILIBRATION HEIGHT h^* PREDICTED BY A COMBINATION OF DARTON'S BUBBLE GROWTH CORRELATION AND WERTHER'S BUBBLE RISE VELOCITY CORRELATION

OTHER DETAILS ARE: H = 2.8 m; D = 0.6 m; A_0 = 56 mm²; U_{mf} = 0.01 m/s

The figure demonstrates the importance of h^* in determining ε_b.

If measurements of ε_b are made for different fluidized bed heights H over a range of superficial gas velocities, then the relation (10) may be used to determine h^*. Data on total bed expansion (ε) and dense phase expansion (ε_{df}) are also obtained simultaneously with ε_b using the dynamic gas disengagement technique. These three parameters are important in determining the efficiency of gas-solid contact, as will be seen later in discussions on conversion calculations.

We now turn our attention to the interphase mass transfer between the bubbles and the dense phase.

Height of Transfer Unit

One of the most important parameters which governs achievable conversion in a fluid bed of height H is the number of (mass) transfer units, NTU, given by

$$NTU = \frac{k_L a H}{U} \tag{11}$$

where k_L is the interphase mass transfer coefficient, m/s; a is the interfacial area per unit reactor volume, m^2/m^3. We deliberately use the common notation of gas-liquid transfer operations because of the remarkable similarities between a gas fluidized bed and a bubble column operating in the "churn-turbulent" regime (Bach, 1977; Vermeer and Krishna, 1979).

The height of transfer unit HTU defined as

$$HTU \equiv H/NTU \tag{12}$$

is thus

$$HTU = U/k_L a \tag{13}$$

Alternatively, we may re-write equation (13) in the form

$$HTU = V_b/k_L a' \tag{14}$$

where a' is the interfacial area per unit bubble volume, m^2/m^3.

Van Swaaij and Zuiderweg (1973) have correlated experimental data for large diameters and A-type fluidization (Geldart classification) by the equation:

$$HTU = \left(1.8 - \frac{1.06}{D_T^{1/4}}\right)\left(3.5 - \frac{2.5}{H^{1/4}}\right) \tag{15}$$

Fig. 7 shows typical calculations using equation (15). Our immediate objective will be to seek a mechanistic explanation for the H and D_T dependence of HTU. A good theoretical model would help to extrapolate to new processes and new operating conditions with confidence.

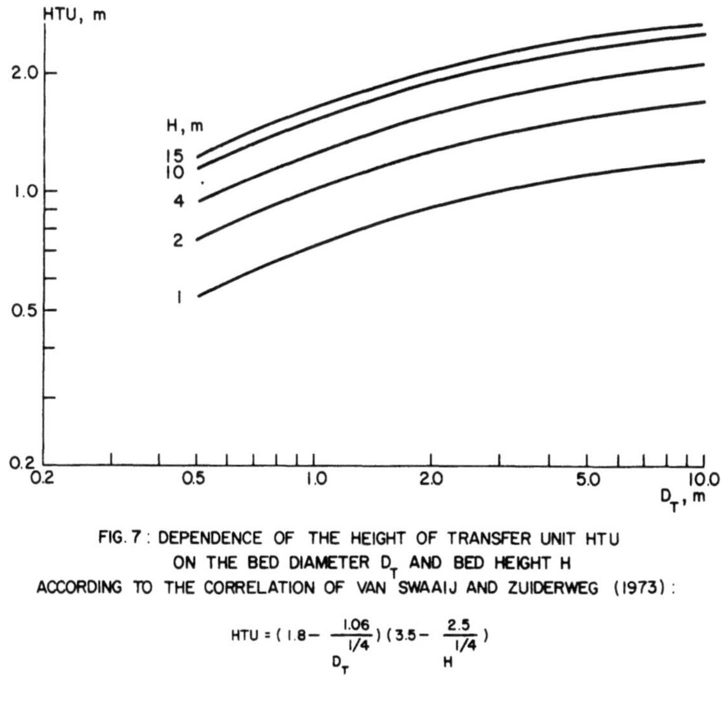

FIG. 7 : DEPENDENCE OF THE HEIGHT OF TRANSFER UNIT HTU ON THE BED DIAMETER D_T AND BED HEIGHT H ACCORDING TO THE CORRELATION OF VAN SWAAIJ AND ZUIDERWEG (1973):

$$HTU = \left(1.8 - \frac{1.06}{D_T^{1/4}}\right)\left(3.5 - \frac{2.5}{H^{1/4}}\right)$$

In analysing the variation of HTU with H and D_T we should appreciate that measured data (e.g. De Groot, 1967; Van Swaaij and Zuiderweg, 1972; De Vries *et al.*, 1972) give integrated average values over the reactor height, in some cases $H < h^*$ and in others $H > h^*$.

If we adopt Davidson's model for transfer from a hemi-spherical gas bubble (see Davidson et al., 1977), there are two contributions to interphase transfer: cross-flow and diffusion. The total amount of gas transferred by these two mechanisms is given by

$$Q = 1.19\ \pi\ d_b^{\,2}\ U_{mf} + 0.91\ \pi\ d_b^{\,7/4}\ D_G^{\frac{1}{2}}\ g^{1/4}\ \frac{\varepsilon_{mf}}{1+\varepsilon_{mf}} \tag{16}$$

where Q, m^3/s, is the flow rate across a single bubble of diameter d_b; D_G is the gas phase diffusion coefficient, m^2/s; ε_{mf} is the porosity at minimum fluidization.

The mass transfer coefficient per unit bubble volume can be readily calculated from

$$k_L a' = \frac{Q}{\pi\ d_b^3\ /6} = 7.14\ U_{mf}\ d_b^{\,-1} + 5.46\ D_G^{\frac{1}{2}}\ g^{\frac{1}{4}}\ \frac{\varepsilon_{mf}}{(1+\varepsilon_{mf})}\ d_b^{\,-5/4} \tag{17}$$

$k_L a'$ varies along the reactor height because of the variation of d_b with H. Close to the distributor the bubbles are small and $k_L a'$ is large. For fast reactions a major portion of the conversion will take place in this zone above the distributor, and a proper account must be taken of the bubble growth in this region. The distributor region is of less importance for slower reactions requiring deep beds.

It is also clear from equation (17) that $k_L a'$ must be independent of the column diameter for freely bubbling (non-slugging) beds.

The coefficient $k_L a'$ calculated from equation (17) gives the local coefficient for transfer from the bubble phase to the cloud surrounding bubbles. Kunii and Levenspiel (1969) argue that the transfer from cloud to emulsion (dense) phase would experience a resistance offered by the cloud itself. They derive an expression for transfer from cloud to emulsion phase as

$$k_L a')_{ce} = 6.78 \left(\frac{\varepsilon_{mf}\ D_G\ V_b}{d_b^{\,3}} \right)^{\frac{1}{2}} \tag{18}$$

If we introduce Werther's bubble rise velocity correlation, equation (7), into equation (18) we find

$$k_L a')_{c-e} = 6.78\ (\varepsilon_{mf}\ D_G\ \phi\ \sqrt{g})^{\frac{1}{2}}\ d_b^{-5/4} \qquad (19)$$

Kunii and Levenspiel further suggest the usual addition-of-resistances rule for combining the mass transfer resistances from bubble to cloud and from cloud to dense phase. Examination of equations (17) and (18) shows that the variation of $k_L a'$ with d_b is similar for these two transfers and we shall proceed further with the Davidson transfer model to develop our scale-up procedure.

The integral averaged mass transfer coefficient $\overline{k_L a'}$ over the reactor height can be calculated from

$$\overline{k_L a'} = \frac{1}{H} \int_0^H k_L a'\ dh \qquad (20)$$

where we should recognise that the integration must be carried out separately over the regions $0-h^*$ and h^*-H (see equations (4) and (6)).

Evaluating the integral with the help of equations (4), (6) and (17) we get

$$\overline{k_L a'} = \frac{5\alpha\ h^{*\,1/5}}{H} + \frac{\beta}{H} \ln \left\{ \frac{h^* + 4\sqrt{A_o}}{4\sqrt{A_o}} \right\} +$$

$$+ \frac{\alpha(1-h^*/H)}{h^{*4/5}} + \frac{\beta(1-h^*/H)}{(h^* + 4\sqrt{A_o})} \qquad \text{if } H > h^* \qquad (21)$$

and

$$\overline{k_L a'} = 5\alpha\ H^{-\frac{4}{5}} + \frac{\beta}{H} \ln \left\{ \frac{H + 4\sqrt{A_o}}{4\sqrt{A_o}} \right\} \qquad \text{if } H \leqslant h^* \qquad (22)$$

In equations (21) and (22) α and β are given by

$$\alpha \equiv \frac{7.14}{0.54}\ g^{0.2}\ U_{mf}\ (U-U_{mf})^{-2/5} \qquad (23)$$

and

$$\beta \equiv \frac{5.46}{(0.54)^{1.25}} g^{0.5} \left(\frac{\varepsilon_{mf}}{1+\varepsilon_{mf}}\right) D_G^{\frac{1}{2}} (U-U_{mf})^{-\frac{1}{2}} \tag{24}$$

Equations (21) - (24) lead to the following observations:

(i) $\overline{k_L a'} \sim D_T^{\,0}$ (25a)

(ii) $\overline{k_L a'} \sim (U-U_{mf})^{-0.4 \text{ to } -0.5}$ (25b)

(iii) $\overline{k_L a'} \sim H^{-\frac{4}{5} \text{ to } -1}$ for $H \leqslant h^*$ (25c)

and (iv) $\overline{k_L a'} = c_1 + c_2/H$ for $H > h^*$ (25d)

where c_1 and c_2 depend on h^* and operating conditions.

Each of the equations (25) contains a scale-up rule, if measurements are made of $\overline{k_L a'}$ in a small-scale cold experimental unit under varying operating conditions U and H. (The equilibration height h^* is known from measurements of ε_b as a function of U and H).

We shall now verify each of the dependences of $\overline{k_L a'}$ given by equation (25).

Van Swaaij and Zuiderweg (1972) report experimental HTU data in fluidized sand beds of diameters 0.1, 0.23, 0.3 and 0.6 m for $H \approx 2$ m; see Fig. 8. The data show that HTU under similar operating conditions increases with increasing reactor diameter, D_T.

From the experimental HTU data of Fig. 8 we may calculate $\overline{k_L a'}$ from:

$$\overline{k_L a'} = \bar{V}_b/\text{HTU} \tag{26}$$

where $\bar{V}_b$ is an integral average bubble rise velocity:

$$\bar{V}_b = \phi \sqrt{g\, \bar{d}_b} \tag{27}$$

with the integral average bubble diameter $\bar{d}_b$ calculated from

$$\bar{d}_b = \frac{1}{H} \int_0^H d_b \, dh \tag{28}$$

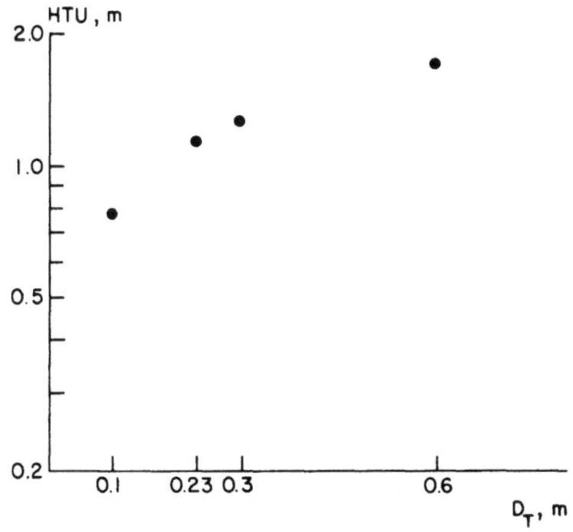

FIG. 8 : DEPENDENCE OF HTU ON COLUMN DIAMETER D_T
DATA OF VAN SWAAIJ AND ZUIDERWEG (1972)
FOR FLUIDIZED SAND BEDS (APPROX. 4% FINES)
OTHER DETAILS : $H \approx 2$ m ; $U \approx 0.14$ m/s $h^* = 0.5$ m

The values of $\overline{k_L a'}$ calculated from equations (26) - (28) are given in Fig. 9, which confirms the independence of $\overline{k_L a'}$ of D_T. Implicitly the Werther correlation, given by equations (7) and (8), is also verified by the Van Swaaij and Zuiderweg data.

Fig. 10 shows the variation of HTU with gas velocity U, for a fluidized sand bed of diameter 0.3 m. Values of $\overline{k_L a'}$ calculated from these data are plotted in Fig. 11 against $(U-U_{mf})$. The data show that $\overline{k_L a'} \sim (U-U_{mf})^{-0.4}$, following equation (25b). Other experimental data of HTU versus U given by Van Swaaij and Zuiderweg (1972) can be similarly interpreted.

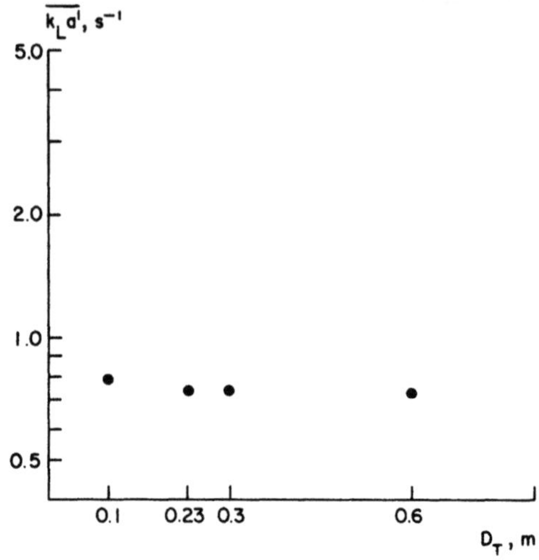

FIG. 9: EXPERIMENTS DEMONSTRATING THE INDEPENDENCE OF $\overline{k_L a'}$ ON COLUMN DIAMETER D_T

DATA TAKEN FROM VAN SWAAIJ AND ZUIDERWEG (1972) FOR FLUIDIZED SAND BEDS (APPROX. 4% FINES) OTHER DETAILS: $h^* = 0.5$ m; $H \approx 2$ m; $U \approx 0.14$ m/s THE EXPERIMENTAL $\overline{k_L a'}$ VALUES WERE CALCULATED FROM $\overline{k_L a'} \equiv \bar{V}_b/\text{HTU}$ WHERE $\bar{V}_b = \phi\sqrt{g\bar{d}_b}$

Another tentative conclusion to be drawn from that -0.4 power dependence of $\overline{k_L a'}$ on $(U-U_{mf})$ is that the cross-flow term, the first term of equation (17), predominates in the interphase transfer process.

A value of $h^* = 0.5$ m was chosen for fluidized sand bed analysis following the calculations for Werther (1977).

Fig. 12 gives the variation of HTU with fluidized bed height H for sand beds with 3-16 % fines. To analyse these data, again $\overline{k_L a'}$ was calculated from equations (26) - (28), taking $h^* = 0.5$ m and $U_{mf} = 0.01$ m/s. Equations (25c) and (25d) suggest plotting of $\overline{k_L a'}$ separately for the two regions $H \leqslant h^*$ and $H > h^* = 0.5$ m.

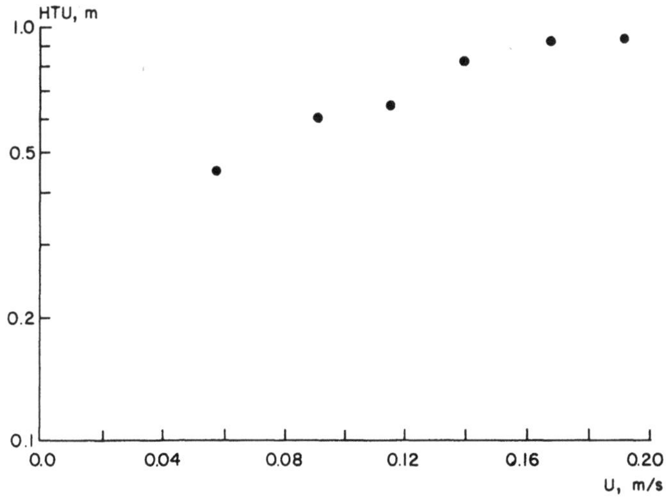

FIG. 10: DEPENDENCE OF HTU ON SUPERFICIAL GAS VELOCITY U

DATA OF VAN SWAAIJ AND ZUIDERWEG (1972)
FOR FLUIDIZED SAND BEDS (APPROX. 3 % FINES)
OF DIAMETER $D_T = 0.3$ m, $H = 1.20 - 1.35$ m AND $h^* = 0.5$ m

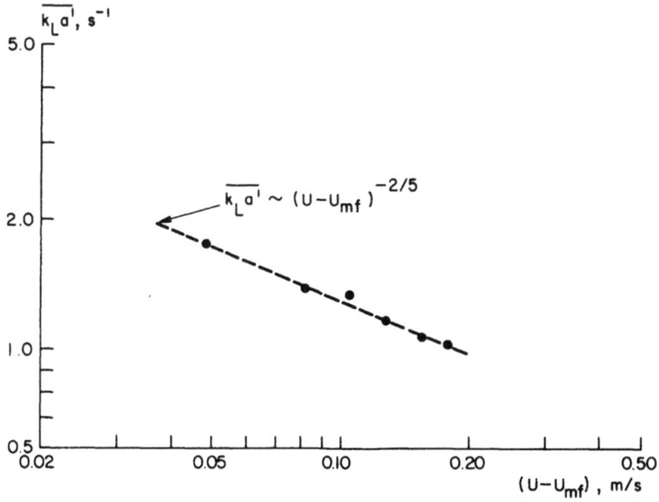

FIG. 11: DEPENDENCE OF $\overline{k_L a'}$ ON $(U - U_{mf})$ FOR FLUIDIZED SAND BED (3% FINES)

DATA FROM VAN SWAAIJ AND ZUIDERWEG (1972) $H = 1.2 - 1.35$ m
A VALUE OF $h^* = 0.5$ m WAS ASSUMED IN CALCULATING $\overline{k_L a'} = \overline{V}_b / HTU$
$U_{mf} = 0.01$ m/s (ASSUMED)

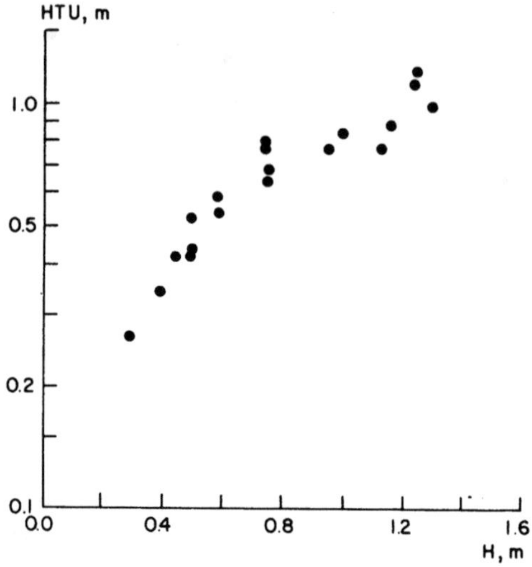

FIG.12: DEPENDENCE OF HTU ON HEIGHT ABOVE DISTRIBUTOR, H
DATA FROM VAN SWAAIJ AND ZUIDERWEG (1972) FOR FLUIDIZED SAND BED OF 0.23 m DIAMETER AND SUPERFICIAL GAS VELOCITY U = 0.15 m/s (FINES CONTENT 3–16%); $h^* = 0.5$ m

Fig. 13 shows that $\overline{k_L a'}$ varies as $H^{-0.8}$ in the region $H \leqslant h^* = 0.5$ m, as would be expected if the cross-flow term in equation (17) predominates. For $H > h^*$ the linearity of $\overline{k_L a'}$ versus $\frac{1}{H}$ (Fig. 14) supports the theoretical trend of equation (25d). It is interesting to note that, as H tends to large values:

$$\underset{H \longrightarrow \infty}{\text{Lt}} \quad \overline{k_L a'} \longrightarrow c_1 = \frac{\alpha}{h^{*\,4/5}} + \frac{\beta}{(h^* + 4\sqrt{A_o})} \qquad (29)$$

which value should be useful for performing "worst case" design calculations for tall industrial units. Equation (29) also indicates the paramount importance of determining the equilibration height h^*, or equivalently the maximum stable bubble size, $d_{b\,max}$.

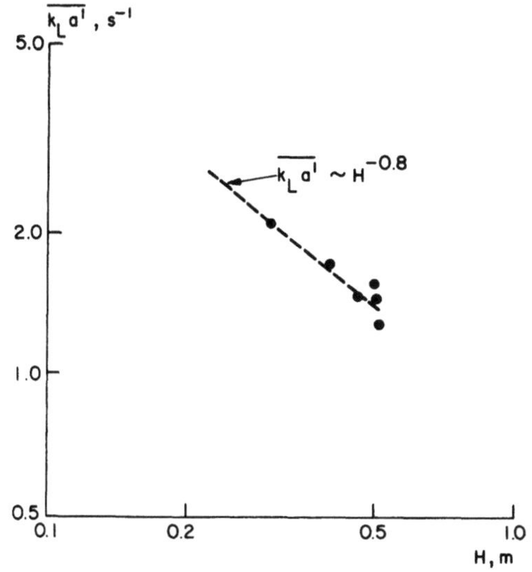

FIG. 13: VARIATION OF $\overline{k_L a'}$ WITH H FOR $H \leq h^* = 0.5$ m FOR FLUIDIZED SAND BED OF 0.23 m DIAMETER

DATA FROM VAN SWAAIJ AND ZUIDERWEG (1972) U = 0.15 m/s

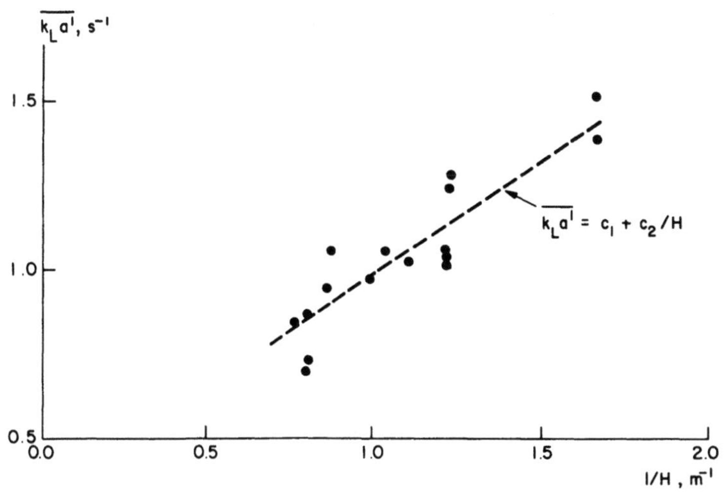

FIG. 14: VARIATION OF $\overline{k_L a'}$ WITH 1/H FOR $H > h^* = 0.5$ m

For fluidized beds of silica and commercial catalyst, the experimental data for HTU reported by De Groot (1967), Van Swaaij and Zuiderweg (1972) and De Vries et al. (1972) have been reanalysed according to the procedure of equations (26) - (28). A value of $h^* = 1.00$ m was assumed (this would vary for the different cases due to the strong influence of particle size distribution on h^*). For $H > h^*$ the data are plotted as $\overline{k_L a'}$ versus $\frac{1}{H}$, as in Fig. 15. The spread in the data is to be attributed to

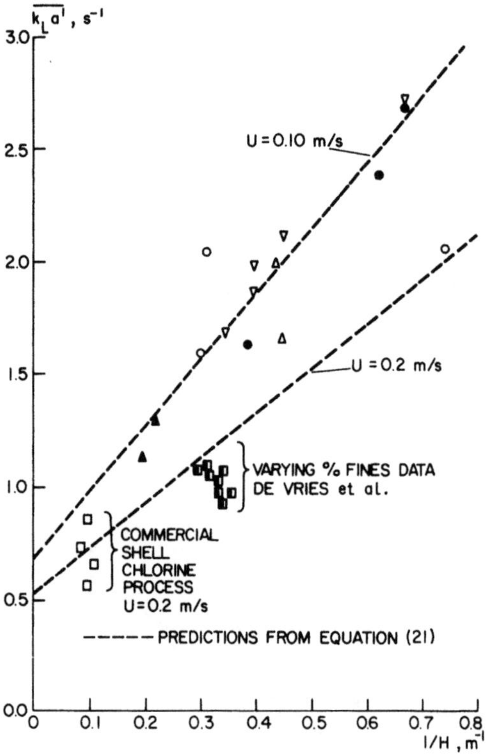

FIG. 15 : VARIATION OF $\overline{k_L a'}$ WITH 1/H FOR FLUIDIZED SILICA BEDS ALSO SHOWN ARE PREDICTIONS OF DAVIDSON - DARTON MODEL

$h^* = 1.0$ m ; $A_0 = 56$ mm^2 ;
$U_{mf} = 0.01$ m/s ; $\epsilon_{mf} = 0.5$;
$D_g = 0.2 \times 10^{-4}$ m^2/s)

○ 0.3 m DIA, △ 0.6 m DIA, ▲ 1.5 m DIA } DE GROOT
▽ 0.3 m DIA, ● 0.6 m DIA } VAN SWAAIJ & ZUIDERWEG
◧ 0.6 m DIA, □ 2.9 m DIA } DE VRIES et al.

varying U and varying percentage of fines (affecting both U_{mf} and h^*).

To test the quantitative predictions of the Davidson-Darton model, calculations of $\overline{k_L a'}$ according to equations (21) - (24) were made for the following chosen parameter values:

$U = 0.1,\ 0.2$ m/s; $h^* = 1.0$ m; $A_o = 56$ mm^2 (valid for porous plate distributors); $D_G = 0.2 \times 10^{-4}$ m^2/s; $U_{mf} = 0.01$ m/s; $\varepsilon_{mf} = 0.5$.

The most important parameters are U_{mf} and h^*. In the classical two-phase model the interstitial flow is assumed to be U_{mf}. There is some evidence to suggest that the actual flow could be much higher than U_{mf} and this would of course lead to an increase in $\overline{k_L a'}$.

The calculations of the Davidson-Darton model for the above chosen parameters are also presented in Fig. 15. By choosing data appropriate to each experiment a much better fit may be expected. Still, it can be seen that the Davidson-Darton model not only predicts the right trends with variations in diameter, gas velocity and height, but also yields the right magnitudes for the transfer coefficient $\overline{k_L a'}$.

It is also heartening to note that data for the commercial scale Shell Chlorine Process reactor, operating at $U \approx 0.20$ m/s, can be predicted by the Davidson-Darton model (Fig. 15). The variation in HTU for the four different runs reported by De Vries et al. (1972) (see Table IV of the paper by De Vries) can be explained by the fact that the varying fines content alters h^*. The De Vries data are given in Fig. 16, along with predictions of the Davidson-Darton model for $h^* = 0.5,\ 1.0,\ 1.5,\ 2.0$ m. The theoretical HTU values were calculated from the relation:

$$HTU = \frac{H}{NTU} = \frac{H}{\int_0^H \frac{k_L a' \, dh}{V_b}} \tag{30}$$

utilizing equations (4), (6), (7) and (8).

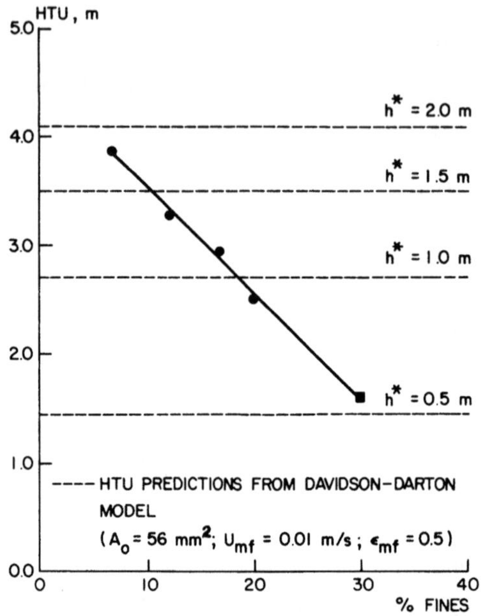

FIG.16: INFLUENCE OF FINES ON HTU IN COMMERCIAL SHELL CHLORINE PROCESS REACTOR ($D_T = 2.9$ m ; $H \approx 10$ m)

DATA OF DE VRIES et al. THEORETICAL PREDICTIONS OF DAVIDSON – DARTON MODEL ARE ALSO SHOWN

The data in Fig. 16 show that the effect of fines addition is to lower h^* and thus increase $\overline{k_L a'}$ and lower HTU. Werther (1977) reaches the same conclusions in his analysis. While Werther has fitted the experimental data with h^* as a variable, we have in the present work retained the physical model till the very end to reach the same conclusions as Werther. Our analysis here gives more information and insight into fluid bed transfer processes and provides us with a scale-up procedure of mass transfer coefficient. It follows from Fig. 15 that for engineering scale-up studies $\overline{k_L a'}$ must be measured in a sufficiently large diameter bed (to avoid slugging) for different heights. A plot of $\overline{k_L a'}$ versus $\frac{1}{H}$ should yield a straight line, which can be extrapolated to higher H-values corresponding to the industrial situation. Care must be taken to ensure that measurements are made with $H > h^*$. For easily fluidizable powders such as silica (with fines addition), $H > 3$ m would usually satisfy this condition.

The techniques for measuring $\overline{k_L a'}$ are adequately discussed in the fluid bed literature and include gas tracing, measurement of conversion of model reaction, and single gas bubble injection.

Effect of Gas Distributor on Interphase Mass Transfer

In the Davidson-Darton model for interphase mass transfer the effect of the method of gas distribution is accounted for in the term A_o in equation (4). This term determines the size of the initial bubbles formed. In experimental units porous plate distributors are common and for this case Darton (1977) has calculated a value $A_o = 56\ mm^2$ from bubble size data fitting. In Fig. 15 this value $A_o = 56\ mm^2$ was used in the prediction calculations for the large scale (2.9 m diameter) industrial Shell chlorine Process reactor. The agreement with the industrial reactor (in which nozzles were employed; see De Vries et al., 1972) data with extrapolated data obtained from units with porous distributor plates is somewhat surprising and could be attributed to enhanced mass transfer from gas jets and break-down of the two-phase theory in this region (see Werther, 1977).

Further work on mass transfer in the region of the distributor may clear this anomaly. For scale-up purposes, industrial type gas distributors (nozzles, bubble caps etc.) could be used in the experimental unit.

Mass Transfer Enhancement due to Fast Reactions

For fast reactions there may be enhancement of the interphase transfer coefficient k_L due to steepening of the composition gradients in the boundary region. The classical Hatta model for mass transfer with chemical reaction was used by Werther (1978) to explain the decrease in HTU with reaction rate as reported by

Van Swaaij and Zuiderweg (1972) (see Fig. 17).

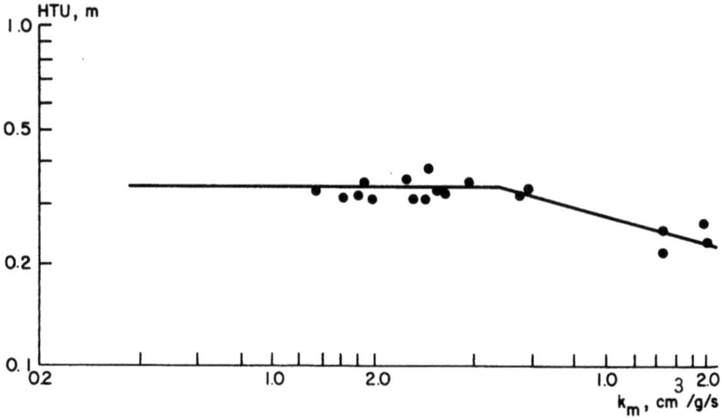

FIG. 17: DEPENDENCE OF HTU ON THE REACTION RATE CONSTANT FOR OZONE DECOMPOSITION IN FLUIDIZED SILICA BED OF DIAMETER D = 0.23 m

OTHER DETAILS : H ≈ 1 m ; U = 0.12 m/s

Use of the enhancement factor:

$$E = \frac{Ha}{\tanh Ha} \tag{31}$$

where Ha is the Hatta number:

$$Ha = \sqrt{k_m \rho_p (1-\varepsilon_{df}) D_G} / k_L \tag{32}$$

enables the measured data for $\frac{c}{c_o}$ (i.e. fraction unreacted gas) to be quantitatively predicted as seen in Fig. 18.

It must be remarked that for reactions fast enough to lead to a mass transfer enhancement factor E significantly greater than unity, the gas/solid contact zone in the distributor region will become of great importance (small bubbles, high mass transfer). Also, for this case, the bed height required will not be large and the coalescence region $h = 0 - h^*$ will be important. A bubble growth model will be absolutely indispensable here.

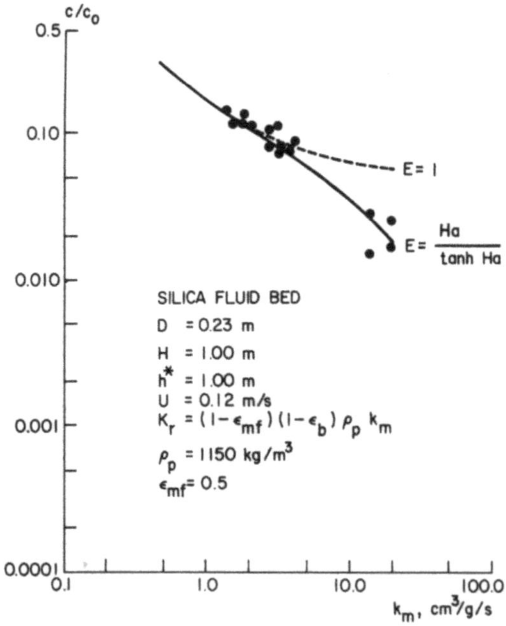

FIG. 18 : DATA OF VAN SWAAIJ AND ZUIDERWEG (1972) FOR OZONE CONVERSION IN FLUIDIZED SILICA BED (D_T = 0.23) INTERPRETED BY WERTHER (1978) USING A HATTA MODEL FOR MASS TRANSFER ENHANCEMENT ; E IS THE ENHANCEMENT FACTOR

Effect of Increased Pressure and Temperature

The gas density and viscosity are both altered under conditions of increased pressure and/or temperature above ambient conditions. Experimental results of various workers presented at the recent fluidization conference at Cambridge (see Davidson and Keairns, 1978) suggest that increased pressure leads to "smoother" fluidization and that the bubbles tend to be smaller and the dense-phase expansion (ε_{df}) higher than under atmospheric pressure conditions. The flow of gas in the dense phase U_{df} is consequently higher than under atmospheric pressure conditions. This increased dense-phase gas flow must of course be taken into account in the mathematical modelling (see below).

In view of the vastly increased interest in high pressure and high temperature fluidization, due to applications to coal conversion processes (e.g. fluid bed gasification, combustion etc.), we may expect a clear picture of the effects of P and T on bubble mechanics and mass transfer to emerge in the near future.

Building a high pressure "cold" engineering experimental unit of large diameter (say, > 1 m) may be prohibitively expensive. An alternative is to use a "model" gas to simulate high P, T conditions. Mobil Research and Development Corporation (Quarterly Report, January-March 1977, ERDA Contract EX-76-C-01-2276) choose model gases SF_6 and Freon-22 to simulate (ρ_g/μ_g) of gas under reaction conditions in the process for direct conversion of synthesis gas to high-octane gasoline. Further theoretical, and experimental, work should reveal whether (ρ_g/μ_g) is the appropriate parameter to mimic hydrodynamic and mass transfer processes.

Effect of Internals

In deep beds involving exothermic, but slow, catalytic chemical reaction vertical tubes may have to be installed to provide cooling. Horizontal heat exchange surfaces are constructionally suitable only for shallow beds, as for example in fluid bed combustion.

The effect of vertical tubes on bubble mechanics and mass transfer is a subject where some contradictory data are present in the literature (see Volk. et al., 1962; Van Swaaij and Zuiderweg, 1973; Botton, 1970).

Botton (1970), who has presented extensive results in beds of diameters 0.065, 0.16, 0.25, 0.5 and 1.8 m for three types of solids, has concluded that for fine, easily fluidizable powders the effect of internals on HTU is negligible. He rejects the "equivalent diameter" concept put forth by Volk et al. (1962).

The fundamental question to be answered is whether vertical inserts alter bubble size or not. With the increased amount of experimental work in this area, we may expect to get a clear answer in the near future.

Dense Phase Backmixing

In the literature there are data on solids mixing (assumed here to be equal to dense gas phase mixing) in fluid beds up to 2.9 m diameter (De Vries *et al.*, (1972). The axial dispersion coefficient $\mathbb{D}$ should be independent of the reactor height, except for shallow beds with $H < D_T$. Using the Kolmogoroff theory of isotropic turbulence, Baird and Rice (1975) showed that the axial dispersion coefficient for sparged columns, spray towers and gas fluidized beds could be correlated in the form

$$\mathbb{D} = \text{const} \ \ell^{4/3} \ P_m^{\ 1/3} \tag{33}$$

where ℓ is characteristic length parameter and P_m is energy dissipation rate per unit mass.

For gas fluidized beds the dense phase mixing is caused by energy input from the gas phase:

$$P_m = g \ U \tag{34}$$

and the length of the largest eddy may be taken as the diameter D_T (provided $H > D_T$):

$$\ell = D_T \tag{35}$$

giving

$$\mathbb{D} = \text{constant} \ (g \ U)^{1/3} \ D_T^{\ 4/3} \tag{36}$$

Baird and Rice (1975) have found the value of the constant to be 0.35. For example, in a reactor of diameter 2.9 m operating at a superficial gas velocity U = 0.175 m/s the coefficient $\mathbb{D}$ is calculated as 1.7 m^2/s. De Vries *et al.* (1972) report an experimentally measured value of 1.5 m^2/s for a commercial reactor operating under these conditions. It appears that equation (36) will be adequate for scaling-up purposes.

The effect of backmixing on the conversion achieved in a reactor is governed by the "number of mixing units", NMU, defined by

$$\mathrm{NMU} = \frac{U\,H}{f\,\mathbb{D}} \tag{37}$$

where f is the fraction of the reactor occupied by the gas in the dense phase. This effect must be discussed in connection with a mathematical model of the reactor.

Mathematical Modelling of Fluidized Bed Reactors

There is a vast amount of literature on this subject, with each author favouring his own model; see the extensive reviews of Potter (1978) and Werther (1978). A realistic, and practical, approach to modelling is to use the two-phase model (May, 1959; Ven Deemter, 1961) and at the same time account for variation of the hydrodynamic and mass transfer parameters along the reactor height. This approach has been used in this paper to develop our scale-up strategy.

If we assume:

(i) solid-free "bubble" phase (superscript ') in plug flow, and

(ii) axially dispersed "dense" phase (superscript "), then the differential equations for the mole fraction x_i of the component i, which undergoes a first-order reaction, are given by

$$\frac{dx_i'}{d\xi} + \frac{1}{\nu}\,\mathrm{NTU}\,(x_i' - x_i'') = 0 \tag{38}$$

and

$$\frac{d^2x_i''}{d\xi^2} - \mathrm{NMU}(1-\nu)\,\frac{dx_i''}{d\xi} - \mathrm{NMU}\,(\mathrm{NTU} + \mathrm{NRU})\,x_i'' + \mathrm{NMU}\,\mathrm{NTU}\,x_i' = 0 \tag{39}$$

where, additionally, ν is the fraction of the incoming gas which travels up the reactor in the bubble phase; $(1-\nu)$ of the incoming gas goes up in the dense phase; $\xi = h/H$ is the fractional distance parameter along the reactor height (inlet: $\xi = 0$; outlet: $\xi = 1$).

The solution to equations (38) and (39), for the appropriate boundary conditions, will yield information on the composition profiles along the reactor, from which the conversions may be calculated.

It must be of course appreciated that ν, NMU, NTU, NRU all vary in the region $0\text{-}h^*$. If integral averaged (constant) values of $\bar{V}_b$, $\bar{\varepsilon}_b$, $\bar{\varepsilon}_{df}$, $\overline{k_L a'}$ etc. are used over the total reactor height, then the differential equations (38) and (39) may be solved analytically.

Fig. 19 presents some calculations to show the effect of dense phase backmixing on $1-\chi$, where χ is the fraction of incoming reactant gas that is converted in the reactor. It is clear that for NMU < 3.0 the effect of backmixing is important. This would be true of shallow beds of $H/D_T \approx 1$. On the other hand, if $H/D_T > 3$, the NMU value will usually exceed 5, approaching plug flow.

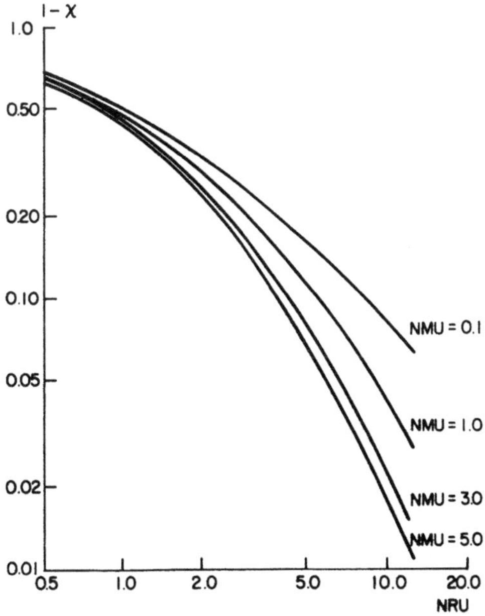

FIG.19: FRACTIONAL UNCONVERTED GAS LEAVING THE REACTOR AS A FUNCTION OF NRU, THE NUMBER OF REACTION UNITS, AND NMU, THE NUMBER OF MIXING UNITS

NTU = 4.0; ν = 1.0

The backmixing then ceases to be of great importance.

Another important parameter, especially for high pressure operations, is the effect of the contribution of dense phase gas flow on conversion. Usually the assumption is made that, for high operating gas velocity $U > 5\ U_{mf}$, the fraction of incoming gas taken up by the bubble phase, ν, is unity. For high pressure operation with high dense gas flow the conversion achieved in the dense phase will not be negligible. Fig. 20 gives the effect of $(1-\nu) = 0.4$ and shows a large dense phase contribution, to the conversion, especially at high reaction rates.

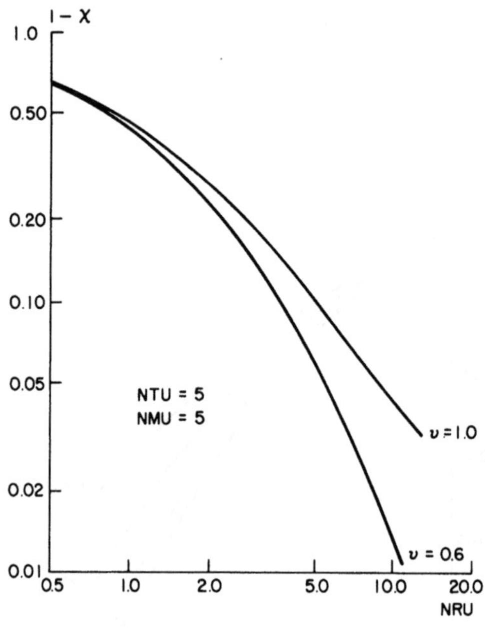

FIG.20: FRACTIONAL UNCONVERTED GAS LEAVING THE REACTOR, 1–X AS A FUNCTION OF THE NUMBER OF REACTION UNITS, NRU, AND ν, THE FRACTION OF INCOMING GAS TRANSPORTED ALONG THE REACTOR BY THE BUBBLES $(1-\nu)$ IS DENSE PHASE FLOW FRACTION

Most mathematical models for fluid bed reactors assume isothermal conditions, not often with justification. Thus, for very tall fluid beds, carrying out exothermic reactions, there may exist temperature gradients along the reactor height (thermal dispersion may be important here) and there may also be temperature differences between the bubble phase and the dense phase at any given height H. The effects of temperature gradients are to alter the physical, thermodynamic and transport properties. Since molar concentrations of any component i, c_i ($\equiv x_i P_t/RT$), depend on the temperature, the commonly used driving force, Δc_i, will be altered by the temperature gradients. The correct way in which to model non-isothermal conditions is to use mole fractions and mole fraction driving forces, as has been done in equations (38) and (39). Use of mole fractions avoids the "hot radiator paradox" of Sherwood, Pigford and Wilke (1975). In fluidized bed gasification and combustion applications the correct application of mole fraction driving forces is especially important because of the existence of large temperature gradients in the boundary layer surrounding a particle.

Concluding Remarks

In this paper we have considered the problem of scaling up gas fluidized bed chemical reactors, especially for cases in which large diameter columns and deep beds are involved. We have discussed criteria for selecting proper dimensions for a "cold" experimental unit for obtaining hydrodynamic and mass transfer scale-up information. By developing a combined Davidson-Darton model for predicting bubble sizes, bubble rise velocities and interphase transfer coefficients we have identified key physical parameters such as the bubble hold-up (ε_b), equilibration height h^*, and mass transfer coefficient per unit bubble volume, $\overline{k_L a'}$. Suitable procedures for estimating, or measuring, these parameters have been discussed at appropriate points in our discussions.

It has been shown that measured experimental data on columns with D_T ranging from 0.1 to 3 m and H varying from 0.5 to 10 m, with fluidized sand and silica beds can be interpreted in terms of the Davidson-Darton model, thus confirming the scale-up potential of the theory.

For example, the Van Swaaij and Zuiderweg (1973) correlation for HTU, equation (15), can be put on a fundamental basis. Fig. 21 shows the dependence of HTU on H and D_T for a gas velocity U = 0.15 m/s, as predicted from the theory. This graph shows similar trends as those given in Fig. 7. The Van Swaaij and Zuiderweg correlation can be seen to be more conservative for $D_T > 5$ m. For new processes involving high pressures and temperature, and with catalysts having properties different from those of silica, a physically based scale-up procedure, as discussed here, will be required.

With the information from the "cold" experimental unit and information on chemical kinetics, the parameters NTU, NRU, NMU and ν of the two-phase theory can be estimated and conversion calculations performed by solution of the appropriate differential equations (38) and (39). The information obtained in the pilot plant may help to further test the accuracy of model predictions.

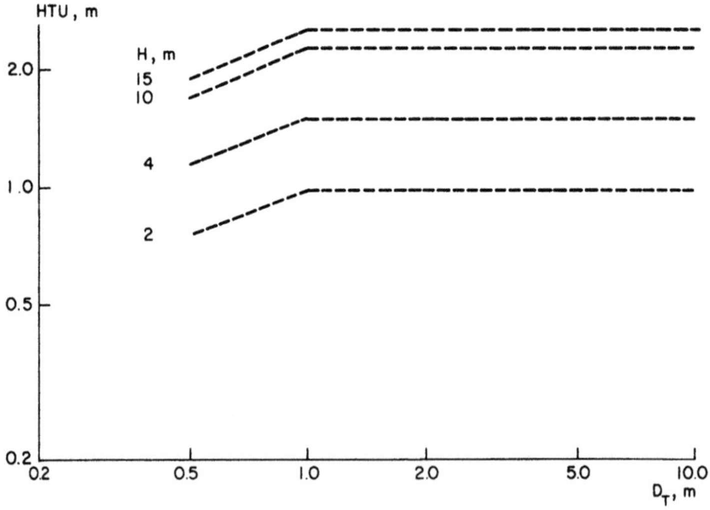

FIG. 21 : DEPENDENCE OF HTU ON COLUMN DIAMETER D_T AND HEIGHT H

PREDICTIONS MADE BY DAVIDSON–DARTON MODEL WITH U = 0.15 m/s; U_{mf} = 0.01 m/s; ϵ_{mf} = 0.5; A_o = 56 mm^2 AND h^* = 1.0 m

Fig. 22 summarizes the overall scale-up procedure for a commercial reactor.

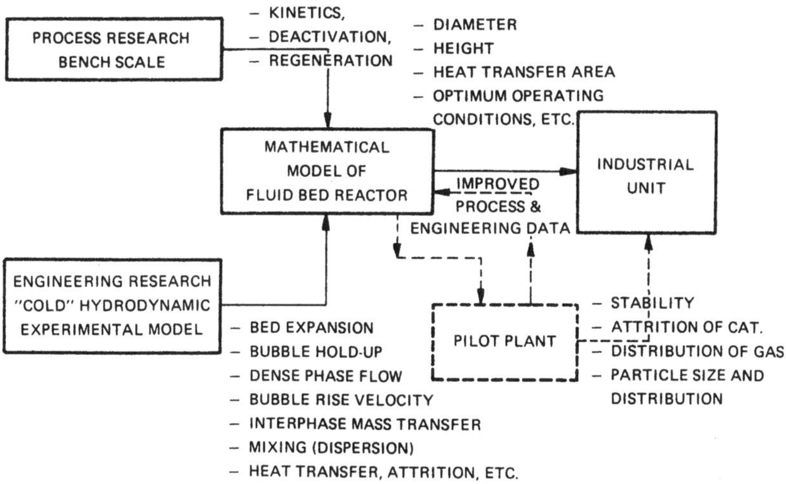

FIG. 22: SCALE-UP STRATEGY FOR
GAS FLUIDIZED BED CHEMICAL REACTOR

LIST OF SYMBOLS

a	interfacial area between the bubble phase and the dense phase, per unit expanded fluidized bed reactor volume	m^2/m^3
a'	interfacial area between the bubble phase and the dense phase, per unit volume of gas dispersed in the form of bubbles in the fluidized bed. $a' = \varepsilon_b\, a$, where ε_b is the bubble phase hold-up	m^2/m^3
A	species A in the reaction sequence: $A \rightarrow$ Products	-
A_o	"catchment area" of the bubble stream at the distributor plate, which characterizes different distributors, and is usually the area of plate per orifice	m^2
A_r	net free cross-sectional area of the fluidized bed reactor available for flow of phases	m^2
c	molar concentration of reacting species in the gas mixture	$kmol/m^3$
c_o	molar concentration of the reacting species entering the fluidized bed reactor	$kmol/m^3$
c_i	molar concentration of species i in the gas mixture	$kmol/m^3$
c_1	constant defined by equations (21) and (25d)	-
c_2	constant defined by equations (21) and (25d)	-
Δc_1	molar concentration difference driving force	$kmol/m^3$
d_b	diameter of a sphere having the same volume as a bubble in the fluidized bed. d_b varies with the height of bed, h	m
$\overline{d_b}$	integral height-averaged value of the bubble diameter over a height H of fluid bed	m

D_T	diameter of fluidized bed column (tower)	m
$\mathbb{D}$	axial dispersion coefficient describing the backmixing characteristics of the dense phase gas	m^2/s
D_G	molecular diffusion coefficient for diffusion of reactant species A in the gas mixture	m^2/s
E	enhancement factor describing mass transfer acceleration due to fast chemical reaction in the dense phase	-
f	the fraction of the reactor occupied by the dense phase gas. $f = (1 + \varepsilon_{df} - \rho_b/\rho_p)/(1 + \varepsilon)$, where ε and ε_{df} are the total bed and dense phase expansions, respectively	-
g	acceleration due to gravity	9.81 m/s^2
h	height along the fluidized bed, measured from the distributor	m
h^*	height above the distributor at which the bubbles reach their equilibrium stable size	m
H	total fluidized bed reactor expanded height	m
Ha	Hatta number defined by equation (32)	-
HTU	height of mass transfer unit, describing the ease (or more strictly, the difficulty) of interphase mass transfer	m
H_o	settled bed height	m
k_L	mass transfer coefficient for interphase (bubble-dense phase) transfer	m/s
$k_L a$	volumetric mass transfer coefficient per unit reactor volume	s^{-1}
$k_L a'$	volumetric mass transfer coefficient per unit volume of gas dispersed in the form of bubbles	s^{-1}

k_m	first order reaction rate constant expressed per unit kg of catalyst	$m^3.kg^{-1}.s^{-1}$
K_r	first order reaction constant expressed per unit volume of expanded fluidized bed volume	s^{-1}
ℓ	characteristic length scale	m
NMU	number of mixing units	-
NRU	number of reaction units	-
NTU	number of mass transfer units	-
P_m	energy dissipation rate per unit mass	m^2/s^3
P_t	total pressure in the system	N/m^2
Q	interfacial flow of gas from bubble to dense phase	m^3/s
R	gas constant	8314.4 $J.kmol^{-1}.K^{-1}$
T	system absolute temperature	K
U	superficial gas velocity. In the presence of internals this velocity is based on the net free cross-sectional area available for flow	m/s
U_{mf}	superficial gas velocity corresponding to minimum fluidization conditions	m/s
V_b	rise velocity of the bubble phase	m/s
$\overline{V_b}$	height averaged rise velocity of the bubble phase over the total fluid bed height H	m/s
x_i	mole fraction of component i in the gas mixture	-
x_i'	mole fraction of component i in the bubble phase	-
x_i''	mole fraction of component i in the dense phase	-

Greek Letters

α	parameter defined by equation (23)	
β	parameter defined by equation (24)	
ε	bed expansion $(= (H_{exp} - H_o)/H_o)$	-
ε_b	hold-up of bubble phase expressed as volume of bubble phase per unit volume of fluid bed	-
ε_{mf}	bed voidage at minimum fluidization (= f at $U = U_{mf}$)	-
ε_{df}	dense phase expansion	-
μ_g	viscosity of gas phase	Ns/m^2
ν	fraction of gas which flows in the bubble phase	-
ξ	fractional height parameter along fluid bed, $\xi = h/H$	-
ρ_g	gas density	kg/m^3
ρ_b	bulk density of catalyst	kg/m^3
ρ_{fb}	kg catalyst per unit expanded fluid bed volume	kg/m^3
ρ_p	particle density of the catalyst	kg/m^3
ϕ	constant defined by equation (8)	-
χ	fractional gas conversion	-

Superscripts

'	per unit dispersed bubble phase volume
'	bubble phase parameter
"	dense phase parameter
—	overbar denoting integral height averaged value

Subscripts

A species A parameter

b referring to the bubble phase

c-e cloud-emulsion transfer

g gas phase parameter

G gas phase parameter

mf minimum fluidization parameter

df dense phase parameter

REFERENCES

H.F. Bach, "Einfluss der Konstruktions-, Betriebs- und Stoffgrössen auf die Fluiddynamik und damit die Funktion von Blasensäulen", Ph. D. Thesis, Technische Universität, München, 1977.

H.M.I. Baird and R.G. Rice, Chem. Eng. J., 9, 171 (1975).

R.J. Botton, Chem. Eng. Progr. Symposium Series No. 101 66, 8 (1970)

R.C. Darton, et al., Trans. Inst. Chem. Engrs, 55, 274 (1977).

R.C. Darton, Trans. Inst. Chem. Engrs. 57, 134 (1979).

J.F. Davidson, et al., in L. Lapidus and N.R. Amudson (eds). Chemical Reactor Theory, A Review, Chapter 10, Prentice-Hall, Englewood Cliffs, New Jersey (1977).

J.F. Davidson and D.L. Keairns (eds), Fluidization, Proceedings of the Second Engineering Foundation Conference, 1978.

J.J. van Deemter, Chem. Eng. Sci., 13, 143 (1961).

J.H. de Groot, Proceedings of the Symposium on Fluidization, Eindhoven, 1967, p. 348.

D. Kunii and O. Levenspiel, Fluidization Engineering, Wiley, New York (1969).

W.G. May, Chem. Eng. Prog. 55, 49 (1959).

O.E. Potter, Catal. Rev. - Sci. Eng. 17, 155 (1978).

P.N. Rowe, Chem. Engng. Sci. 31, 285 (1976).

T.K. Sherwood, R.L. Pigford and C.R. Wilke, Mass Transfer, McGraw-Hill (1975), p. 15.

W.P.M. van Swaaij and F.J. Zuiderweg, Proceedings of the Fifth European/Second International Symposium on Chemical Reaction Engineering, Amsterdam, 1972.

W.P.M. van Swaaij and F.J. Zuiderweg, Proceedings of the International Symposium on Fluidization and its Applications, Toulouse, 1973.

D.J. Vermeer and R. Krishna, the paper presented at the 72nd Annual AIChE meeting, San Francisco, November 25-29, 1979.

W. Volk, C.A. Johnson and H.H. Stotler, Chem. Eng. Progr. 58, 44 (1962).

R.J. de Vries, *et al*., Proceedings of the Fifth European/Second International Symposium on Chemical Reaction Engineering, Amsterdam, 1972.

J. Werther, Chem. Ing. Tech., 49, 777 (1977).

J. Werther, Chem. Ing. Tech., 50, 850 (1978).

SCALE-UP CONSIDERATIONS FOR MIXED PHASE CATALYTIC REACTORS

R. M. Koros

Exxon Research and Engineering Company
Florham Park, New Jersey

1.0 SCALING STRATEGIES

The interactions of physical factors -- catalyst contacting, vapor/liquid equilibrium, and mass and heat transfer -- with the reactions commonly carried out in multi-phase reactors are not well quantified. Because of this difficulty, a priori design of multiphase reactors can seldom be done; therefore, the design engineer depends on pilot plant or other test data for guidance. One of the important tasks of the design engineer is to choose a scaling strategy. His choice will depend on a number of factors, such as:

- The process involves new chemistry.
- The chemistry involves an extension of the known. For example, the conversion of a commercial reactor must be increased from 90 to 95%.
- The chemistry is amenable to parallelism. For example, the physical and chemical characteristics of the feed material are similar to that in an existing commercial reactor.

To focus our discussion on multiphase reactor design and scale up, we will deal with the most complex situation in which the process involves new chemistry. Further, we will concentrate on only one multiphase reactor type: the trickle bed reactor. In this reactor, vapor and liquid flow concurrently downward through a fixed bed of catalyst. The trickle bed reactor serves as a good vehicle to discuss the universal concerns in the design and scale up of other multiphase reactors such as co-current upflow fixed beds, countercurrent fixed beds, mixed phase moving beds, gas-liquid fluidized beds, and slurry beds.

Three-phase reactor systems share the following relationships that must be considered in scale up and design:

1. The reactants must transfer from the gas to the liquid phase.
2. The reactants must transfer from the liquid to the solid catalyst phase.

3. The reactants may transfer directly from the gas to the solid catalyst phase.

4. The liquid phase reacts on the catalyst surface.

5. The gaseous reactants may directly react on the catalyst surface.

6. Diffusion may limit the gaseous reactants within the catalyst pores.

7. The liquid reactants may be diffusion limited in the catalyst pores.

8. The heat of reaction has to be dissipated from the catalyst into the liquid and/or gaseous stream.

9. The gaseous and liquid reactants must be effectively contacted with the catalyst.

10. The reactant gases and liquids must be transported through the reactor in a predictable fashion, i.e., plug flow, back mixed.

11. The partition of the gas, liquid and catalyst holdups in the reactor may be critical.

12. The distribution pattern of the gas, liquid and catalyst within the reactor volume in the radial and axial direction may be critical.

The designer's task is to determine which of these effects are important for his reaction and how the controlling effects scale from a practically sized pilot plant reactor to a commercial reactor. The principal difference between a trickle bed reactor and the other reactor types are the flow patterns and the phenomological relationships that allow a designer to quantify the relationships listed above. The metholology of diagnosing the important factors in the operation of the reactor in question is similar for each of the reactor types.

The focus on trickle beds was selected because this reactor type is industrially important and is the subject of a large body of literature. Trickle beds have the advantage of simplicity and high capacity. Cocurrent downward flow of gas and liquid through packing is only limited by the pressure drop that one is willing to expend on flowing through the bed and perhaps by the crushing strength of the catalyst. Also, as discussed later, trickle beds behave essentially as plug flow reactors, which is advantageous for typical reactions carried out in trickle beds.

1.1 The Basic Design Equations

Most reactions of commercial interest tend to be described by either first order or second order kinetics. Fortunately, for the reactor engineer, most hydrotreating and other reactions of commercial interest are well described by first order kinetics or by a combination of first order reactions. For first order kinetics, plug flow will give the highest conversion possible. Plug flow and full catalyst utilization are the ideals that a design engineer should strive for.

The mass balance for a first order reaction in plug flow gives a very simple expression in terms of the apparent first order reaction rate, k_{app}; the space velocity, SV, defined as the flow of reactants, F, divided by the amount of catalyst, W, in the reactor; and the inlet, C_i, and outlet, C_o, concentration of the reactant.

$$k_{app} = \frac{F}{W} \ln \frac{C_i}{C_o} = SV \ln \frac{C_i}{C_o} \qquad (1.1\text{-}1)$$

Since mass transfer processes are also first order, much of the modelling and analyses of real reacting system lend themselves to much simplification and tractable mathematical modelling. For example, for first order systems, the apparent kinetic constant, k_{app}, can be defined as an intrinsic kinetic constant, k_i, multiplied by an effectiveness factor. This effectiveness factor in the simplest situation can take into account possible diffusion resistances of the reactants within the catalyst pores. In more complex situations, it can also take into account the mass transfer resistance and contacting resistances from the flowing reactants into the catalyst. This k_{app} can then be considered as a lumped kinetic constant and used for scale up and design.

In many situations, there is not enough information to be able to calculate these various effectiveness factors, so that one must rely on experimental data and diagnostics to discover the magnitude of each of the factors affecting k_{app}. We will now focus on how in a trickle bed reactor design, we can evaluate this k_{app} and how it can be used for scale up and design. To attach physical significance to the k_{app}, we will use mathematical models which reflect perceptions of the realities of the flows within the reactor. As we shall see, because the flows in trickle beds are complex, a complete mathematical model would also be complex, but reasonable simplifications can make the mathematical models tractable. Useful diagnostics and criteria can be based on these simplfied models.

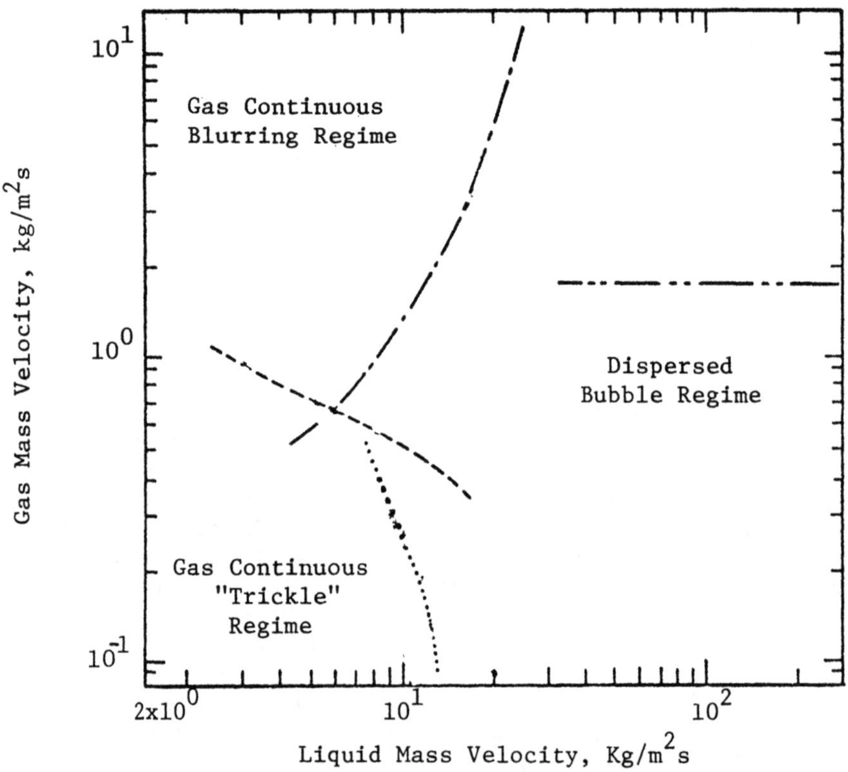

Figure 1 Flow Regimes in Multiphase Flow in Fixed Beds

2.0 FLOW DYNAMICS

In spite of the fact that mixed-phase downflow fixed bed reactors are commonly referred to as trickle bed reactors, the "trickle" flow regime, i.e., gas continuous, corresponds to only one of several possible flow regimes. The other flow regimes found in a mixed-phase downflow fixed bed reactor are bubble flow, pulsed flow, spray flow and slug flow, much the same as in two-phase flow in pipes. Various studies (1-8) have identified the boundaries between these flow regimes and have suggested correlating parameters to define the boundaries between these regimes. Each of these regimes exhibit their own characteristic patterns for holdup, pressure drop, mass transfer and heat transfer.

2.1 Flow Maps

In general, at low gas and liquid flow rates, the flow will be in the gas continuous channeled regime. With increasing liquid flow rate, the flow will shift to pulsed and then to dispersed bubble flow. With increasing gas rate, the flow pattern transitions are more complex. At low liquid rate, the pattern changes from gas continuous channeled to gas continuous blurring flow. At an intermediate liquid rate, the flow changes from gas continuous channeled flow, to pulsed flow and then to gas continuous blurring flow. At higher liquid rates, the pattern can shift from dispersed bubble to pulsed flow. Flow maps such as that presented by Sato (6) and by Shah (38) show these transitions. (Figure 1)

Talmor (4) proposed flow map coordinates for both foaming and non-foaming systems consisting of ratios of inertial and gravity forces to interfacial and viscous resistances (evaluated in terms of the Froude, Reynolds and Weber numbers). These maps (Figures 2 and 3) encompass data for a large range of flow rates, packing properties and fluid systems for columns of 3.4 cm to 29 cm. Foaminess was measured by the time of collapse of the foam produced by shaking. The principal difference between the two maps are the coordinate positions for the transitions between the various regimes.

One of the weaknesses of these maps is the arbitrariness of the measurement of foaminess. A foamy liquid was taken to be that which had a time of collapse of 3 seconds or more. This is clearly an arbitrary choice -- both of the method of measurement and criteria. However, a distinction between foaming and non-foaming fluids is necessary, because the fluid dynamic behavior is different for the two situations. A better definition is certainly necessary for a priori design.

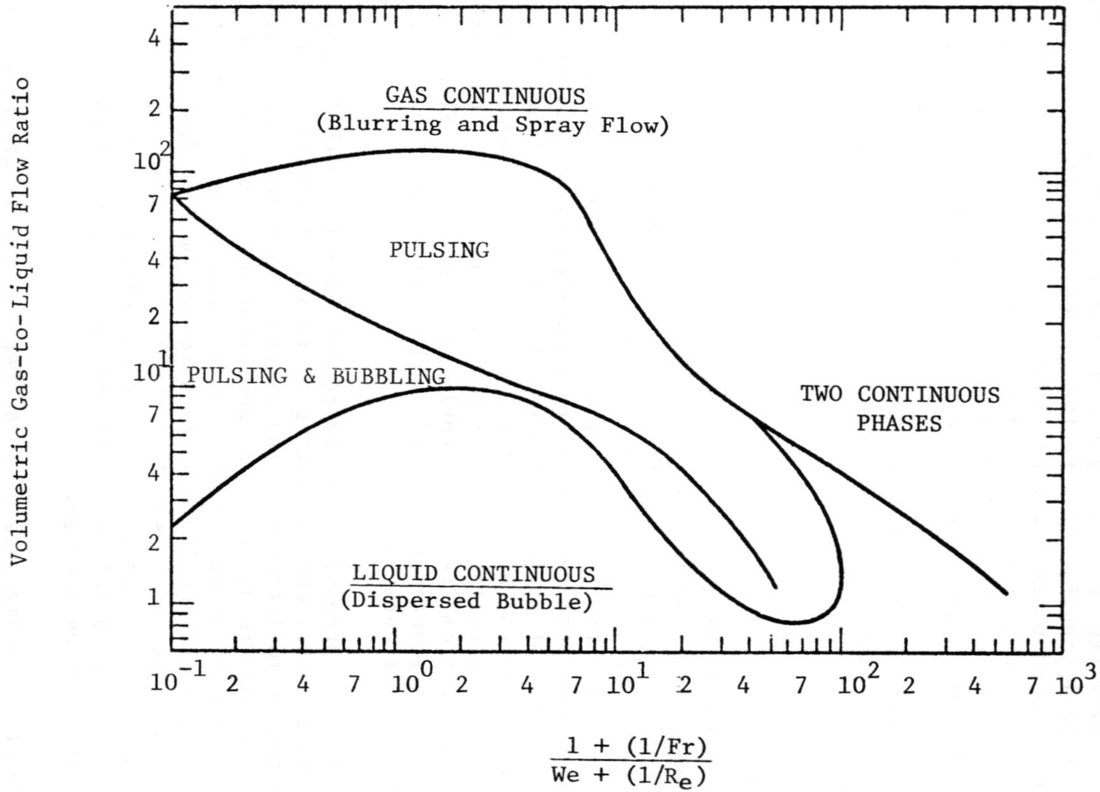

Figure 2 Flow Map after Talmor[4] for Non-Foaming Liquids in Two-Phase Flow in Fixed Beds

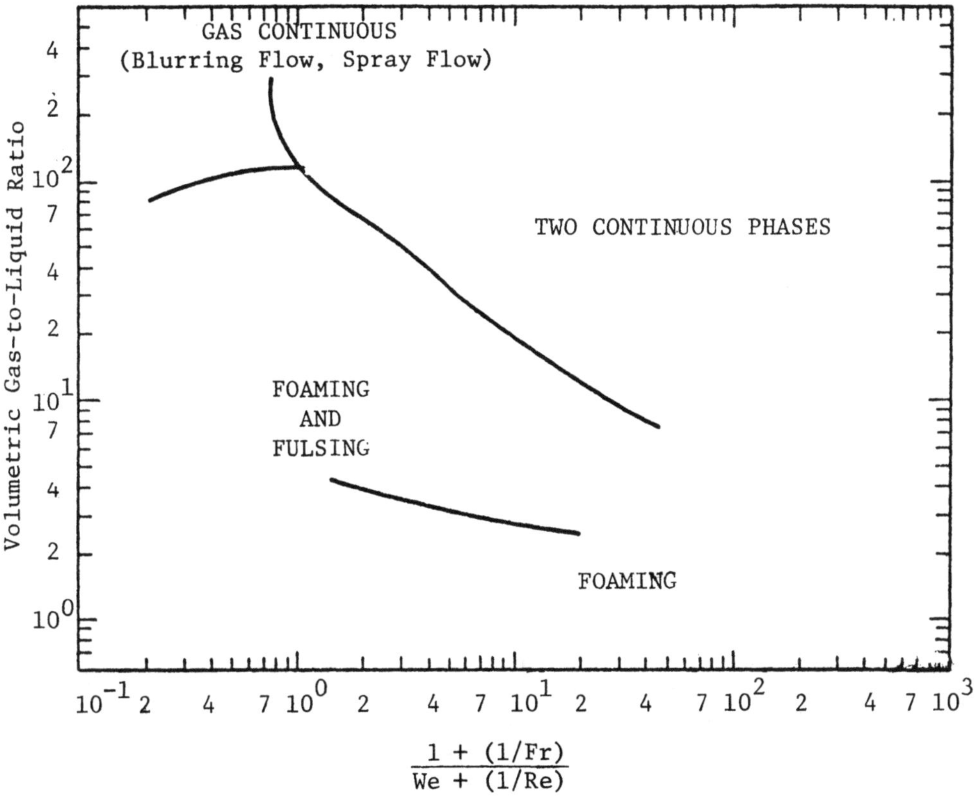

Figure 3 Flow Map after Talmor[(4)] for Foaming Liquids in Two-Phase Flow in Fixed Beds

Chou, Worley, and Luss (2) showed that a decrease in the surface tension shifted the transition of the gas continuous flow to pulsed flow to lower gas and liquid velocities. Also, using the flow map coordinates of Charpentier and Favier (3), they showed that besides foaminess and liquid phase surface tension, the interfacial tension between the liquid with the solid should also be considered in the correlations.

These flow regimes have an impact on how the liquid is distributed among the pores of the catalyst and the interstitial volume between the catalyst and how this interstitial volume is divided between stagnant zones and the flowing liquid. As

Beimesch and Kessler (1) pointed out, in the gas continuous regime liquid flows via rivulets that are fairly steady and well defined. In the pulsing regime, the rivulets merge into liquid slugs with gas "holes". In foamy flow, the gas-liquid mixture has the appearance of an emulsion and moves through the bed either in steady fashion or in pulses depending on the gas and liquid flow rates. Foamy flows are characterized by high pressure drops in comparison to gas continuous and pulsing regimes (7, 8).

A diagnostic that a designer can use to determine the flow regime in his commercial reactor involves measuring the pressure drop of the actual vapor and liquid feeds and products through a test catalyst bed section using transducers that have a rapid response, for example, with a first order of response of a few seconds. The pulsing regime can easily be detected by observing oscillatory pressure drops. Foaming systems can be detected by larger than normal pressure drops. Pressure drop equations that can be used as a basis for this diagnostic and for design will be discussed in a later section. It should be pointed out that this diagnostic requires running a pilot plant reactor at flow rates envisioned for the commercial units, since pulsing and foamy flow are strong functions of flow rate.

Small pilot plant tends to run in the gas continuous regime and efforts to provide scale up rules have centered principally on situations where the flow remains in the gas continuous regime. Directional effects for situations where the flow regime shifts into the pulsing and foaming regime upon scale-up have been suggested and will be discussed in later sections.

2.2 Pressure Drop

In the design and operation of pilot plant reactors, pressure drop is usually not a major issue unless the reaction or feed is fouling in nature. As a matter of fact, at the low liquid gas flow rates typical of small pilot plants, pressure drop is usually difficult to measure, particularly if the reaction conditions are at elevated temperature and pressure. In commercial reactor design, however, it is important to be able to predict the pressure drop in the reactor being that it is important in the design of compressors and pumps necessary for the operation of these units.

Several pressure drop correlations are available in the literature. One popular one is by Larkins, White and Jeffrey, (7) which is an empirical correlation based on the work of

Lockhart and Martinelli (9) for flows in pipes. The form of this correlation has been supported by several subsequent workers as for example by Weekman and Myers (8) and Sato, et al (10). The coordinates of the correlation are Ø, the square root of the ratio of the mixed phase pressure drop, δ_{LG}, to the pressure drop of the gas phase, δ_G, if it were to flow alone in the bed. X is the square root of the ratio of the pressure drop of the liquid, δ_L, divided by the pressure drop of the gas, δ_G, if each were to flow alone in the bed. The correlation below was found adequate for both trickle and pulsed flow:

$$\log_{10} \frac{\delta_{LG}}{\delta_L + \delta_G} = \frac{0.416}{(\log_{10} X)^2 + 0.666} \qquad (2.2\text{-}1)$$

However, in foaming systems, the pressure drop was underpredicted by factors of 50 to 40-fold depending on the surfactant level used.

Weekman and Myers (8) found that for foamy systems, a modified X parameter allowed correlation of each set of runs with a given surfactant concentration. This modified X, which they called X^1 is given by the equation below,

$$X^1 = \frac{\delta_L}{\delta_G}^{0.5} \qquad (2.2\text{-}2)$$

Another correlation follows the work of Turpin and Huntington (11) in which the pressure drop is based on a friction coefficient which is a function of the ratio of the packing Reynolds numbers of the gas, Re_G, and the liquid Re_L. Specchia and Baldi (12) modified this correlation to take into account foaming and non-foaming systems via parameters suggested by Charpentier and Favier (3) that include surface tension, σ, and viscosity, μ. Also, these workers distinguished between the low interaction and high interaction regime which corresponds to the continuous gas flow regime and the pulsed flow regime. The correlations given by Specchia and Baldi for the high interaction regime are shown below,

$$\delta_{LG} = \frac{3\, f_{LG}\, \rho_G\, U_G^2}{d} \left(\frac{1-\varepsilon}{\varepsilon}\right) \qquad (2.2\text{-}3a)$$

$$f_{LG} = f(Z^1, \psi) \qquad (2.2\text{-}3b)$$

$$z^1 = \frac{(Re_G)^{1.167}}{(Re_L)^{0.767}} \qquad (2.2\text{-}3c)$$

$$\psi = \frac{\sigma_w}{\sigma_L}\left[\frac{\mu_L}{\mu_w}\left(\frac{\rho_w}{\rho_L}\right)^2\right]^{1/3} \qquad (2.2\text{-}3d)$$

The subscripts w and L refer to water and the actual liquid respectively, ρ is the density of the fluids; ε, the void fraction and U_G the gas velocity.

Another correlation was proposed by Clements and Schmidt (13) which is based on silicone fluids of viscosities ranging from 1.5 to 8 centipoises and surface tensions of 18 to 20 dynes/cm which they state to be typical of hydrocarbon systems. They varied gas and liquid flow rates to cover the gas continuous and pulsing regimes. They found that there was no break in the curves between these two regimes and that their data were well correlated by the following expression:

$$[\delta_{LG}/\delta_G] = 190\ [\mu_L d \left(\frac{\varepsilon}{1-\varepsilon}\right)^3\left(\frac{Re_G\ We_G}{Re_L}\right)^{-1/3} \qquad (2.2\text{-}4)$$

Their data using 1 mm extrudates and 1.6 and 2.9 mm spheres showed less deviation with this equation than when fitted with the equation by Larkins.

Charpentier (14), in a recent study, showed that the propensity for foam to collapse can be correlated to the pressure drop increase shown by the liquid. This was shown by measuring the rate of coalescence of gas bubbles in mixtures of cyclohexane and gas oil and measuring the pressure drop of a trickle bed using the same mixtures. Small amounts of gas oil added to cyclohexane did not appreciably change the viscosity or surface tension, however, it changed markedly the propensity to foam. The pressure drop of cyclohexane was increased 5-fold by introducing approximately 15% gas oil into it. The large influence of percent gas oil and cyclohexane on the rate of coalescence of the mixture was shown in a test apparatus that measures the coalescence of bubbles. This appears to be a promising technique but it needs further confirmation. For high temperature and pressure reactions, foam collapse data would probably have to be obtained at the reactor conditions. Extrapolation of foaminess from room temperature to elevated temperatures is not feasible.

Being that there are still some very important, unresolved questions with respect to these various correlations, all of which are claimed by their authors to follow fairly closely their own data and that of other workers in the field, it is best that a designer, when handling an unfamiliar feed obtain data using the actual reactor feed at the liquid mass velocity and gas rates of the commercial unit design. The pilot plant should be run at the temperature and pressure conditions of the commercial reactor. For greater assurance, both feed and product material should be tested and checked against these correlations to see which one fits the best for design purposes.

2.3 Flow Distribution

The importance of having adequate flow distribution at the top of a fixed bed of catalyst was pointed out in a classic paper by Ross (15). By using radioactive tracers, he showed that the better performance of a pilot plant reactor than a commercial hydrotreater could be explained in terms of inadequate feed distribution over the catalyst. Improved distributor geometries (unfortunately not described) were shown to improve both the liquid hold up and conversion indicating a connection.

Most flow distribution studies for trickle beds have concentrated on the texture of the flow under conditions of thorough irrigation at the top of the column. These cannot be used for distributor design. However, a few studies (16-20) are available dealing with radial dispersion from single points and other geometries. Generally, these studies showed that dispersion of the flow is by rivulets flowing over the packing particles and that distributions tended to even out to an equilibrium good distribution. Some disagreement exists over the importance of the wall flow. Herskowitz (17) gives a criterion to avoid excessive wall flow and proposes that the reactor diameter size should be at least 16 catalyst particle diameters. Koros (21) showed that the criterion is a function of liquid mass velocity and that reactors as small as 1.4 particle diameters in size can operate at full catalyst utilization if the liquid mass velocity is maintained above about 10 kg/m^2s. This does not require large flow rates since the reactor diameter for such a case would be small.

Hoftyzer (16) outlines generalized mathematical expressions needed to calculate dispersion from a variety of distributors -- a central stream, disc, ring, annular disc and an eccentric stream. This study showed that the characteristic dimensionless group for all geometries is given by

$$\frac{LZ}{Y^2} \tag{2.3-1}$$

where L(in cm) is the lateral dispersion coefficient, Z is the distance (in cm) down the column at which a given quality of distribution is achieved and Y is the diameter (in cm) of the column or the zone associated with a single injector. Hoftyzer and subsequent workers found that L increased slightly with increasing liquid and gas rates and was directly proportional to the packing size, L = d/7 where d (in cm) is the equivalent diameter of the packing. Subsequent work by Onda suggested $L \propto d^{0.5}$. Using this relationship and incorporating the data of Herskowitz (17), Onda (18), Stanek (19) and Zarzycki (20) into the Hoftyzer relationship, one obtains the following criterion:

$$K_{crit} = \frac{Y^2}{d^{0.5} Z} \leq 4 \tag{2.3-2}$$

The quality of distribution that was used to obtain a $K_{crit} = 4$ was judgemental and was arbitrarily chosen to be acceptable when the flows in all of the measured column sections reported by these studies became equilibrated. Different values for the constant would be obtained for other criteria. Also, the data used to obtain K_{crit} only included results obtained in columns ranging in diameters between 4 and 40 cm. The relationship between $K_{critical}$ and column diameter is not known but the value is expected to decrease with increasing diameter and feed injector spacing.

It is interesting to note that for a typical small pilot plant reactor, 2.5 cm in diameter, using catalyst with an equivalent diameter of 0.16 cm, the above relationship predicts that a single point injector would produce even distribution at about 4 cm from the top of the bed. This is an insignificant distance since most pilot plant reactors are packed with inert beds above the bed. Hence, we can assume that flow distribution in pilot plants is not an issue. The findings of Koros (21) confirmed this by showing that single point and spray nozzle injectors gave the same conversions in a 2.5 cm diameter reactor.

Equation (2.3-2) also suggests a criterion for maintaining similarity of flow distribution for scale up. For a constant fraction of bed height devoted to distribution of the flow, the separation distance between distribution points should increase no more than with the square root of the ratio of the commercial to pilot plant reactor heights. Within the limitations of the data base for the criterion, this scale-up estimate is conservative since it does not take into account the increase in dispersion with increased liquid and gas velocities upon scale up and the decreased surface per unit volume of reactor available for wall flow.

For a priori design, the relationship given in Equation (2.3-2) would have to be quantitatively related to the efficiency of catalyst utilization in the distributing zone. This relationship is not available in the literature at the present time.

Because of the limited spreading ability of liquid in a trickle bed reactor relative to a single phase reactor, extra precautions for trickle beds have to be taken particularly for exothermic reactions to assure that hot spots are not generated due to flow maldistributions. A study by Jaffe (34) showed that the shadowing effect from large reactor internals in a trickle bed reactor can cause low flow zones to persist for a large enough distance that the low flow rate generated within that distance can build up to very high temperatures. This study was done for a hydrocracking reactor in which commercially hot spots were observed.

The model for these calculations was that of Deans and Lapidus (35) in which a packed bed was simulated using a parallel and series network of stirred tank reactors. Deans showed that if each stage is assumed to be 0.845 d in width and 1.0 d in length that the radial Peclet number would be 12 and the axial Peclet number would equal 2 which is what is expected from a single phase reactor. Jaffe compared calculated and measured temperature profiles showing a very good match. The calculations were supported by kinetics obtained in a small pilot plant reactor over a wide space velocity range.

Being that in a mixed phase downflow reactor, the rivulet dispersion of liquid is less than in a single phase reactor, the effects estimated by Jaffe should be, for equal kinetics and heat release, even larger and more significant for mixed phase system and therefore should be kept in mind in the design of internals for trickle bed systems.

2.4 Scale Up By Fluid Dynamic Similarity

This is by far the most assured way of scaling up a trickle bed reactor. It consists of running a pilot plant at the same liquid superficial velocity and space velocity as the commercial unit to be designed. The pilot plant would also use the same catalyst type and particle size as the commercial reactor.

Using this technique, questions arising from possible changes in catalyst contacting efficiencies associated with changes in flow regime upon scale up are avoided. Based on data obtained in the gas continuous regime, both the pilot plant and the commercial reactor would be designed to run at a liquid mass velocity exceeding 4 kg/m^2s, which has been shown to yield maximum catalyst utilization (Satterfield (5), Koros (21)). Gianetto et al (52) recommends a liquid velocity greater than 6 x 10^{-3} m/s which is approximately the same as 4 kg/m^2s for hydrocarbon systems.

To check that indeed the maximum catalyst utilization had been achieved in the pilot plant, the pilot plant would have to be run over a large range of velocities while maintaining the space velocity constant. If the conversion remains constant, over say a range of 4 to 1 in liquid velocity, maximum catalyst utilization was probably achieved. For these tests, of course, the gas/liquid ratio and the temperatures must be held constant also to maintain fluid dynamic and reaction rate similarities. Changing both the space velocity and liquid mass velocity at the same time should be avoided because this introduces the reaction kinetics into the interpretation of the data and the kinetics of the reaction are not always well known. The assumption of first order reaction, which is satisfactory for many reactions, is not necessarily true in the general case.

The technique of scaleup by fluid dynamic similarity is not restricted to the situation where the commercial reactor is designed at the maximum contacting efficiency possible for the catalyst, however desirable that may be. The procedure should also be satisfactory for situations where the contacting efficiency is less than maximum. In either case, it is assumed that the designer will be able to provide, both for the pilot plant and the commercial reactor, a liquid and gas distribution system which will properly distribute the two phases over the catalyst and maintain them in good distribution throughout the reactor length.

3.0 PHENOMOLOGICAL APPROACHES FOR PERFORMANCE SCALE UP

Since a validated mathematical model of a trickle bed reactor does not exist, a priori design of trickle beds is not yet possible. However, a phenomological approach to scale up from small pilot plants is feasible in most cases. In this approach, the principal steps affecting the apparent reaction are identified and applicable phenomological correlations are used for scale-up.

In a trickle bed reactor, the intrinsic reaction rate can be modified by limitations that can come about due to resistances in the transfer through one or several of the following interfaces:

- Flowing liquid to catalyst pellet
- Flowing liquid to stagnant liquid
- Flowing gas to flowing liquid
- Flowing gas to stagnant liquid
- Flowing gas to catalyst pellet which may be partially wetted
- Stagnant liquid to catalyst pellet

The magnitude of the influence of these resistances on the apparent reaction rate depends on the magnitude of the following:

- Liquid and/or gas phase homogeneous reactions rates
- Vapor phase reaction rates on the catalyst surface
- Liquid phase reaction rates on the catalyst surface
- Reactant concentrations in the fluids

- Diffusional resistances within the catalyst pores
- Catalyst pellet wetting
- Stagnancies surrounding the catalyst pellet
- Heat of reaction

The task of the reactor engineer in scaling up a pilot plant reactor which is run at lower velocities than the anticipated commercial reactor is to ascertain how the apparent reaction rate measured in the pilot plant will change upon scale up due to changes in mass transfer and contacting associated with the fluid dynamic changes inherent in increasing the fluid velocities. As we shall see in the discussion that follows, scale-up concerns intensify as the reaction rates increase and as vapor phase reactants become more dilute. Criteria and diagnostics will be presented that can be used to determine whether or not scale up will affect the apparent rate of the reaction.

3.1 Axial Dispersion Effects on Conversion

For reaction kinetics greater than zero order, axial dispersion of reactants in the reactor can decrease the observed conversion in short reactors, particularly at high conversion levels, resulting in a lowered apparent rate of reaction. Mears (40) using a dispersion model for the trickle bed reactor obtained the following criterion to assure less than 5% deviation of the apparent reaction rate from that measured with plug flow:

$$\frac{Z}{d} > \frac{20\ n}{Pe_L} \ln \frac{C_i}{C_o}$$

where Z is the reactor length; d, the catalyst particle diameter; n, the reaction order; Pe_L, the axial Peclet number; and C_i and C_o, the inlet and outlet concentrations. Using the axial dispersion data in the gas continuous regime obtained by Hochman and Effron (36), it can be shown that for liquid velocities that yield good catalyst utilization, i.e., a liquid mass velocity of 4 kg/m^2s as suggested by Koros (21), plug flow is approximated by beds greater than 10-20 catalyst particles long for a conversion level of 85% -- a condition which is surely met by any practical commercial reactor. Hence, axial dispersion of the liquid in a trickle bed operating at elevated liquid velocities is of no concern.

Recent measurements by Lerou, Glasser and Luss (42) in the pulsing regime showed that axial dispersion in the liquid phase is low just as in the gas continuous regime. Measurements in the frothing/bubbling regime do not exist to the knowledge of the author, but it is expected that in this regime also, the dispersion will be low also. On the other hand, in the mist flow regime, it would be expected that liquid dispersion could be quite high, since the gas could carry a fraction of the liquid flow faster than the mean flow of the liquid down the reactor.

The axial Peclet number of the liquid decreases with decreasing liquid velocity; therefore, for small pilot plants, the criteria shows that the reactor should be longer -- about 2000 particles for 1.6 mm diameter catalyst. This represents a minimum length of about 3 m. Higher conversion than 85% would require longer beds. Mears (40) and Koros (21) show data that supports this concern.

Also, it should be noted that the value of the Peclet number measured by Hochman and Effron varied over a wide range for any given liquid velocity. (Figure 4) They varied from run to run and were also affected by the start up conditions of the bed. For example, pre-flooding the bed gave different results than starting with a dry bed. This may be one of the causes for larger variabilities in conversion observed when using short trickle beds (21, 43). For the estimate of the minimum reactor length given above, a conservative extrapolation of the data was used.

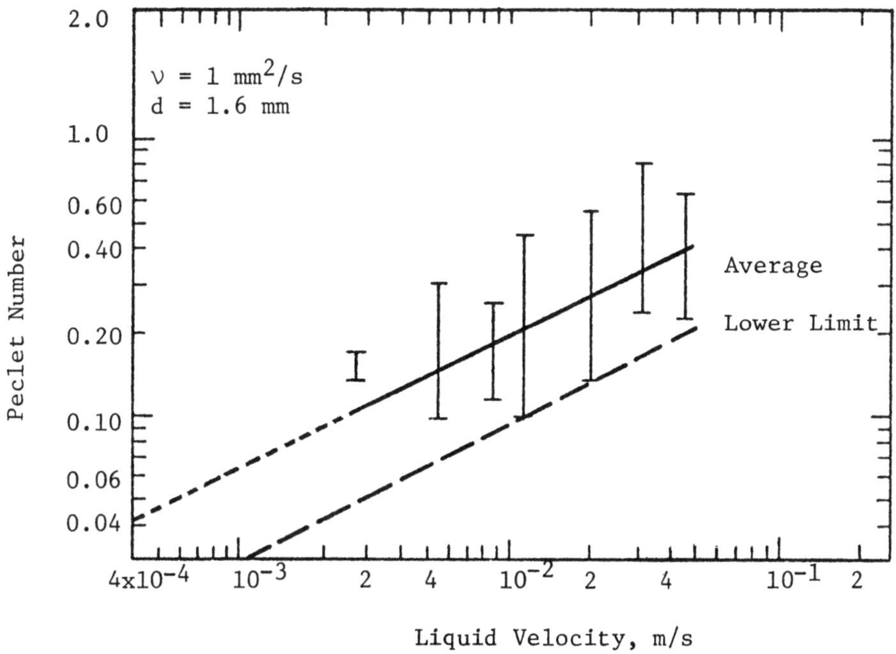

Figure 4. Liquid Peclet Number in Two-Phase Flow in Fixed Beds Using Data of Hochman and Effron[36]

Values for the Peclet number for the gas phase were found to be about one order of magnitude larger than for a single phase reactor (36), i.e., Pe $\simeq 0.2$ but varied inversely with liquid velocity. Increased dispersion with increasing flow rate was interpreted as being due to increased bridging by the liquid. The effect of gas dispersion on conversion for a reactor run at high liquid mass velocity is, however, not significant. For pilot plants, the Peclet number for the gas is higher than for the liquid.

3.2 Scale up of the Homogeneous Reaction Component Based on Liquid Hold Up

As discussd earlier, Ross (15) used liquid hold up as an indicator of the quality of distribution and found that catalyst utilization for pilot plant and commercial unit data were brought in line by assuming that contacting was proportional to total liquid hold up. Henry and Gilbert (41) developed correlations based on this finding, however, in the design equation, (e.g., Equation 1.0-1) for a catalytic trickle bed reactor, hold up does not enter into the equation. Henry and Gilbert proposed that liquid hold up represents a direct measure of the catalyst that is being contacted. Much of the limited success in using this concept is circumstantial because many other important factors other than liquid hold up influence contacting.

For example, Hoogendoon and Lips (37), Hochman and Effron (36) and others have shown that liquid hold up in a trickle bed is composed of several types. There is liquid in the pores, in stagnant zones among the catalyst, and there is flowing liquid. The hold up scaling model lumps the effectiveness of the stagnant and flowing hold ups. As discussed later, this is not universally true.

In a situation, however, where the reaction also occurs in a liquid phase in a homogeneous manner, hold up would be an important scaling parameter. One can postulate a whole range of situations from the one extreme in which the liquid reactions are insignificant through one in which the liquid phase reaction is just as important as the catalytic one and may even interact with a catalytic reaction. For example, one might have a thermal cracking reaction which is being modified by catalytic hydrogenation. Or, one might have the situation in which the packing is inert and only a homogeneous reaction is occurring.

In general, liquid hold up increases with increasing liquid velocity and decreases with increasing gas velocity. Gianetto et al (52) have reviewed the various available correlations. A recent correlation by Specchia and Baldi (12) gives the hold up in both the low and high interaction regimes as a function of the liquid and gas Reynolds′ number and the Galileo number.

Since liquid hold up tends to vary with the same parameters as contacting efficiency, it is not feasible to test the sensitivity of the reaction to the homogeneous component by varying, for instance, the liquid mass velocity. Therefore, to test for the influence of homogeneous reactions, inert zones should be provided within the catalyst bed. These inert zones should be spread throughout the bed by diluting the catalyst with inerts. These inerts, however, should be of the same size and shape as the catalyst, otherwise, the contacting efficiency of the reactor can be affected also and confound the test. To proceed with this approach, the designer must be sure that the catalyst bed is already long enough to avoid possible axial mixing effects as discussed in Section 3.1.

If it is discovered that a homogeneous reaction component is important, several alternatives may be followed. One is to design a pilot plant reactor so that the liquid mass velocity and gas velocities are comparable to the commercial unit, i.e., same liquid hold up. Second, introduce the right amount of inert material in the lower velocity pilot plant to simulate the additional liquid hold up of the commercial reactor. The second approach could be quite unreliable since the correlations have a large variability. If the homogeneous reaction is important, scale up based on constant space velocity based on the catalyst will tend to be conservative since liquid hold up increases equally or more greatly with gas and liquid velocity than catalyst contacting.

3.3 Mass Transfer From Flowing Liquid to Catalyst Pellet

The mass transfer coefficient, K_s, in the correlations for mass transfer from the liquid to solids are based on a total interfacial area of the catalyst within the reactor and should be used accordingly. Wetting effects are included within the value of the coefficient.

In general, these correlations show that the mass transfer coefficient increases with increasing liquid flow rate and is relatively insensitive to the gas flow rate. As the flow regime changes from the gas continuous to pulsed flow, the mass transfer coefficient does not show marked increases. However, the sensitivity to the gas rate increases from practically nil in the gas continuous flow regime to more marked sensitivity in the pulsed and dispersed bubble regime. These are illustrated by the study of Hirose, More and Sato (26). They correlated their mass transfer data using an equation of the form

$$\varepsilon H\ Sh(Sc)^{-1/3} \propto f\ (Re_L) \qquad (3.3\text{-}1)$$

where: Sh is the Sherwood number ($=K_S d/D$); d, is the equivalent catalyst diameter; D, the diffusion coefficient; ε, the void fraction; Sc, the Schmidt number ($= \nu/D$) where ν is the kinematic viscosity, and Re_L, the liquid phase particle Reynolds number (=Ud/ν) where U is the liquid superficial velocity; H, the liquid hold up.

$$H \propto X^{0.22}$$

where X is the square root of the ratio of the pressure drop of the liquid to the gas phase if they were flowing alone in the bed. Figure 5 shows their correlation in terms of the liquid mass velocity for a bed using 1.16 mm diameter catalyst and with ν = 1.0 centistokes.

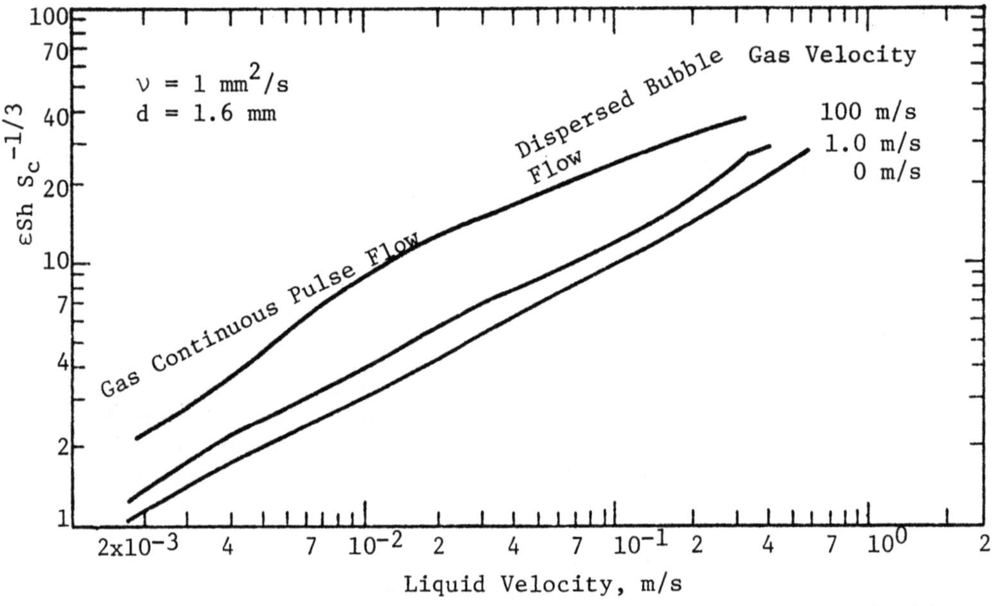

Figure 5. Effect of Liquid Velocity and Flow Regime on Liquid-Particle Mass Transfer in Two-Phase Flow in Fixed Beds. After Hirose(26)

This correlation can be used to check whether or not mass transfer might limit the rate of the reaction in a trickle bed reactor. For this, a criterion of the form suggested by Satterfield (27) can be used as shown below,

$$C_L > \left(\frac{10d}{3\ K_S}\right) R_P \tag{3.3-2}$$

where C_L is the concentration of the limiting reactant; d, the particle size; and R_P, the intrinsic reaction rate per catalyst pellet. If the criterion is met, the concentration of the limiting reactant in the bulk flow will be within 5% of its concentration on the surface. This relationship shows that for a given set of reactor conditions there is a minimum reactant concentration required to avoid mass transfer limitations. This reactant can be a liquid phase reactant or a dissolved gaseous reactant, such as hydrogen. For this situation, an overall mass transfer coefficient, K_{LS} can be used lumping the gas to liquid mass transfer coefficient, K_L, with K_S,

$$\frac{1}{K_{LS}} = \frac{1}{K_S} + \frac{1}{K_L} \tag{3.3-3}$$

This procedure assumes that the interfacial area for mass transfer of the gas to liquid and liquid to the solid are equal. This is generally not true; the gas-liquid interfacial area being a factor of 2 to 3 lower than the liquid-solid interfacial area. This correction can be avoided by using $K_S a_S$ and $K_L a_L$, where a_S and a_L are the liquid/catalyst and gas/liquid interfacial areas respectively.

Satterfield, using very conservative estimates of mass transfer coefficients, and trickle bed hydrodesulfurization data presented by Nobel and Choufoer for the Shell Hydrodesulfurization Process, concluded that liquid film resistance for that reaction is negligible. He took the reaction rate per pellet to be 1×10^{-6} gram-mol hydrogen/cm^{3}s and the hydrogen solubility as being 4.8×10^{-4} g-mol/cm^3. In another example, Satterfield (27) takes the data of Satterfield and Pelosoff (28) for the hydrogenation of liquid alpha-methyl styrene to cumene over a palladium catalyst and shows that whereas the reaction rates are about one order of a magnitude slower than in the previous example, that because the solubility of hydrogen is two orders of magnitude lower, mass transfer limitations of the hydrogen should be important. And, indeed, this was shown to be true in the experimental work of Satterfield and Pelosoff (28).

These examples show that it is not only important to consider the reaction rate but also the concentrations of the limiting reactant when considering mass transfer effects.

Another useful criterion can be developed by recasting equation (3.3-2) in terms of the intrinsic reaction rate defined in terms of the concentration on the surface of the catalyst, i.e., $R_p = k_i C_S$.

Again, using the criteria that for mass transfer limitations to be significant, C_S has to be less than 0.95 C_L, we obtain the following criterion:

$$\frac{K S}{d k_i} < 3 \qquad (3.3\text{-}4)$$

When this criterion is met, mass transfer should not be limiting. k_i can be obtained in a stirred tank reactor or any other device in which the pellet intrinsic reaction rate can be measured directly. It is not an observable quantity in a trickle bed reactor. However, for a conservative estimate, the observed rate constant may be inserted since this will be equal to or less than the intrinsic rate constant. For the data of Pelosoff, the value of this ratio is about 1000 showing that the reaction was very much controlled by mass transfer. For hydrodesulfurization in trickle bed reactors, Satterfield and Pelosoff estimated intrinsic reaction rates of the order of 10^{-3} per second, for which this criterion gives a value of approximately 3 showing that mass transfer limitations in such a reactor is not an important consideration.

A useful criteria given by Froment and Bischoff, based on earlier work by Petersen, shows that mass transfer through the liquid film to the catalyst cannot be limiting unless the reaction is limited by pore diffusion. The analysis was developed for single phase systems but is also applicable to mixed phase systems. The requirement is that the Sherwood number (Sh) for mass transfer be 1-10. This is the case for trickle beds operated at velocities in the range of 10^{-2} - 10^{-3} m/sec superficial liquid velocity.

Calculations presented by Satterfield, et al, (28) show that mass transfer limitations will be important for reaction rates about one order of magnitude higher than for those that just exhibit pore diffusion limitations, i.e., for reactions giving and effectiveness factors, η, < 0.1.

Hence, checking for possible mass transfer limitations may be done by determining the effectiveness factor of the catalyst. This may be done in a stirred tank reactor by classical methods (27, 29), or, in the trickle bed, using a diagnostic suggested by Koros and Nowak (30). In this latter technique, the intrinsic activity of the catalyst is changed in known fashion while keeping the particle size and fluid dynamic conditions in the reactor constant. This may be achieved for example by varying the metal content of a supported catalyst. This diagnostic involves observing the apparent reaction rate as a function of catalyst activity. A linear relationship indicates that the catalyst is in the kinetic control regime and no mass transfer limitations are important. A square root relationship of apparent vs. intrinsic activities indicates a pore diffusion limitation. No effect of catalyst concentration on activity indicates mass transfer limitation on the outside of the catalyst. This technique has been analyzed by Maddon (31, 32).

3.4 Catalyst Wetting as a Scaling Factor

Mears (49) suggested that the contacting efficiency of a trickle bed reactor is proportional to the fraction of the total catalyst area that is wetted by the liquid, a_{fw}, as given by the equation below by Onda (50).

$$a_{fw} = \left[1 - \exp -1.36\ Ga_L^{0.05}\ \ We_L^{0.2} \left(\frac{\sigma_c}{\sigma_L}\right)^{0.75}\right] \qquad (3.4\text{-}1)$$

where Ga is the Galileo number ($=d^3g/\ _L{}^2$); We, the Weber number ($=U^2d/\sigma_L\rho_L{}^3$); σ_c, the surface tension value of the liquid over the packing that gives a zero contact angle; and σ_L, the liquid surface tension.

The form of the equation shows an increase in wetted area with increasing mass velocity reaching an asymptote at high mass velocities. This is shown by the form of the equation below in which the Galileo and Weber numbers have been replaced by a lumped constant, W.

$$a_{fw} = 1 - e^{-WU^{0.4}} \qquad (3.4\text{-}2)$$

W lumps the physical factor such as surface tension, density and particle diameter which in a scale-up situation remain constant. Combining this relationship with the plug flow equation for a trickle bed reactor for a first order reaction by assuming that the apparent rate of reaction is proportional to a_{fw}, Mears obtained:

$$\ln \frac{C_i}{C_o} = \frac{k_{app}}{(SV)} [1 - \exp \ (-WU^{\ 0.4} \ (SV)^{0.4})] \qquad (3.4\text{-}3)$$

For low mass velocities where full wetting is not expected, Mears used the correlation of Puranik and Vogelpohl (51) for catalyst wetting:

$$\ln \frac{C_i}{C_o} \ \alpha \ \frac{L^{0.32}}{SV^{0.68}} \ d^{0.18} \ (\nu_L)^{-0.05} \left(\frac{\sigma_c}{\sigma_L}\right)^{0.2} \qquad (3.4\text{-}4)$$

This relationship predicts that log-log plots of the logarithim of the ratio of the inlet to the outlet concentration vs the reciprocal space velocity should give lines of slope 0.68 for any given reactor length.

Shah (38) reviews recent data and concludes that the slopes varied from 0.53 to 0.92 and were dependent on the temperature. This would indicate that for certain reactions, the contacting efficiency depends on the catalyst wetting to a greater and lesser extent. This cannot be if indeed only catalyst wetting is important and it implies that contacting efficiency for these reactions involves more complex phenomena.

Clearly, using this model a universal exponent for the liquid hourly space velocity and the reactor length does not exist. Consequently, the use of this type of equation is limited and requires pilot plant work over a wide range of liquid mass velocities and lengths to determine the actual coefficients for a particular feed stock, reaction type and rate, catalyst size and type.

This approach can be used as a diagnostic using the generalized form of equation (3.3-1) as given by Gianetto, Baldi, Specchia and Sicardi (52)

$$\ln \frac{C_i}{C_o} \ \alpha \ \frac{L^w}{(SV)^{1-w}} \qquad (3.4\text{-}5)$$

where w is a fitted constant, i.e., 0.32 for the Mears model. At constant L, a log-log plot of ln C_o/C_i vs reciprocal SV, would give a slope of 1-w. If w = 0, there is no mass velocity effect. Increase values of w would indicate increasing mass velocity effects on the conversion, a value of w = 1 signifies that the catalyst contacting efficiency is increasing proportionally to the liquid mass velocity.

This diagnostic is confounded by the effects of direct vapor phase reactions, therefore, it cannot be used for reactions in which the direct vapor phase reaction is shown to be important. Instead, the diagnostics discussed in Section 3.7 should be used for those instances.

Bondi (53) suggested another equation relating the contacting efficiency to gas and liquid rates as shown below,

$$\frac{1}{k_{app}} - \frac{1}{k_i} = \frac{A}{U^b} \qquad (3.4\text{-}6)$$

where U is the liquid velocity and A, b are fitted constants. Bondi estimated $b \simeq 2/3$ but commented that it varies widely from reaction to reaction. This is an entirely empirical correlation but it does have the feature that at high liquid velocity it correctly asymptotes to $k_{app} \rightarrow k_i$. This relationship offers little advantages over equation (3.4-5).

Another approach which expands on these concepts was proposed by Colombo, Baldi and Sicardi (54). These workers proposed that the external effective wetting theory is too simplistic and does not take into account the important situation where reaction is occurring within the pores of a catalyst which only requires partial wetting to attain full utilization. This situation can occur when the reaction rate is slow. On the other hand, interfacial contact area would be most important in situations where the reaction is fast and where only the surface of the catalyst enters into the reaction. Thus, they point out that the approach to contacting efficiency through only considering effective wetting is too restrictive and would yield conservative results for slow reactions.

Colombo et al (54) adopt the definition of Satterfield for the contacting efficiency as being

$$\frac{k_{app}}{k_i} \times 100 \qquad (3.4\text{-}7)$$

where the k_{app} represents the apparent measured rate and the k_i represents the intrinsic rate of the catalyst. They equate this ratio with the ratio of the effectiveness factor for partially and totally wetted catalyst. The effectiveness factors are calculated using the conventional relationships, however, for the partially wetted catalyst, measured apparent diffusivities are used. Colombo et al used KCl and $ZnSO_4$ tracers to calculate the apparent diffusivities D_{app}, from measurements made in a trickle bed over a range of liquid velocities.

Their measurements show that D_{app}/D reaches unity at a velocity of 4 to 8 x 10^{-3} m/s which is in the range shown by Koros (21) to be needed for full catalyst utilization. (Figure 6)

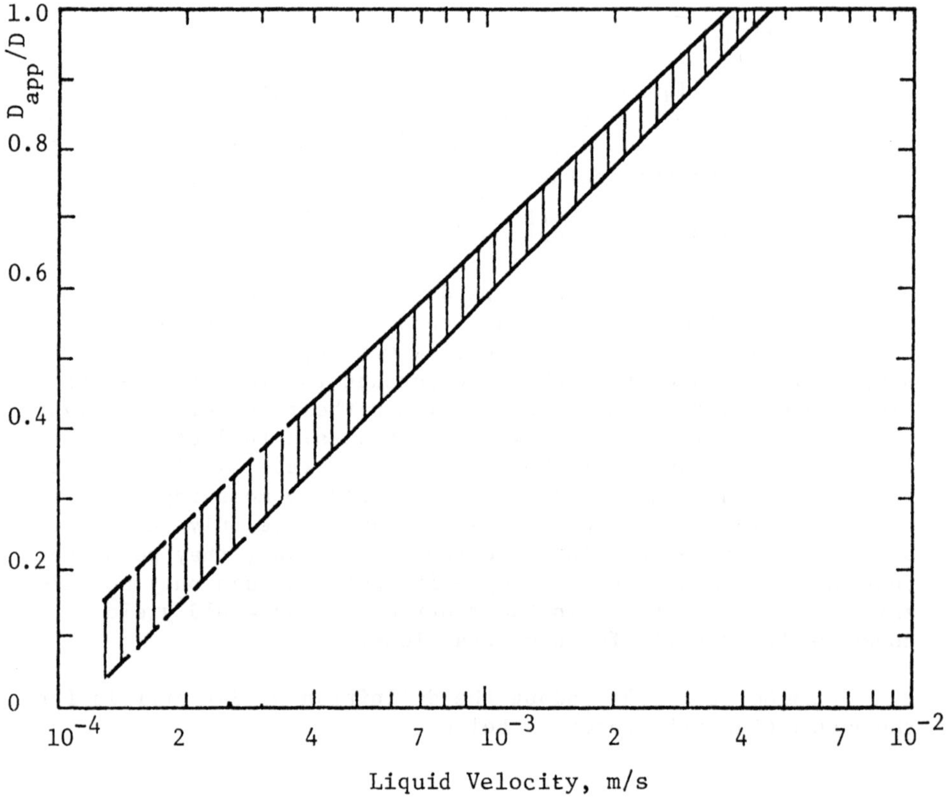

Figure 6 Effect of Liquid Velocity on the Ratio of D_{app} to D

The authors took their measurements of apparent and internal diffusivity and calculated k_{app}/k_i in the style of Satterfield (5) using typical hydrosulfurization kinetics. Their calculations show good agreement with the chart proposed by Satterfield[5] and the data of Koros (21) (Figure 7).

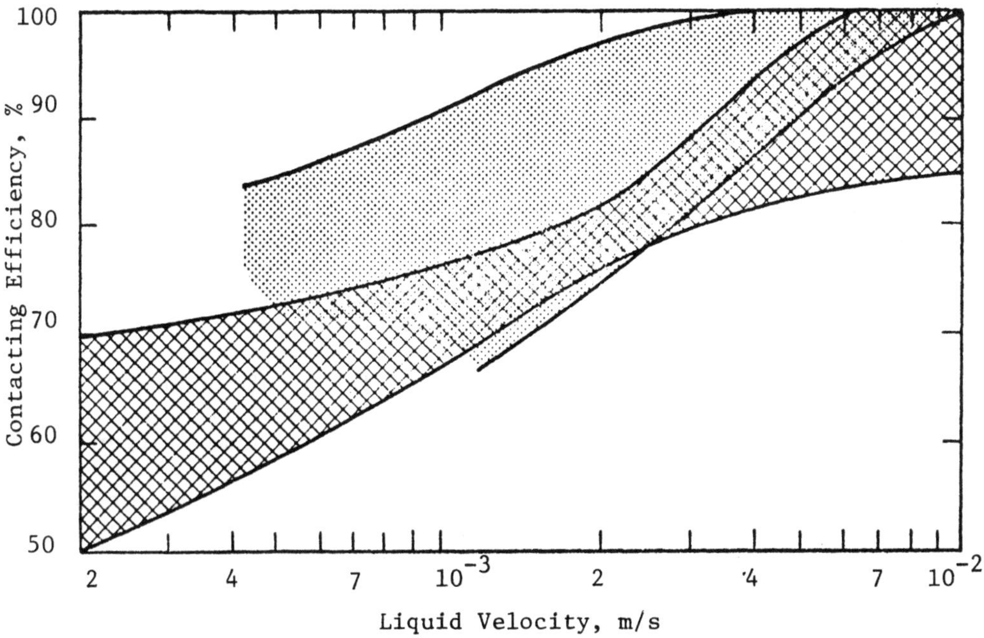

Figure 7 Comparison of contacting efficiency in trickle beds as per Colombo et al (54) and Koros (21)

In all of these calculations total internal wetting was assumed because of the strong capillary effects in the catalyst. Also, the assumption was made that mass transfer of gas to liquid is not limiting and that the mass transfer from the liquid to the solid is also not limiting. Lastly, the possible effects of bed stagnancies were not taken into account.

The results of this study can be used to develop a criterion for whether or not a mass velocity effect is to be expected for a given reaction. The basis for the criterion is that for the maximum change in D_{app}/D that has been measured, the reaction should not be diffusion controlled. The data of Colombo, et al at low velocity, approaches 0.1 - 0.2 for D_{app}/D. Therefore, using the form of the Weisz-Prater criterion, (29)

$$\phi = \frac{d^2 R_p}{36 D_{app} C_L} \ll 1$$

$$\frac{5\ d^2 R_p}{36\ DC_L} = \frac{5}{36}\ \frac{d^2}{D}\ k_i << 1 \qquad (3.4\text{-}8)$$

For this criterion to be useful, the generality of the relationship of D_{app}/D vs. liquid velocity has to be demonstrated. Some variability with packing was exhibited in the data of Colombo et al and only water was used, however, the approach appears promising.

3.5 Mass Transfer of Flowing Liquid to Stagnant Liquid

A characteristic of trickle beds is that a fraction of the liquid hold up is in the form of stagnant zones (36, 37) that interchange with the flowing liquid. Tracer studies by Hochman and Effron (36) using the cross-flow model quantified the exchange coefficient for the gas continuous regime and Lerou, Glasser and Luss (42) for pulsed flow. Hochman and Effron showed that the mass transfer coefficient between the flowing and stagnant pool varies with the particle Reynolds number to the 0.6 power, much like the liquid to solid mass transfer coefficient, except that its value is about ten times larger. For pulsed flow, Lerou (42) found $K_e \simeq 20\ s^{-1}$ at $Re/(1-\varepsilon)$ of 43 which is 200 times larger than that found by Hochman for the gas continuous measurements. So it appears that both in the continuous gas regime and the pulsed flow regime the flowing liquid to stagnant liquid mass transfer would limit only at very high reaction rates.

3.6 Mass Transfer from Flowing Gas to Flowing and Stagnant Liquid

Correlations for mass transfer coefficients from the flowing gas to the flowing liquid are summarized by Charpentier (33), Shah (38) and Hirose, et al (39). Those correlations show that the gas-liquid mass transfer coefficient is usually larger than the liquid-solid mass transfer coefficient, except as Hirose points out for small particles when the resistances are of the same order of magnitude.

Diagnosis for gas mass-transfer resistance follows the procedure outlined in the section on liquid-solid mass transfer.

3.7 Contacting of Flowing Gas with the Catalyst Pellet

Direct gas-catalyst contacting can be important in the scale up of a trickle bed reactor. Its importance is a strong function of the influence of catalyst wetting on the kinetics of the reaction, the relative rates of the vapor and liquid phase reactions, and whether the limiting component in the reaction is in the gas phase or in the liquid phase. Indeed, situations can arise where an increase in wetting increases the rate of the apparent reaction whereas there are cases in which just the opposite is true. Therefore, unless the designer knows which of the two cases he is dealing with, he should scale up at constant wetting. This limits the designer to a situation where the pilot plant and the commercial reactor are run at the same liquid and gas mass flow rates, the same catalyst size and the same reaction conditions which is basically scaling by fluid dynamic similarity. If, on the other hand, the designer wants to be able to use data from a pilot plant that has lower feed rates, he must determine first the sensitivity of the reaction to direct gas phase contacting with the catalyst.

The work of Herskowitz, Carbonell and Smith, (44) Sedriks and Kenney, (45) and Satterfield and Ozel (46) suggest several diagnostic procedures that a design engineer can use to determine whether or not gas phase contacting and wetting of the catalyst by the liquid phase reaction are important issues.

In these analyses, mass transfer of the gaseous components to the catalyst surface that is not covered by trickling liquid was assumed to be governed by a mechanism similar to a vapor phase fixed bed which is packed with particles which have partially blocked (i.e., by liquid rivulets) surfaces. Hence, j-factor correlations could be applied and these show that direct mass transfer to the surface from the gas phase is rapid and not likely to limit the rate of reaction. Therefore, changes in the gas velocity should not affect the observed rate of reactions. However, as discussed in Section 3.3, liquid to solid mass transfer can be limiting and, hence, if a catalyst particle is fully wet, the mass transfer rate of the gas may be slowed because the gas will have to reach the catalyst dissolved in the liquid. Thus, a lowered reaction rate could be observed. In both cases, the catalyst pores are assumed to be liquid filled by capillary action.

The work of Herskowitz, et al, suggests that the importance of gas phase contacting and wetting on the global reaction rate can be determined by measuring the global reaction rate over a wide range of liquid flow rates. In their experiments, the rate of hydrogenation of alpha methyl styrene on palladium catalyst was measured as a function of liquid flow rate. For these runs, the hydrogen and the liquid was in equilibrium. The data for two catalysts concentrations, 0.75% palladium and 2.5% palladium on alumina, showed that the behavior of the two catalysts were different, one anomalous and the other one normal. (Figure 8)

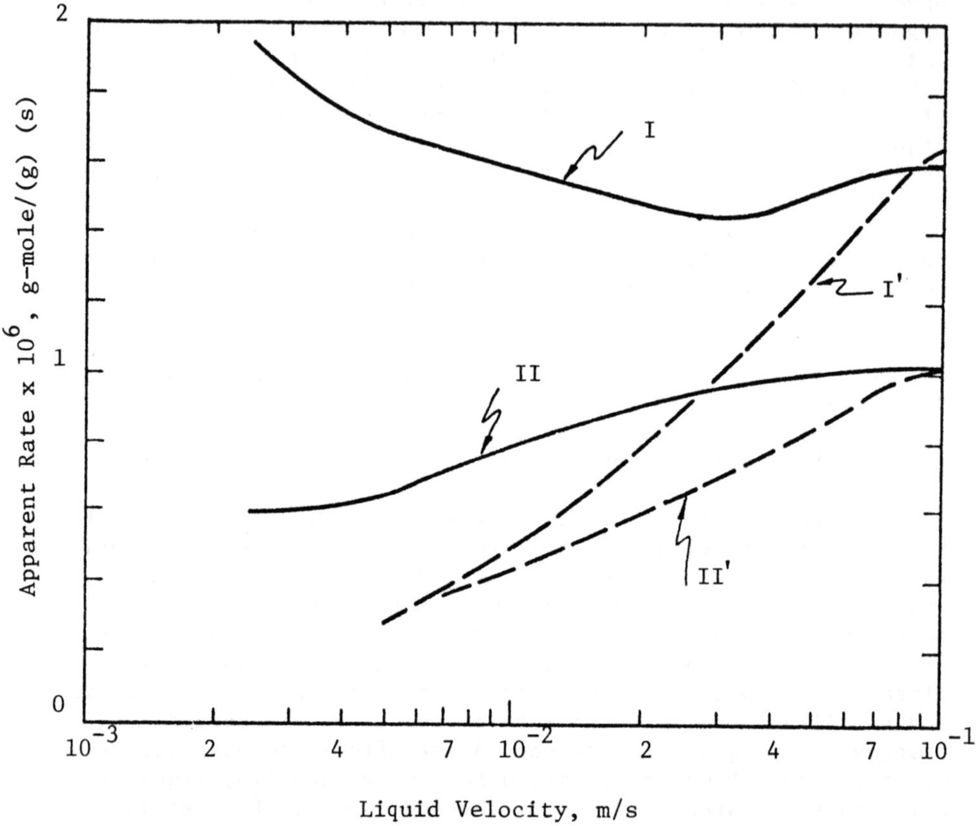

Figure 8 Effect of liquid velocity on the apparent reaction rate for a reaction in a fixed bed reactor limited by gas/liquid mass transfer. Hydrogenation of alpha methyl styrene on 2.5% Pd/Al_2O_3 (I) and 0.75% Pd/Al_2O_3 (II) shown with the corresponding rates assuming full wetting (I', II')

Namely, the more active catalyst showed a minimum in the global rate vs. liquid mass velocity plot; whereas, the lower activity catalyst showed an increasing global rate with liquid mass velocity. Variation of the gas rate over a ten-fold range did not affect the results.

This behavior can be explained by assuming that the external particle surface at the low liquid rates was covered by liquid only partially. The argument goes as follows: since the reaction rate is first order in hydrogen, the concentration of hydrogen at the catalyst surface is important. Hydrogen can reach the surface either directly from the gas phase to the particle through a very thin liquid film or by first transferring from the gas to the liquid and then in solution from the liquid to the particle.

The authors calculated that the gas-particle mass transfer coefficients at the reactor conditions were much greater than the liquid-particle mass transfer coefficients by using the correlations of Dwivedi and Upadhyay (47) for the gas mass transfer coefficient and that of Goto and Smith (48) for the liquid phase transfer. The high value for the gas-particle mass transfer coefficient was also demonstrated by the laboratory data which showed that a change in gas velocity over a ten-fold range did not affect the reaction rate.

Therefore, Herskowitz et al argued that at the low flow rates, the high global rate was evidenced because of partial wetting of the catalyst which allowed for higher mass transfer rates by direct gas to particle transfer. As the liquid flow rate was increased, wetting increased, decreasing the contribution of this fast path. However, with further increases in the liquid mass velocity, the liquid to solid mass transfer rate increased offsetting the negative effect of the increased wetting on the gas mass transfer rate. Thus, a minimum was produced. At the lower catalyst activity, the minimum in the apparent reaction rate was not observed, indicating that the influence of wetting was not important.

This sensitivity to gas and liquid flow rates and wetting was increased by decreasing the concentration of hydrogen in the feed liquid below the equilibrium value. The rates of reaction were lower at the highly wetted condition (i.e., at high liquid flow rate) as would be expected by the lowered hydrogen concentration in the liquid. Also, at the low liquid flow rate, the sensitivity of the system was increased showing now an increase in global rate for both the high and low activity catalyst.

Thus, a diagnostic suggested by this work involves running the reactor at a wide range of liquid flow rates and at various levels of presaturation of the liquid and observing the global rate as a function of these parameters.

Another diagnostic that is suggested by the work of Herskowitz, et al, uses intrinsic reaction rate data from a liquid full reaction system such as an upflow reactor or a catalyst basket reactor. With this technique, the fully wetted global reaction rate is measured and then is compared to the apparent rates measured in the trickle bed reactor. With an upflow liquid full reactor, the liquid phase mass transfer is part of the measurement of the global rate. Thus, a direct comparison between the upflow data and the trickle bed data shows directly the fraction of the global rate attributable to the liquid phase reaction and the gas phase reactions. As expected, Herskowitz showed that at high flow rates, the two global rates merge with the trickle bed performance for both the high and the low catalytic rates. This technique is restricted to low conversions of gas since there is no gas-liquid contact within the catalyst bed.

A diagnostic technique based on the work of Sedriks and Kenney (45) and Satterfield and Ozel (46) involves using the transient behavior during the start up of a trickle bed reactor to determine the relative importance of the gas and liquid phase reactions. Sedriks and Kenney used the hydrogenation of crotonaldehyde for their study and Satterfield and Özel used the hydrogenation of benzene to cyclohexane. The studies of Sedriks and Kenney were done at low liquid superficial velocities whereas the studies of Satterfield and Ozel were done at much higher liquid velocities. The former studies cover liquid flow rates ranging from 0.0045 cm/s to 0.0158 cm/s at a hydrogen flow rate of 0.57 cm/s calculated at STP. The latter study covered the range of 0.11 to 0.37 cm/s liquid velocity and a vapor velocity of 1.17 cm/s. Both studies showed that starting with a dry bed, the conversion of the reaction decreased with time on stream. This observation can be interpretted to indicate that the gas phase reaction on the catalyst has a higher rate than the liquid phase because after the bed is put into service wetting of the bed increased as a function of time until all the catalyst pores were saturated.

Thus, another diagnostic for the relative importance of the gas phase and the liquid phase reaction on wetting is to observe the apparent reaction rate during start up of a reactor from a pre-wetted reaction. Clearly, in this situation, the catalyst has to be pre-treated properly and stabilized before these experiments can be carried out. Pre-wetting is necessary because this diagnostic may give erroneous results if the bed is started from a completely dry state. For example, Satterfield (27) reports that the operation of a trickle bed reactor using a nickel kieselgur catalyst, reaction rates differed by approximately one order of magnitude depending on the start-up conditions. This was interpretted to be caused by changes in diffusivity associated with whether or not the pores were filled with liquid. The increased reaction rate that would be observed on start-up from a dry catalyst are not of interest for this diagnostic. Hence, the pores of the catalyst should be saturated before performing this diagnostic.

A third diagnostic, which is suggested by the work of Satterfield and Özel involves determining the concentration of reactants both in the gas phase and the liquid phase and establishing whether or not the concentration of the gas phase is in equilibrium with the liquid phase at the exit of the reactor. If the product concentrations in the gas phase are greater than that based on equilibrium with the liquid phase, this would be an indication that there is a significant gas phase reaction occurring. Since the gasliquid mass transfer is relatively rapid for most situations, this is not a sensitive diagnostic technique and certainly obtaining a correspondence between the equilibrium values and the actual concentrations in the product gas is not an acceptable test for the abscence of this effect.

In a laboratory test, this third diagnostic could be run by running the reaction with no inerts below the bed and collecting the liquid and the gas phase separately immediately after exiting from the catalyst bed and comparing these results with results obtained with the catalyst bed followed by inert packing in which the gas and liquid equilibrate. If these two runs show a difference, the diagnostic is positive -- a direct gas-catalyst reaction exists. By making these two back-to-back runs, one avoids the problem of trying to determine equilibrium concentrations from vapor-liquid equilibrium data.

Herskowitz, et al (44) gives a procedure to analyze the importance of the direct gas phase reaction. To use their theory, it is first required to determine the intrinsic rate of the reaction with a fully wetted catalyst. This intrinsic rate is used to calculate the catalyst effectiveness factor which is then used to calculate the effect of wetting efficiency on the overall effectiveness factor for the catalyst activities in question using the liquid mass transfer coefficient as a parameter. The model used for this calculation involves calculating the parallel contributions through direct gas-catalyst contact and indirectly through the liquid. A cubic catalyst geometry was used and one or more faces assumed to be wetted to simplify the calculations.

A trial and error computation is used to solve for the wetting efficiency and the mass transfer coefficient. A minimum of two catalyst activities measurements are required being that two factors are being evaluated. A principal assumption is that the mass transfer coefficient is a function of the wetting efficiency. The mass transfer coefficient of Goto and Smith (48) which takes wetting into account was used.

Plots of wetting efficiency vs. superficial liquid velocity obtained by the procedures outlined above show that the data fall well within the range of the correlation that Satterfield suggested in his 1975 review (5) and those presented by Koros (21). (Figure 9)

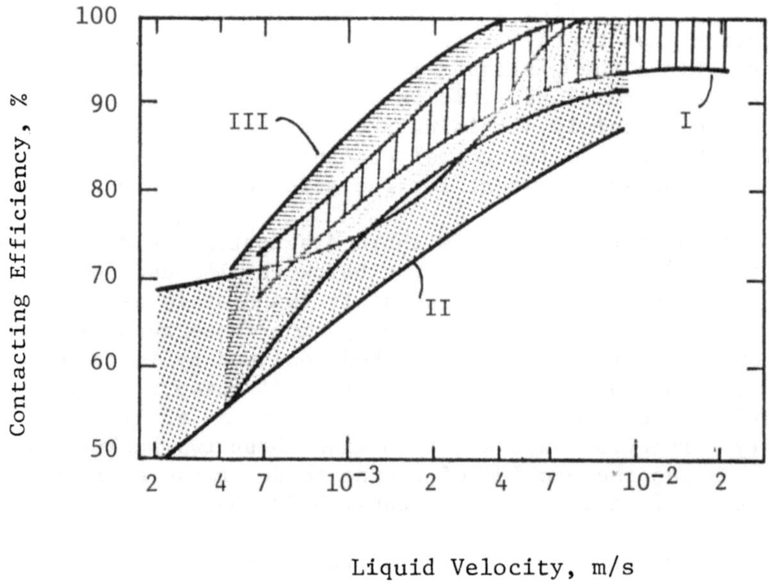

Figure 9 Comparison of contacting efficiency as calculated by Herskowitz (56) using wetting model (I), Koros (21) using a liquid phase reaction (II), and the estimate by Satterfield (5) (III).

A fourth diagnostic, which was suggested by Schwartz, et al (49), involves the use of multiple tracers in the reactor. The tracers are in the liquid phase and only give information on the wetting efficiency of the system.

Using the diagnostic of Koros and Nowak (30) described in Section 3.3, we note that at high wetting, the ratio of the global rates is approximately proportional to the square root of the Pd contents. Assuming that the reaction is first order in Pd content, this signifies pore diffusion limitations for the reaction, and no liquid phase mass transfer limitation the same as concluded by Herskowitz. At low wetting, the ratio of the rates approached unity, i.e., kinetic control. This is a misdiagnosis, hence, this diagnostic must not be used if there is suspicion that a direct gas-catalyst reaction can affect the apparent rate.

In the theory of Herskowitz et al, the role of stagnancies is entirely neglected, however, as discussed in Section 3.8 at low mass velocities stagnancies play an important role in determining the contacting efficiency.

Clearly, it is important to determine if the direct gas reaction plays a significant role in determining the apparent rate since scale up from a small low mass velocity pilot plant to a higher velocity commercial unit could result in considerable lowering of the apparent rate. Equation (3.3-2) can be used to check if this factor should be of concern. If this mass transfer criterion indicates no mass transfer limitations for the gas phase reaction, then the effects discussed here should not be significant. The data of Herskowitz does not meet the criterion as expected.

3.8 The Effect of Liquid Stagnancies on Trickle Bed Performance

In a recent paper, Sicardi, Baldi and Gianetto (55) considered the effect of stagnant pockets between the catalyst particles on the overall efficiency of a trickle bed reactor. They found that the reaction effectiveness of the catalyst areas covered by the flowing liquid and the stagnant liquid varied with the Thiele modulus. They found that the two effectivenesses approach each other at a low Thiele modulus (no diffusion limitation in the catalyst) and at a high modulus where the reaction is entirely mass transfer limited. In between, they found a large ratio between the two rates (a factor of two for their example) and they proposed that the variability in the mass velocity effect from reactor to reactor is explained by variability in the stagnancies within these reactors.

The work of Koros (21) indicates that the most important factor in a design of trickle bed reactors is the value of the liquid mass velocity, and that this is due to the strong influence of liquid mass velocity on the size of stagnant pockets within the bed. For these experiments, small pilot plant reactors, varying in size from 0.46 cm to 2.54 cm, were used packed with granules, pills and extrudate catalysts. The test reaction was the decomposition of hydrogen peroxide in a methanol and water solution. This reaction exhibits first order kinetics and has a rate which is comparable to reactions in commercial hydrotreater reactors used in the petroleum industry. In the study, the catalyst was calibrated using a well stirred tank reactor in which the catalyst was suspended in a basket. The efficiency of catalyst utilization was defined in the same manner as Satterfield[5], as shown below.

$$\text{Cat. Util, \%} = \frac{k_{app}}{k_i} \times 100 \qquad (3.8\text{-}1)$$

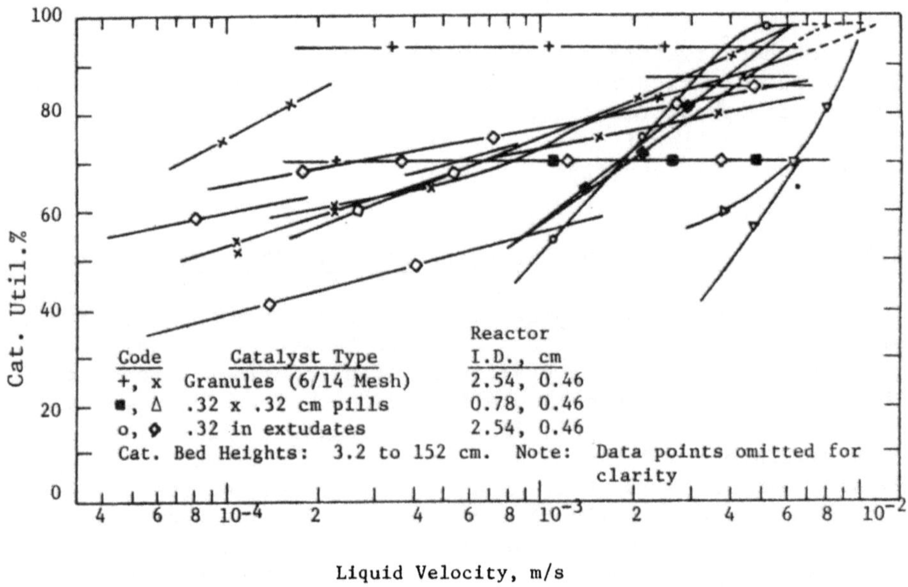

Figure 10 Effect of liquid velocity on the efficiency of catalyst utilization. After Koros (21)

The test trickle bed reactors were operated over a large range of liquid rates from 0.041 to 8.2 kg/m^2s and the gas that is typically used in a hydrotreater reaction was simulated using nitrogen. The gas to liquid ratio was varied from 5 to 100 volume/volume. The efficiency of catalyst utilization measured as a function of liquid mass velocity showed that there is a critical mass velocity above which full catalyst utilization was achieved. (Figure 10) The value of the critical mass velocity does not appear to be a universal constant. It is a function of reactor geometry and catalyst bed packing. It was also observed that in some instances mass velocity had no effect on the efficiency of catalyst utilization. Thus, there is an uncontrolled variable in the operation of these reactors which is a function of bed packing parameters.

In this study, it was shown that the effect of the liquid mass velocity on the contacting efficiency in trickle bed reactors is not associated with liquid-catalyst mass transfer resistance, nor with the flow non-uniformity, nor with mass transfer limitations from the flowing liquid to the stagnant pools, nor with radial mixing limitations, nor with wall flow. The absence of liquid catalyst mass transfer resistance was demonstrated by changing the reaction rate over a factor of two from 3.3 to 7.5 W/H/W and finding that the efficiency of catalyst utilization is not affected by this change. Furthermore, a high value for the activation energy was observed, 42 kJ/mol, which was identical for both the trickle bed reactor and the stirred tank reactor. The stirred tank reactor was demonstrated not to have mass transfer limitations. Consequently, this demonstrated that the trickle bed was not limited by mass transfer either. Also, the activation energy of 42 kJ/mol is considerably larger than the 13 to 17 kJ/mol expected for a purely diffusive mass transfer limited situation.

Using the criterion (Equation (3.3-4)), described previously it can be shown that the reaction rate, indeed, should not be limited by mass transfer at the high liquid mass velocity range and perhaps slightly at the lower range. The modulus varies from about 100 for the high mass velocity tests to about unity for the low range. This compares with the value of 3 which would be required for having negligible mass transfer resistance and indicates that at very low mass velocity, at about 0.04 to 0.14 kg/m s, a small portion of the debit in catalyst utilization was perhaps associated with some mass transfer resistance.

Catalyst utilization and flow uniformity was shown not to go hand in hand. Whereas with increasing mass velocity, the catalyst utilization increases and then plateaus, the flow distribution measured in the annular and central portion of the pilot plant reactor showed that the flow uniformity was poor over most of the range of the experiments. Also, it was observed that the conversion uniformity between the central and annular region was excellent, showing that interchange between the flowing rivulets throughout the cross section of the reactor was very good. The data of Herskowitz and Smith (56) also indicates that this should be the case. They showed that effective flow distribution and tracer spreading by rivulet mixing exists within the bed and also with the flow at the wall.

Further evidence that intraparticle flow patterns and stagnancies are important in determining the efficiency of catalyst utilization in trickle beds was obtained by comparing data between a bed in which the catalyst was packed normally with beds in which (a) the interstices of the catalyst were filled with sand and (b) the catalyst was imbedded in a bed of sand. (Figure 11)

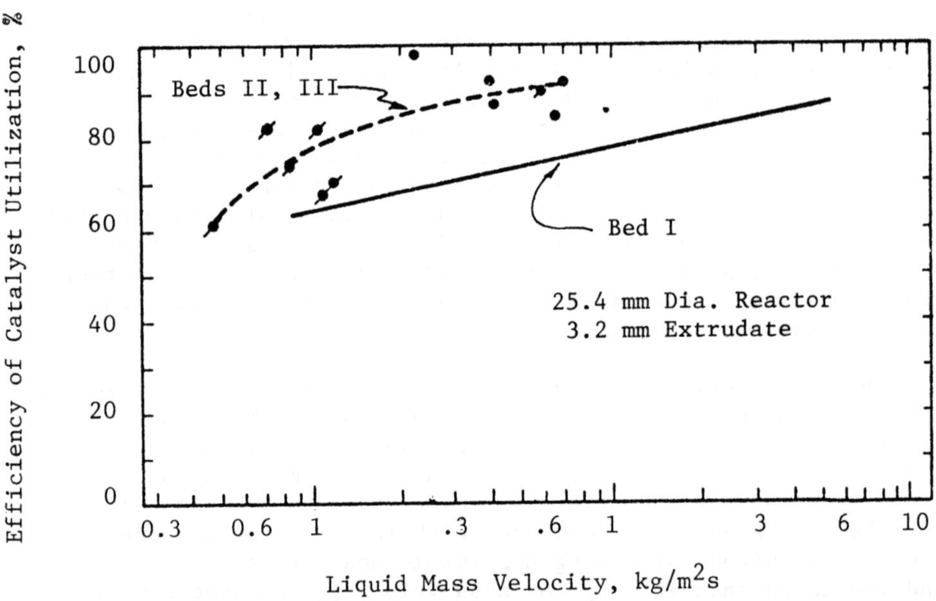

Figure 11 Effect of imbedding the catalyst bed in sand after Koros (21). Bed I packed with vibration. Bed II (•) packed with interstices filled with somesh sand. Bed III (ø) catalyst imbedded in a 50 mesh sand bed

The data showed that the catalyst efficiency was raised by surrounding or imbedding the catalyst in sand, demonstrating that interestitial bed geometry is, indeed, an important factor.

The improvement in catalyst utilization with introduction of sand was also reported by van Klinken and van Dongen (43). The work of van Klinken and van Dongen also supports the view that catalyst utilization is tied closely to the stagnancies in the bed. These workers diluted a bed of spherical 1.5 mm catalyst with 0.2 mm silicon carbide powder and found by the use of tracers that the dynamic hold up was increased from the undiluted to the diluted bed. Also, the ratio of the dynamic hold up to the total liquid hold up increased from 40% to 77%. The apparent rate of reaction was increased 3 to 10-fold by dilution. The diluted bed gave a reproducible apparent reaction rate whereas the undiluted bed fluctuated over about a factor of 3.

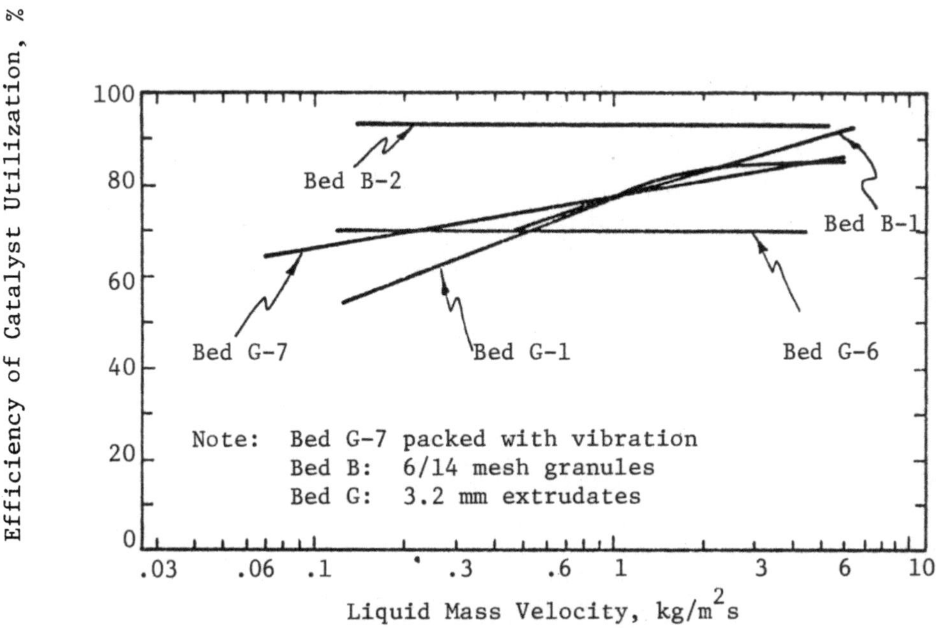

Figure 12 Effect of packing procedure on catalyst utilization after Koros (21)

Being that mass transfer was shown to be not a limiting factor, the only effect left appears to be interstitial flow effects, that is the effect of stagnancies. In this case, sand provided a very uniform flow environment in which the number of stagnancies were minimized and their sizes reduced because of the small size of the sand.

The importance of the interstitial bed geometries is also shown by the variability and contacting efficiency measured with beds which presumably were packed in the same way. Beds packed with vibration and no vibration showed that sometimes the performances were reproducible and sometimes they were not. (Figure 12) This is an effect also observed by other workers in the field. The variabilities of trickle beds were also shown by the irreproducibility of measurements from run to run and day to day as shown by the data scatter. Thus, one of the prime parameters in trickle bed design is the liquid mass velocity and some packing parameter.

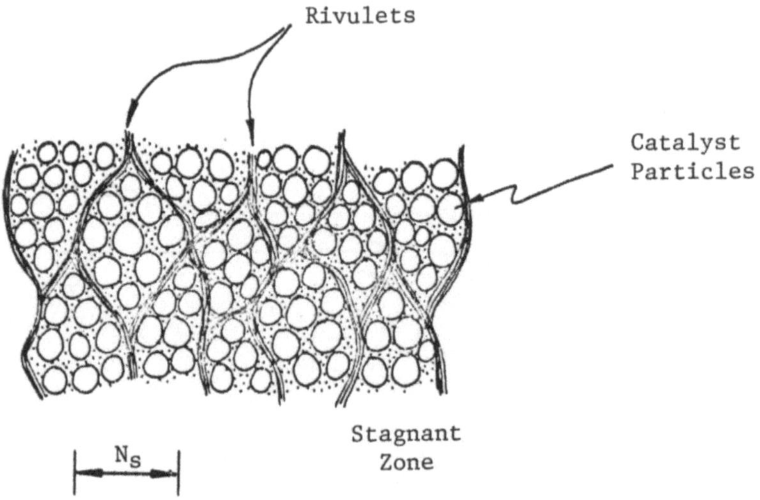

Figure 13 Stagnant zone model for contacting efficiency in a trickle bed

Experimental evidence that stagnancies in trickle beds affect the performance of the bed was given by Van Swaaij, Charpentier and Villermaux (58). In a paper in which they described the effect of wetting on the performance of a packed bed contactor, they reported that as the ratio of dynamic to static holdup becomes less than 8, the spreading mechanism in the bed is best explained by a stagnancy cross-flow model. They found that this ratio is affected by liquid mass velocity and the wettability of the packing. Decreasing mass velocity and increasing the wettability of the packing decreases this ratio making the stagnant zones more important. They found that small packing diameter also increased the importance of the stagnancies.

The work of Michell and Furzer (59) also supports this model. Michell and coworker, much like Hochman and Effron, found that dispersion is due to stagnancies which are distributed across the bed.

It is concluded, therefore, that the factor that determines the efficiency of the trickle bed is the size of the stagnant zones and that these zones could be thought of as being clumps of catalysts which act together much as if they were a larger size catalyst. Using this model, the contacting efficiency of a trickle bed would be affected by the effectiveness of the fraction of the catalyst volume enveloped by the stagnant zones.

The importance of these zones on the contacting efficiency of a bed could be estimated using a modified Weisz and Prater criterion (65). Thus, the stagnant zones could be viewed as a pseudo pellet enveloped by flowing rivulets (Figure 13) with a global diffusivity which combines the diffusivity in the particles and in the stagnant liquid surrounding it. Consequently, this problem can be treated with a Thiele model in which the total effectiveness of a catalyst pellet is affected by its size. The size and number of these clumps are a function of the interstitial geometry, liquid mass velocity, flow regime, catalyst particle size and the physical characteristics of the fluids (i.e., viscosity, surface tension, etc.). These relationships are not quantified at this time, however, estimates of the size of the stagnancies can be made based on available flow measurements and, as we shall see, some useful criteria and diagnostics can be developed using this model.

We will use a modified form of the Weiz and Prater criterion

$$\emptyset_{mv} = \frac{S^2}{36D} k_{app} \ll 1 \qquad (3.8\text{-}2)$$

where S is the size of the stagnant clump of particles, D is the overall diffusivity of the liquid in the clump (molecular diffusivity), and $\emptyset_{mv}$ is the criterion for the absence of an effect of stagnancies on trickle bed performance.

An estimate for S can be obtained observing that at low liquid mass velocity, Hochman and Effron found that the axial Peclet number was about an order of magnitude larger for mixed phase flow than for single phase flow, and that this was probably due to bridging of the liquid flow over the particles. In single phase flow, each particle is viewed to act singly, therefore, the ratio of the two Peclet numbers is a measure of the size of the stagnant clumps. Substituting for

$$S = n_s \cdot d \cong 10d \qquad (3.8\text{-}3)$$

where n_s is the ratio of the Peclet numbers for trickle and single flow and is a measure of the number of particles diameters of the stagnant zones. It must be kept in mind that the measurements of Hochman and Effron were made in a 15 cm diameter column and that for small reactors, Y/d 8-16, smaller values of n_s are more reasonable, say a value of about 3-5, then,

$$\emptyset_{mv} = \frac{n_s^2 d^2}{36D} k_{app} \cong \frac{d^2}{2D} k_{app} \ll 1 \qquad (3.8\text{-}4)$$

Another estimate of n_s can be obtained by comparing equations (3.4-8) and (3.8-4). From this comparison $n_s = \sqrt{5} = 2.25$ or about 2-3 particles.

For the hydrogen peroxide decomposition experiments, the value for the criterion was 1.5-4 confirming that a mass velocity effect should have been observed. We would expect that n_s approaches unity at a high mass velocity. Thus, the contacting efficiency would increase with mass velocity as observed.

Since high temperature diffusivity data for complex mixtures of hydrocarbons are not available, this criterion is difficult to use except for rough screening. However, the form and concept of the criterion suggests that if the reaction can be experimentally shown not to be diffusion limited such that even a three to ten fold increase of particle size will not make the reaction diffusion limited, then we should expect that the reaction rate will not be sensitive to changes in liquid mass velocity.

Past fluid dynamic studies have not quantified the size of the stagnancies as a function of particle size, packing arrangement and mass velocity. However, using intuitive arguments, one can infer the functionality of the size of the stagnancies with respect to variables such as catalyst size and mass velocity. The "S" saturation shape curve was chosen for these conjectural relationships because from intuitive arguments for small particle sizes and high mass velocities, the stagnancies should disappear or be minimized and the particles will asymptote to acting essentially as single particles, ie. $n_s=1$. (Figure 14)

A supporting argument for n_s approaching unity as particles size decreases is that imbedding catalyst in small sand particles improves catalyst efficiency to full-utilization. For large particles or low mass velocities, an upper limit for the size of the stagnancies is the size of the reactor. However, in large

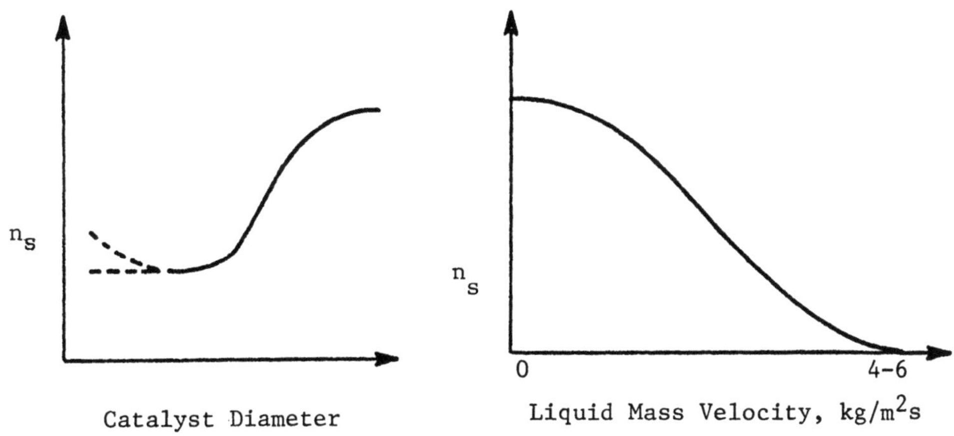

Figure 14 Conjectural relationships of the stagnant zone size and liquid mass velocity and catalyst diameter

diameter vessels, one would expect that equilibration of the size of the stagnancies will occur well below the size of the reactor and that the size of the stagnancy will asymptote to a certain value. On the basis of the axial gaseous dispersion data of Hochman and Effron (36) discussed earlier, the value for n_s should asymptote to a value in the range of about 5 to 10. Complex interactions between the various parameters will give rise to various shapes for these basic curves. For instance, at high liquid mass velocities one would expect a weaker influence of particle size than at low mass velocities.

Introducing the relationship of $n_s=f(d)$ into the Thiele parameter ($=Sk_i/6D$) gives a new diagnostic that can be used to ascertain if a mass velocity effect on contacting is to be expected for a reaction. On a Thiele coordinate plot of catalyst effectiveness vs the Thiele modulus, the slope for a large value of the modulus will be greater than -1. The slope will vary with the value of the Thiele modulus depending on f(d). A plot of the apparent rate of reaction or ln C_i/C_o at a constant mass velocity vs the catalyst particle diameter will give increasingly negative slopes past unity with increasing particle diameters values if stagnancies are limiting. (Figure 15)

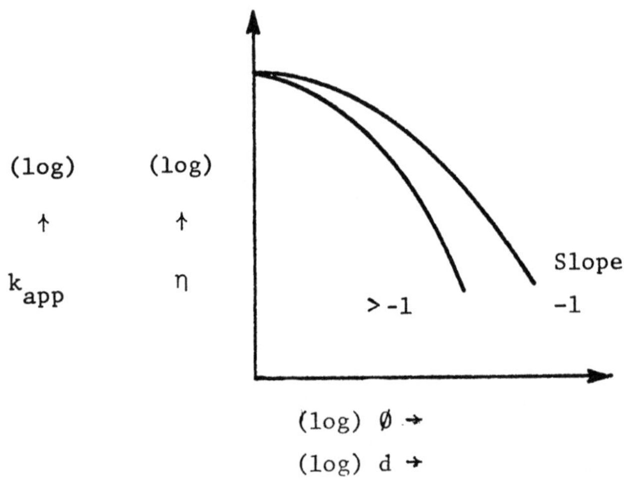

Figure 15. Predicted relationship between k_{app} and catalyst particle size, d, based on the relationship between the effectiveness factor, η, and the Thiele modulus Ø.

Thus, a laboratory diagnostic test would consist of measuring the conversion for a series of catalyst sizes at a constant mass velocity. The slope of the plot of ln C_i/C_o vs the catalyst size serves as the diagnostic. A slope of zero would denote no diffusional limitations and hence no effect of mass velocity on catalyst contacting efficiency should be observed in the pilot plant. A limiting slope of -1 would indicate that the potential for a mass velocity effect exists particularly at a lower mass velocity since the catalyst is diffusion limited. A limiting slope greater than -1 or an increasing negative slope with increasing particles size would denote that for that range of particle sizes an effect of mass velocity does exist and full catalyst utilization is not being obtained.

This model predicts that for a given feed and catalyst combination a mass velocity effect may be observed for a large catalyst particle size but not for smaller ones. Also, that the mass velocity affect will be a function of catalyst pore structure (i.e. D) and catalytic activity.

Support for the first predictions is given by the data of Montagna, Shah and Paraskos (60) combined with the data of Montagna and Shah (61) and Paraskos, Frayer and Shah (62). In these studies, a variety of reduced Kuwait crudes were hydrodesulfurized over a hydrotreating catalyst consisting of a group 6 and group 7 metal combination on alumina. The same reactor and experimental procedures were used in these two studies and it appears that the catalyst in both cases was the same. Montagna, Shah and Paraskos (60) studied the effect of particle size from 0.3 to 3 mm on the desulfurization conversion, metals and asphaltene removed for 22% and 36% reduced Kuwait crude. Montagna and Shah (61) studied the effect of liquid velocity and reactor length on these reactions for 36% and 53% reduced Kuwait crude. Paraskos, Frayer and Shah (62) measured the effect of liquid velocity and reactor length for a 53% reduced Kuwait crude and other lighter oils.

Plots (60) of the logarithm of the ratio of the inlet to outlet concentration of the reacting species vs. the particle diameter (Figure 16) showed a very strong dependence of the observed rate with particle size in most cases. Exceptions were runs using the 22% reduced crude with particle sizes below about 1 mm in diameter. The authors of the paper argued that the data supports the wetting model of Mears (49). Alternatively the data can be explained using the stagnancy model. Recognizing that the form of their plots follow the format for the stagnancy diagnostic for the mass velocity effect, we can use the slopes directly.

The slopes of the observed reaction rate vs. particle size of the data of Montagna (60) show slopes steeper than -1 suggesting the existence of a mass velocity effect for the large particle size (1-3 mm). The results of Montagna and Shah (61) for the 36% reduced crude showed that indeed there was an effect

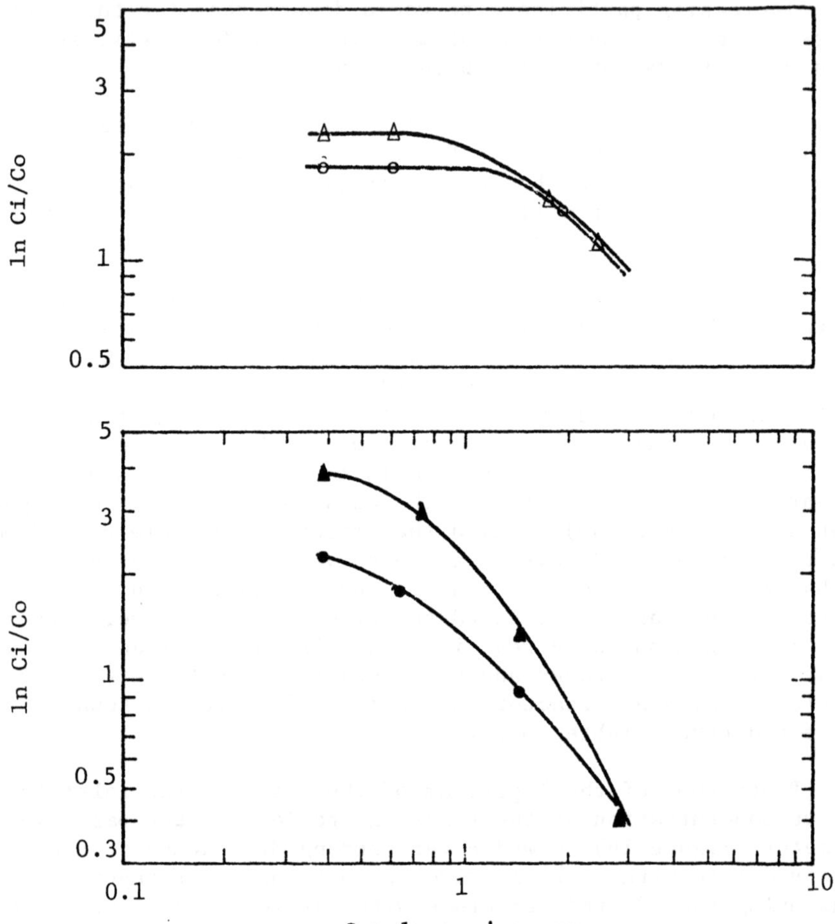

Figure 16 Effect of catalyst size on sulfur removal (o, Δ) and nickel removal (•, ▲) reactions for 36% Reduced Kuwait Crude (o, •) and 22% Reduced Crude (Δ, ▲) after Montagna et al (60)

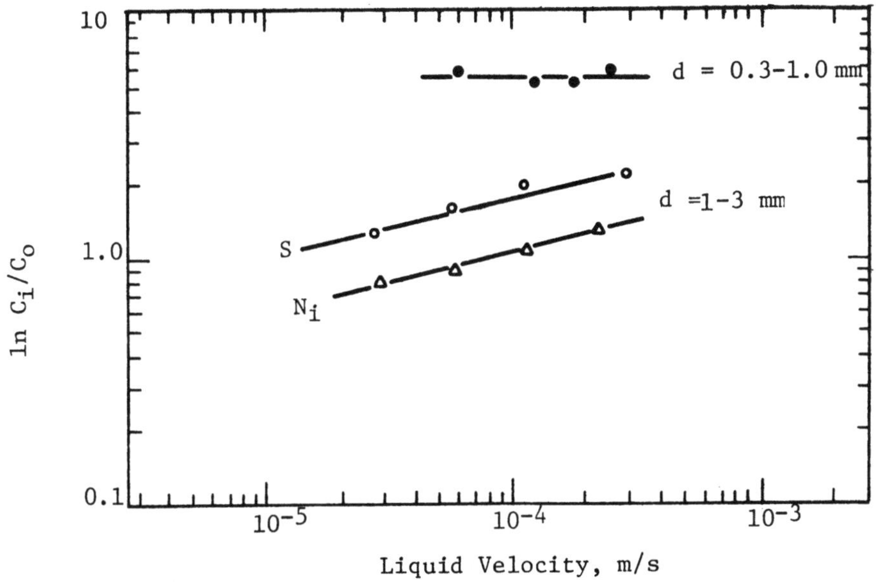

Figure 17 Effect of liquid velocity on the removal of sulfur (O) and nickel (Δ) using 36% Kuwait Reduced Crude and on sulfur removal using 53% Kuwait Reduced Crude (•). After Montagne et al (61)

of mass velocity on the apparent rate when using an 8 to 14 mesh catalyst particle size (corresponding to about 1 to 3 mm). (Figure 17)

For the 53% reduced crude, there was no mass velocity effect measured (60). For these runs, a 20 to 30 mesh size catalyst was used which corresponds to 0.5 to 0.8 mm, a particular size These results are consistent with the stagnancy model diagnostic as given above.

Now if we assume that the 53% reduced crude, which is lighter than the 36% reduced crude, has a diffusivity coefficient which is the same or higher than the 36% reduced crude, and if we note that the reactivity of the two feeds are about the same as shown by the equality of the logarithm of their inlet and outlet concentrations measured using the same small particle sizes (kinetic regime), we can infer that both feeds would behave similarly as a function of particle size.

Then observing that the 36% reduced crude exhibited slopes less than unity for large particles (> 1 mm) in the ln C_i/C_o vs. particle size plots, the diagnostic predicts a mass velocity effect for large particles. A mass velocity effect was observed (62) for runs with 53% reduced crude using a 1.6 mm catalyst.

Summarizing then, to check experimentally for the possibility of a mass velocity effect on the contacting efficiency in a trickle bed reactor in which direct gas/catalyst interactions are not important, one should measure conversions at various mass velocities in the reactor at constant particle size or with varying particle size at constant mass velocity. In a log-log plot of the observed rate vs. the particle size, a slope of zero would signify that a mass velocity effect is not likely. An increasing slope towards -1 or a steeper slope would indicate increasingly stronger influence of mass velocity on the observed rate. Since data on the value of n_s for the stagnancies as a function of reactor size and liquid mass velocity and catalyst particles size are not available, scale up based on the stagnancy model is not recommended. A better policy is to design pilot plant and commercial units outside the range where mass velocity effects are important to be able to take advantage of the full catalyst activity and also avoid working with "masked" kinetics.

3.9 Scale-Up of Highly Exothermic Reactions

For highly exothermic reactions, the designer must also concern himself with the possibility that the kinetics of the reaction may change upon vaporization of the reactants. This can give rise to sharp temperature increases due to increases in the rate of heat release and changes in the heat transfer coefficient. An example of a situation where the kinetics was changed by vaporization of the reactants is given in the paper by Hanika and coworkers (57). Here they studied the cyclohexene hydrogenation to cyclohexane in a trickle bed reactor. The reaction was carried out over a 3% palladium on charcoal and it was found in a stirred tank reactor that the effectiveness factor for the catalyst in a liquid phase was low, approximately 0.014 and that this resulted in a very low efficiency measured in the trickle bed reactor, an overall efficiency of about 0.0061. In the vapor phase, however, the reaction rate accelerated considerably.

Hanika (57) observed that when the reactor was run with cyclohexene diluted with product cyclohexane to slow down the reaction, the temperature increase with respect to axial distance was moderate, about 5°C. Whereas, when the liquid contained

75% cyclohexene the reaction rate was accelerated to provide enough heat to vaporize the cyclohexene/cyclohexane mixture and once vaporized, the temperature rose very quickly because the reaction rate now became accelerated in the vapor phase. There was a dip in the temperature curve at the back end of the reactor due to heat transfer through the reactor wall. The observation that the reaction rate in a liquid phase diffusion limited reaction can increase as the material vaporizes was also reported by Satterfield and Ozel (46) who studied the hydrogenation of benzene to cyclohexane.

Acceleration of the reaction rate upon vaporization is not a general phenomena. Reaction rates can remain the same or decrease when going from the liquid to the vapor phase depending on the concentration distribution of the reactants between the two phases and the kinetics of the liquid and vapor phase reactions. Therefore, each reaction that has the potential for going through a "dry point" should be kinetically tested in both phases.

In the scale up of a highly exothermic reaction from a small diameter pilot plant with heat dissipation through the reactor wall, to a large diameter adiabatic reactor, the designer must take into account that the heat effects and possible hot spotting may be more pronounced in the large scale reactor. To test for this concern, the pilot plant should be operated so that the reaction passes through the "dry point" and detailed axial temperature measurements should be taken. The pilot plant should be of a large enough diameter to retain essentially all of the reaction heat and not dissipate it through the wall.

Estimates of the requisite size can be made using the heat transfer correlations proposed be Weekman and Myers (63). When handling a sensitive reaction, care must be taken to assure that good distibution is obtained both in the pilot plant and in the commercial reactor to prevent shadowing effects which have a potential for localized hot spotting such as described by Jaffe (34) and discussed in Section 2.3.

4.0 MATHEMATICAL SCALE-UP MODELS

The complexity of the gas and liquid flows and contacting mechanisms between these phases and the catalyst in a trickle bed reactor have not been appropriately described by a universal mathematical model. Two types of models have been proposed: a cross-flow model and a dispersive model. These models have been shown to yield similar answers for axial dispersion (23), however, the cross-flow model offers greater potential as a predictive model.

The model has to be able to describe a gas-liquid flowing system in which the liquid flows in rivulets (23, 24) and in which all catalyst is not contacted evenly by the rivulets and the vapor. Also, the model must take into account the existence of stagnant zones which are next to and surround catalyst particles and which exchange with the flowing streams. In addition, the model must account for contacting between the vapor and the flowing and stagnant liquid and with the catalyst directly. To the knowledge of the writer, such a universal model has not been written. Pieces of this model do exist and have been discussed in the sections dealing with scale up by the use of phenomological correlations.

The cross-flow model is the most attractive model because it best describes the importance of the stagnant zones in a trickle bed reactor both by appropriately describing tracer responses (24, 36) and by explaining the observed catalyst contacting behavior (21) of these reactors. Various forms of the cross-flow models have been proposed but have yet to be demonstrated as useful for reactor design. A model proposed recently by Hinduja, Sundaresan and Jackson (23) has most of the required elements but this model has not been adapted to mixed phase flow. Giannetto, Baldi, Specchia and Sicardi (52) and Shah (38) have recently reviewed the cross-flow models used for interpreting tracer data.

Sato, Hirose, Takahashi and Toda (24) proposed a model based on an adsorption column design approach which lumps the effects of stagnancies and wetting into the mass transfer expressions. They demonstrated the validity of their model for the oxidation of ethanol.

Ramachandran and Smith (25) proposed a mixing cell in series model in which they use a mass transfer lumping technique like Sato, et al (24). Design calculations are simplified by the staging approach. They demonstrated that the model approximates plug flow with about ten mixing cells, however, the effect of stagnant zones and cross flow which are important for the modeling of pilot plants is not included.

It appears to the writer that a combination of the model elements of Ramachamdran and Smith (25) and that of Hinduja, Sundaresan and Jackson (23) would provide a general broad based model for scale up but this task is yet to be done.

The final story on trickle bed reactor design is far from having been told. Much progress has been made in developing an understanding of trickle bed reactors, but a priori design and scale up procedures have so far eluded us. However, much progress has been made in the understanding of multi-phase reaction systems such as trickle beds and many processes have successfully progressed from the bench to the pilot plant and into commercial use.

Aside from the inherent complexity caused by the presence of three phases (gas, liquid and the solid catalyst) among which contacting can permute in many ways, the definition of where each of the two phases (gas and liquid) are in any one time and point in the reactor has caused major difficulties in modeling of multiphase reactors. Hence, a complete mathematical model would have to describe not only the continuity equations for each of the phases but also will have to describe how these phases interact at any point within the reactor. Consequently, these equations will also have to describe the distribution of the two phases within the reactor. A general and accepted model of this type does not exist today.

This situation however, is not hopeless. The reactor engineer has available elements of the necessary model which he can use to design his pilot and commercial reactors and scale them from one to the other. One of the principal tasks of the reactor engineer is to diagnose his reaction system properly so that he may design his pilot plant tests to minimize the risks in scale-up.

NOMENCLATURE

a_{fw}	Fraction wetted area in the bed
C_i, C_o	Inlet and outlet concentration
C_L	Concentration in the liquid phase
C_S	Concentration on the catalyst surface
D	Effective molecular diffusion within catalyst pores
d	Effective particle size
f_{LG}	Mixed phase friction factor in pressure drop correlation, (Equation 2.2-3)
g	Gravitational constant
Ga	Galileo number ($=d^3g/\nu_L^2$)
H	Liquid hold up, dimensionless
k_{app}, k_i	Apparent and intrinsic reaction rate constants
$K_{crit.}$	Dispersion constant
K_e	Mass transfer coefficient between flowing liquid and stagnant liquid
K_g	Mass transfer coefficient between gas and liquid
K_s	Mass transfer coefficient, liquid to catalyst pellet
L	Reactor length
n_s	Number of particle diameters in a stagnant clump of catalyst
R_e	Reynolds number ($=Ud/\nu$)
R_p	Rate of reaction per catalyst pellet volume
S	Size of stagnancy in a bed
Sc	Schmidt number ($= \nu/D$)
Sh	Sherwood number ($= K_s d/D$)

T	Tortuosity factor for diffusion in catalyst
V	Liquid velocity
W	Lumped parameter defined by Equation (3.4-2)
w	Fitting parameter for Equation (3.4-5)
We	Weber number $(= V^2 d/\sigma_L \rho_L^3)$
X	Square root of the ratio of the pressure drop of the liquid to the gas phase if they were flowing alone in the bed
X^1	Defined by Equation (2.2-2)
Y	Diameter of reactor or effective spreading diameter from a point source
Z	Axial distance in the reactor
Z^1	A factor in pressure drop correlation given by Equation (2.2-3)
δ_{LG}, δ_L, δ_G	Pressure drop per unit length of bed
ε	Catalyst void fraction
ν	Kinematic viscosity
ρ	Density
σ	Surface tension
η	Catalyst effectiveness factor
μ	Viscosity
ψ	A factor in pressure drop correlation given by Equation (2.2-3)

REFERENCES

1. Beimesch, W. E. and D. P. Kessler, AIChE J., 17, 1160 (1971).

2. Chou, T. S., F. L. Worley, Jr., and D. Luss, I&EC Proc. Des. Dev., 16, 424 (1977).

3. Charpentier, J. C, and M. F. Favier, AIChE J, 21, 1213 (1975).

4. Talmor, E., AIChE J, 23, 868 (1977).

5. Satterfield, C. N., AIChE J, 21, 209 (1975).

6. Sato, Y., P. Hirose, F. Takahashi, M. Toda, and Y. Hashiguchi, J. Chem. Eng. Japan, 6, 315 (1973a).

7. Larkins, R. P., R. R. White, and P. W. Jeffrey, AIChE Journal, 7, 231 (1961).

8. Weekman, V. W., Jr., and J. E. Meyers, AIChE J, 10, 1951 (1964).

9. Lockhart, R. W. and R. C. Martinelli, Chem. Eng. Prog., 45, 39 (1949).

10. Sato, Y., Hirose, T., Takahashi, F. and Toda, N., J. of Chem. Eng. of Japan, 6, 147 (1973).

11. Turpin, J. L. and R. L. Huntington, AIChE J, 13, 1196 (1967).

12. Specchia, V. and Baldi, G. Chem. Eng. Sci., 32, 515 (1977).

13. Clements, L. D. and P. C. Schmidt, Paper Presented at AIChE 69th Annual Meeting, November, 1976, Chicago, Illinois.

14. Charpentier, J. C., "Hydrodynamic of Two-Phase Flow Through Porous Media", Chemical Engineering of Gas-Liquid-Solid Catalyst Reactions, G. A. L'Homme, Editor, Proceedings of an International Symposium held at the University of Liege, March, 1978.

15. Ross, L. D., Chem. Eng. Prog., 61, 77 (1965).

16. Hoftyzer, P. J., Trans. Inst. Chem. Engrs. (London) 42, T109 (1964).

17. Herskowitz, M., "Performance of Trickle Bed Reactors", PhD Thesis, Univ. of Calif., Davis, 1978.

18. Onda, K., H. Takeuchi, Y. Naeda and N. Takeuchi, Chem. Eng. Sci., 28, 1677 (1973).

19. Stanek, V. and N. Kolev, Chem. Eng. Sci., 33, 1049 (1978).

20. Zarzycki, R., Inst. Chem. Eng., 12, 88 (1972).

21. Koros, R. M., Proceedings of the 4th International 16th European Symposium on Chemical Reaction Engineering, Page 372, Heidleberg, Federal Republic of Germany, April 6-8, 1976.

22. Schwartz, J. G. and Roberts, G. W., Ind. Eng. Chem. Process Develop., 12, 262 (1973).

23. Hinduja, M. J., Sundaresan, S., and Jackson, R., AIChE J, 26, 274 (1980).

24. Sato, Y., P. Hirose, D. Takahashi, and M. Toda, Pacific Chemical Engineering Congree, Keyota, Japan, October 10-14, 1972, Part 2, Paper 8-3, Page 187.

25. Ramachandran, P. A. and J. M. Smith, The Chem. Eng. J., 17, 91 (1979).

26. Hirose, T., Y. Mori, and Y. Sato, J. of Chem. Eng. Japan, 9, 220 (1976).

27. Satterfield, C. N., "Mass Transfer In Heterogeneous Catalysis", MIT Press, Cambridge, 1970.

28. Satterfield, C. N., A. A. Pelossof and T. K. Sherwood, AIChE J, 15, 226 (1969).

29. Froment, G. F., Bischoff, K. B., "Chemical Reactor Analysis and Design", John Wiley and Sons, New York, 1979.

30. Koros, R. M., and E. J. Nowak, Chem. Eng. Sci., 22, 470 (1976).

31. Maddon, R. J., "Gradientless Hydrogenation Kinetics in a Slurry Reactor" PhD Disertation, Stanford Univ., Calif. (1974).

32. Maddon, R. J., J. P. O'Connell and M. Boudart, AIChE J, 24, 904 (1978).

33. Charpentier, J. C., The. Chem. Eng. J., 11, 161 (1976).

34. Jaffe, S. D., Ind. Eng. Chem. Process Des. Dev., 15, 410 (1976).

35. Deans, H. A. and Lapidus, L., AIChE J, 6, 656, (1960).

36. Hochman, J. M. and E. Effron, Ind. Eng. Chem. Fund., 8, 63 (1969).

37. Hoogendoon, C. J. and J. Lipps, Can. J. Chem. Eng., 43, 125 (1965).

38. Shah, Y. T., "Gas-Liquid Solid Reactor Design", McGraw-Hill, New York, 1979.

39. Hirose, T. M., M. Toda and Y. Sato, J. Chem. Eng. (Japan) 7, 187 (1974).

40. Mears, D. E., Chem. Eng. Sci., 26, 361 (1971).

41. Henry, H. C., Gilbert, J. B., Ind. Eng. Chem. Process Design Develop., 12, 328 (1973).

42. Lerou, J. J., D. Glasser, and D. Luss, Ind. Eng. Chem. Fundam., 19, 66 (1980).

43. van Klinken, J. and R. H. van Dongen, Chem. Eng. Sci., 35, 59 (1980).

44. N. Herskowitz, R. G. Carbonell, and J. M. Smith, AIChE J, 25, 272 (1979).

45. Sedriks, W. and C. N. Kenney, Chem. Eng. Sci., 28, 559 (1973).

46. Satterfield, C. N. and F. Ozel, AIChE J, 19, 1259 (1973).

47. Dwivedi, P. N. and S. N. Upadhyay, Ind. Eng. Chem. Process Design Develop., 16, 157 (1977).

48. Goto, S. and J. M. Smith, AIChE J, 21, 706 (1975).

49. Mears, D. E., Adv. Chem. Ser., 133, 218 (1974).

50. Onda, K., H. Takeuchi and H. Koyana, Kagaku Kogaku, 32, 121 (1976).

51. Puranik, S. A. and A. Vogelpohl, Chem. Eng. Sci., 29, 501 (1974).

52. Gianetto, A., G. Baldi, V. Specchia and S. Sicardi, AIChE J, 24, 1087, (1978).

53. Bondi, A., Chem. Technol., 1, 185 (March, 1971).

54. Colombo, A. J., G. Baldi, and S. Sicardi, Chem. Eng. Sci., 31, 1101 (1976).

55. Sicardi, S., G. Baldi, A. Gianetto and V. Specchia, Chem. Eng. Sci., 35, 67 (1980).

56. Herskowitz, N. and J. M. Smith, AIChE J, 24, 439, 450 (1978).

57. Hanika, J., K. Sporka, V. Ruzicka, and J. Hrstka, The Chem. Eng. J., 12, 193 (1976).

58. Van Swaaij, W. P. N., J. C. Charpentier and J. Villermaux, Chem. Eng. Sci., 124, 1083 (1969).

59. Michell, R. W. and I. A. Furzer, Trans. Instn. Chem. Engineer, 50, 334 (1972).

60. Montagna, A. A., Y. P. Shah and J. A. Paraskos, Ind. Eng. Chem. Process Des. Dev., 16, 152 (1977).

61. Montagna, A. A. and Y. P. Shah, Ind. Eng. Chem. Process Des. Dev., 14, 479 (1975).

62. Paraskos, J. A., J. A. Frayer and Y. P. Shah, I&EC Process Design Dev., 14, 315 (1975).

63. Weekman, Jr., E. W. and J. E. Myers, AIChE J, 11, 13 (1965).

64. Schwartz, J. G., E. Weger, and N. P. Dudukovic, AIChE J, 22, 894 (1976).

65. Koros, R. M. To be published.

LABORATORY REACTORS FOR MIXED PHASE CATALYTIC STUDIES

J.A.Mahoney

Amoco Oil Co.
Naperville,Illinois 60566,USA

1. INTRODUCTION

In the design of any commercial catalytic reactor system evaluation of the reaction kinetics is the crucial step in the overall process development program. Equations to describe the physical rate processes (heat and mass transfer) can usually be predicted from existing correlations. However,the prediction of chemical rate parameters has not reach the "a priori" stage and thus experimental determination of these constants is unavoidable.

Choosing the type of laboratory reactor for evaluating process kinetics is extremely complex and an important factor of any project. A poor choice will result in data that would scale-up erroneously leading to a disastrous design of a commercial unit. Selection of a laboratory catalytic reactor has to be made on the basis of several considerations.

This review will focus mainly on some of the important types of laboratory reactors and to discuss both the strong points and limitations of each.

The review will also briefly cover some indirect methods of determining catalyst activity. Included in the discussion will be the theory in linking the measurement to activity as well as actual or potential applications to real systems.

The paper will include a discussion of the transport resistances commonly encountered in actual commercial catalytic systems and commonly used criteria for diagnosing the importance of these phenomena.A discussion of important criteria for testing catalysts and some pitfalls commonly encountered in actual laboratory situations are presented. A case history of the development of a new generation of coal liquefaction and reforming catalysts will be reviewed.

2. DIRECT ACTIVITY MEASUREMENTS

2.1. Catalytic reactor types

Reactors encountered in the chemical process industry are commonly divided into the following classifications: homogeneous, or single phase,and heterogeneous reactors.Heterogeneous reactors can be further subdivided into gas-solid and multiphase systems. The occurence of multiphase (gas-liquid,liquid-liquid,liquid-solid, gas-liquid-solid) in industrial operation is very common.

There are many types of reactors employed in industrial practice and in this overview we classify them as follows (1):

a) conventional reactors - these reactors are used for large as well as small scale operations for a wide range of chemical reactions.

b) specific purpose and novel reactors - these are largely laboratory scale reactors.

Conventional reactors are usually designed to optimize the profits of a chemical process but are often encountered in the laboratory because of the inherent simplicity of these reactor systems as well as the intrinsec confidence people have in them because of years of industrial experience and exposure.

Conversely,specific purpose reactors,are information producing devices and are designed for specific information purposes such as the measurement of intrinsec kinetics and transport resistances.

The emphasis in this review will be placed on laboratory-scale specific purpose reactors.In particular we will limit ourselves to reaction systems employing solid catalysts,which is the system type most frequently encountered in industry and which the maximum profit potential can be generated from precise design.

2.2 Objectives of catalyst testing programs

The heart of practically all chemical processes is the catalyst itself.Laboratory reactors are commonly designed to generate data to optimize catalyst formulations and operating conditions. Obviously,transport resistances can be a profound influence on the performance of a catalyticreactor.The path a reactant molecule must follow to get from the bulk phase to an active catalytic site within a porous catalyst particle as well as the reverse path for a product molecule is commonly described in eight steps:

(a) transport of reactants from the bulk gas to the bulk liquid
(b) transport of reactants from the bulk liquid to the liquid-solid interface
(c) intraparticle transport of reactants into the catalyst particle
(d) adsorption of reactants at the interior sites of the catalyst particles
(e) chemical reaction of adsorbed reactants to adsorbed products
(f) desorption of adsorbed products
(g) transport of products from the interior sites to the outer surface of the catalyst particle
(h) transport of products from the fluid-solid interface into the bulk-fluid stream.

Any one of the above steps can cloud information we are trying to develop about the chemical reaction step (e).The catalytic researcher is faced with the task of designing a reactor to either eliminate the mass transfer steps,clouding his catalytic kinetic measurements,or control them to such an extent that their influence on the catalytic reaction can be precisely measured.

The objectives of catalyst testing programs usually include three areas,viz. catalyst development,mechanistic studies,and process research.

Catalyst development includes screening of new formulations for activity,selectivity,and deactivation rate measurements;optimizing promising new catalyst formulations;and modelling sucessful commercial catalytic systems.

Kinetic mechanism studies of industrially important catalysts are designed to answer the question of "how and why" a particular catalyst functions.This information aids in modelling the reaction system (but is not essential if a purely empirical rather than theoretical model to describe the reaction system).Such information is also used in developing a second or improved generation catalyst for a given chemical reaction.

Finally,process research is a key element of any sucessful catalyst testing program so that the industrial reactors will be operated at the optimal temperature and pressure to insure maximum product generation.

Several practices are essential in catalyst testing programs. These include:

- flow systems- these eliminate or minimize transients and result in greater kinetic precision

- kinetic measurements of conversion- greater precision over non-kinetic measurements

. gradientless laboratory reactors- these minimize heat and mass transfer effects while generating kinetic data of maximum precision

. complete mass balances- essentially for precise selectivity measurements,particularly in gas-liquid systems

. in-lab analytical capability- essential to constantly monitor data throughout a given run

. reproducibility- between experiments and reactors

. consistent catalyst pretreatment- extremely important in deactivation measurements

. fresh vs. aged catalyst- normally catalysts are fresh for only a short period of their industrial life.The catalyst should be studied in the form it will be used industrially

. initial or steady-state activity- initial activity measurements alone do not give practical information on long-lived reaction systems.

2.3. Reactor operation

Laboratory-reaction studies and reaction system in general are conducted in three general modes (2):batch operation,semibatch operation,and continuous-flow operation.

In the batch operation a closed reactor system is used ,and the progress of reaction is monitored as a function of time.

In semibatch operation one reactant is charged and the coreactant is added during the course of reaction.

In contrast to batch and semibatch operation,continuous-flow reactors can be operated in the steady state.Reaction progress is followed as a function of residence time in the reactor rather than real time. Two limiting situations exist for continuous-flow operation: pulg-flow reactors were all molecules have the same residence time and concentrations vary only along the length of the tubular reactor,and continuous-flow stirred-tank reactor where due to vigorous agitation,the reactor contents are well mixed,so that the effluent composition equals that in the tank.As effluent and reactor contents are of uniform composition,progress of reaction is monitored by measuring the exit composition vs. average residence time behavior in the steady state.

In the plug-flow reactor the effluent composition is an average value based on the overall,integral conversion.On the other hand as the composition is uniform throughout the steady-state continuous-flow stirred-tank reactor,the effluent composition vs. average residence time represent point or local values.Plug flow reactors can be operated to give either integral or differential conversions.Small conversions obtained in differential reactors can cause analytical problems resulting in loss of precision.

2.3.1.Reactor environments

In a kinetic study the temperature field is of great importance.Reactors are commonly operated between the extremes of isothermal and adiabatic cases.Adiabatic operation on a laboratory scale is difficult to achieve by design and although it is not as complex to analyze as nonadiabatic,nonisothermal data,it is much preferred to follow laboratory procedures that generate isothermal data.

2.3.2. Data analysis

Benson (3) provides a summary to guide for procurement of precise rate parameters.To obtain a rate constant with an error of $\pm$ ε,it is necessary:

(a) To measure concentration with an accuracy of

$$\pm \frac{\text{Change in concentration}}{\text{Largest concentration}} \quad \frac{\varepsilon}{1.4\%}$$

(b) To measure time (or residence time) with an accuracy of

$$\pm \frac{\text{Time interval}}{\text{Largest time}} \quad \frac{\varepsilon}{1.4\%}$$

(c) To measure temperature with an accuracy of

$$\pm \frac{\varepsilon}{35\%} = \frac{\varepsilon T}{3500}$$

The consequent accuracy of E,the activation energy,for two measurements of k,each of accuracy $\pm\varepsilon$ percent,will be

$$\overline{\Delta^2 E}^{1/2} = \pm \frac{\text{Highest temperature}}{\text{Temperature interval}} \quad \frac{\varepsilon}{18}$$

Benson declares these to be rough but reliable rules,accurate for reactions where E/RT is about 35.

Benson also notes that when k is computed from concentrations measured at two time intervals,the error in k,the rate constant,is reduced when relative time and concentration intervals are large, from the Arrhenius relation:

$$k = A\exp(-E/RT)$$

$$d(\ln k) = \frac{E}{RT^2}\,dT$$

$$\frac{dk}{k} = \frac{E}{RT}\,\frac{dT}{T}$$

The error in T produces a consequent error in k,the magnitude of which depends on the activation energy,E. The error in E increases as the temperature interval over which k is measured is reduced.

2.4. Laboratory catalytic reactor types

As discussed earlier steady-state flow reactors are preferred for precise measurement of catalytic rate parameters.The five main types of reactors employed in the laboratory are:

(a) differential plug-flow reactor

(b) integral plug-flow reactor

(c) continuously-stirred tank reactor

(d) recycle reactors

(e) transport reactors

These reactors are sketched in Fig.2.4-1.

Differential reactors yield unreliable kinetic data because of the high degree of analytical precision needed with small conversions.Also,synthetic feeds must be employed in order to measure reaction rates as a function of composition.One way of overcoming this difficulty is to use an integral reactor to provide feed for the differential reactor (4).

Integral reactor data are intrinsecally more accurate than differential reactor data because of the high conversions involved. However,integral reactors have a serious disadvantage.The effect of temperature on the rate constant,k,is exponential as described by the Arrhenius law.Thus relatively small errors in temperature measurements and deviations from isothermality can lead to large errors in measured rates for reactions; large reactors have substantial concentration and temperature gradients that are caused by large conversions,and their operation is often restricted by excessive pressure drop.

In Fig. 2.4-2 we have illustrated the gradients present in a fixed bed catalytic reformer operating at integral conversion le-

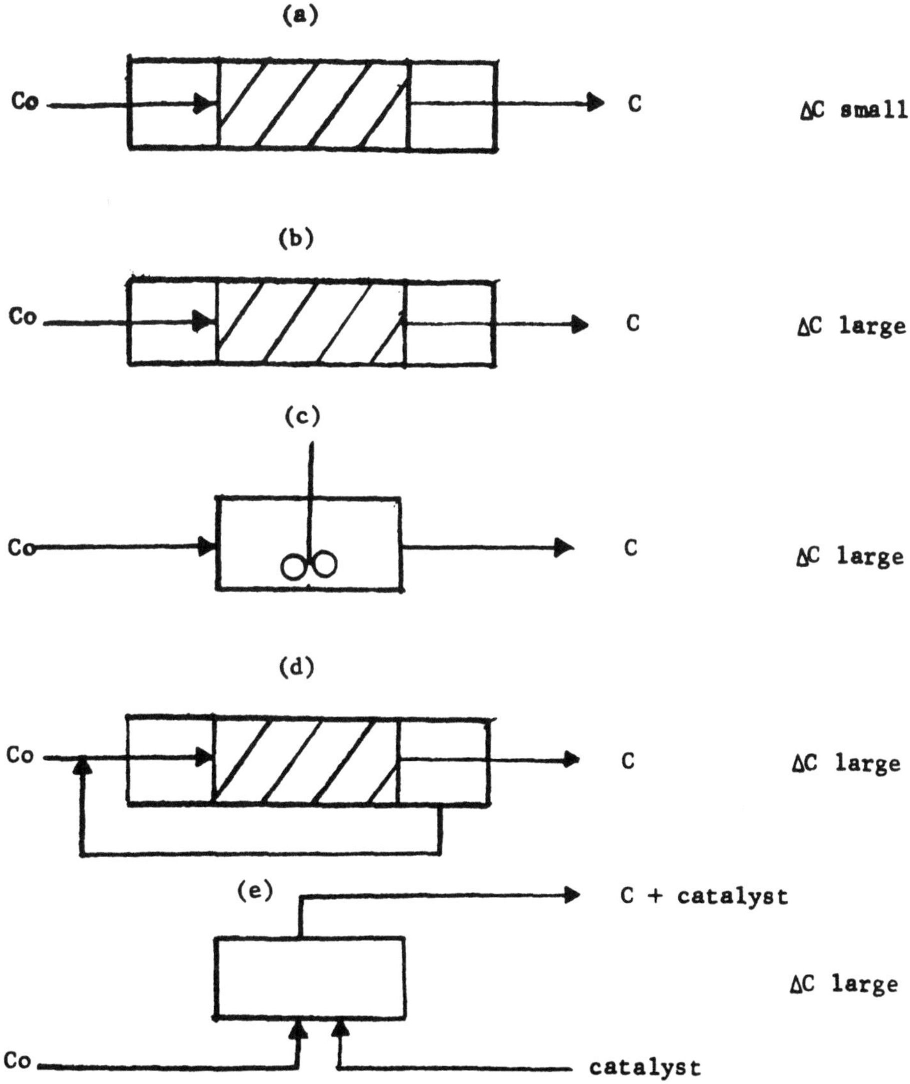

Figure 2.4-1 Laboratory Catalytic Reactor Types

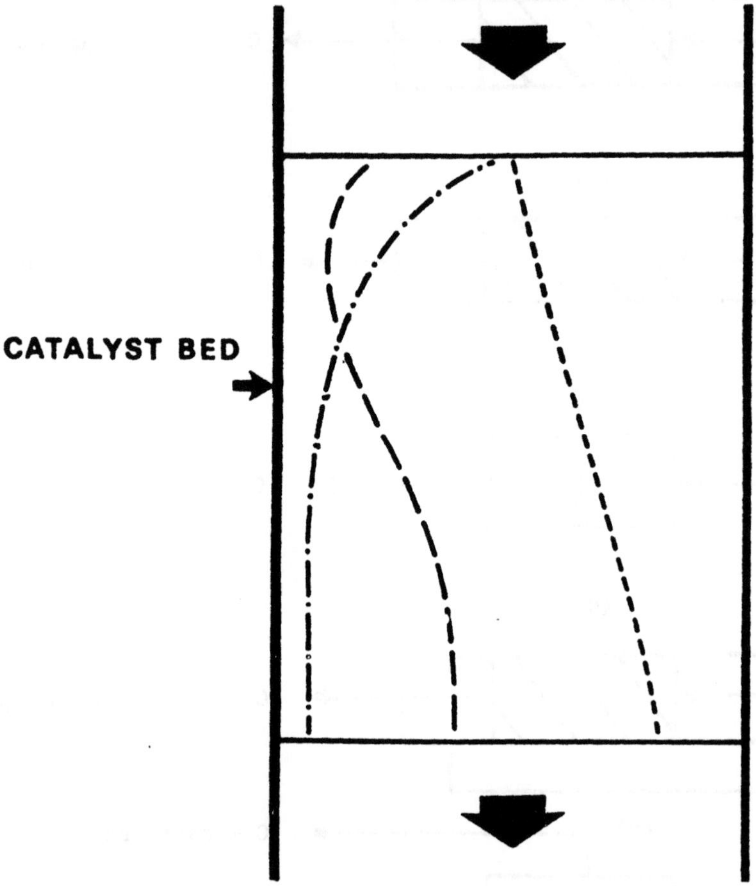

— — Temperature
------Coke level
-.-.-.Reactant concentration (first order)

Figure 2.4-2 Gradients in fixed bed tubular reactors with plug flow conditions vels.The temperature varies due to the different heat effects of the complex reaction network.The reactant concentration decreases while passing through the bed as the reactant is converted to products.The undesirable side reaction is the production of coke,which, in this illustration,the concentration increases as the reactant passes down the bed. A precise mathematical design relationship between all these nonlinear interacted parameters is difficult to obtain.

The stirred reactor and recycle reactors are fundamentally similar in that both give a direct measure of rate from the measured difference between inlet and outlet concentration.The principle feature of both these reactors is that mass and heat transfer resistances are almost eliminated and perfect mixing can be achieved. Thus,both can be operated under practically gradientless conditions.

Transport reactors have had limited use in the laboratory but are widely employed commercially in the catalytic cracking of heavier petroleum fractions to gasoline.

Recycle reactor schemes,with either external or internal circulation,are the most useful for obtaining catalytic kinetic data. Nevertheless,external circulation is plagued by mechanical pumps, which require cooling and subsequent reheating of the recycle stream; also,perfect mixing,which depends on the recycle rate,is hard to achieve. CSTR behavior (perfect mixing) can be achieved in a recycle reactor if the recycle ratio is greater than 25 (5).

2.4.1. Recycle reactors

If a differential reactor is operated such that part of the effluent stream is recycled to the feed stream,a reactor system is obtained with several advantages over the conventional differential reactor.The high recirculation velocities reduce or eliminate heat and mass transfer resistances.Thus,we are able to obtain integral conversions in this system even though the conversion per pass is differential.Because of the problems involved with the external recirculation pump,reactors employing internal recycle are preferred.

In an internal recycled gradientless reactor with a stationary basket the reactant mixture enters the vessel through the inlet connection and is driven up the draft tube by the motion of the impeller as seen in Fig. 2.4-3.The vapor passes through the catalyst bed,moves outside the draft tube,and is then recycled through the catalyst bed.After some 20 to 100 passes,the mixture is expelled through the outlet connection by pressure difference.The high recycle rate insures that the vapor-phase mixture inside the reactor is completely backmixed,i.e.,at a uniform temperature and composition.

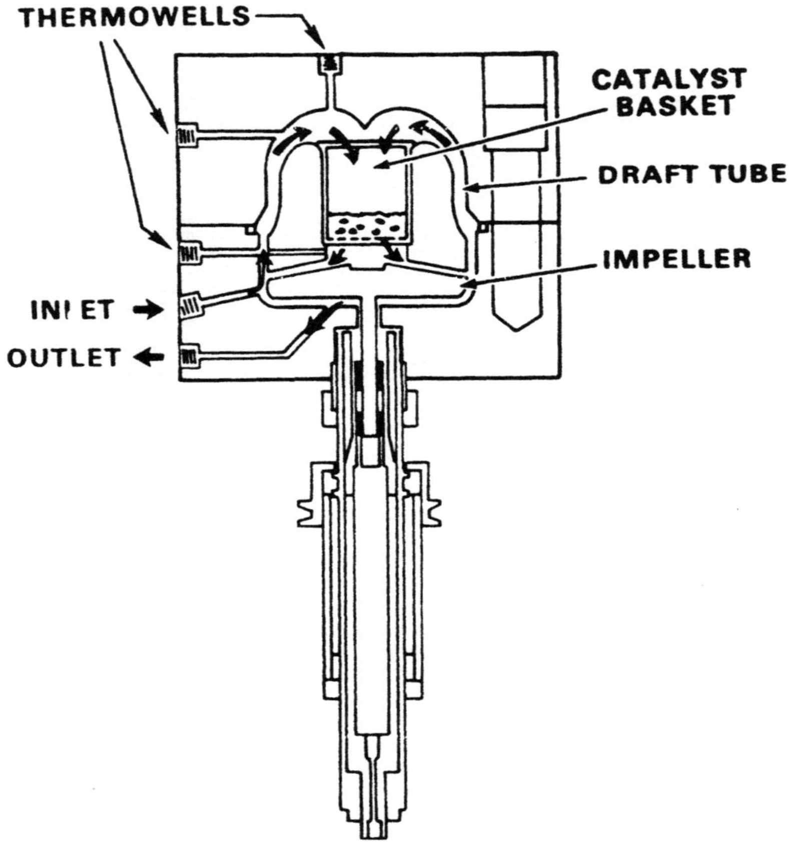

Figure 2.4-3 **INTERNALLY RECYCLED CATALYTIC REACTOR**

Because the gradientless reactor permits reaction studies at isothermal conditions with uniform concentrations,it eliminates the need for integral or partial-differential equations,whose solution often requires complex computer programming and processing. Instead,each steady-state experiment with integral conversion yields a reaction rate that is simply calculated from the ordinary difference equation:

$$\frac{W}{F} = \frac{X_o - X_i}{r_p}$$

instead of the integral equation

$$\frac{W}{F} = \int_{X_i}^{X_o} \frac{dX}{r_p}$$

where r_p is the global rate of reaction per unit mass of catalyst, F is the feed rate of reactant,W is the mass of catalyst,X is the concentration of reactant,and subscripts o and i refer to outlet and inlet concentration.

In the rotating basket reactor the catalyst is held in a wire mesh basket which is attached to a stirrer rotating inside a reactor pot.Basket design and baffling are chief variation in individual design.When the basket is rotated at high speeds the contents of the reactor are mixed to uniform composition and the geometry of the basket causes the contents of the reactor to circulate through the basket.This reactor is superior to the external recycle reactor in that very high speeds of rotation (2500 RPM) ensure the absence of heat and mass transfer effects.

Two types of rotating basket designs have been employed in our research.The first type,a four rectangular paddle basket , was modelled after the original design of Carberry and Tajbl (6). This design was soon abandoned in favor of a second type of multiphase gradientless reactor that consisted of a spinning annular basket held in a 300 cc stirred autoclave (7).Baffles around the basket insure good mixing and contacting of the liquid and gas and also eliminates vortexing.

A disadvantage of rotating basket gradientless reactors is that the actual temperature at the surface of the catalyst cannot be measured.The temperature measured is that of the fluid in the reactor and not the temperature at the surface of the catalyst. Additionally,the measurement of the fluid velocity past the catalyst surface cannot be measured with this reactor configuration. There is also some question whether all the catalyst in the basket is exposed uniformly to the reactor contents.

Gradientless reactors with stationary catalysts are preferred for several reasons.The most important feature of these types of reactors is that the actual temperature of the catalyst,the crucial parameter in any kinetic study,can be measured directly. These reactors minimize bypassing and contacting problems and thus can be used as single or multi pellet reactors.In some designs,such as the Berty reactors (see Fig. 2.4-3),pressure drop and hence flow rate across the catalyst bed can be measured (8).Determinations of flow rate past the catalyst pellet permits this type of reactor to be used in mass transfer as well as kinetic studies.Reactors with the catalyst constrained in a stationary basket are preferred over versions with the catalyst impregnated on the reactor walls or on internal cylinder because the low amount of catalytically active surface area maximizes the flow volume to catalyst volume ratio thereby limiting its use in kinetic studies of noncatalytic or homogeneous side reactions are occuring simultaneously with the catalytic reaction.The Berty design minimizes the free volume to catalyst volume ratio by proper design of the catalyst basket and draft tube.Data obtained from reactors employing stationary catalyst baskets will be discussed in the Case History section.

2.4.2. Microreactors

Microreactors are characterized by the use of a very small quantity of catalyst.Since quantities of reactants and catalyst are small,heat effects are negligible and hence they can be operated almost isothermally.The reactor can be of a relatively simple design and reactant quantities can be minimized.The reactors can be operated in either the steady-state or transient flow regime.Such reactors are usually limited to low pressure and temperature studies and thus cannot be reliably used to study most catalytic systems of industrial interest.Good mass balances are difficult if not impossible to obtain in these reactors and selectivity measurements,particularly of mixed phase systems,would contain considerable error.A good discussion of microreactors is contained in Doraiswany and Tajbl's review article (9).

These reactors have not been employed in mixed phase catalytic studies to any great extent.

3. INDIRECT ACTIVITY MEASUREMENTS

Indirect activity measurements are an important source of supplemental information for those involved in the study of catalytic systems.Obviously,qualitative,rather than quantitative information is the goal of most indirect activity studies.A detailed description of available indirect activity techniques is beyond the scope of this paper,however,a short discussion of some of the more useful techniques is appropriate.

Diffusion-adsorption characterization is important in any heterogeneous catalytic system.Surface area measurements,pore size distributions,and pore shape measurements are needed to taylor make catalysts for a given reaction system.Such measurements assume increased importance in gas-liquid reaction systems because of the increased importance of intraparticle diffusion resistances.

Surface chemistry characterization is a useful guide for optimizing the catalytic components contained on the surface of the catalyst.Measurement of acid-base strengths is extremely important in synthetic fuel catalysis.Infra-red studies yield information on the oxidation state of catalytically active metals.Such information is essential in pretreatment studies.The field of surface chemistry characterization has expanded rapidly in recent years with the introduction of ESCA,Auger spectroscopy,and a wide variety of other techniques.

X-ray characterization has always been extremely important in the study of catalytically reacting systems.This is the primary technique for measurement of the size and amount of catalytically active metal clusters.Catalytic activity is directly related to the molecular size of the catalytic active material.This technique in conjunction with electron microscope has been used in recent years to study concentration gradients of catalytically active metals across extrudates as well as deposition of metals in resid and synthetic fuels processing.

Other techniques such as the measurement of magnetic properties of catalysts and their semi-conductor character have been widely used.

4. MASS TRANSFER RESISTANCE CRITERIA

Kinetic data disguised by diffusion effects is not useful for formulating mathematical expressions that can be used in the design of mixed phase catalytic reactors.The chief characteristic of the various laboratory reactors centers around the need for either eliminating heat and mass transfer resistances or accounting for them in a precise manner.

Diffusional effects have been well reviewed by others (2). General guidelines are given for minimizing heat and mass resistances and literature equations are shown in Table 4.0-1 (10).In operating any reactor it should be ensured that the criteria in Table 4.0-1 are met if the kinetic data obtained are to reflect the true chemical behavior of the reactions involved.

TABLE 4.0-1

Criteria for Absence of Heat and Mass Resistances

1. For the absence of concentration gradients in an isothermal spherical particle,

$$\frac{r R_p^2}{C_o D_{eff}} \left\} \begin{array}{l} < 6.0 \text{ zero order} \\ < 0.6 \text{ first order} \\ < 0.3 \text{ second order} \end{array} \right.$$

2. For the absence of temperature gradients

$$\frac{\Delta H \, r \, R_p^2}{C_o D_{eff}} < \frac{T_o R}{E}$$

3. For the absence of interphase gradients in isothermal systems; power law kinetics

$$\frac{R_p}{C_o k_m} < \frac{0.15}{\eta}$$

4. For the minimum reactor length required for neglecting axial dispersion effects

$$\frac{L}{d_p} > \frac{20 \eta}{Pe_a} \ln \frac{C_o}{C_f}$$

5. For control of bypassing effects due to dilution

$$\frac{L}{d_p} > 250 \frac{b}{\Delta}$$

5. REACTOR SELECTION

Weekman addressed himself to the general problem of evaluating reactor types for a mixed phase catalytic system (11). Weekman rated each reactor with respect to such characteristics as construction, sampling, isothermality, contact time distribution, and selectivity disguise-decay. His conclusions for non-decaying systems are shown in Table 5.0-1. In non-decaying or slowly deactivating reaction networks, the continuous stirred tank catalytic reactor (CSTCR) stands out above the other reactor types. As we dis-

TABLE 5.0-1 REACTOR RATINGS
Gas-Liquid, Powdered Catalyst, Nondecaying Catalyst System

Reactor type	Sampling and analysis	Isothermality	Residence contact time	Selectivity disguise-decay	Construction problems
Differential	P-F	F-G	F	G	G
Fixed bed	G	P-F	F	G	G
Stirred batch	F	G	G	G	G
Continuous stirred tank catalytic reactor	G	G	F-G	G	F-G
Continuous stirred tank	F	G	F-G	G	P-F
Straight-through transport	F-G	P-F	F-G	G	F-G
Recirculating transport	F-G	G	G	G	P-F
Pulse	G	F-G	P	G	G

G = good, F = fair, P = poor.

cussed earlier,this reactor is preferred because integral conversion are obtained under uniform concentrations and temperature. External film transport coefficients can be measured in stationary basket CSTCR as well as intraparticle diffusion influences.

6. CASE HISTORIES

6.1 Vapor phase gradientless reactor

Our aim was to design a bench-scale catalytic reactor to operate at commercial petroleum processing conditions and also eliminate the effect of external heat and mass transfer on the reaction rates.To satisfy this requirement each catalyst particle inside the reactor must be surrounded by a vapor at the same temperature and composition with film gradients around the pellet eliminated. Also, the physical environment around the catalyst particle must approach that in a full-scale commercial reactor.

Unit description

The Union Carbide gradientless reactor met all of the above criteria with the single exception of being able to operate at petroleum processing temperatures,viz.,in the range of 700-1100 F. The Teflon bearing used in their standard Magne-drive assembly could withstand a maximum reactor temperature of about 600 F,which could not be exceeded merely by substituting a graphite bearing for Teflon.Graphite is not a suitable bearing material in a dry,high temperature,hydrogen-rich atmosphere. The problem was solved by adding an additional bearing made of Rulon,a ceramic material impregnated with Teflon,farther from the reactor body and surrounding it with a water cooling jacket (see Fig. 2.4-3 and 6.1-1). Without the jacket,any bearing in this position or closer to the reactor (as in the Union Carbide reactor) would be destroyed at 1100 F. Rulon was also used as the bearing material throughout the drive shaft and Magne-drive assembly.(Additional details are available from Autoclave Engineers,Inc.,Erie,PA,USA).

The degree of mixing is a function of the impeller speed and the gas-phase density,and perfect mixing is easily achieved under most operating conditions.However,at high temperature and low pressure,some difficulty does arise.The criterion for complete mixing is the insensitivity of conversion to turbine speed at constant process conditions.

Recycle rates and hence linear velocity can be determined by measuring either the small pressure or temperature drop across the catalyst bed.The pressure drop is measured by sensitive manometers and them correlated with linear velocity by means of Leva's

FIGURE 6.1-1

MODIFIED DRIVE SHAFT ASSEMBLY

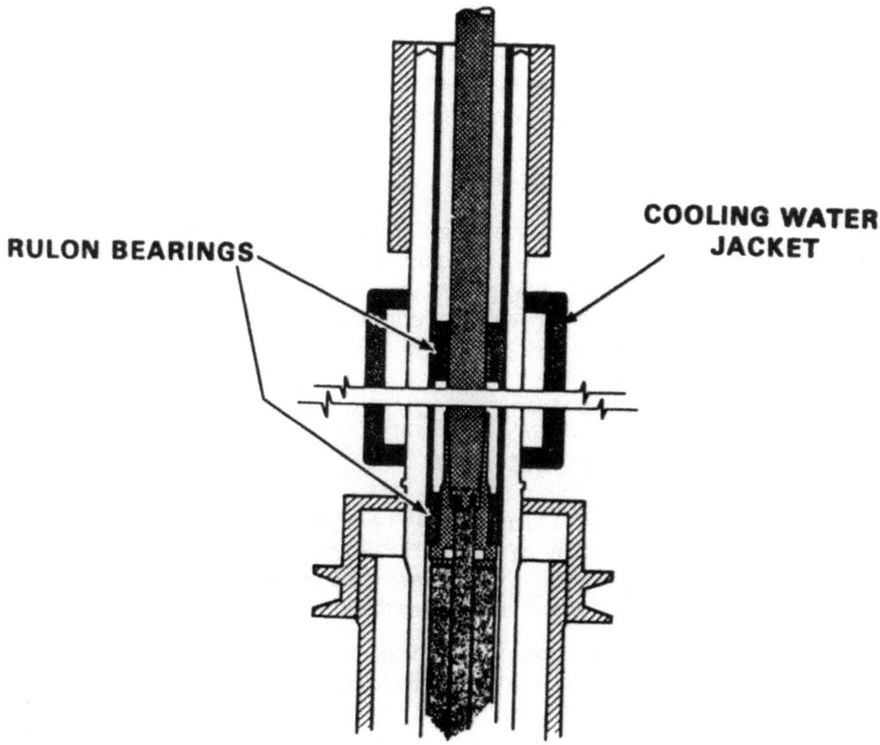

correlation.The temperature difference across the catalyst bed is determined by strategically placed thermocouples.The recycle rate can then be determined by the ratio of the adiabatic temperature change and the measured temperature difference

$$\frac{\text{mass flow through the catalyst bed}}{\text{mass flow out of the reactor}} = \frac{\Delta T \text{ adiabatic}}{\Delta T \text{ measured}}$$

The total adiabatic temperature change is calculated from heat and material balance.A typical value of the recycle rate in our experiments is 60,based on a thermodynamic calculation.

Ultraforming kinetic studies

Ultraforming,which is a high-severity catalytic naphta reforming process,operates at high conversions,and the catalyst deactivates rapidly between regenerations.Thus,kinetic studies are essential in order to model this process.These studies can be divided into two main areas:

1. Reaction mechanism

2. Catalyst deactivation

The discussion will focus on the catalyst deactivation phenomenon.Dehydrocyclization of paraffins,the slowest of the major types of reforming reactions,requires effective promoters to accelerate the reaction rate and decrease its decline with time.

Our studies were concerned with the reactions of n-heptane catalyzed by a commercial platinum-on-alumina catalyst which had been activated by a halide promoter.The n-heptane was dried with 3 Å molecular sieve.No impurities could be detected on a Hewlett--Packard 5750 gas chromatograph.

In a typical run,prepurified hydrogen was mixed with the hydrocarbon in the prescribed mole ratio (see Figure 6.1-2).The reformate product was collected in an iced liquid-gas separator which was at reactor pressure.The overhead gas passed through a gas sampling bomb and a wet test meter.The liquid and gas samples were analyzed by gas chromatography and the results were combined to give the final values of the rate constants.A typical run lasted from a few hours to a few days.Process conditions were constant so that the catalyst deactivated at a constant temperature and pressure.

Figure 6.1-3 presents some experimental results from our n-heptane catalytic reforming experiments.The data are presented on a semilog plot of the rate constant versus the square root of time on stream,the familiar Voorhies relationship.The rate cons-

FIGURE 6.1-2

FLOW DIAGRAM OF GRADIENTLESS REACTOR

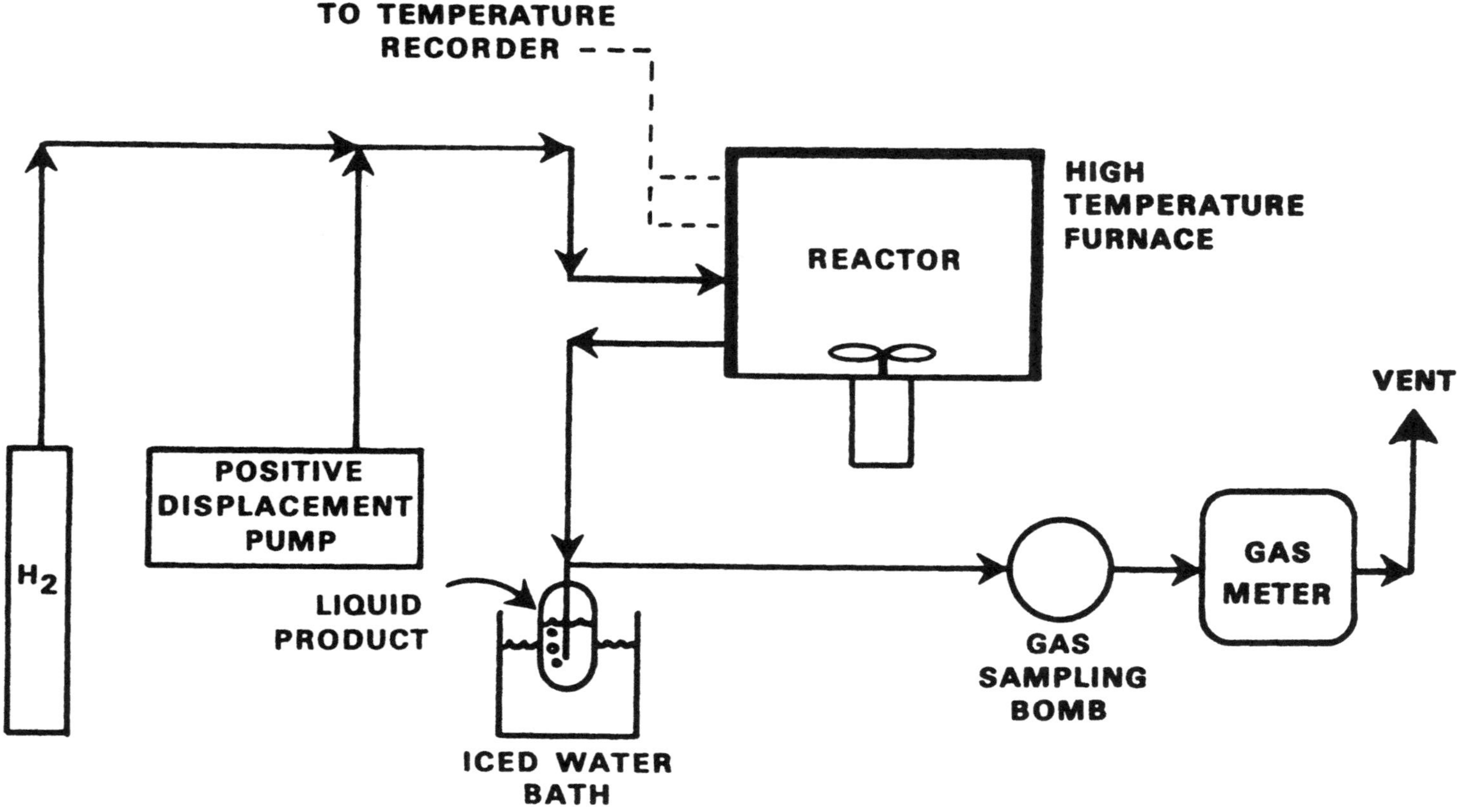

tant is in units of g-moles per hour per gram of catalyst. The abscissa is in units of the square root of time. The dehydrocyclization deactivation is seen to take place in two distinct regimes, a rapid, initial step followed by a slower, more gradual decline. It should be pointed out that Arrhenius activation energies are normally based on initial values of the rate constant, which are determined by extrapolation back to zero time. Thus, both the initial, extrapolated value of the rate constant and the value of ΔE depend on which deactivation regime is chosen for extrapolation. Considerable confusion in the literature could have been avoided if the precise means of determining the value of the rate constants and ΔE had been disclosed.

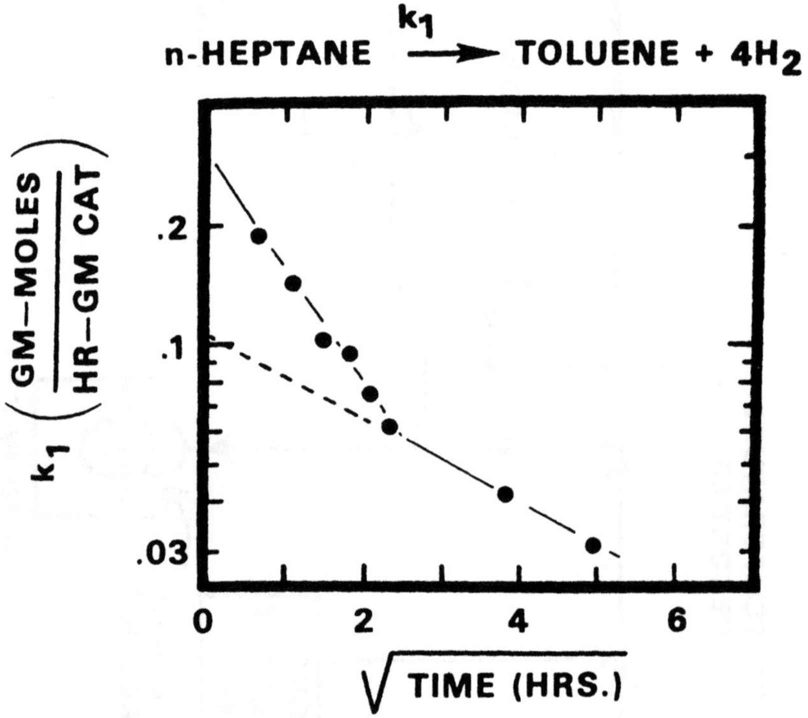

FIGURE 6.1-3

DEHYDROCYCLIZATION DEACTIVATION

6.2 Multiphase gradientless reactor

Another application where gradientless reactors are employed at Amoco Oil is to test catalysts for coal hydroliquefaction.Intended application of the catalysts is for liquefaction processes in advanced stages of development such as H-Coal from Hydrocarbon Research,Inc.

Unit description

Figure 6.2-1 shows a schematic flow diagram of the pilot plant.The coal slurry is pumped with a metering pump to reactor pressure (137 atm) and joined with the regulated inlet hydrogen stream.The combined stream of coal slurry and hydrogen is continuously introduced to the reactor bottom and the reaction products are withdrawn through a vertical overflow tube to either one of the two high pressure product separator/receivers.The reactor holdup can be varied by changing the overflow tube height so that the degree of thermal reaction taking place in the bulk phase,relative to the catalytic reaction,can be adjusted.No preheater is employed and all the transfer lines are electrically heated to less than 70 C. A bypass line is provided directly to the reactor top to perform catalyst pretreatments.

The product gas exiting from the high pressure separator/ receiver is scrubbed of ammonium salts in a gas scrubber filled with water to prevent downstream plugging,passed through a water vapor trap,and depressured by a pressure letdown valve.The depressured product gas is metered through a wet test meter before it is vented out or sampled for gas chromatographic analysis.The slurry product collected in the higher pressure receiver is drained either manually or automatically through a two-step pressure letdown loop.The operation is once through with respect to both hydrogen and slurry feed.

Reactor description

A 1-liter stirred autoclave is used as a reactor,where an annular catalyst basket can be placed in a fixed position.Figure 6.2-2 shows a schematic drawing of the reactor internals with a catalyst basket and a top view of the basket itself.The autoclave internal dimensions are 7.6 cm in diameter and 22.9 cm deep.The catalyst basket and impeller were designed based on a glass mockup study to give adequate mixing and contacting of the reaction mixture with the catalyst. The reactor is heated by a split-type electric furnace surrounding the reactor wall.The reactor internal temperature is controlled and measured through a pair of thermocouple wells immersed in the slurry phase.

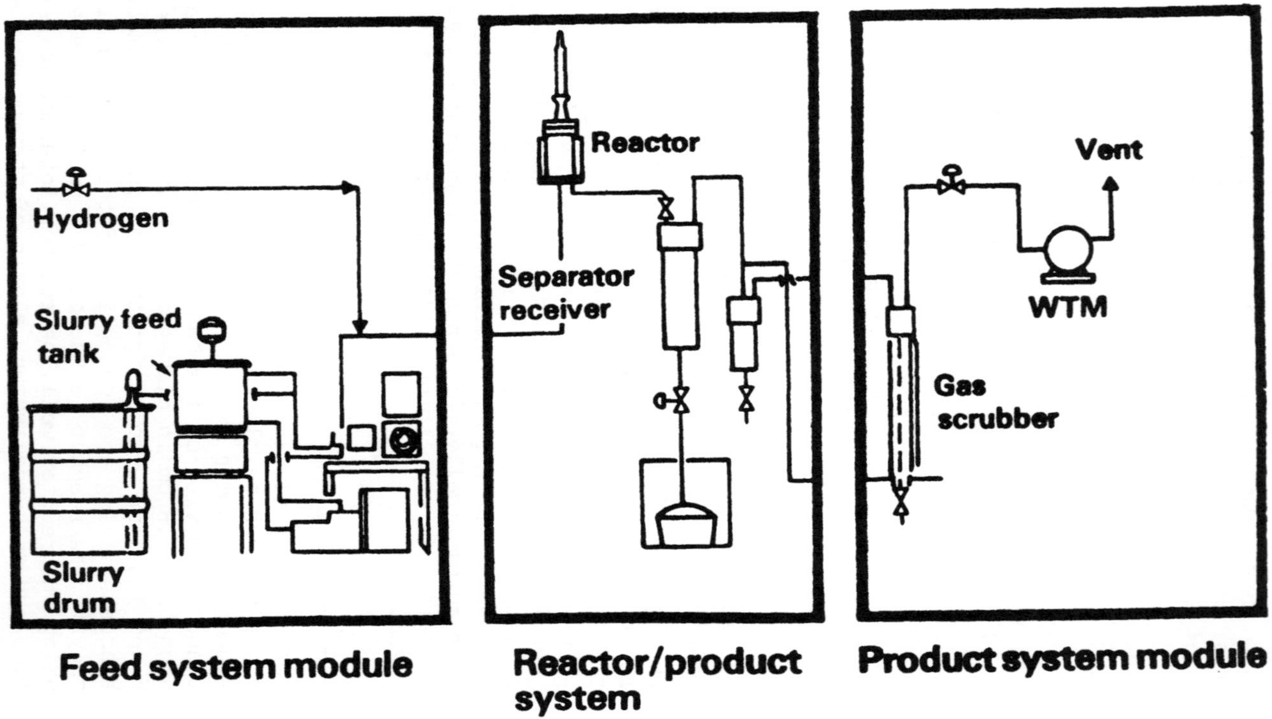

Figure 6.2-1 Flow Diagram

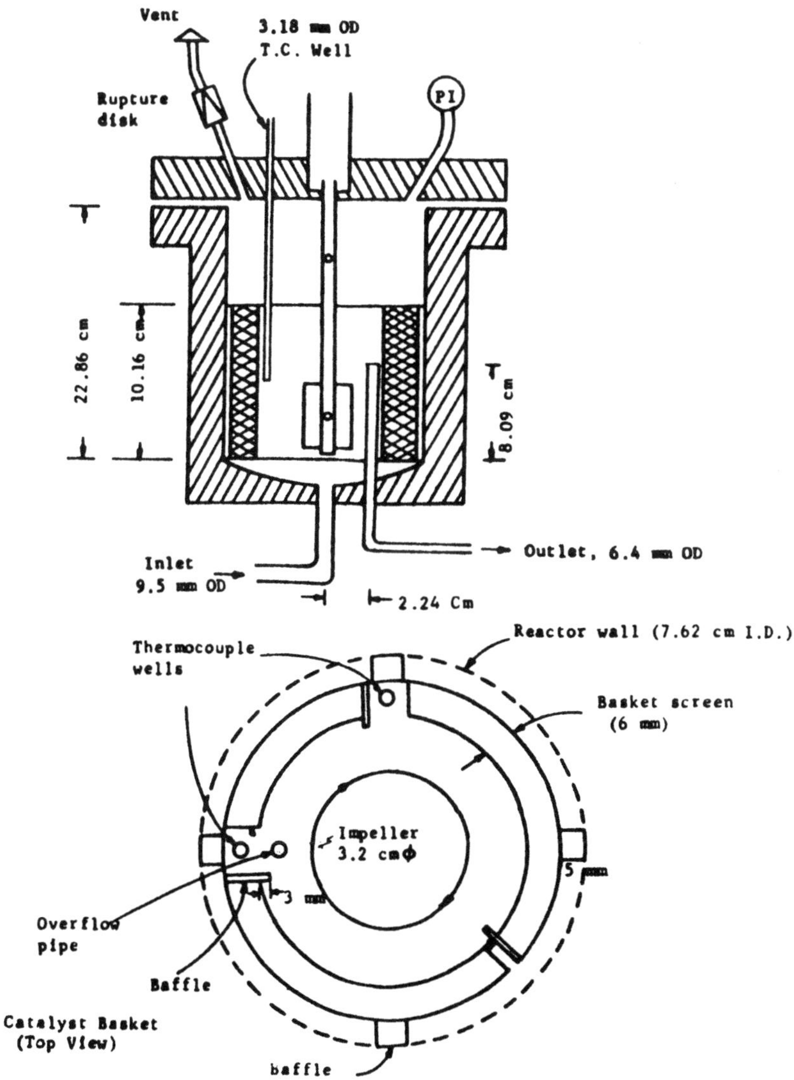

Figure 6.2-2 Reactor Internal

The original autoclave impeller,Dispersimax,was not adequate for slurry mixing.Three different types of impellers were designed,and the one that provided the best mixing was chosen from glass mockup studies.The improved impeller,which has six flat blades,is 3.8 cm long and 3.2 cm in diameter.The hollow stirrer shaft can deliver up to 0.17 horsepower and the maximum mixing speed is 1500 rpm.The pilot plant runs are usually conducted at the maximum agitation rate.

The annular catalyst basket,made of 14 U.S. mesh screen with 1.4 mm openings,is 6 mm thick and 10.2 cm higher,allowing a maximum of 88 cc catalyst charge to be loaded in 2-3 pellet thickness.The thin basket pockets are intended to provide intimate contact of the reaction mixture with the catalyst at the expense of reduced catalyst loading.The baffles inside and outside of the annulus prevent vortexing of the slurry and result in a well-defined,gas-liquid interface at the level of the overflow pipe.The narrow gap between the reactor wall and basket creates both upward and downward circulation.No direct assessment was made on the mixing efficiency and bulk mass transfer effects.A suggestion to spin the basket was not attempted due to mechanical limitations of the autoclave stirrer mechanism.No plugging or bridging of the basket screen was observed,and contacting was assumed adequate since the catalyst desulfurization performance was high.

Residence time and the feed rate to catalyst ratio (1/LHSV) are the two time parameters which respectively correspond to the thermal and catalytic reactions taking place in coal liquefaction. The residence time reported is a nominal value calculated from reactor holdup divided by the flowrate of slurry feed.The overflow pipe typically extends to 2.1 cm below the catalyst basket tip to give about 320 cc void volume,i.e.,reactor holdup.However,the actual volume occupied by the slurry is slightly less than the measured void volume because of gas holdup.Observations with glass mockup studies show approximately 5 percent volume expansion of the bulk phase due to gas bubbles.The gas bubbles are constantly dispersed through the slurry and rapidly disengaged into the free space above the overflow pipe.The nominal residence time give a relative measure of the holding time for a given slurry density.

Catalyst testing

Catalysts were tested for short term aging behavior with typical test periods of one week.The catalyst loading was 60 cc of whole pellets.The annular catalyst basket was loaded with catalyst in 2-3 pellet thickness and the remaining space was filled with 3 mm glass beads and a layer of glass wool to prevent catalyst spillover.

The standard test conditions for measuring catalyst activities are as follows:

STANDARD TEST CONDITIONS

Catalyst Charge:	**60 cc**
Reactor Holdup:	**320 cc**
Slurry Composition:	**20 wt% Ill. #6 coal (-400 mesh) in anthracene oil**
Pressure:	**137 atm (2000 psig)**
Temperature:	**427°C (800°F)**
Inlet Hydrogen:	**170 liters/hr (6 SCFH)**
Slurry Feed Rate:	**400 g/hr**
Residence Time:	**48 minutes**
LHSV:	**1.33 g coal/hr/cc catalyst**
Mixing Speed:	**1500 rpm**

The hydroliquefaction of coal is a complex process which involves close interaction between the solubilized coal,hydrogen donor solvent,and catalyst.An improved catalyst for coal liquefaction must satisfy several requirements,but two key aspects are high initial activity for liquefaction-desulfurization and good aging characteristics.The results are evaluated for liquefaction conversion,desulfurization,and hydrogen consuption.

Three commercial and one experimental catalysts with varying surface area and pore size are shown in the following table.All of these are about 1/16" extrudates containing cobalt-molybdenum on alumina support.

CATALYST INSPECTION DATA

Catalyst	Wt% CoO	Wt% MoO3	SA m^2/g	APD Å	PV cc/g	ABD g/cc	Size Inches
Cyanamid HDS-1442A	3.1	13.2	323	58	.64	0.57	1/16
Cyanamid HDS-20A	4.9	16.2	229	72	.54	0.72	1/20
Filtrol HPC-5	3.1	15.0	181	125	.56	0.77	1/16
Amoco KSA-LP (1)	3.1	14.7	157	201	.63	0.62	1/12

(1) Contains 1.2 Wt% P

Test results

In coal liquefaction,the conversion measurement depends on the product workup solvent and method used to determine insoluble residue.For this study benzene and tetrahydrofuran (THF) were used as the workup solvents.THF extraction has been previously used to

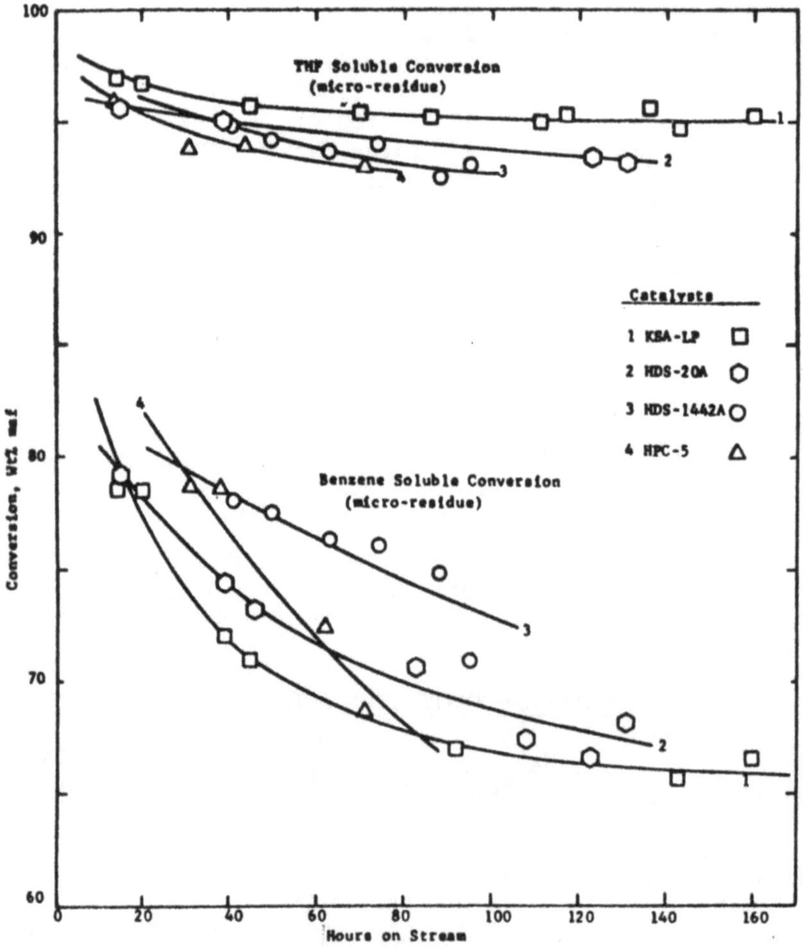

Figure 6.2-3 Liquefaction performance(based on micro residue method)

approximate boiler fuel product for the H-Coal process.On the other hand,benzene soluble material represents the upgraded fraction of boiler fuel.

The THF soluble and benzene soluble conversions for the four catalysts are given in Figure 6.2-3.The THF soluble conversion does not change much from one catalyst to the next.Amoco KSA-LP,which has large pores,shows the highest THF soluble conversion;it also maintains conversion fairly well.The two Cyanamid catalysts follow. Finally,Filtrol HPC-5 shows the lowest conversion to THF soluble oil oil,i.e.,boiler fuel as defined in this work.

The benzene soluble conversion in the bottom half of Figure 6.2-3 represents the upgraded boiler fuel.The difference between the THF soluble and benzene soluble conversions gives some measure of product selectivity.Cyanamid HDS-1442A,the H-Coal catalyst, shows the best performance in this case,which corresponds to more hydrogen consumption.Amoco KSA-LP declines very rapidly to a lower conversion level.The Filtrol HPC-5 catalyst also declines rapidly and would eventually level off as Amoco KSA-LP.

REFERENCES

1. Y.T.Shah,Gas-liquid-solid reactor design,McGraw-Hill(1979)
2. J.Carberry,Chemical and catalytic reaction engineering,McGraw Hill(1976)
3. S.Benson,Foundations of chemical kinetics,McGraw Hill(1960)
4. P.Lunde and F.Kester,Ind.Eng.Chem.Proc.Des.Dev.,11,350(1972)
5. J.Carberry,Ind.Eng.Chem.,56,39(1964)
6. D.Tajbl,J.Simmons and J.Carberry,Ind.Eng.Chem.Fund.,5,171(1966)
7. E.Meyer and K.Robinson,"Multiphase kinetic studies in a spinning basket reactor",Chemical Reaction Engineering-Houston, ACS Symposium Series,G5,1978
8. J.Mahoney,Journal of Catalysis,32,247(1974)
9. L.Doraiswamy and D.Tajbl,Cat.Rev.Sci.,20,177(1975)
10. J.Butt and V.Weekman,AIChE Symp.Series Nº143,70,27.
11. V.Weekman,AIChEJ,20,833(1974).

MIX
Papier aus verantwortungsvollen Quellen
Paper from responsible sources
FSC® C105338

If you have any concerns about our products,
you can contact us on
ProductSafety@springernature.com

In case Publisher is established outside the EU,
the EU authorized representative is:
Springer Nature Customer Service Center GmbH
Europaplatz 3, 69115 Heidelberg, Germany

Printed by Libri Plureos GmbH
in Hamburg, Germany